COMPUTATIONAL VISION AND MEDICAL IMAGE PROCESSING

BALKEMA – Proceedings and Monographs
in Engineering, Water and Earth Sciences

PROCEEDINGS OF VIPIMAGE 2007 – FIRST ECCOMAS THEMATIC CONFERENCE ON COMPUTATIONAL VISION AND MEDICAL IMAGE PROCESSING, PORTO, PORTUGAL, 17–19 OCTOBER 2007

Computational Vision and Medical Image Processing

VipIMAGE 2007

Edited by

João Manuel R. S. Tavares & R. M. Natal Jorge
Faculdade de Engenharia da Universidade do Porto, Porto, Portugal

Taylor & Francis
Taylor & Francis Group

LONDON / LEIDEN / NEW YORK / PHILADELPHIA / SINGAPORE

Taylor & Francis is an imprint of the Taylor & Francis Group, an informa business

© 2008 Taylor & Francis Group, London, UK

Typeset by Charon Tec Ltd (A Macmillan Company), Chennai, India
Printed and bound in Great Britain by Antony Rowe (A CPI-group Company), Chippenham, Wiltshire

Published by: Taylor & Francis/Balkema
                P.O. Box 447, 2300 AK Leiden, The Netherlands
                e-mail: Pub.NL@tandf.co.uk
                www.taylorandfrancis.co.uk/engineering, www.crcpress.com

ISBN: 978-0-415-45777-4

Computational Vision and Medical Image Processing – João Tavares & Natal Jorge (eds)
© 2008 Taylor & Francis Group, London, ISBN 978-0-415-45777-4

Table of Contents

Invited Keynotes/Lectures

Contributed Papers

Acknowledgements

The editors and the Conference co-chairs acknowledge the support towards the publication of the Book of Proceedings and the organization of the First ECCOMAS Thematic Conference VIPIMAGE to the following organizations:

– Universidade do Porto (UP)
– Faculdade de Engenharia da Universidade do Porto (FEUP)
– Fundação para a Ciência e a Tecnologia (FCT)
– Instituto de Engenharia Mecânica – Pólo FEUP (IDMEC-Polo FEUP)
– Instituto de Engenharia Mecânica e Gestão Industrial (INEGI)
– European Community on Computational Methods in Applied Sciences (ECCOMAS)
– International Association for Computational Mechanics (IACM)
– Associação Portuguesa de Mecânica Teórica Aplicada e Computacional (APMTAC)
– TecnoHospital – Revista de Instalações e Equipamentos de Saúde
– Unicer – Bebidas de Portugal, S.A.
– Sociedade Portuguesa de Neurorradiologia
– Sociedade Portuguesa de Engenharia Biomédica (SPEB)

Computational Vision and Medical Image Processing – João Tavares & Natal Jorge (eds)
© 2008 Taylor & Francis Group, London, ISBN 978-0-415-45777-4

Preface

This book contains the keynote lectures and full papers presented at VIPIMAGE 2007 - First ECCOMAS Thematic Conference on Computational Vision and Medical Image Processing, held in Porto, Portugal, during the period 17–19 October 2007. The event had 10 invited lectures, 44 oral presentations distributed by thirteen sessions and 20 posters. The contributions came from sixteen countries: Algeria, Belgium, Brazil, Canada, Colombia, Cuba, France, Germany, Italy, Japan, Portugal, Spain, Switzerland, The Netherlands, UK and USA.

In last decades the research related with object modeling has been a source of hard work in several distinct areas of science, such as mechanical, physics, informatics and mathematics. One major application of object modeling is in medical area. For instance, it is possible to consider the use of statistical or physical procedures on medical images in order to model the represented objects. Its modeling can have different goals as, for example: 3D shape reconstruction, object segmentation in 3D or 2D images, data interpolation, data registration, data compression, simulation, temporal tracking, motion and deformation analysis, computer-assisted therapy and therapy planning and tissue characterization.

The main goals of the ECCOMAS Thematic Conference on Computational Vision and Medical Image Processing consisted in the provision of a comprehensive forum for discussion on the current state-of-the-art in the associated/related fields.

The Conference brought together several researchers representing several fields related to Computational Vision, Computer Graphics, Computational Mechanical, Mathematics, Statistics, Medical Imaging and Medicine. The expertises spanned a broad range of techniques, such as finite element method, modal analyses, stochastic methods, principal components analyses, independent components analyses and distribution models.

The Conference co-chairs would like to take this opportunity to thank to The International European Community on Computational Methods in Applied Sciences, to The Portuguese Association of Theoretical, Applied and Computational Mechanics, to all sponsors, to all members of the Scientific Committee, to all Invited Lecturers, to all Session-Chairs to the Conference Organizers and to all Authors for submitting their contributions.

João Manuel R.S. Tavares
Renato M. Natal Jorge
(Conference Co-Chairs)

Computational Vision and Medical Image Processing – João Tavares & Natal Jorge (eds)
© 2008 Taylor & Francis Group, London, ISBN 978-0-415-45777-4

Invited Lectures

During the Conference VIPIMAGE 2007 were presented Invited Lectures by ten Expertises from five countries:

– Automatic Generation of Computer Models from Multi-modal Bio-medical Imaging, by Chandrajit Bajaj – University of Texas, USA
– Computational Bioimaging and Visualization, by Chris Johnson – University of Utah, USA
– From Geometrical Models to Physiological Models of the Human Body, by Hervé Delingette – Institut National de Recherche en Informatique, France
– Latest Advances in Cardiovascular Informatics, by Ioannis A. Kakadiaris – University of Houston, USA
– Robust Algorithms for Deformable Contours, by Jorge S. Marques – Instituto Superior Técnico, Portugal
– Image Sequence Evaluation, by Juan J. Villanueva – Autonomous University of Barcelona, Spain
– Fast Surface Segmentation and Remeshing by Finding Geodesics, by Laurent Cohen – Université Paris Dauphine, France
– Processing of Simultaneous Acquisition of EEG and fMRI, by Mario Forjaz Secca – Universidade Nova de Lisboa, Portugal
– Automatic Construction of Statistical Shape Models using Non-Rigid Registration, by Tim Cootes – University of Manchester, UK
– Theory of Digital Manifolds and its Applications to Medical Imaging, by Valentin E. Brimkov – State University of New York, USA

Computational Vision and Medical Image Processing – João Tavares & Natal Jorge (eds)
© 2008 Taylor & Francis Group, London, ISBN 978-0-415-45777-4

Scientific Committee

All works were submitted to an International Scientific Committee composed by eighty four expert researchers from recognized institutions of sixteen countries:

Adelino F. Leite-Moreira	Faculty of Medicine of University of Porto, Portugal
Adérito Marcos	University of Minho, Portugal
Alberto De Santis	Università degli Studi di Roma "La Sapienza", Italy
Ana Mafalda Reis	Instituto de Ciências Biomédicas Abel Salazar, Portugal
Antonio Susín Sánchez	Universitat Politècnica de Catalunya, Spain
Arrate Muñoz Barrutia	University of Navarra, Spain
Augusto Goulão	Garcia de Orta Hospital, Portugal
Bernard Gosselin	Faculte Polytechnique de Mons, Belgium
Bogdan Raducanu	Computer Vision Center, Spain
Chandrajit Bajaj	University of Texas, USA
Chang-Tsun Li	University of Warwick, UK
Charles A. Taylor	Stanford University, USA
Chris Johnson	University of Utah, USA
Constantine Kotropoulos	Aristotle University of Thessaloniki, Greece
Daniela Iacoviello	Università degli Studi di Roma "La Sapienza", Italy
Demetri Terzopoulos	University of California, Los Angeles, USA
Dinggang Shen	University of Pennsylvania, USA
Djemel Ziou	University of Sherbrooke, Canada
DRJ Owen	University of Wales Swansea, UK
Eduardo Borges Pires	Instituto Superior Técnico, Portugal
Enrique Alegre Gutiérrez	University of León, Spain
Eugenio Oñate	Universitat Politècnica de Catalunya, Spain
Francisco Perales	Balearic Islands University, Spain
Georgeta Oliveira	Pedro Hispano Hospital, Portugal
Gerald Schaefer	Aston University, UK
Gerhard A. Holzapfel	Royal Institute of Technology, Sweden
Helcio R.B. Orlande	Federal University of Rio de Janeiro, Brazil
Hélder Araújo	University of Coimbra, Portugal
Hélder C. Rodrigues	Instituto Superior Técnico, Portugal
Hemerson Pistori	Dom Bosco Catholic University, Brazil
Henrik Aanæs	Technical University of Denmark
Hervé Delingette	Institut National de Recherche en Informatique, France
Ian Jermyn	Institut National de Recherche en Informatique, France
Ioannis A. Kakadiaris	University of Houston, USA
Ioannis Pitas	Aristotle University of Thessaloniki, Greece
Isaac Cohen	Honeywell, Advanced Technology Lab, USA
Isabel M.A.P. Ramos	Faculty of Medicine of University of Porto, Portugal
J. Paulo Vilas-Boas	Faculty of Sport of University of Porto, Portugal
Jan C. De Munck	Vrije Universiteit Amsterdam, The Netherlands
Jan Koenderink	Utrecht University, The Netherlands
João A.C. Martins	Instituto Superior Técnico, Portugal
João M.C.S. Abrantes	Technical University of Lisbon, Portugal
João M.R.S. Tavares	Faculty of Engineering of University of Porto, Portugal
João Martins Pisco	Universidade Nova de Lisboa, Portugal
João Paulo Costeira	Instituto Superior Técnico, Portugal

Invited Keynotes/Lectures

Computational Vision and Medical Image Processing – João Tavares & Natal Jorge (eds)
© 2008 Taylor & Francis Group, London, ISBN 978-0-415-45777-4

Automatic construction of statistical shape models using non-rigid registration

Tim Cootes

The University of Manchester, England

ABSTRACT: Statistical models of object shape and appearance have been widely used for interpretting medical images. In early work, such models were constructed by manually annotating training images, which is both time-consuming and potentially error-prone. There is considerable demand for algorithms to automatically construct such models from training data, with minimal human intervention. Building on work on registering 2D boundaries and 3D surfaces, we have developed methods for registering unlabelled images so as to construct compact models. This short paper gives a list of some useful references in the area.

1 OVERVIEW

Statistical models of shape (Cootes et al. 1995) and appearance (Cootes et al. 2001) have been widely used for image interpretation, particularly when analysising images of faces or in the medical domain. Such models can be rapidly matched to new images using the Active Shape Model (Cootes et al. 1995) or the Active Appearance Model (Cootes et al. 2001) algorithms, or their more recent derivatives, such as (van Ginneken et al. 2002), (Matthews and Baker 2004) to name but two.

Such models require training sets in which landmark points are annotated, defining the *correspondence* between equivalent points across the images. In early work such points were manually or semi-automatically marked, which is time consuming, and almost completely impractical for large 3D datasets.

We have thus developed algorithms which can automatically find correspondences given sets of 2D curves (Kotcheff and Taylor 1997), (Davies et al. 2002) or 3D surfaces (Davies et al. 2002). These have demonstrated the utility of expressing the correspondence problem as an optimisation problem in which we seek to find the correspondences which lead to a model which can encode the data most compactly – a minimum description length (MDL) approach.

However, though such approaches have proved very useful, they still require the object(s) of interest to have been segmented from the training images. In more recent work we have developed methods in which the models are constructed directly from images, with minimal manual intervention.

This can be thought of as an image correspondence problem, in which we wish to find a set of mappings which warp each image in the set to every other image in an 'optimal' way. By analogy with our work on surfaces, we treat it as an image coding problem. We can represent the mappings using a dense set of points defining correspondences. Given a set of correspondences across a set of images, we can construct an appearance model (Cootes et al. 2001), which can be used to encode the image set. By modifying the correspondences, we change the model and the cost of the encoding. This leads to an optimisation problem, in which we modify the point positions on each image, one at a time, so as to optimise the model coding cost (Cootes et al. 2005).

More sophisticated models of anatomical structure can be constructed by computing the statistics of estimates of tissue class fraction, rather than image intensity (Petrovic et al. 2006). This leads to models which are more independent of imaging modality.

2 RELATED WORK

Finding mappings between structures across a set of images can facilitate many image analysis tasks. One particular area of importance is in medical image interpretation, where image registration can help in tasks as diverse as anatomical atlas matching and labelling, image classification, and data fusion. Many researchers have investigated image registration methods and the use of deformable models, for overviews see for example (Zitová and Flusser 2003; Maintz and Viergever 1998).

Baker *et al.* (Baker et al. 2004) considered building an appearance model as an image coding problem.

The model parameters are iteratively re-estimated after fitting the current model to the images, leading to an implicit correspondence defined across the data set.

Other related work on image registration for model building includes (Marsland et al. 2003), (Jones and Poggio 1998), (Rueckert et al. 2001), (S. Duchesne 2002), (Jebara 2003), (Cootes et al. 2004), (S. Joshi et al. 2004), (E. G. Miller et al. 2000), (Miller 2004).

3 CONCLUSION

Methods for constructing statistical models of shape and appearance from sets of minimally labelled images are very useful for practical applications, and are currently the subject of much research.

REFERENCES

Baker, S., I. Matthews, and J. Schneider (2004). Automatic construction of active appearance models as an image coding problem. *IEEE Transactions on Pattern Analysis and Machine Intelligence 26*(10), 1380–84.

Cootes, T., S. Marsland, C. Twining, K. Smith, and C. Taylor (2004). Groupwise diffeomorphic non-rigid registration for automatic model building. In *8th European Conference on Computer Vision*, Volume 4, pp. 316–327. Springer.

Cootes, T., C. Twining, V.Petrović, R.Schestowitz, and C. Taylor (2005). Groupwise construction of appearance models using piece-wise affine deformations. In *16th British Machine Vision Conference*, Volume 2, pp. 879–888.

Cootes, T. F., G.J. Edwards, and C.J. Taylor (2001). Active appearance models. *IEEE Transactions on Pattern Analysis and Machine Intelligence 23*(6), 681–685.

Cootes, T. F., C. J. Taylor, D. Cooper, and J. Graham (1995, January). Active shape models - their training and application. *Computer Vision and Image Understanding 61*(1), 38–59.

Davies, R., C. Twining, T. Cootes, and C. Taylor (2002). A minimum description length approach to statistical shape modelling. *IEEE Transactions on Medical Imaging 21*, 525–537.

Davies, R., C. Twining, T. Cootes, J. Waterton, and C. Taylor (2002). 3D statistical shape models using direct optimisation of description length. In *European Conference on Computer Vision*, Volume 3, pp. 3–20. Springer.

E. G. Miller, M. E. Matsakis, and P. A. Viola (2000). Learning from one example through shared densities on transforms. In *IEEE Proc Computer Vision and Pattern Recognition*, Volume 1, pp. 464–471.

Jebara, T. (2003). Images as bags of pixels. In *ICCV*, Volume 1, pp. 265–272.

Jones, M. J. and T. Poggio (1998). Multidimensional morphable models: A framework for representing and matching object classes. *International Journal of Computer Vision 2*(29), 107–131.

Kotcheff, A. and C. J. Taylor (1997). Automatic construction of eigen-shape models by genetic algorithm. In *15th Conference on Information Processing in Medical Imaging*, pp. 1–14.

Maintz, J. B. A. and M. A. Viergever (1998). A survey of medical image registration. *Medical Image Analysis 2*(1), 1–36.

Marsland, S., C. Twining, and C. Taylor (2003). Groupwise non-rigid registration using polyharmonic clamped-plate splines. In *MICCAI*, Lecture Notes in Computer Science.

Matthews, I. and S. Baker (2004, November). Active appearance models revisited. *International Journal of Computer Vision 60*(2), 135–164.

Miller, M. (2004). Computational anatomy: shape, growth, and atrophy comparison via diffeomorphisms. *NeuroImage 23*, S19–S33.

Petrovic, V. S., T. F. Cootes, C. J. Twining, and C. Taylor (2006). Automatic framework for medical image registration, segmentation and modeling. In *Proc. Medical Image Understanding and Analysis*, Volume 2, pp. 141–5.

Rueckert, D., A. Frangi, and J. Schnabel (2001). Automatic construction of 3D statistical deformation models using non-rigid registration. In *MICCAI*, pp. 77–84.

S. Duchesne, J. Pruessner, D. C. (2002). Appearance-based segmentation of medial temporal lobe structures. *NeuroImage 17*, 515–531.

S.Joshi, B.Davis, M.Jomier, and G.Gerig (2004). Unbiased diffeomorphic atlas construction for computational anatomy. *NeuroImage 23*, S151–S160.

van Ginneken, B., A.F.Frangi, J.J.Stall, and B. ter Haar Romeny (2002). Active shape model segmentation with optimal features. *IEEE-TMI 21*, 924–933.

Zitová, B. and J. Flusser (2003). Image registration methods: A survey. *Image and Vision Computing 21*, 977–1000.

CardioSense3D: Towards patient-specific cardiac simulation for clinical applications

H. Delingette, M. Sermesant, J.-M Peyrat & N. Ayache,
Asclepios Team, INRIA Sophia-Antipolis, France

D. Chapelle, J. Sainte-Marie & Ph. Moireau
Macs Team, INRIA Rocquencourt, France

M. Fernandez, J-F. Gerbeau
Reo Team, INRIA Rocquencourt, France

K. Djabella, M. Sorine
Sisyphe Team, INRIA Rocquencourt, France

ABSTRACT: In this paper, we overview the objectives and achievements of the CardioSense3D project dedicated to the construction of an electro-mechanical model of the heart.

1 TOWARDS A VIRTUAL PHYSIOLOGICAL HUMAN

Geometrical and physical/biomechanical modeling of the human body provide representations allowing geometric reasoning, navigation, and various forms of simulated interactions including deforming and cutting soft tissues with realistic visual and haptic feedback. However the tissues and organs still behave like passive material, and it is necessary to go one step further to model the active properties of living tissues and the dynamic nature of normal or pathological evolving processes. A grand challenge for the coming years is thus to build a multiscale virtual physiological human where the interaction between all biological systems are models from cells to organs. In another words the objectives of such initiative would be to replace a qualitative and phenomenological knowledge of the human body with a quantitative and multiscale description of the human body. The European commission has chosen to fund research on this topic in the 7th framework programm, directing more than 100 million euros over the next 5 years.

An important example of this virtual physiological human is the modeling of the electro-mechanical activity of the heart. The simulation of the heart has received a growing attention due to the importance (Hunter, Nash, and sands 1997; Noble and Rudy 2001; Bassingthwaighte 2000; Sachse, Seemann, Werner, Riedel and Dössel 2001; Frangi, Niessen, and Viergever 2001;

Wong, Zhang, Liu, and Shi 2006) of cardiovascular diseases in industrialized nations[1] and to the high complexity of the cardiac function. Indeed, formulating a computational model of the cardiac function of a specific patient represents a great challenge due to :

- the intrinsic physiological complexity of the underlying phenomena which combine tissue mechanics, fluid dynamics, electro-physiology, energetic metabolism and cardiovascular regulation;
- the partial information available for a specific patient and the variety of the objectives of data processing ranging from global detection of pathological situations to local diagnosis and personalized therapy planning.

2 CARDIOSENSE3D

CardioSense3D is a 4-year Large Initiative Action launched in 2005 and funded by the French national research center INRIA which focuses on the electro-mechanical modeling of the heart.

The **objectives** of **CardioSense3D** are threefold :

1. To **build a cardiac simulator**, with identifiable parameters, that couples 4 different physiological

---

[1] With 180 000 deaths per year, cardiovascular diseases represent the leading cause of death in France before cancer. In the United States more than 1 million deaths occur every year caused by cardio-vascular diseases.

phenomena: electrophysiology, mechanical contraction and relaxation, myocardium perfusion and cardiac metabolism,

2. To **build data assimilation software** that can estimate patient specific parameters and state variables from given observations of the cardiac activity,

3. To **build several application softwares** based on this simulator and data assimilation techniques to solve clinical problems related to the diagnosis or therapy of cardiac pathologies.

in order to:

- **significantly improve medical practice** in terms of better prevention, diagnosis, quantitative follow-up, simulation and guidance of therapy,
- **support biomedical research** in the preparation and evaluation of new diagnostic and therapeutic tools,
- **advance** the **fundamental knowledge** of the integrative *physiology* of the **heart**.

CardioSense3D relies on the expertise of four INRIA research teams (resp. Asclepios, Reo, Macs, Sosso2) covering the fields of medical image analysis, computational structural and fluid dynamics, numerical analysis and control. But is also a collaborative framework that involves clinical centers such as the Guy's Hospital London, the Laboratory of Cardio-Energetics at the National Institutes of Health, the Hospital Henri Mondor (J. Garot), and other partners listed in the web site of the project[2].

Reaching those three objectives requires to tackle the following challenges :

1. The introduction of models and related numerical procedures to represent some important physiological phenomena still not considered, in particular: cardiac metabolism, perfusion and tissue remodeling. The extended models must remain identifiable with the available data and computationally tractable. This sets the limits of the otherwise endless quest for model fidelity and simulation accuracy.

2. The formulation of effective data assimilation methodologies associated with those models, that can estimate patient-specific indicators from actual measurements of the cardiac activity. Major shortcomings of existing methods include robustness and computational cost (the "curse of dimensionality").

3. The adaptation and optimization of the cardiac simulator (including both direct and inverse approaches) to some targeted clinical applications. For each application, specific problems connected with clinical science will be considered.

Figure 1. Fiber tracking performed on an average Canine heart build from nine canine images.

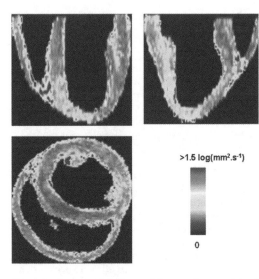

>1.5 log(mm².s⁻¹)

0

Figure 2. Images of the trace of the covariance matrix of diffusion tensors from nine canine hearts. The variability of those tensors seems to be low in most part of the myocardium.

3 SOME CARDIOSENSE 3D RESEARCH ACTIVITIES

We illustrate below some of recent research advances performed within CardioSense3D.

3.1 *Statistical analysis of diffusion tensor imaging of canine hearts*

In Figures 1 and 2, we show some recent results (Peyrat, Sermesant, Delingette, Pennec, Xu, McVeigh,

---
[2] www.inria.fr/CardioSense3D/

6

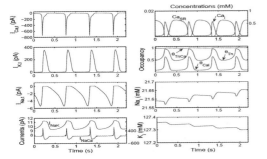

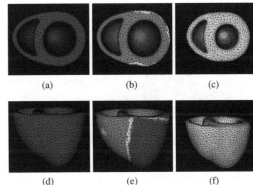

(a)   (b)   (c)

Figure 3. Computed spontaneous action potential and ionic currents (left) and intracellular Ca^{++} dynamics (from (Djabella and Sorine 2006a)).

(d)   (e)   (f)

Figure 4. Short axis (top row) and long axis (bottom row) views of an electromechanical heart model during end diastole (left column), ventricular depolarization (middle column) and end systole (right column).

and Ayache 2006) concerning the statistical analysis of Diffusion Tensor Imaging (DTI) of nine canine hearts. Diffusion imaging helps to reveal the fine structure of the myocardium such as the fiber orientation and possibly the location and orientation of laminar sheets. This structural information is crucial for modeling both the mechanical and electrophysiological function of the heart.

3.2 *Electro-physiology modeling*

Several electrophysiological models have been proposed within CardioSense3D, including front propagation techniques (Sermesant, Coudière, Moreau-Villéger, Rhode, Hill, and Ravazi 2005) phenomenological models (Pop, Sermesant, Coudière, Graham, Bronskill, Dick, and Wright 2006) and a 8-variable cardiac cell model describing the dynamics of calcium (Djabella, and Sorine 2006b) Model-based ECG processing for identification of restitution curves have also been proposed in (Illanes Manriquez, Zhang, Medigue, Papelier, and Sorine 2006).

3.3 *Electro-mechanical model of the heart*

The coupling between electrophysiology and mechanics (Bestel, Clément, and Sorine 2001) is governed by a chemically-controlled constitutive law which is consistent with general thermodynamics and with the behavior of myosin molecular motors. The biomechanical model is based on a Hill-Maxwell rheological scheme (Krejci, Sainte-Marie, Sorine, and Urquiza 2006), pressure boundary conditions being controlled by valve and Winkessel models (Sainte-Marie, Chapelle, Cimrman, and Sorine 2006; Sermesant, Delingette, and Ayache 2006). Figure 4 shows the biventricular model at end diastole and end systole (Sermesant, Delingette, and Ayache 2006).

3.4 *Coupling models with observations*

The objective of estimating model parameters from observations is a key aspect of CardioSense3D.

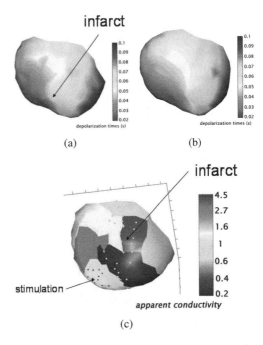

(a)   (b)

(c)

Figure 5. (a) Measured depolarization isochrones of a canine heart with an infarcted region; (b) Simulated depolarization isochrones based on a phenomelogical model after the automatic estimation of regional appearent conductivities; (c) View of the apparent conductivity map where regions of low conductivities matches infarcted regions.

Preliminary results have been obtained on this front, by globally integrating the electromechanical heart models with clinical datasets (Sermesant, Rhode, Sanchez-Ortiz, Camara, Andriantsimiavona, Hegde, Rueckert, Lambiase, Bucknall, Rosenthal, Delingette, Hill,

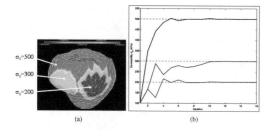

(a)   (b)

Figure 6.   (a) Three regions of an electro-mechanical model of the heart have been set with different contractility parameters; (b) A data assimilation technique has been used to recover those parameters.

Ayache, and Razavi 2005; Rhode, Sermesant, Brogan, Hegde, Hipwell, Lambiase, Rosenthal, Bucknall, Qureshi, Gill, Razavi, and Hill 2005) (3D endocardial mapping with tagged, SSFP and late enhancement MR images), by estimating apparent electrical conductivities from electrophysiological mappings (Moreau-Villéger, Delingette, Sermesant, Ashikaga, Faris, McVeigh, and Ayache 2006) (see Figure 5) and by estimating regional contractilities (Sermesant, Moireau, Camara, Sainte-Marie, Andriantsimiavona, Cimrman, Hill, Chapelle, and Razavi 2006; Sainte-Marie, Chapelle, Cimrman, and Sorine 2006) from motion information (see Figure 6).

REFERENCES

Bassingthwaighte, J. B. (2000). Strategies for the physiome project. *Annals of Biomedical Engineering 28*, 1043–1058.

Bestel, J., F. Clément, and M. Sorine (2001). A biomechanical model of muscle contraction. In W. Niessen and M. Viergever (Eds.), *Medical Image Computing and Computer-Assisted intervention (MICCAI'01)*, Volume 2208 of *Lecture Notes in Computer Science (LNCS)*, pp. 1159–1161. Springer.

Djabella, K. and M. Sorine (2006a). A differential model of controlled cardiac pacemaker cell. In *Proc. of the 6th IFAC Symposium on Modelling and Control in Biomedical Systems*.

Djabella, K. and M. Sorine (2006b, July 31-August 4). A reduced differential model for cardiac action potentials. In *SIAM Conference on the Life Sciences*, Raleigh, USA.

Frangi, A., W. Niessen, and M. Viergever (2001). Three-dimensional modeling for functional analysis of cardiac images: A review. *IEEE Transactions on Medical Imaging 1*(20), 2–25.

Hunter, P., M. Nash, and G. Sands (1997). Computational electromechanics of the heart. In *Computational biology of the heart*, pp. 345–407. A.V. Panfilov and A.V. Holden Eds, John Wiley & Sons.

Illanes Manriquez, A., Q. Zhang, C. Medigue, Y. Papelier, and M. Sorine (2006). Electrocardiogram-based restitution curve. In *Computers in cardiology*, Valencia, Spain.

Krejci, P., J. Sainte-Marie, M. Sorine, and J. Urquiza (2006, September). Solutions to muscle fiber equations and their long time behaviour. *Nonlinear Analysis: Real World Applications 7*(4).

Moreau-Villéger, V., H. Delingette, M. Sermesant, H. Ashikaga, O. Faris, E. McVeigh, and N. Ayache (2006, August). Building maps of local apparent conductivity of the epicardium with a 2D electrophysiological model of the heart. *IEEE Transactions on Biomedical Engineering 53*(8), 1457–1466.

Noble, D. and Y. Rudy (2001) Models of cardiac ventricular action potentials: iterative interaction between experiment and simulation. *Phil. Trans. R. Soc. Lond. A*, 1127–1142.

Peyrat, J.-M., M. Sermesant, H. Delingette, X. Pennec, C. Xu, E. McVeigh, and N. Ayache (2006, 2–4 October). Towards a statistical atlas of cardiac fiber structure. In *Proc. of MICCAI'06, Part I*, Number 4190 in LNCS, pp. 297–304.

Pop, M., M. Sermesant, Y. Coudière, J. Graham, M. Bronskill, A. Dick, and G. Wright (2006). A theoretical model of ventricular reentry and its radiofrequency ablation therapy. In *3rd IEEE International Symposium on Biomedical Imaging: Macro to Nano (ISBI'06)*, pp. 33–36.

Rhode, K., M. Sermesant, D. Brogan, S. Hegde, J. Hipwell, P. Lambiase, E. Rosenthal, C. Bucknall, S. Qureshi, J. Gill, R. Razavi, and D. Hill (2005). A system for real-time XMR guided cardiovascular intervention. *IEEE Transactions on Medical Imaging 24*(11), 1428–1440.

Sachse, F., G. Seemann, C. Werner, C. Riedel, and O. Dössel (2001). Electro-mechanical modeling of the myocardium: Coupling and feedback mechanisms. In *Computers in Cardiology*, Volume 28, pp. 161–164.

Sainte-Marie, J., D. Chapelle, R. Cimrman, and M. Sorine (2006). Modeling and estimation of the cardiac electromechanical activity. *Computers & Structures 84*, 1743–1759.

Sermesant, M., Y. Coudière, V. Moreau-Villéger, K. Rhode, D. Hill, and R. Ravazi (2005). A fast-marching approach to cardiac electrophysiology simulation for XMR interventional imaging. In *Proceedings of MICCAI'05*, Volume 3750 of *LNCS*, Palm Springs, California, pp. 607–615. Springer Verlag.

Sermesant, M., H. Delingette, and N. Ayache (2006). An electromechanical model of the heart for image analysis and simulation. *IEEE Transactions in Medical Imaging 25*(5), 612–625.

Sermesant, M., P. Moireau, O. Camara, J. Sainte-Marie, R. Andriantsimiavona, R. Cimrman, D. Hill, D. Chapelle, and R. Razavi (2006). Cardiac function estimation from MRI using a heart model and data assimilation: Advances and difficulties. *Medical Image Analysis 10*(4), 642–656.

Sermesant, M., K. Rhode, G. Sanchez-Ortiz, O. Camara, R. Andriantsimiavona, S. Hegde, D. Rueckert, P. Lambiase, C. Bucknall, E. Rosenthal, H. Delingette, D. Hill, N. Ayache, and R. Razavi (2005). Simulation of cardiac pathologies using an electromechanical biventricular model and XMR interventional imaging. *Medical Image Analysis 9*(5), 467–480.

Wong, K. C. L., H. Zhang, H. Liu, and P. Shi (2006). Physiome model based state-space framework for cardiac kinematics recovery. In *MICCAI (1)*, pp. 720–727.

Computational Vision and Medical Image Processing – João Tavares & Natal Jorge (eds)
© *2008 Taylor & Francis Group, London, ISBN 978-0-415-45777-4*

Finding a closed boundary by growing minimal paths from a single point

Fethallah Benmansour, Stéphane Bonneau & Laurent Cohen
CEREMADE, Place du Marchal De Lattre De Tassigny, Paris, France

ABSTRACT: In this paper, we present a new method for segmenting closed contours. Our work builds on a variant of the Fast Marching algorithm. First, an initial point on the desired contour is chosen by the user. Next, new keypoints are detected automatically using a front propagation approach. We assume that the desired object has a closed boundary. This a-priori knowledge on the topology is used to devise a relevant criterion for stopping the keypoint detection and front propagation. The final domain visited by the front will yield a band surrounding the object of interest. Linking pairs of neighboring keypoints with minimal paths allows us to extract a closed contour from a 2D image. Detection of a variety of objects on real images is demonstrated.

1 INTRODUCTION

Energy minimization techniques have been applied to a broad variety of problems in image processing and computer vision. Since the original work on snakes (Kass, Witkin, and Terzopoulos 1988), they have notably been used for boundary detection. An active contour model, or snake, is a curve that deforms its shape in order to minimize an energy combining an internal part which smooths the curve and an external part which guides the curve toward particular image features. For instance, the geodesic active contour model (Caselles, Kimmel, and Sapiro 1997) relies on the minimization of a geometric energy functional that deforms an initial curve toward local geodesics in a Riemannian metric derived from the image. Whereas the geodesic active contour model presents significant improvements compared to the original snake model, the energy minimization process is still prone to local minima. Consequently, results strongly depend on the model initialization.

To avoid local minima, Cohen and Kimmel (Cohen and Kimmel 1997) introduced an approach to globally minimize the geodesic active contour energy, provided that two endpoints of the curve are initially supplied by the user. This energy is of the form $\int_{\gamma} \tilde{\mathcal{P}}$ where the incremental cost $\tilde{\mathcal{P}}$ is chosen to take lower values on the contour of the image, and γ is a path joining the two points. The solution of this minimization problem is obtained through the computation of *the minimal action map* associated to a source point. The minimal action map can be regarded as the arrival times of a front propagating from the source point with velocity $(1/\tilde{\mathcal{P}})$, and it satisfies the Eikonal equation. We can compute simultaneously, and efficiently, the minimal action map and its Euclidean path length with the *Fast Marching Method* as will be detailed in section 2.2.

In section 3, we introduce a novel segmentation approach, based on the Fast Marching Method, to distribute a set of points on a closed curve that is not known a priori. We only assume the user provides a single point (or more if desired) initialized on the desired object boundary. Each newly detected keypoint is immediately defined as a new source of propagation, and keypoints are detected with a criterion based on the Euclidean length of the minimal paths. Since the front propagates faster on the object boundary, the first point for which the length λ is reached, is located in this area (of small values of $\tilde{\mathcal{P}}$) and is a valuable choice as a new keypoint. By using the a-priori knowledge on the topology of the manifold, we devise a relevant criterion for stopping the keypoint detection and front propagation. The criterion is general for any dimension.

In section 4, we explain how to extract a boundary curve using the previous results. The main idea is to link pairs of neighboring keypoints with minimal paths via gradient descent on the minimal action map. Segmentation results on a set of 2D images are presented in section 5. Finally conclusions and perspectives follow in section 6.

2 BACKGROUND ON MINIMAL PATHS

2.1 *Definitions*

Given a 2D image $I : \Omega \rightarrow \mathbb{R}^+$ and two points $\mathbf{p}_1$ and $\mathbf{p}_2$, the underlying idea introduced in (Cohen and

Kimmel 1997) is to build a potential $\mathcal{P} : \Omega \to \mathbb{R}^{*+}$ which takes lower values near desired features of the image I. The choice of the potential $\mathcal{P}$ depends on the application. For example, one can define $\mathcal{P}$ as a decreasing function of $\|\nabla I\|$ to extract edges by finding a curve that globally minimizes the energy functional $E : \mathcal{A}_{\mathbf{p}_1,\mathbf{p}_2} \to \mathbb{R}^+$

$$E(\gamma) = \int_\gamma \left\{ \mathcal{P}\big(\gamma(s) + w\big) \right\} \mathrm{d}s = \int_\gamma \tilde{\mathcal{P}}\big(\gamma(s)\big) \mathrm{d}s, \quad (1)$$

where $\mathcal{A}_{\mathbf{p}_1,\mathbf{p}_2}$ is the set of all paths connecting $\mathbf{p}_1$ to $\mathbf{p}_2$, s is the arc-length parameter, $w > 0$ is a regularization term and $\tilde{\mathcal{P}} = (\mathcal{P} + w)$. A curve connecting $\mathbf{p}_1$ to $\mathbf{p}_2$ that globally minimizes the energy (1) is a *minimal path* between $\mathbf{p}_1$ and $\mathbf{p}_2$, noted $\mathcal{C}_{\mathbf{p}_1,\mathbf{p}_2}$. The solution of this minimization problem is obtained through the computation of the *minimal action map* $\mathcal{U}_1 : \Omega \to \mathbb{R}^+$ associated to $\mathbf{p}_1$. The minimal action is the minimal energy integrated along a path between $\mathbf{p}_1$ and any point $\mathbf{x}$ of the domain Ω:

$$\forall \mathbf{x} \in \Omega, \; \mathcal{U}_1(\mathbf{x}) = \min_{\gamma \in \mathcal{A}_{\mathbf{p}_1,\mathbf{x}}} \left\{ \int_\gamma \tilde{\mathcal{P}}\big(\gamma(s)\big) \mathrm{d}s \right\}. \quad (2)$$

The values of $\mathcal{U}_1$ may be regarded as the arrival times of a front propagating from the source $\mathbf{p}_1$ with velocity $(1/\tilde{\mathcal{P}})$. $\mathcal{U}_1$ satisfies the Eikonal equation

$$\begin{cases} \|\nabla \mathcal{U}_1(\mathbf{x})\| &= \tilde{\mathcal{P}}(\mathbf{x}) \quad \text{for } \mathbf{x} \in \Omega, \\ \mathcal{U}_1(\mathbf{p}_1) &= 0. \end{cases} \quad (3)$$

The map $\mathcal{U}_1$ has only one local minimum, the point $\mathbf{p}_1$, and its flow lines satisfy the Euler-Lagrange equation of functional (1). Thus, the minimal path $\mathcal{C}_{\mathbf{p}_1,\mathbf{p}_2}$ can be retrieved with a simple gradient descent on $\mathcal{U}_1$ from $\mathbf{p}_2$ to $\mathbf{p}_1$ (see Fig. 1), solving the following ordinary differential equation with standard numerical methods like Heun's or Runge-Kutta's:

$$\begin{cases} \dfrac{\mathrm{d}\mathcal{C}_{\mathbf{p}_1,\mathbf{p}_2}(s)}{\mathrm{d}s} &= -\nabla \mathcal{U}_1\big(\mathcal{C}_{\mathbf{p}_1,\mathbf{p}_2}(s)\big), \\ \mathcal{C}_{\mathbf{p}_1,\mathbf{p}_2}(0) &= \mathbf{p}_2. \end{cases} \quad (4)$$

Let us extend the definitions given so far to the case of multiple sources and introduce other definitions which will be useful hereinafter. These definitions hold in dimension 2 and higher. The *minimal action map* associated to the potential $\tilde{\mathcal{P}} : \Omega \to \mathbb{R}^{*+}$ and the set of n sources $\mathcal{S} = \{\mathbf{p}_1, \dots, \mathbf{p}_n\}$ is the function $\mathcal{U} : \Omega \to \mathbb{R}^+$ defined by

$$\forall \mathbf{x} \in \Omega, \; \mathcal{U}(\mathbf{x}) = \min_{1 \le j \le n} \{\mathcal{U}_j(\mathbf{x})\},$$

where $\mathcal{U}_j(\mathbf{x}) = \min_{\gamma \in \mathcal{A}_{\mathbf{p}_j,\mathbf{x}}} \left\{ \int_\gamma \tilde{\mathcal{P}}\big(\gamma(s)\big) \mathrm{d}s \right\}. \quad (5)$

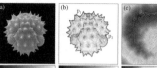

Figure 1. Extraction of an open contour from an electron microscopy image. (a) Original image I. (b) Potential $\mathcal{P} = (\|\nabla I\| + \varepsilon)^{-3}$, where ε is a small positive constant, and user-supplied points $\mathbf{p}_1$ and $\mathbf{p}_2$. (c) Minimal action map $\mathcal{U}_1$ and minimal path $\mathcal{C}_{\mathbf{p}_1,\mathbf{p}_2}$ between $\mathbf{p}_1$ and $\mathbf{p}_2$. (d) Image I and minimal path $\mathcal{C}_{\mathbf{p}_1,\mathbf{p}_2}$.

The map $\mathcal{U}$ is a weighted distance map to the set of sources $\mathcal{S}$, and it satisfies the Eikonal equation

$$\begin{cases} \|\nabla \mathcal{U}(\mathbf{x})\| &= \tilde{\mathcal{P}}(\mathbf{x}) \quad \text{for } \mathbf{x} \in \Omega, \\ \mathcal{U}(\mathbf{p}_j) &= 0 \qquad \text{for } \mathbf{p}_j \in \mathcal{S}. \end{cases} \quad (6)$$

The *Voronoi region* associated to the source $\mathbf{p}_j \in \mathcal{S}$, noted $\mathcal{R}_j$, is the locus of points of the domain Ω which are closer (in the sense of a weighted distance) to $\mathbf{p}_j$ than to any other source of $\mathcal{S}$. The region $\mathcal{R}_j$ is a connected subset of the domain Ω, and its boundary is noted $\partial \mathcal{R}_j$. The union of Voronoi regions and its complementary set, the *Voronoi diagram*, leads to a tessellation of the domain Ω, called the *Voronoi partition*. The *Voronoi index map* is the function $\mathcal{V} : \Omega \to \{1, \dots, n\}$ that assigns to any point of the domain Ω the index of its Voronoi region:

$$\forall \mathbf{x} \in \mathcal{R}_j, \; \mathcal{V}(\mathbf{x}) = j. \quad (7)$$

If two Voronoi regions $\mathcal{R}_i$ and $\mathcal{R}_j$ are adjacent (i.e. if $\partial \mathcal{R}_i \cap \partial \mathcal{R}_j$ is a non-empty set), then the minimal path $\mathcal{C}_{\mathbf{p}_i,\mathbf{p}_j}$ passes through the point of $\partial \mathcal{R}_i \cap \partial \mathcal{R}_j$ which has the smallest $\mathcal{U}$ value. This point, noted $\mathbf{m}_{i|j}$, is the *midpoint* of the minimal path $\mathcal{C}_{\mathbf{p}_i,\mathbf{p}_j}$ since it is equidistant to $\mathbf{p}_i$ and $\mathbf{p}_j$ in the sense of a weighted distance. This is a saddle point of $\mathcal{U}$.

The *Euclidean path length map* is the function $\mathcal{L} : \Omega \to \mathbb{R}^+$ that assigns to any point $\mathbf{x}$ of the domain Ω the Euclidean length of the minimal path between $\mathbf{x}$ and the source which is the closest in the sense of a weighted distance :

$$\forall \mathbf{x} \in \mathcal{R}_j, \; \mathcal{L}(\mathbf{x}) = \int_{\mathcal{C}_{\mathbf{p}_j,\mathbf{x}}} \mathrm{d}s. \quad (8)$$

Note that if $\tilde{\mathcal{P}}(\mathbf{x}) = 1$ for all $\mathbf{x} \in \Omega$, then the maps $\mathcal{U}$ and $\mathcal{L}$ are equal and both correspond to the Euclidean distance map to the set of sources $\mathcal{S}$.

2.2 Fast Marching Method

The *Fast Marching Method* (FMM) is a numerical method introduced in (Sethian 1999b; Sethian 1999a;

Table 1. Fast Marching Method for solving equation (6).	Table 2. FMM with keypoint detection.

- **Notation.**
  $\mathcal{N}_M(\mathbf{x})$ is the set of M neighbors of a grid point $\mathbf{x}$,
  where M $= 4$ in 2D and M $= 6$ in 3D.
- **Initialization.**
  For each grid point $\mathbf{x}$, do
  Set $\mathcal{U}(\mathbf{x}) : = +\infty$, $\mathcal{V}(\mathbf{x}) : = 0$ and $\mathcal{L}(\mathbf{x}) : = +\infty$.
  Tag $\mathbf{x}$ as *Far*.
  For each source $\mathbf{p}_j \in \mathcal{S}$, do
  Set $\mathcal{U}(\mathbf{p}_j) : = 0$, $\mathcal{V}(\mathbf{p}_j) := j$ and $\mathcal{L}(\mathbf{p}_j) := 0$.
  Tag $\mathbf{p}_j$ as *Trial*.
- **Marching loop**.
  While the set of *Trial* points is non-empty, do
  Find $\mathbf{x}_{min}$, a *Trial* point with the smallest $\mathcal{U}$ value.
  Tag $\mathbf{x}_{min}$ as *Alive*.
  For each point $\mathbf{x}_n \in \mathcal{N}_M(\mathbf{x}_{min})$ which is not *Alive*, do
  $\{u, v, \ell\} : = $ UpdateSchemeFMM $(\mathbf{x}_n, \mathcal{N}_M(\mathbf{x}_n))$.
  Set $\mathcal{U}(\mathbf{x}_n) : = u$, $\mathcal{V}(\mathbf{x}_n) : = v$ and $\mathcal{L}(\mathbf{x}_n) : = \ell$.
  If $\mathbf{x}_n$ is *Far*, tag $\mathbf{x}_n$ as *Trial*.

- **Notation.**
  $\mathcal{N}_M(\mathbf{x})$ is the set of M neighbors of a grid point $\mathbf{x}$,
  where M $= 4$ in 2D and M $= 6$ in 3D.
  $\mathcal{N}_{M^+}(\mathbf{x})$ is the set of M^+ neighbors of a point $\mathbf{x}$,
  where $M^+ = 8$ in 2D and $M^+ = 26$ in 3D.
- **Initialization.**
  For each grid point $\mathbf{x}$, do
  Set $\mathcal{U}(\mathbf{x}) : = +\infty$, $\mathcal{V}(\mathbf{x}) : = 0$ and $\mathcal{L}(\mathbf{x}) : = +\infty$.
  Tag $\mathbf{x}$ as *Far*.
  For each source $\mathbf{p}_j \in \mathcal{S}$, do
  Set $\mathcal{U}(\mathbf{p}_j) : = 0$, $\mathcal{V}(\mathbf{p}_j) : = j$ and $\mathcal{L}(\mathbf{p}_j) : = 0$.
  Tag $\mathbf{p}_j$ as *Trial* and as *Boundary*.
  $m : = 1$, *StopDetection*: $= FALSE$.
- **Marching loop**.
  While the set of *Trial* points is non-empty, do
  Find $\mathbf{x}_{min}$, a *Trial* point with the smallest $\mathcal{U}$ value.
  If (*StopDetection* $= FALSE$) and ($\mathcal{L}(\mathbf{x}_{min}) \geq \lambda$), do
  Here, $\mathbf{x}_{min}$ *is defined as the keypoint* $\mathbf{p}^*_{n+m}$.
  Set $\mathcal{U}(\mathbf{x}_{min}) : = 0$, $\mathcal{V}(\mathbf{x}_{min}) : = n + m$, $\mathcal{L}(\mathbf{x}_{min}) : = 0$.
  $m : = m + 1$.
  Else, do
  Tag $\mathbf{x}_{min}$ as *Alive*.
  For each grid point $\mathbf{x}_n \in \mathcal{N}_M(\mathbf{x}_{min})$, do
  If $\mathbf{x}_n$ is not *Alive*, do
  $\{u, v, \ell\} : = $ UpdateSchemeFMM $(\mathbf{x}_n, \mathcal{N}_M(\mathbf{x}_n))$
  Set $\mathcal{U}(\mathbf{x}_n) : = u$, $\mathcal{V}(\mathbf{x}_n) : = v$ & $\mathcal{L}(\mathbf{x}_n) : = \ell$.
  If (*StopDetection* $= FALSE$) & ($\mathbf{x}_n$ is *Far*), do
  Tag $\mathbf{x}_n$ as *Trial* and as *Boundary*.
  Else if $\mathcal{V}(\mathbf{x}_n) \neq \mathcal{V}(\mathbf{x}_{min})$, do
  $\{u, v, \ell\} : = $ UpdateSchemeFMM $(\mathbf{x}_n, \mathcal{N}_M(\mathbf{x}_n))$
  If $u < \mathcal{U}(\mathbf{x}_n)$, do
  Set $\mathcal{U}(\mathbf{x}_n) : = u$, $\mathcal{V}(\mathbf{x}_n) : = v$ & $\mathcal{L}(\mathbf{x}_n) : = \ell$.
  Tag $\mathbf{x}_n$ as *Trial*.
  If $\mathbf{x}_{min}$ is *Boundary*, do
  Tag $\mathbf{x}_{min}$ as *Interior*.
  If *StopDetection* $= FALSE$, do
  StopDetection: $=$
  IsBoundarySplit $(\mathbf{x}_{min}, \mathcal{N}_{M^+}(\mathbf{x}_{min}))$.

Sethian 1996) and (Tsitsiklis 1995) for efficiently solving the isotropic Eikonal equation on a cartesian grid. In equation (6), the values of $\mathcal{U}$ may be regarded as the arrival times of wavefronts propagating from each point of $\mathcal{S}$ with velocity $(1/\tilde{\mathcal{P}})$. The central idea behind the FMM is to visit grid points in an order consistent with the way wavefronts propagate, i.e. with the Huygens principle. It leads to a single-pass algorithm for solving equation (6) and computing the maps $\mathcal{U}$, $\mathcal{V}$ and $\mathcal{L}$ in a common computational framework (see Table 1). The FMM is a front propagation approach that computes the values of $\mathcal{U}$ in increasing order, and the structure of the algorithm is almost identical to Dijkstra's algorithm for computing shortest paths on graphs (Dijkstra 1959). In the course of the algorithm, each grid point is tagged as either *Alive* (point for which $\mathcal{U}$ has been computed and frozen), *Trial* (point for which $\mathcal{U}$ has been estimated but not frozen) or *Far* (point for which $\mathcal{U}$ is unknown). The set of *Trial* points forms an interface between the set of grid points for which $\mathcal{U}$ has been frozen (the *Alive* points) and the set of other grid points (the *Far* points). This interface may be regarded as a set of fronts expanding from each source until every grid point has been reached (see Table 1). The key to the speed of the FMM is the use of a priority queue to quickly find the *Trial* point with the smallest $\mathcal{U}$ value. If *Trial* points are ordered in a min-heap data structure, the computational complexity of the FMM is $\mathcal{O}(N \log_2 N)$, where N is the total number of grid points.

Outputs of the routine UpdateSchemeFMM in Table 1 are estimated using a correct first order accurate scheme, for equation 6, given by Rouy an Tourin in (Rouy and Tourin 1992). The scheme is an upwind scheme: the forward and backward differences are chosen to follow the direction of the flow of information. The Euclidian length ℓ is computed in the same

manner as the minimal action map by solving the equation $\|\nabla \mathcal{L}\| = 1$ by using the same neighbors as used to solve 6 (see (Deschamps and Cohen 2002)).

3 DISTRIBUTION OF A SET OF POINTS ON A CLOSED CURVE

First, we consider the case where the domain Ω is a 2D domain. We assume that we are given an initial set $\mathcal{S} = \{\mathbf{p}_1, \ldots, \mathbf{p}_n\}$ of points on a closed curve along which a potential $\tilde{\mathcal{P}} : \Omega \to \mathbb{R}^{*+}$ takes lower values. Note that the set $\mathcal{S}$ may contain only one point.

We propose here a variant of the FMM, called the *Fast Marching Method With keypoint Detection* (FMMWKD, see Table 2), to propagate fronts from each point of $\mathcal{S}$ with velocity $(1/\tilde{\mathcal{P}})$ and sequentially detect, during the front propagation, a set of

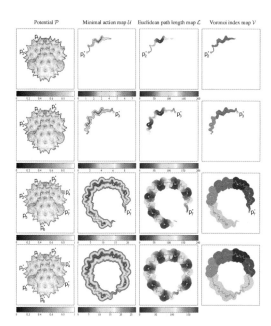

Potential $\mathcal{P}$ Minimal action map $\mathcal{U}$ Euclidean path length map $\mathcal{L}$ Voronoi index map $\mathcal{V}$

Figure 2. Intermediate and final results for the FMMWKD applied to the 2D potential of the Figure 2.**b**, with $\mathcal{S} = \{\mathbf{p}_1\}$ and $\lambda = 200$. The first, second and third rows show intermediate results obtained when are detected, respectively, $\mathbf{p}_2^*$ (the first keypoint), $\mathbf{p}_3^*$ (the second keypoint) and $\mathbf{p}_7^*$ (the last keypoint). The last row shows final results.

keypoints $\mathcal{S}^* = \{\mathbf{p}_{n+1}^*, \ldots, \mathbf{p}_{n+m}^*\}$ on the closed curve along which $\tilde{\mathcal{P}}$ takes low values. Each newly detected keypoint is immediately defined as a new source of propagation, and keypoints are detected with a criterion based on the Euclidean length of minimal paths. This criterion depends on only one parameter, denoted λ. Front propagation and keypoint detection ceases as soon as the domain visited by the fronts contains the whole curve of interest.

The final domain visited by the fronts, denoted Ω_F, correspond to a band surrounding the curve of interest. Furthermore, the FMMWKD also enables the computation of the minimal action map $\mathcal{U} : \Omega_F \to \mathbb{R}^+$, the Voronoi index map $\mathcal{V} : \Omega_F \to \{1, \ldots, n + m\}$ and the Euclidean path length map $\mathcal{L} : \Omega_F \to \mathbb{R}^+$ associated to the potential $\tilde{\mathcal{P}}$ and the set of sources $\mathcal{S} \cup \mathcal{S}^*$.

3.1 *Keypoint detection and local correction of maps $\mathcal{U}$, $\mathcal{V}$ and $\mathcal{L}$*

Initially, fronts are propagated from each point of $\mathcal{S}$ with velocity $(1/\tilde{\mathcal{P}})$, until a grid point $\mathbf{x}$ such that $\mathcal{L}(\mathbf{x}) \geq \lambda$ is tagged as *Alive*. This point is then defined as the first keypoint, denoted $\mathbf{p}_{n+1}^*$ (see Fig. 2). Such a criterion has already been used in (Deschamps and Cohen 2001) to find a minimal path given only one

endpoint and also to adapt front propagation for segmentation of tubular shapes (Deschamps and Cohen 2002). Assuming that the point $\mathbf{p}_{n+1}^*$ belongs to the Voronoi region $\mathcal{R}_j$ when it is detected, this criterion ensures that the minimal path $\mathcal{C}_{\mathbf{p}_j, \mathbf{p}_{n+1}^*}$ minimizes the integral of $\tilde{\mathcal{P}}$ (along itself) over all open curves with Euclidean lengths greater than or equal to λ and with endpoints in $\mathcal{S}$. Therefore, $\mathbf{p}_{n+1}^*$ is likely to belong to the curve along which the values of $\tilde{\mathcal{P}}$ are low.

Once the first keypoint has been detected, it is defined as a new source of propagation. It is unnecessary to restart the overall algorithm since values of $\mathcal{U}$, $\mathcal{V}$ and $\mathcal{L}$ which have already been estimated would not differ in the vicinity of initial sources (i.e. in the vicinity of points of $\mathcal{S}$). In order to limit the computational cost, one just needs to update $\mathcal{U}$, $\mathcal{V}$ and $\mathcal{L}$ in the following manner :

$$\mathcal{U}(\mathbf{p}_{n+1}^*) := 0, \quad \mathcal{V}(\mathbf{p}_{n+1}^*) := n + 1, \quad \mathcal{L}(\mathbf{p}_{n+1}^*) := 0,$$

tag $\mathbf{p}_{n+1}^*$ as *Trial* and continue front propagation. However, without any additional modification of the original FMM, final values of $\mathcal{U}$, $\mathcal{V}$ and $\mathcal{L}$ would be incorrect for grid points which are tagged as *Alive* when $\mathbf{p}_{n+1}^*$ is detected and closer (in the sense of a weighted distance) to $\mathbf{p}_{n+1}^*$ than to the initial sources. These errors would be solely due to the fact that, in the original FMM, values of $\mathcal{U}$, $\mathcal{V}$ and $\mathcal{L}$ are frozen for *Alive* points. An easy way to avoid this problem is just to let an *Alive* point be tagged as *Trial* again if it is closer to the new source of propagation than to initial sources. This algorithmic trick enables the local correction of $\mathcal{U}$, $\mathcal{V}$ and $\mathcal{L}$ in the neighborhood of $\mathbf{p}_{n+1}^*$.

Next, front propagation is continued until a grid point $\mathbf{x}$ such that $\mathcal{L}(\mathbf{x}) \geq \lambda$ is tagged as *Alive*. This point is defined as the second keypoint, denoted $\mathbf{p}_{n+2}^*$, and is added to the set of sources. Afterward, front propagation is continued, and so on. Thus, during the front propagation, keypoints are sequentially detected on the curve along which $\tilde{\mathcal{P}}$ takes low values (see Fig. 2).

3.2 *Stopping criterion for keypoint detection and front propagation*

In order to prevent the algorithm from distributing keypoints over the whole domain Ω, one needs to stop the keypoint detection as soon as the domain visited by the fronts contains the curve of interest. Note that even if this curve is unknown, we assume that it is closed. This topological assumption is used to devise a relevant criterion for stopping keypoint detection and front propagation.

One possible strategy is to take into account the Voronoi partition, and to stop keypoint detection as soon as each Voronoi region is adjacent to at least two other Voronoi regions (i.e. as soon as there exists a cycle of Voronoi regions). This strategy, although correct, is limited to the 2D case. To get a scheme which may be extended to higher dimensions, another strategy is employed in the FMMWKD. Let us denote by Ω_F the domain visited by the propagating fronts, defined as the set of grid points which are not *Far* (i.e. the set of grid points which are either *Alive* or *Trial*). In the FMMWKD, keypoint detection is stopped as soon as Ω_F becomes a simply connected subset of Ω delimited by exactly two simply connected boundaries.

The set Ω_F may be divided into two subsets : the set of interior points, denoted $int(\Omega_F)$, and the set of boundary points, denoted $\partial\Omega_F$. In the original FMM, $int(\Omega_F)$ and $\partial\Omega_F$ respectively correspond to the set of *Alive* points and the set of *Trial* points. This is no longer true in the FMMWKD because of the local correction of $\mathcal{U}$, $\mathcal{V}$ and $\mathcal{L}$ in the neighborhood of a keypoint. That is why a second labelling is introduced in the FMMWKD: each grid point which is not *Far*, in addition to being tagged as *Alive* or *Trial*, is also tagged as *Interior* or *Boundary* depending on whether it belongs to $int(\Omega_F)$ or $\partial\Omega_F$. Noting that the iteration of the marching loop at which $int(\Omega_F)$ becomes a simply connected subset of Ω is also the iteration at which the number of simply connected components of $\partial\Omega_F$ increases for the first time, we just need to monitor the topological changes of $\partial\Omega_F$.

In the algorithm detailed in Table 2, the stopping criterion for keypoint detection is satisfied as soon as the routine `IsBoundarySplit` returns *TRUE*. This routine is called after the grid point $\mathbf{x}_{min}$ is moved from the set of *Trial* points to the set of *Alive* points, once some of the $M=4$ neighbors of $\mathbf{x}_{min}$ have been tagged as *Boundary*. The routine `IsBoundarySplit` returns *TRUE* if both of the following tests are satisfied:

- *Local test for detecting a front collision.*
  First, we check if some fronts collide in the vicinity of $\mathbf{x}_{min}$. Let us denote by $\mathcal{N}_{M^+}(\mathbf{x}_{min})$ the set of $M^+ = 8$ neighbors of $\mathbf{x}_{min}$, and by $\partial\Omega_F \cap \mathcal{N}_{M^+}(\mathbf{x}_{min})$ the set of points of $\mathcal{N}_{M^+}(\mathbf{x}_{min})$ which are tagged as *Boundary*. The local test simply relies on the computation of the number of 8-connected components of $\partial\Omega_F \cap \mathcal{N}_{M^+}(\mathbf{x}_{min})$, denoted $\#C(\partial\Omega_F \cap \mathcal{N}_{M^+}(\mathbf{x}_{min}))$. Most of the time, $\mathbf{x}_{min}$ is a simple point of $int(\Omega_F)$, and $\#C(\partial\Omega_F \cap \mathcal{N}_{M^+}(\mathbf{x}_{min})) = 1$ (see Fig. 3a). The local test is satisfied if $\#C(\partial\Omega_F \cap \mathcal{N}_{M^+}(\mathbf{x}_{min})) > 1$, i.e. when there is a shock between some propagating fronts (see Fig. 3b).
- *Global test for detecting a topological change of $\partial\Omega_F$.*
  When the local test is satisfied, we need to check if the different components of $\partial\Omega_F \cap \mathcal{N}_{M^+}(\mathbf{x}_{min})$ are also disconnected at a global scale. The global test is satisfied if the front collision has split an 8-connected component of $\partial\Omega_F$ into several 8-connected components. Such a test is easy to

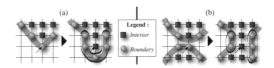

Figure 3. Local test applied in the vicinity of a grid point $\mathbf{x}_{min}$ (the point marked with an arrow) to detect a front collision. (a) $\mathbf{x}_{min}$ is a simple point of $int(\Omega_F)$ and $\#C(\partial\Omega_F \cap \mathcal{N}_{M^+}(\mathbf{x}_{min})) = 1$. (b) Two fronts have collided in the neighborhood of $\mathbf{x}_{min}$ and $\#C(\partial\Omega_F \cap \mathcal{N}_{M^+}(\mathbf{x}_{min})) = 2$.

implement. For instance, consider the case where $\#C(\partial\Omega_F \cap \mathcal{N}_{M^+}(\mathbf{x}_{min})) = 2$. Let $\mathbf{x}_1$ and $\mathbf{x}_2$ be two grid points such that $\mathbf{x}_1$ belongs to the first component of $\partial\Omega_F \cap \mathcal{N}_{M^+}(\mathbf{x}_{min})$ and $\mathbf{x}_2$ to the second. We just have to visit all grid points which belong to the same 8-connected component of $\partial\Omega_F$ as $\mathbf{x}_1$, and assign to each visited point a temporary label. Then, the global test is satisfied if $\mathbf{x}_2$ has not been labeled.

Since the scheme used to detect the iteration at which the keypoint detection has to be stopped mainly requires tests at a local scale, it is considerably less computationally expensive than globally counting the number of connected components of $int(\Omega_F)$ and $\partial\Omega_F$ at each iteration of the marching loop. Moreover, note that special care is required to deal with the fact that a propagating front may reach the border of the domain Ω. We suggest adding virtual points along each border of the discrete grid and tagging as *Boundary* every virtual point in the neighborhood of an *Interior* point lying on the border of the grid. This ensures that any connected component of $int(\Omega_F)$ is completely delimited by a connected set of *Boundary* points.

Once the keypoing stopping criterion is satisfied, no more grid points are moved from the set of *Far* points to the set Ω_F, and computations are continued until correct values of $\mathcal{U}$, $\mathcal{V}$ and $\mathcal{L}$ have been assigned to each point of Ω_F.

4 BUILDING A CYCLIC SEQUENCE OF MINIMAL PATHS TO EXTRACT A CLOSED CONTOUR

The FMMWKD may be used to extract a closed contour from a 2D image I given a single contour point $\mathbf{p}_1$ in an easy and fast manner. Once a potential $\tilde{\mathcal{P}}$ has been chosen to drive the front propagation, applying the FMMWKD with $S = \{\mathbf{p}_1\}$ gives a set of points $S \cup S^*$, but also the maps $\mathcal{U}$ and $\mathcal{V}$. We exploit the Voronoi diagramm to decide if two sources $\mathbf{p}_i$ and $\mathbf{p}_j$ of $S \cup S^*$ are adjacent. Then we look for the associated saddle point $\mathbf{m}_{i|j}$ as described in (Cohen 2001) to make two gradient descent to $\mathbf{p}_i$ and $\mathbf{p}_j$. Linking pairs of neighboring points of $S \cup S^*$ by minimal paths enable the extraction of the desired contour.

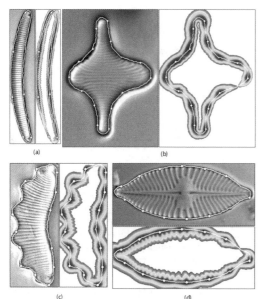

(a)   (b)

(c)   (d)

Figure 4. Extraction of a closed contour from a 2D microscopy image. Potential $\mathcal{P}$, set of sources $\mathcal{S} \cup \mathcal{S}^*$, Minimal action map and cyclic sequence of minimal paths. (a) Image size 101×521, $\lambda = 180$. (b) 385×532, $\lambda = 80$. (c) 153×380, $\lambda = 60$. (d) 1032×435, $\lambda = 160$.

5 RESULTS AND DISCUSSION

The way the FMMWKD is built ensures that λ is an upper bound of the Euclidean path length map $\mathcal{L}$ whenever a new keypoint is detected. Thus, the smaller the value given to λ is, the smaller the number of grid points visited during the front propagation is. In a sense, the FMMWKD may be regarded as a way to limit the front propagation to a small neighborhood around the manifold of interest. Better still, FMMWKD speeds up the segmentation process.

In figure 4, we show segmentation results on microscopy images obtained on a commercial computer in under a second.

6 CONCLUSION

We have presented a new fast front propagation approach for closed contour segmentation. Our method is interactive. At least one keypoint and the Euclidean length parameter λ have to be given by the user. Extending our method to higher dimensions is straightforward, but one may only get a mesh of minimal paths on a closed surface called *geodesic meshing*. Future work will include a new step based on an implicit method, to obtain a complete closed surface.

REFERENCES

Caselles, V., R. Kimmel, and G. Sapiro (1997). Geodesic active contours. *International Journal of Computer Vision 22*, 61–79.

Cohen, L. D. (2001). Multiple contour finding and perceptual grouping using minimal paths. *Journal of Mathematical Imaging and Vision 14*, 225–236.

Cohen, L. D. and R. Kimmel (1997). Global minimum for active contour models: a minimal path approach. *International Journal of Computer Vision 24*, 57–78.

Deschamps, T. and L. Cohen (2002, August). Fast extraction of tubular and tree 3d surfaces with front propagation methods. In *16th International Conference on Pattern Recognition, ICPR'02*, Quebec, Canada.

Deschamps, T. and L. D. Cohen (2001). Fast extraction of minimal paths in 3D images and applications to virtual endoscopy. *Medical Image Analysis 5*, 281–299.

Dijkstra, E. W. (1959). A note on two problems in connection with graphs. *Numerische Mathematic 1*, 269–271.

Kass, M., A. Witkin, and D. Terzopoulos (1988). Snakes: active contour models. *International Journal of Computer Vision 1*, 321–331.

Rouy, E. and A. Tourin (1992). A viscosity solution approach to shape from shading. *SIAM Journal on Numerical Analysis 29*, 867–884.

Sethian, J. A. (1996). A fast marching level set for monotonically advancing fronts. *Proceedings of the National Academy of Sciences 93*, 1591–1595.

Sethian, J. A. (1999a). Fast marching methods. *SIAM Review 41*, 199–235.

Sethian, J. A. (1999b). *Level Set Methods and Fast Marching Methods*. Cambridge University Press.

Tsitsiklis, J. N. (1995). Efficient algorithms for globally optimal trajectories. *IEEE Transactions on Automatic Control 40*, 1528–1538.

Computational Vision and Medical Image Processing – João Tavares & Natal Jorge (eds)
© 2008 Taylor & Francis Group, London, ISBN 978-0-415-45777-4

HERMES: A research project on human sequence evaluation

J. Gonzàlez
IRI-UPC, Barcelona, Spain

X. Roca & J. Villanueva
CVC/Dept. Computer Science, UAB, Bellaterra, Spain

ABSTRACT: Human Sequence Evaluation concentrates on how to extract descriptions of human behaviour from videos in a restricted discourse domain, such as (i) pedestrians crossing inner-city roads where pedestrians appear approaching or waiting at stops of busses or trams, and (ii) humans in indoor worlds like an airport hall, a train station, or a lobby. These discourse domains allow to explore a coherent evaluation of human movements and facial expressions across a wide variation of scale. This general approach lends itself to various cognitive surveillance scenarios at varying degrees of resolution: from wide-field-of-view multiple-agent scenes, through to more specific inferences of emotional state that could be elicited from high resolution imagery of faces. The true challenge of the HERMES project will consist in the development of a system facility which starts with basic knowledge about pedestrian behaviour in the chosen discourse domain, but could cluster evaluation results into semantically meaningful subsets of behaviours. The envisaged system will comprise an internal logic-based representation which enables it to comment each individual subset, giving natural language explanations of why the system has created the subset in question.

1 INTRODUCTION

Hermeneutics, according to Wilhelm Dilthey, is the art of interpretation of hidden meanings. The name comes from HERMES, the God known as the messenger of the intentions of the Gods to the human beings. In particular, interpretation in cultural sciences requires to *know* its object, a human being, from the inside. That means, we can infer the intentions of a person because we also are persons. Towards this end, the HERMES project will address basic methods for the extraction, description and animation of human motion in the same scenario (indoor or outdoor), and new methods for the interpretation of dynamic scenes.

The design and implementation of such a cognitive system still constitutes a challenge, even if the discourse domain will be drastically constrained within which it is expected to operate. An algorithmic system with analogous capabilities can be considered an instantiation of a 'cognitive system'. In particular, the term Human Sequence Evaluation (HSE) denotes the design, implementation and evaluation of such a cognitive system (Gonzàlez 2004). In general terms, we proposed to develop towards weakly embodied cognition within a system for understanding an environment containing autonomous agents. By understanding, we mean that the system must move beyond merely describing the scene: in addition it must be able to reason about the scene and give suitable explanations for various events and behaviours.

Thus, the generation of semantic descriptions conveys the meaning of motion, i.e. *where, when, what, how* and also *why* the motion is being detected. As a result, this high-level understanding provide a richer, broader and even more challenging domains of research, which will encompass not only research in Computer Vision, but also in Pattern Recognition, Artificial Intelligence and Computer Animation, to cite few.

At present, few video surveillance systems exploits all these aspects of cognition: in the HERMES project, we restrict cognition to assure HSE, that means, on the one hand, to develop transformation processes to perform human motion understanding and, on the other hand, to convey inferred interpretations to human operators by means of natural language texts or synthesized agents in virtual environments.

This paper presents how HSE considers the interpretation of human motion as a transformation process between raw video signals and high-level, qualitative descriptions. At least, this process will involve (i) the extraction of relevant visual information from a video sequence, (ii) the representation of that information in a suitable form, and (iii) the interpretation of visual information for the purpose of recognition and learning about human behaviour.

2   STATE OF THE ART

During the past three decades, important efforts in Computer Vision research have been focused on developing theories, methods and systems applied to video sequences. Broadly speaking, research is focused on describing *where and when* motion is being detected by camera sensors. For this purpose, the goal is set to describe motion using quantitative values, such as the spatial position of a given agent over time, for example.

Suitable discourse domains are, e.g., well-frequented streets, pedestrian-crossings, bus-stops, reception desks of public buildings, railway platforms. This demand in surveillance systems is due to the huge amount of video which should be selected, watched, and analyzed by a small number of operators in real time. Current textual descriptions generated automatically from surveillance sequences helps to detect abnormal and dangerous situations on-line. As a long-term result of HSE, surveillance systems will not only recognize and describe, but also *predict* abnormal or dangerous behaviours on-line, instead of merely record video sequences.

The basis of current research in any of the aforementioned domains is the detection of agents within the scene. Two different approaches are found in the literature, namely, *background modeling/substraction* and *motion detection*. The former necessitates implementing a suitable background model of the scene to determine foreground regions. Most referred publications use a background modeling-based approach (Haritaoglu 2000, Stauffer 2000, Li 2004). On the other hand, motion detection computes motion information from consecutive frames. Consequently, an action can be described in terms of a proper motion characterization (Lipton 1998, Ricquebourg 2000, Masoud 2003).

Additionally, *tracking* procedures are usually incorporated in order to reduce segmentation errors (Sanfeliu 2005) . In recent years, new tracking techniques are defined based on a hypothesis/validation principle (Comaniciu 2003). Thus, the tracking process is modeled using a probabilistic scheme, which is based on the Bayes' rule (Isard 1998, Sidenbladh 2002, Nummiaro 2003, Bullock 2004).

Tracking techniques should embed knowledge about the human agent, such as its observed motion, appearance, or shape. This knowledge can be based on *image features* or *predefined body models*. On the one hand, the spatial information of the agent state in video surveillance systems is often represented using simple image features, such as points, lines, or regions. Most popular representations are blobs (Li 1998,) or blob attributes, such as the centroid, median or bounding box. On the other hand, model-driven approaches incorporate known physical constraints of limbs and extremities of the body to help both localisation and tracking. By providing a synthetic body model, anatomical information and kinematic constraints are incorporated into the action model, thus allowing tracking of limbs, synthesis of motion, and performance analysis. Most referred models are those based on stick figures (Dockstader 2003, Karaulova 2002, Deutscher 2005), 2-D contours (Yamada 1998, Wagg 2004) and volumetric models (Gavrila 1996, Ben Aire 2002, Ning 2004).

Once the body model is properly tracked over time, it is possible to recognize predefined motion patterns and to produce high-level descriptions. In fact, the basis of motion understanding is *action recognition*. In order to deal with the inherent temporal and spatial variability of human performances, suitable analytical methods have been used in the literature for matching time-varying data. Most referred algorithms are Dynamic Time Warping (DTW), Hidden Markov Models (HMM) and Neural Networks (NN) (Galata 2001, Wang 2003).

Subsequently, human motion information is then combined with the known information about the environment in order to derive complex semantic descriptions (Gonzàlez 2004). From a semantic perspective, conceptual predicates extracted from video sequences are classified according to different criteria, such as *specialization relationship* (Karaulova 2002), *semantic nature* (Remagnino 1998) or *temporal ordering* (Intille 2001). Likewise, suitable behaviour models explicitly represent and combine the specialization, semantic and temporal relationships of their constituent semantic predicates (Nagel 1988). For this purpose, semantic primitives involved in a particular behaviour are organized into hierarchical structures, such as networks (Sagerer 1997) or trees (Wachter 1999, Kojima 2002) which allow motion understanding.

On the one hand, semantic interpretation is still mostly restricted to express the *relationships* of an agent with respect to its environment. However, the *internal state* of the agent has traditionally received little (or none) attention in human motion understanding. But human agents have inner states (based on emotions, personality, feelings, goals and beliefs) which may determine and modify the execution of their movement. These inner states are hard to be derived from a single picture. Instead, we need image sequences to evaluate emotions, like *sad*, *happy* or *angry*, in a temporal context.

Emotion descriptions will require high-detailed images which will be obtained by means of active cameras. In fact, camera's zoom are controlled to supply imagery at the appropriate resolution for motion analysis of the human face, thus facilitating emotion analysis (Cohen 2003, Zhang 2001). Current state-of-the-art is mainly concerned with posed facial expression recognition. In the proposed scenario, we would encounter spontaneous expressions that are considerably more difficult to handle. Only few publications can be found on spontaneous facial expression recognition and are mostly limited to very specific facial motions such as eye blinking.

On the other hand, semantic interpretation also leads to *uncertainty*, due to the vagueness of the semantic concepts utilized, and the incompleteness, errors and noise in the agent state's parameters (Ma 2004). In order to cope with the uncertainty aspects, integration and fusion methods can be learnt using a probabilistic framework: PCA, Mixtures of Gaussians (MoG) (Morris 2000, Fod 2002) and Belief Networks (BN) (Intille 2001, Remagnino 1998) provide examples. Alternatively, Fuzzy Metric Temporal Logic (FMTL) can also cope with the temporal and uncertainty aspects of integration in a goal-oriented manner (Schäfer 1997).

3 APPROACH TO RESEARCH

The main objective of HSE is to develop a cognitive artificial system based in a framework model which allows both recognition and description of a particular set of human behaviors arising from real-world events. Specifically, we propose to model the knowledge about the environment in order to make or suggest interpretations from motion events, and to communicate with people using natural language texts, audio or synthetic films. These events will be detected in image data-streams obtained from arrays of multiple active cameras (including zoom, pan and tilt).

HSE thus aims to design a Cognitive Vision System for human motion and behaviour understanding, followed by communication of the system results to end-users, based on two main goals. We assume that three different types of descriptions can be obtained, which depend on the resolution of the acquisition process: facial expressions, body postures and agent trajectories, where each topic demands its own specific requirements and computational models for a proper representation.

So the first goal is to determine which interpretations are feasible to be derived in each category of human motion, see Fig. 1. Consequently, for each category, suitable human-expressive representations of motion will be developed and tested. In particular, HSE will interpret and combine the knowledge inferred from three different categories of human motion, namely the motion of agent, body and face, in the same discourse domain.

The distinction between these motion categories is due to the fact that knowledge of different nature is required to interpret agent trajectories, body poses and facial expressions, since these types of interpretations strongly depend on the details of motion which can be inferred from active video cameras. The strategy is to obtain the available information at a particular level (i.e., agent), thereby providing this incomplete

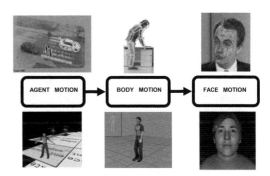

Figure 1. Human-expressive representations of motion.

knowledge to higher levels (i.e., body and face) which can update their representations as more information becomes available, and which can feedback the new information to the lower level.

The second objective of HSE is set to establish how these three types of interpretations can be linked together in order to coherently evaluate the human motion as a whole in image sequences. Such evaluation will require, at the very least, to *acquire* human motion from video cameras, to *represent* the recorded human motion using computational models, to *understand* the developments observed within a scene using high-level descriptions, and to *communicate* the inferred interpretations to a human operator by means of natural language texts or synthesized virtual agents as a visual language.

Thus, the main procedure of HSE will be the combination of:

- detection and tracking of agents while they are still some distance away from a particular location (for example a bus station, a pedestrian crossing, or a passenger in an airport, or a guest in a lobby);
- when these agents come closer to the camera, or when the active camera zooms in on these agents, their body posture will be evaluated to check for compatibility with behaviour hypotheses generated so far;
- if they are even closer and their face can be resolved sufficiently well, facial emotions will be checked in order to see whether these again are compatible with what one expects from their movements and posture in the observational and locational context which has been accumulated so far by the system.

Naturally, the interest is greater to integrate the three different components of human motion for someone approaching than someone leaving the camera. In addition, the most complex task (emotion evaluation) will come last, when the most is known already about the person in question from the preceding observations. Moreover, emotion recognition will become more specific because it can be embedded into the

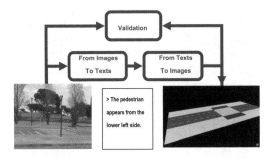

Figure 2.   Evaluation of image sequences.

context of the preceding observations and it can exploit the rigid and non-rigid motion of the face.

A suitable discourse domain comprise two types of scenarios: (i) open worlds (such as well-frequented streets, pedestrian-crossings and bus-stops), and (ii) indoor worlds (airports, train stations or lobbies, for example). Multiple active cameras will record people to infer what the humans intend to do. The main objective is the characterisation of humans to study the behaviour of people in these domains. It will be interesting whether the abilities to detect, track and characterize pedestrians would already be sufficiently advanced to reliably detect regional differences within the EU.

By implementing the aforementioned tasks, HSE will fulfill two main objectives, see Fig. 2: on the one hand, the goal will be *description*, or the generation of conceptual descriptions based on acquired and analysed motion patterns. On the other hand, the aim will be communication using *visualization*, or the generation of synthetic motion patterns based on textual descriptions.

Firstly, natural language text generation will be accommodated within HSE based on the following considerations:

- Semantic descriptions will enable researchers to check details of the conceptual knowledge base.
- Semantic descriptions will allow communication with end-users of HSE in a most natural manner.
- Semantic descriptions will support conceptual abstraction, thereby facilitating the communication of short messages or essential details, possibly in response to inquiries communicated by a microphone near the recording camera or by an UMTS mobile phone, for example for blind people.

Descriptive texts will be applied to outdoor or indoor scenes from different parts of the EU. The inclusion of videos from different parts of Europe will also constitute a means to prevent over adaptation of HSE to a small set of learning videos. In addition, once a system-internal conceptual representation has been built, it will be possible to enlarge this for natural language text generation in the languages of all groups

cooperating within HSE. Also, we will test whether the same video recordings are interpreted in different manners in different parts of Europe (or similar situations just happen in a different manner, for example people nicely queuing up at a bus station in one country and habitually cluster around the bus doors in another). Thus, on the one hand, HSE will achieve automatic translation of visual information and, on the other hand, it will be able to investigate how and why human motion may produce different descriptions, due to the cultural characteristics of the areas where a given language is spoken.

Secondly, animation will be accommodated within HSE based on the following considerations:

- Analysis-by-synthesis at the three stages of human behaviour, i.e. motion of people, their posture analysis, and their face characterisation.
- Animated computer graphics as a visual language to quickly communicate essential aspects to involved people like bus-drivers, policemen for helping people at pedestrian crossings, waiters in a lobby, etc.
- Animated computer graphics, again at three motion categories, for checking the conceptual knowledge base underlying the entire approach. Since this knowledge base is expected to grow or need adaptation throughout the project, animated computer graphics will provide the means to quickly check larger parts.

Using both natural language text generation and animation, quantitative measures and qualitative descriptions will be developed to analyze the robustness and the efficiency of the proposed cognitive system. In fact, the performance of the system will be studied by considering the following strategy (Arens 2003, Nagel 2004): let the system generate a synthetic image sequence using the textual descriptions obtained from a previously recorded image sequence. Both synthetic and original sequences can be compared to evaluate the suitability and correctness of the knowledge being considered so far.

Additionally, it will be possible to assess the results of the system by controlling the inference processes which are applied. The objectives will be:

- to trace the computational process which generates the result,
- to determine the internal information requested by the system, and
- to assess the selection of a particular interpretation.

As a result, it will become easier to debug the system. Therefore, the designer can decide to incorporate extra knowledge (by means of models, restrictions, and default options) for improving the performance of the system in terms of reliability. Also, there will be an increase of confidence of end-users in the results reported by the system: evaluation, in the sense of explanation, will ease the understanding of the results by non-expert users.

4 INNOVATION BROUGHT BY HSE RESEARCH

As an innovation, HSE proposes to develop an unified framework for human motion analysis which will be applied to confront both animation and description. Our basis is that procedures for synthetic video generation should rely on knowledge very similar to the knowledge required for textual description.

Image-sequence evaluation will be driven to incorporate assessment strategies to guide and validate the system results by:

- presenting the results of cognition using natural language texts or virtual animations, and
- arguing about inferred interpretations in order to assist and validate the system processes.

Using this know-how, we will be able to look for characterizing the behaviour of pedestrians approaching to a traffic-light-controlled pedestrian-crossing of a well-frequented inner-city street, for example. In this particular domain, a pedestrian-crossing could switch to green without grossly interfering with vehicle traffic by preparing the transition phase (green-yellow-red for vehicles) while vehicles are still some distance away. Also, switching back to green earlier, even saving gas, thus helping the environment, compared with stopping a cavalcade of vehicles in full drive after having had the pedestrians waiting for several minutes. A similar idea could survey the environment around bus-stops with an associated gain in efficiency and comfort for all involved agents. Furthermore, provided one can extend this characterization of pedestrians reliably enough, it might become possible to design special help for handicapped people.

The basic procedure of HSE will be the combination of detection and tracking of agents while they are still some distance away from a particular location. Detecting and tracking people in crowded scenes is a challenging problem as people differ in their appearance caused by various types and styles of clothing and occluding accessories, undergo a large range of movements and moreover occlude each other. Previous approaches have either used appearance-based models or local features to detect people while a majority of trackers is still based on interactive initialization.

In HSE, cooperating pan-tilt-zoom sensors will also enhance this process of cognition via controlled responses to uncertain or ambiguous interpretations. Therefore, the challenge will be to provide sensor data for each of the modules by coupling the modules together in a sensor perception/action cycle (Nakazawa 2002, Ukita 2005). These cooperating pan-tilt-zoom sensors involved in acquisition will also serve the

purpose of providing sensor data for each of the modules, but more importantly couple the other work-packages together in a sensor perception/action cycle. The use of zoom will provide a unification for interpretations of different resolution imagery, and will bestow the ability to switch the sensing process between different streams in a controlled fashion.

5 CONCLUSIONS

Multiple issues will be contemplated to perform HSE, such as detection and localization; tracking; classification; prediction; concept formation and visualization; communication and expression, etc. And this is reflected in the literature: a huge number of papers confront some of the levels, but rarely all of them. Summarizing, agent motion will allow HSE to infer behaviour descriptions. The term behaviour will refer to one or several actions which acquire a meaning in a particular context. Body motion will allow HSE to describe action descriptions. We define an action as a motion pattern which represents the style of variation of a body posture during a predefined interval of time. Therefore, body motion will be used to recognize style parameters (such as age, gender, handicapped, identification, etc.). Lastly, face motion will lead to emotion descriptions. The emotional characteristics of facial expressions will allow HSE to confront personality modeling, which would enable us to carry out multiple studies and researches on advanced human-computer interfaces. So these issues will require, additionally, assessing how, and by which means, the knowledge of context and a plausible hypothesis about he internal state of the agent may influence and support the interpretation processes.

REFERENCES

M. Arens and H.-H. Nagel. "*Behavioural Knowledge Representation for the Understanding and Creation of Video Sequences*". In Proceedings of the 26th German Conference on Artificial Intelligence (KI-2003), Hamburg, Germany; LNAI 2821, Springer–Verlag, 2003, pp. 149–163.

J. Ben-Aire, Z. Wang, P. Pandit, S. Rajaram, "*Human activity recognition using multidimensional indexing*", IEEE Trans. Pattern Analysis and Machine Intelligence 24(8) (2002) 1091–1104.

D. Bullock, J. Zelek, "*Real-time tracking for visual interface applications in cluttered and occluding situations*", Image and Vision Computing 22(12) (2004) 1083–1091.

I. Cohen, N. Sebe, A. Garg, L. Chen, and T.S. Huang. "*Facial expression recognition from video sequences: temporal and static modeling*". Computer Vision and Image Understanding, 91(1–2):160–187, 2003.

D. Comaniciu, V. Ramesh and P. Meer, "*Kernel-based object tracking*", IEEE Trans. Pattern Analysis and Machine Intelligence 25(5) (2003) 564–577.

J. Deutscher, I. Reid, "*Articulated body motion capture by stochastic search*", International Journal of Computer Vision 61(2) (2005) 185–205.

S. Dockstader, M. Berg, A. Tekalp, "*Stochastic kinematic modeling and feature extraction for gait analysis*", IEEE Trans. Pattern Analysis and Machine Intelligence 12(8) (2003) 962–976.

A. Fod, M. Mataric, O. Jenkins, "*Automated derivation of primitives for movement classification*", Autonomous Robots 12(1) (2002) 39–54.

A. Galata, N. Johnson, D. Hogg, "*Learning variable-length markov models of behaviour*", Computer Vision and Image Understanding 81(3) (2001) 398–413.

D. Gavrila, L. Davis, "*3D model-based tracking of humans in action: A multiview approach*", in: Proceedings of IEEE Conference on Computer Vision and Pattern Recognition (CVPR'96), 1996, pp. 73–80.

J. Gonzàlez. "*Human Sequence Evaluation: the Keyframe Approach*". PhD Thesis. Universitat Autònoma de Barcelona. October 2004.

Haritaoglu I., Harwood D., Davis L.S. "*W^4: Real-Time Surveillance of People and their Activities*". IEEE Transactions on Pattern Analysis and Machine Intelligence, Vol. 22, No.8, pp. 809–830, 2000.

S. Intille, A. Bobick, "*Recognized planned, multiperson action*", International Journal of Computer Vision 81(3) (2001) 414–445.

M. Isard, A. Blake, "*Condensation: Conditional density propagation for visual tracking*", International Journal of Computer Vision 29(1) (1998) 5–28.

I. Karaulova, P. Hall, A. Marshall, "*Tracking people in three dimensions using a hierarchical model of dynamics*", Image and Vision Computing 20 (2002) 691–700.

A. Kojima, T. Tamura, K. Fukunaga, "*Natural language description of human activities from video images based on concept hierarchy of actions*", International Journal of Computer Vision 50(2) (2002) 171–184.

Y. Li, S. Ma, H. Lu, "*A multiscale morphological method for human posture recognition*", in: Proceedings of Third Int. Conference on Automatic Face and Gesture Recognition, Nara, Japan, 1998, pp. 56–61.

L. Li, W. Huang, I. Gu, Q. Tian, "*Statistical modeling of complex backgrounds for foreground object detection*", IEEE Transactions on Image Processing 11(13) (2004) 1459–1472.

A. Lipton, H. Fujiyoshi, R. Patil, "*Moving target classification and tracking from real-video*", in: IEEE Workshop on Applications of Computer Vision (WACV'98), Princeton, NJ, 1998, pp. 8–14.

M. Ma, P. McKevitt, "*Interval relations in lexical semantics of verbs*", Artificial Intelligence Review 21(3–4) (2004) 293–316.

O. Masoud, N. Papanikolopoulos, "*A method for human action recognition*", Image and Vision Computing 21(8) (2003) 729–743.

T. Moeslund, E. Granum, "*A survey of computer vision based human motion capture*", Computer Vision and Image Understanding 81(3) (2001) 231–268.

R. Morris, D. Hogg, "*Statistical models of object interaction*", International Journal of Computer Vision 37(2) (2000) 209–215.

H.-H. Nagel, "*From image sequences towards conceptual descriptions*", Image and Vision Computing 6(2) (1988) 59–74.

H.-H. Nagel, "*Steps toward a Cognitive Vision System*". AI Magazine, Cognitive Vision 25(2):31–50, 2004.

A. Nakazawa, H. Kato, S. Hiura, S. Inokuchi, "*Tracking multiple people using distributed vision systems*", IEEE Int. Conf. on Robotics and Automation 2002, pp. 2974–2981.

H. Ning, T. Tan, L. Wang, W. Hu, "*People tracking based on motion model and motion constraints with automatic initialization*", Pattern Recognition 37 (2004) 1423–1440.

K. Nummiaro and E. Koller-Meier, L.J. Van Gool. "*An adaptive color-based particle filter*". Image Vision Computing 21(1), pp. 99–110, 2003.

P. Remagnino, T. Tan, K. Baker, "*Agent oriented annotation in model based visual surveillance*", in: Proceedings of International Conference on Computer Vision (ICCV'98), Mumbai, India, 1998, pp. 857–862.

Y. Ricquebourg, P. Bouthemy, "*Real-time tracking of moving persons by exploiting spatio-temporal image slices*", IEEE Trans. Pattern Analysis and Machine Intelligence 22(8) (2000) 797–808.

G. Sagerer, H. Niemann, "*Semantic networks for understanding scenes*", in: M. Levine (Ed.), Advances in Computer Vision and Machine Intelligence, Plenum Press, New York, 1997.

A. Sanfeliu and J.J. Villanueva, "*An approach of visual motion analysis*", Pattern Recognition Letters 26(3), pp. 355–368, 2005.

K. Schäfer, "*Fuzzy spatio-temporal logic programming*", in: C. Brzoska (Ed.), Proceedings of 7th Workshop in Temporal and Non-Classical Logics – IJCAI'97, Nagoya, Japan, 1997, pp. 23–28.

H. Sidenbladh, M. Black, L. Sigal, "*Implicit probabilistic models of human motion for synthesis and tracking*", in: A. Heyden, G. Sparr, M. Nielsen, P. Johansen (Eds), Proceedings European Conference on Computer Vision (ECCV), Vol. 1, LNCS 2353, Springer-Verlag, Denmark, 2002, pp. 784–800.

C. Stauffer, W. Eric L. Grimson, "*Learning patterns of activity using real-time tracking*", IEEE Trans. Pattern Analysis and Machine Intelligence 22(8) (2000) 747–757.

N. Ukita, T. Matsuyama, "*Real-time cooperative multiple-target tracking by communicating active vision agents*", Computer Vision and Iage Understanding 97(2), 2005, pp. 137–179.

S. Wachter, H.-H. Nagel, "*Tracking persons in monocular image sequences*", Computer Vision and Image Understanding 74(3) (1999) 174–192.

D. Wagg, M. Nixon, "*Automated markerless extraction of walking people using deformable contour models*", Computer Animation and Virtual Worlds 15(3–4) (2004) 399–406.

L. Wang, W. Hu, T. Tan, "*Recent developments in human motion analysis*", Pattern Recognition 36(3) (2003) 585–601.

M. Yamada, K. Ebihara, J. Ohya, "*A new robust real-time method for extracting human silhouettes from color images*", in: Proceedings of Third International Conference on Automatic Face and Gesture Recognition, Nara, Japan, 1998, pp. 528–533.

Y. Zhang, E. Sung, E. C. Prakash "*3D modeling of dynamic facial expressions for face image analysis and synthesis*". International Conference on Vision Interface, Canada, 2001.

Computational Vision and Medical Image Processing – João Tavares & Natal Jorge (eds)
© 2008 Taylor & Francis Group, London, ISBN 978-0-415-45777-4

Theory of digital manifolds and its application to medical imaging

Valentin E. Brimkov
Mathematics Department, Buffalo State College, State University of New York

Gisela Klette
Computer Science Department, University of Auckland

Reneta Barneva
Department of Computer Science, SUNY Fredonia

Reinhard Klette
Computer Science Department, University of Auckland

ABSTRACT: Digital manifolds play an important role in computer graphics, 3D image analysis, volume modeling, process visualization, and so forth – in short, in all areas where discrete multidimensional data need to be represented, visualized, processed, or analyzed. The objects in these areas often represent surfaces and volumes of real objects. In this paper we discuss some applications of digital curves and surfaces to medical imaging, implied by theoretical results on digital manifolds.

1 INTRODUCTION

Digital manifolds play an important role in various facets of the modern information society. By becoming a "digital society," the complexity of synthetic digital worlds is increasing. They often represent surfaces and volumes of real objects. This is for example the case in such fields like medicine (e.g., organ and tumor measurements in CT images, beating heart, or lung simulations), bioinformatics (e.g., protein binding simulations), robotics (e.g., motion planning), engineering (e.g., finite elements stress simulations), and security (biometrics). With the rapidly growing variety of synthetic surfaces and volumes, it is becoming critical to develop a relevant theory of digital manifolds and based on it methods for resolving a wide range of problems.

In this paper we review some actual or possible applications in medical imaging implied by some theoretical results on digital manifolds. These applications include visualization of a digitized real object, identification of its topological or geometric properties (such as its tunnels, gaps, skeleton, or boundary), as well as certain metric properties. In Section 2 we refer to some works providing a theoretical basis for the above-mentioned applications that are discussed in Section 3. We conclude with some remarks in Section 4. An extensive bibliography is provided to facilitate interested readers.

2 THEORETICAL FOUNDATIONS

Theory of digital manifolds is a vivid topic of research that is mainly driven by numerous practical applications. In this section we first briefly list and comment some literature sources containing recent developments on the subject. Then we introduce several notions playing an important role in research and related to applications presented in the subsequent sections.

2.1 Research on digital curves and surfaces

Before providing a brief overview of results on digital curves and surfaces, we recall a few basic notions. Two 3-cells (voxels) c_1 and c_2 are called *α-adjacent* iff their intersection $c_1 \cap c_2$ contains an *α-cell*, where $\alpha \in \{0, 1, 2\}$. Alternatively, two grid points $p_1, p_2 \in \mathbb{Z}^3$ are called *6-adjacent* iff $0 < d_e(p_1, p_2) \le 1$, *18-adjacent* iff $0 < d_e(p_1, p_2) \le \sqrt{2}$, and *26-adjacent* iff $0 < d_e(p_1, p_2) \le \sqrt{3}$, where d_e is the Euclidean distance.

Digital surfaces have been studied frequently over the years. For example, (Kim 1983) defines digital surfaces in $\mathbb{Z}^3$ based on adjacencies of 3-cells. A mathematical framework (based on a notion of "moves") for defining and processing digital manifolds is proposed in (Chen & Zhang 1993).

For obtaining α-surfaces by digitization of surfaces in $\mathbb{R}^3$, see (Cohen-Or et al. 1996). It is proved in (Malgouyres 1997) that there is no local characterization of a 26-connected subset S of $\mathbb{Z}^3$ such that its complement $\overline{S}$ consists of two 6-components and every voxel of S is adjacent to both of these components. (Malgouyres 1997) defines a class of 18-connected surfaces in $\mathbb{Z}^3$, proves a surface separation theorem for those surfaces, and studies their relationship to the surfaces defined in (Morgenthaler & Rosenfeld 1981). (Bertrand & Malgouyres 1999) introduces a class of "strong" surfaces and proves that both the 26-connected surfaces of (Morgenthaler & Rosenfield 1981) and the 18-connected surfaces of (Malgouyres 1997) are strong. For further studies on 6-surfaces, see (Chen et al. 1999). Digital surfaces in the context of arithmetic geometry are studied in (Brimkov et al. 2002). For various other topics related to digital manifolds we also refer to (Chen 2004; Chen 2005).

In a recent paper (Brimkov & Klette 2004) two of the authors provided the first definition of digital manifolds of involving the notion of dimension in discrete spaces (Mylopoulos & Pavlidis 1971). Accordingly, a digital curve is one dimensional while a digital surface is $(n-1)$-dimensional set of voxels, where n is the dimension of the considered discrete space. The definition allows classification of all digital manifolds with respect to the type of their "gaps." The concepts of tunnels and gaps and their relevance to certain practical problems is discussed next.

2.2 Tunnels, gaps, and skeletons

A *gap* is an important notion in discrete geometry and topology. Usually, gaps are defined through separability as follows: Let a digital object M be m-separating but not $(m-1)$-separating in a digital object D. Then M is said to have k-*gaps* for any $k < m$. A digital object without m-gaps is called m-*gapfree*. See Figure 1.

Homology groups in topology define *tunnels*, and 2-gaps are sometimes also discussed as being tunnels. Information about the number of gaps or tunnels has been a subject of interest in various disciplines, such as digital topology (Fourey & Malgouyres 2002; Ma & Wan 2000; Nakamura 2006; Srihari 1981), image analysis (Kong & Rosenfeld 1989; Lohmann 1988; Saha & Chaudhuri 1996), graph theory (White 1972), and computational modeling of 3D forms (Desburg et al. 2005). Gaps or tunnels are related to important topological concepts such as Euler characteristic and Betti numbers. See (Klette & Rosenfeld 2004) for more details.

For various applications it is useful to obtain the *skeleton* of a digital set. Skeletons represent the basic topological features of the considered object while being easier to study. They are obtained by thinning algorithms. For more details refer, e.g., to (Klette, G. 2006; Klette, G. & Pan 2004; Klette, G. & Pan 2005;

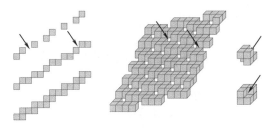

Figure 1. *Left:* From top to bottom: portions of arithmetic lines defined by $0 \le 3x - 5y < 3$, $0 \le 3x - 5y < 5$ (naive line), and $0 \le 3x - 5y < 8$ (standard line). The first one has 1-gaps (and, therefore, also 0-gaps; a 1-gap is pointed out by an arrow), the second one has 0-gaps (one of them pointed out by an arrow) but no 1-gaps, and the third one is gap-free. *Middle:* Portion of an arithmetic plane defined by $0 \le 2x + 5y + 9z < 7$. It has 2-gaps (and, therefore, also 1- and 0-gaps). A 2-gap and a 1-gap are pointed out by arrows. *Right:* Configuration of voxels (in two different orientations) that features a 0-gap (pointed out by an arrow).

Kong 2004; Palagyi & Kuba 2003; Palagyi & Kuba 1998; Palagyi et al. 2001).

3 APPLICATIONS TO VIZUALIZATION, PROCESSING, AND STRUCTURAL ANALYSIS OF DIGITIZED OBJECTS

In this section we briefly discuss possible applications of digital manifolds, mainly in the area of medical imaging.

3.1 Finding and counting gaps

Knowledge about gaps is important for ray tracing or understanding of the topology of digitized 3D sets. See (Kaufman 1987; Kaufman 1993; Kaufman 1987a; Kaufman & Shimony 1986). Assume, for example, that an unknown closed continuous surface Γ has been digitized, e.g., by a tomography scanner. Let M be the resulting digital set of voxels. If now the border $\partial(M)$ of M is determined in a way to constitute a digital surface satisfying the proposed definitions, one will have information about the type of possible gaps in that surface. The requirement for gap-freeness of $\partial(M)$ is important when a discrete model of a surface is traced through digital rays (e.g., for visualization or illumination purposes), since the penetration of a ray through the surface causes a false hole in it. Knowledge about the type of gaps of $\partial(M)$ may predetermine the usage of an appropriate type of digital rays for tracing the border in order to avoid wrong conclusions about the topology of the original continuous 3D set having the frontier Γ. Then, for the purposes of surface reconstruction, one will be able to faithfully model the geometry of the original 3D set. This is of importance for 3D imaging, e.g., in medicine.

Information about gaps is also important for ensuring correctness of representation for simulation purposes. For example, a small hole in a heart surface created by imperfections of the synthetic representation, while possibly insignificant (or simply unnoticeable) for visualization, renders the synthetic surface useless for blood flow simulation. Further, finite element simulations may yield incorrect results if surfaces have singularities. Therefore, it is of primary importance to have sound mathematical methods that can assure correctness of key topological, geometric, and metric properties of synthetic surfaces and volumes.

In (Brimkov et al. 2006) the notion of gap was generalized to higher dimensions and the following formula for the number of $(n-2)$-dimensional gaps in a digital object S has been obtained. Let S_k be the set of k-cells of S and $s_i = |S_i|$, $0 \leq k \leq n$. Then

$$g_{n-2} = -2n(n-1)s_n + 2(n-1)s_{n-1} - s_{n-2} + b,$$

where b is the number of $2^2 1^{n-2}$-blocks of S (see Brimkov et al. 2006) for denotations, definitions, and other details). In particular, the above formula counts the number of 0-gaps and 1-gaps in digital 2D/3D digital objects. A computer program (based on simple linear time algorithm) has been designed to compute the number of 0- and 1-gaps as well as other object parameters. The program also allows to visualize the digital picture S and interactively rotate it along the Ox-, Oy-, and Oz- axes so that the object can be seen from different viewpoints.

3.2 Number of tunnels

Several works address the more difficult and equally important problem of computing the number of tunnels in a digital object. An algorithm from (Saha & Chaudhuri 1996) computes the number of tunnels in a $3 \times 3 \times 3$ neighborhood of a point but not for the whole region. Several other works (Basu 2005; Basu et al. 2005; CHomP & CAPD; De Silva; Kaczynski et al. 2004; Peltier et al. 2005) provide algorithms for the problem, however, with no estimation of the computational complexity.

Using a graph-theoretical approach, in (Li & Klette 2006) the authors present a computationally efficient algorithm with a guaranteed polynomial worst case running time. There is evidence that the same approach could provide an algorithm to compute homology for digitized sets in arbitrary dimension.

3.3 Visualization, skeletonization, and measurements

Some theoretical developments related to digital manifolds are particularly relevant to the analysis

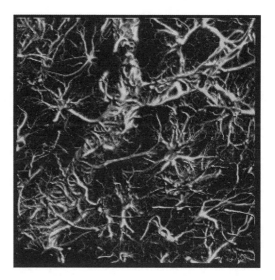

Figure 2. Example of an input data set composed of 42 slices of 256×256 density images generated by confocal microscopy from a sample of human brain tissue.

of curve-like structures in biomedical images. An ongoing research project (Klette, G. 2006) at the University of Auckland aims at analyzing confocal microscope images of human brain tissue (which contain cells called astrocytes, see Figure 2, left). These images have been taken layer by layer and constitute a volume defined on a 3D regular orthogonal grid. The curve-like structures have been obtained by applying a thinning algorithm (see Figure 2, right). (Klette, G. 2006) proposes a classification of voxels in 3D skeletons of binarized volumes for subsequent structural analysis and length measurements of digital arcs. For the former, a specific graph is associated with the skeleton (see Figure 3). The nodes of the graph, called *junctions*, exhibit certain interesting properties. However, within the proposed model they are considered as singletons that constitute the set of graph vertices. For the purposes of length measurements, the digital curves are segmented into subsequent maximum-length digital straight-line segments, and the total length of those is used to evaluate the length of the curves. For more details we refer to (Klette, G. 2006).

Note that the arcs of the skeleton form one-dimensional digital curves and as a whole the skeleton is a digital curve satisfying recent definitions from (Brimkov & Klette 2004). These properties support the segmentation process through a number of available efficient algorithms and, in turn, the curve length measurements. Note that curve-like structures appear also in other biomedical images, for example in 3D scans of blood vessels or in 3D ultrasound images.

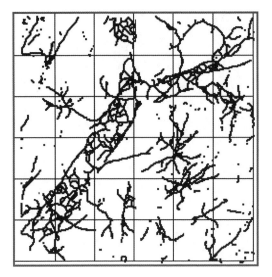

Figure 3. *Right:* A skeleton of the binarized volume shown on the left.

3.4 *Determination of object boundary*

Another possible application of the theory of digital manifolds is seen in designing new algorithms for determining the border of a digital object. Because of its importance, this problem has attracted considerable attention (see, e.g., (Daragon 2005; Kovalevsky 1989; Latecki 1988) and the bibliographies in those).

Our hypothesis is that one would benefit from an algorithm that constructs the border as a digital surface as defined in (Brimkov & Klette 2004). As already discussed earlier, the reason for this is the knowledge about the gaps in the surface.

If a digital object has been obtained by digitizing a set with a "regular" shape (e.g., featuring convexity), then, in practice, the border voxels indeed constitute a digital surface satisfying those definitions. Moreover, for data compression purposes the obtained digital surface can be "linearized" by partitioning it into polygonal portions of digital planes. The fact that any digital plane is a digital surface explains why in practice the

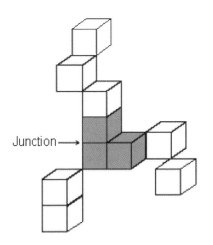

Figure 4. Grey voxels constitute a junction.

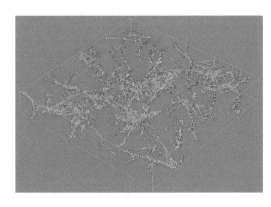

Figure 6. Large view of a sample of human brain tissue, studied within the astrocyte project. The data have been obtained by confocal microscopy and visualized in voxel view mode.

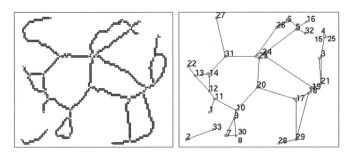

Figure 5. *Left:* Portion of the skeleton from Figure 2 (junctions are shown as small black squares). *Right:* A graph associated with the skeleton from the left. Nodes are labeled by positive integers).

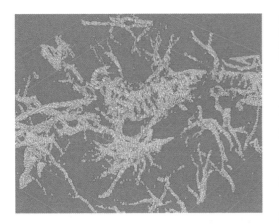

Figure 7. Enlarged view of a detail of the volume in Figure 6.

Figure 8. Further enlargements of subvolumes of the digital image of Figure 2.

requirement for two-dimensionality supports the minimization of the number of digital plane patches. For more details we refer to (Klette & Sun 2001).

In some cases however it is possible that the border voxels of a digital set do not constitute a digital surface. This usually happens when the digital object has a very complex and irregular structure. An illustration of such a complexity is provided in Figures 6, 7, and 8. They present digitized images of a human brain tissues, studied within the previously mentioned astrocyte project. In such cases, one possibility is to algorithmically "repair" the set of border voxels in order to make it two-dimensional. Some theoretical results from (Brimkov & Klette 2004) suggest that such a digitization always exists. Repairing digital objects in order to achieve desired properties has been already used by some researchers (e.g., (Siguara et al. 2005; Latechki 1988)).

4 CONCLUDING REMARKS

The purpose of this paper was to introduce the reader to ongoing research on properties of digital manifolds and related applications to medical imaging. Mathematically sound foundations may guarantee high quality rendering of objects, faultless simulations (e.g., of organ functions), and computational efficiency of the image analysis and processing. In order to achieve optimal effect, theoretical research should go in parallel with applied work. Close collaboration between specialist with diverse expertise will become increasingly important.

REFERENCES

Basu,S. 2005. Computing the first few Betti numbers of semi-algebraic sets in single exponential time. Preprint submitted to Elsevier Science, see www.math.gatech.edu/šaugata.

Basu, S., Pollack, R. & Roy, M. 2005. Computing the first Betti number and describing the connected components of semi-algebraic sets. In Proc. *STOC'05*: 304–312.

Bertrand, G., Malgouyres, R. 1999. Some topological properties of surfaces in $\mathbb{Z}^3$. *J. Mathematical Imaging Vision* 11: 207–221.

Brimkov, V.E., Andres, E., & Barneva, R.P. 2002. Object discretizations in higher dimensions. *Pattern Recognition Letters* 23: 623–636.

Brimkov, V. E., Klette, R. 2004. Curves, hypersurfaces, and good pairs of adjacency relations. In Proc. Int. Workshop Combinatorial Image Analysis: 270–284. LNCS 3322, Springer. Extended version submitted to PAMI.

Brimkov, V. E., Moroni, D. & Barneva, R. 2006. Combinatorial relations for digital pictures. In Kuba, A. et al. (eds), *Discrete Geometry for Computer Imagery*, LNCS 4245: 189–198.

Chen, L. 2004. *Discrete Surfaces and Manifolds: A Theory of Digital-Discrete Geometry and Topology*. Scientific & Practical Computing, Rockville.

Chen, L. 2005. Gradually varied surfaces and gradually varied functions. CITR-TR 156, The University of Auckland.

Chen, L., Cooley, D.H. & Zhang, J. 1999. The equivalence between two definitions of digital surfaces. *Information Sciences* 115: 201–220.

Chen, L. & Zhang, J. 1993. Digital manifolds: an intuitive definition and some properties. In Proc. *Symp. Solid Modeling Applications*: 459–460. ACM/SIGGRAPH.

CHomP (Atlanta) & CAPD (Kraków). Homology algorithms and software. www.math.gatech.edu/chomp/homology/.

Cohen-Or, D. Kaufman, A. & Kong. T.Y. 1996. On the soundness of surface voxelizations. In T.Y. Kong & A. Rosenfeld (eds), *Topological Algorithms for Digital Image Processing*: 181–204. Amsterdam: Elsevier.

Daragon, X., Couprie, M. & Bertrand, G. 2005. Discrete surfaces and frontier orders. *Journal of Mathematical Imaging and Vision*, 23(3):379–399, 2005.

Desbrun, M., Kanso, E. & Kong, Y. 2005. Discrete differential forms for computational modeling. In: *ACM SIGGRAPH 2005 Course Notes on Discrete Differential Geometry*, Chapter 7.

De Silva, V. Plex - A Mathlab library for studying simplicial homology. math.stanford.edu/comptop/programs/plex/plexintro.pdf.

Fourey, S. & Malgouyres, R. 2002. A consise characterization of 3D simple points. *Discrete Applied Mathematics* 125: 59–80.

Kaczynski, T., Mischaikow, K. & Mrozek, M. 2004. *Computational Chomology*. Applied Mathematical Sciences, Vol. 157, Berlin: Springer.

Kaufman, A. 1987. An algorithm for 3D scan-conversion of polygons. In Proc. *Eurographics*: 197–208.

Kaufman, A. 1993. Volume graphics, *Volume Graphics* 26(7): 51–64.

Kaufman, A. 1987. Efficient algorithm for 3D scan-conversion of parametric curves, surfaces, and volumes. *Computer Graphics* 21(4): 171–179.

Kaufman, A. & Shimony, E. 1986. 3D scan-conversion algorithms for voxel-based graphics. In Proc. *Workshop on Interactive 3D Graphics*: 45–75. New York: ACM.

Klette, G. 2006. Branch voxels and junctions in 3D skeletons. In R. Reulke et al. (eds), *Combinatorial Image Analysis*, LNCS 4040: 34–44.

Klette, G., Pan, M. 2004. 3D topological thinning by identifying non-simple voxels. In Proc. *Int. Workshop Combinatorial Image Analysis*, LNCS 3322, 164–175.

Klette, G. & Pan, M. 2005. Characterization of curve-like structures in 3D medical images. In Proc. *Image Vision Computing New Zealand*: 164–175.

Klette, R. & Rosenfeld, A. 2004. *Digital Geometry – Geometric Methods for Digital Picture Analysis*. San Francisco: Morgan Kaufmann.

Klette, R. & Sun, H.-J. 2001. Digital planar segment based polyhedrization for surface area estimation. In C. Arcelli, L.P. Cordella, & G. Sanniti di Baja (eds), *Visual Form*: 356–366. Berlin: Springer.

Kim, C.E. 1983. Three-dimensional digital line segments. *IEEE Trans. Pattern Analysis Machine Intelligence* 5: 231–234.

Kong, T.Y. 2004. On topology preservation in 2-D and 3-D thinning. *Int. J. Pattern Recognition Artificial Intelligence* 9: 813–844.

Kong, T.Y. & Rosenfeld, A. 1989. Digital topology: introduction and survey. *Computer Vision Graphics Image Processing* 48: 357–393.

Kovalevsky, V.A. 1989. Finite topology as applied to image analysis. *Computer Vision, Graphics, and Image Processing* 46(2): 141–161.

Latecki, L.J. 1998. *Discrete Representations of Spatial Objects in Computer Vision*. Dordrecht: Kluwer Academic Publisher.

Li, F. & Klette, R. 2006. Calculation of the number of tunnels. IMA Preprint Series 2113, April 2006.

Lohmann, G. 1988. *Volumetric Image Analysis*, Chichester: Wiley & Teubner.

Ma, C.-M. & Wan, S.-Y. 2000. Parallel thinning algorithms on 3D (18,6) binary images. *Computer Vision Image Understanding* 80: 364–378.

Malgouyres, R. 1997. A definition of surfaces of $\mathbb{Z}^3$: A new 3D discrete Jordan theorem. *Theoretical Computer Science* 186: 1–41.

Morgenthaler, D.G. & Rosenfeld, A. 1981. Surfaces in three-dimensional digital images. *Information Control* 51: 227–247.

Mylopoulos, J.P. & Pavlidis, T. 1971. On the topological properties of quantized spaces. I. The notion of dimension. *J. ACM* 18: 239–246.

Nakamura, A., Morita, K. & Imai, K. 2006. B-problem. CITR-TR-180, The University of Auckland, Computer Science Department.

Palagyi, K. & Kuba, A. 2003. Directional 3D thinning using 8 subiterations. In Proc. *Discrete Geometry Computational Imaging*, LNCS 1568: 325–336.

Palagyi, K. & Kuba, A. 1998. A 3D 6-subiteration thinning algorithm for extracting medial lines. *Pattern Recognition Letters* 19: 613–627.

Palagyi, K., Sorantin, E., Balogh, E., Kuba, A., Halmai, C., Erdohelyi, B., & Hausegger, K. 2001. A sequential 3D thinning algorithm and its medical applications. In *Proc. Information Processing Medical Imaging*, LNCS 2082: 409–415.

Peltier, S., Alayrangues, S., Fuchs, L. & Lachaud, J. 2005. Computation of homology groups and generators. In Proc. *DGCI*, LNCS 3429: 195–205.

Saha, P.K. & Chaudhuri, B.B. 1996. 3D digital topology under binary transformation with applications. *Computer Vision Image Understanding* 63: 418–429.

Siguera, M., Latecki, L.J., & Gallier, J. 2005. Making 3D binary digital images well-composed. In Proc. *Vision Geometry*, SPIE 5675: 150–163.

Srihari, S.N. 1981. Representation of three-dimensional digital images. *ACM Comput. Surveys* 13: 399–424.

White, A.T. 1972. On the genus of the composition of two graphs. *Pacific J. Math.* 41: 275–279.

Contributed Papers

Computational Vision and Medical Image Processing – João Tavares & Natal Jorge (eds)
© 2008 Taylor & Francis Group, London, ISBN 978-0-415-45777-4

Video sequences analysis for eye tracking

A. De Santis & D. Iacoviello

Dip. Informatica e Sistemistica "A.Ruberti", Università "Sapienza" of Rome, Italy

ABSTRACT: An efficient eye tracking procedure is presented providing a non-invasive method for real time detection of a subject eyes in a sequence of frames captured by low cost equipment. The procedure can be easily adapted to any subject and is adequately insensitive to the illumination changes. The eye identification is performed by an optimal approximation procedure of the frames based on a discrete level set formulation of the variational approach to the optimal segmentation problem. The segmentation yields a simplified version of the original data retaining all the information relevant to the application. No eye movement model is required being the procedure fast enough to obtain the current frame segmentation as one step update from the previous frame segmentation.

1 INTRODUCTION

A Human Computer Interface (HCI) is a complex system whose purpose consists in determining, with sufficient accuracy, the point of gaze of a subject exploring a PC screen to provide a smart tool improving the subject autonomy both in terms of communication and environment interaction, Majaranta 2002, Heckenberg 2006. The various solutions proposed have different degree of invasiveness, depending on the technology involved: generally speaking the more sophisticated the devices the more invasive and higher cost, Yu 2004, Ishima 2003, Glenstrup 1995; correspondingly signal analysis is quite simple. The use of non-invasive low cost technology demands complex algorithms to deal with the real time constraint. We consider indeed the latter situation: the subject is positioned in front of the screen of a general purpose PC endowed with a low cost video camera.

In this paper we present a robust and efficient procedure of image analysis to identify the eyes in a video sequence; this is the first block in the functional flow chart shown in Fig. 1. The other blocks are just standard and a possible implementation can be found in Villanueva 2006.

Eye detection procedures usually exploit the pupil reflectance power to perform image zoning to separate the subject head from the environment and therefore identify the pupil shape and center by template matching. As suggested in Zhu 2005, there are mainly three kinds of methods: template, appearance and feature based methods. In the first one, a deformable template is designed to fit at best the eye shape in the image; the eye identification is usually accurate but needs a

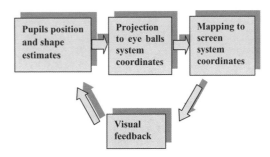

Figure 1. Flow chart of the eye tracking system.

proper initialization, it is computationally expensive and requires a good image contrast, Yuille 1992, Lam 1996. The appearance methods require large amounts of data to train a classifier with eye photometric appearance in different subjects under different face orientations, Huang 1999, Huang 2000; a good performance can be obtained provided that the classifier has a good generalization property, that depends on the completeness of the training set. Feature based methods rely on characteristics such as dark pupil, white sclera, circular iris, eye corners, etc, to distinguish the human eye from the context. In Morris 2002, for instance, a picture zoning is performed by the Turk method Turk 1992 and eye blinking detection is accomplished to initialize the location of the eyes providing the position of the feature points for the eye tracking by means of the Lucas Kanade feature tracker, Xie 1994. All these steps require a high contrast image to detect and track eye corners. In this paper a new method for eye identification in video sequences is proposed; it is an appearance-based method that relies on a novel image segmentation technique. The image segmentation process is accomplished by a region-growing algorithm based on a discrete level set formulation, De Santis 2006. This method was adapted to the case of video sequences by considering a simpler cost functional to derive the optimal segmentation, consisting just in the fit error between data and the piece-wise constant approximation. Moreover the interconnection between frames is obtained by updating the current frame segmentation starting from the previous frame optimal segmentation. Therefore we do not need to use any predictor for the eye position for picture zoning, since the discrete segmentation procedure is very fast and can easily handle the real time constraint with a general purpose PC. Moreover the usual drawbacks, mainly due to illumination changes, that may result in a fail of the predictor are avoided, due to the robustness of the region growing segmentation procedure, De Santis 2006.

The paper is organized as follows. The optimal segmentation procedure is in section 2, while the eye tracking and feature extraction procedure is presented in section 3. In section 4 the application of the proposed procedure to real video sequences is performed. Conclusions and further developments can be found in section 5.

2 THE OPTIMAL SEGMENTATION PROCEDURE

Consider a simple 2D monochromatic image I with just one object over the background; the object boundary can be represented by the *boundary set* ϕ_0 of a function $\phi : D \to \mathbb{R}$, where $D \subset \mathbb{N}^2$ is a grid of points (pixels) representing the image domain. The boundary set is defined as follows

$$\phi_0 = \left\{(i,j) : sign(\phi_{h,k}) \neq sign(\phi_{i,j}), \text{ for at least one}(h,k) \in [i \pm 1, j \pm 1]\right\}$$

Let us assume that the region $\{(i,j) : \phi_{i,j} \geq 0\}$ coincides with the object; then the pixels not belonging to ϕ_0 are either in the interior of the object or in the background. In this case it is easy to obtain a binary representation I_s of the original picture

$$I_s = c_1 \mathcal{X}_{(\phi \geq 0)} + c_2 \mathcal{X}_{(\phi < 0)} \qquad (1)$$

where

$$\mathcal{X}_{(\phi \geq 0)} = \begin{cases} 1 & \phi_{i,j} \geq 0 \\ 0 & otherwise \end{cases}, \quad \mathcal{X}_{(\phi < 0)} = \begin{cases} 1 & \phi_{i,j} < 0 \\ 0 & otherwise \end{cases} \qquad (2)$$

and $c_1, c_2 \in [0,1]$ are two different constant gray level values. Image I_s provides a *piece-wise segmentation*

of the original picture I: it is a simpler representation of the data with a clear-cut between the object and the background; the information about the object shape is preserved. Function ϕ is called *level set function* and, according to (1), operates the image segmentation. Should we need to preserve other object appearance details, the number of gray levels would be increased to four by further segmenting the regions defined by (1), by the use of two more level set functions ϕ_+, ϕ_-

$$I_s = c_{11} \mathcal{X}_{(\phi \geq 0)} \mathcal{X}_{(\phi_+ \geq 0)} + c_{12} \mathcal{X}_{(\phi \geq 0)} \mathcal{X}_{(\phi_+ < 0)} +$$
$$c_{21} \mathcal{X}_{(\phi < 0)} \mathcal{X}_{(\phi_- \geq 0)} + c_{22} \mathcal{X}_{(\phi < 0)} \mathcal{X}_{(\phi_- < 0)} \qquad (3)$$

By the same argument, the number of gray levels can be further increased, but for the eye-tracking problem the four levels segmentation (3) proved to be sufficient to represent all the information relevant to the features extraction process.

Segmentation (1) can be obtained by solving an optimal approximation problem defined as follows:

$$\min_{(c_1, c_2, \phi)} E(c_1, c_2, \phi) = \min_{(c_1, c_2, \phi)} \left[\lambda \|I - I_s\|^2 + \alpha \|\phi\|^2 \right]$$

$$= \min_{(c_1, c_2, \phi)} \left[\lambda \sum_{i,j} \left(I_{i,j} - c_1\right)^2 H\left(\phi_{i,j}\right) + \right. \qquad (4)$$

$$\left. \lambda \sum_{i,j} \left(I_{i,j} - c_2\right)^2 \left(1 - H\left(\phi_{i,j}\right)\right) + \alpha \sum_{i,j} \phi_{i,j}^2 \right]$$

where H is the Heaviside function, λ and α are two positive parameters.

The optimal solution of Problem (4) can be found by the following iterative scheme:

$$\alpha \phi_{i,j}^{n+1} + \lambda \left[\left(I_{i,j} - c_1^n\right)^2 - \left(I_{i,j} - c_2^n\right)^2 \right] \delta\left(\phi_{i,j}^n\right) = 0$$

$$c_1^n = \frac{\sum_{i,j} H\left(\phi_{i,j}^n\right) I_{i,j}}{\sum_{i,j} H\left(\phi_{i,j}^n\right)}, \quad c_2^n = \frac{\sum_{i,j}\left[1 - H\left(\phi_{i,j}^n\right)\right] I_{i,j}}{\sum_{i,j}\left[1 - H\left(\phi_{i,j}^n\right)\right]}$$

$$(5)$$

This scheme has a fast convergence to the unique optimal solution of (4) starting from any initial configuration ϕ^0, De Santis 2005. The level set method has the amenable property that during the level set evolution (5) the boundary set ϕ_0^n, starting from *any* initial shape ϕ_0^0, can *merge* and *split* in order to easily deal with the complex topology of the real world images.

Segmentation (3) is obtained by a hierarchical processing, by simply applying algorithm (5) to the regions individuated by (1), see Fig. 2.

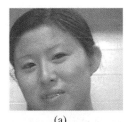

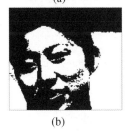

(a)

(b)   (c)

Figure 2. Image segmentation by the discrete level set procedure; a) original; b) binarization; c) four levels segmentation.

3   FEATURES EXTRACTION AND EYE TRACKING

To the purpose of selecting a set of eye features that can be reliably tracked in a video sequence, a four levels segmentation suffices for two reasons: first, the four levels provide a higher contrast image than the original data (with 256 levels) easier to analyze since the principal elements of a human face are preserved and well separated one to another and from the rest of the scene. Then single elements, such as the eye brows, the eyelids, the irises, the pupils, the nostrils, the mouth and so for, can be individually addressed and their shape characterized by a proper set of quantities. The pupils need to be identified and their position and deformation measured: these data is what the eye tracking system needs to correctly determine the line-of-gaze of the subject and therefore the point of the screen the subject is looking at. Among the parameters that can be used to describe the position and deformation of the pupil we can consider: the centroid coordinates, the area, the major and minor axes and the inclination of the ellipse that best fits the pupil, the distance between the centroids of the right and left pupil. These quantities can be accurately estimated from the first frame of the video and stacked in two features vectors that represent the *eyes signature,* Fig. 3. Since even in low cost video cameras the acquisition rate is at least 15 frames per second, the scenes captured in adjacent frames do not differ significantly; therefore, moving from one frame to another, false detections can be avoided by simply matching the current frame features vectors to those of the eyes signatures of the previous frame.

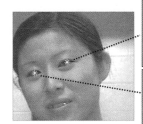

Centroid = (72.9, 65.8)
Area = 82
Major Axis = 17.02
Minor Axis = 6.7
Orientation = 15.03

Centroid = (29.7, 79.6)
Area = 89
Major Axis = 19.6
Minor Axis = 7.28
Orientation = 15.03

Figure 3. Estimation of eyes position and some shape parameters.

The eye-tracking algorithm requires that the current frame is quickly segmented so to obtain in real time the features vectors to be matched to the eyes signature vectors of the previous frame. To this aim we exploit the efficiency of the procedure described in Section 2; the current frame segmentation can be accurately and quickly obtained starting, as initial configuration of the level set function, from the configuration giving the previous frame optimal segmentation.

4   EXPERIMENTS

The standard set-up we refer to considers a person that sists in front of a PC and watches the screen. The screen is endowed with a single low cost video camera pointed on the subject face. The videotaping occurs in a regular room illumination; no additional light sources are considered, like IR lamps. To test the proposed algorithm in critical conditions, we did not produce any ad hoc video sequence, but rather used two sequences taken from the web, courtesy of Prof. Qiang Ji, http://www.ecse.rpi.edu/homepages/cvrl/database/database.html

The tracking procedure we designed is summarized in the flow chart of Fig. 4.

The signal processing step may be required for signal equalization, like gamma correction or histogram equalization. This is useful when the gray level range of the data is too compressed, as in the case of the sequences we analyzed, where a gamma equal to 0.5 was used. Then a four levels segmentation was performed according to the hierarchical processing described in Section 2, with the following choice of the parameters: $\lambda = 10^3, \alpha = 1$. These choices proved to be suitable for both sequences and did not require any special calibration. The high value of λ ensures the segmentation accuracy, especially for low contrast images, and a fast convergence of equation (5).

The difference in magnitude between λ and α makes the two terms of the cost functional E comparable; indeed, the value of the level set function ϕ are of two

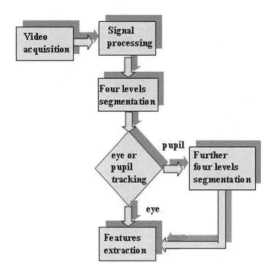

Figure 4. Video sequence segmentation flow chart.

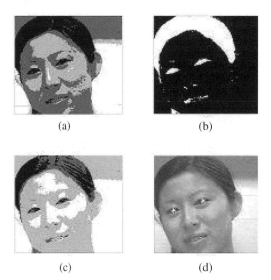

Figure 5. a) four levels segmentation; b) binary image obtained by Boolean operation on Fig. 5a; c) four levels segmentation of the two regions of Fig. 5b); d) detected pupils marked on the original picture.

orders greater than those of the fit error. On the other hand, the α term ensures the existence and uniqueness of the optimal solution by making problem (4) convex.

The four levels segmentation provides a *cartoon image* of the original data where the eyes always belong to the darkest part, whatever the colour of the eyes, the race of the subject. Should we be interested just in the position of the eyes, this segmentation would suffice to identify all the eye parameters of interest, see Fig. 5a.

On the contrary, to track the position of the pupils we need a more detailed segmentation to distinguish the pupil from the other eye elements. To this aim a binary image can be obtained from the previous four levels segmentation (Fig. 5a), by a simple boolean operation, thus highlighting the dark elements (and therefore the eyes) and leaving the rest of the picture in the black background, see Fig.5b. The two regions of this binary image are then further segmented obtaining again a four levels segmentation (Fig. 5c) where, this time, the pupil is well separated from the other elements and its position and shape properties can be easily measured, Fig. 5d. Once the elements of interest, either the eyes or the pupils, are well separated, the features vectors can be built. Many other anthropometric parameters could be determined on the segmented image to increase the information extracted from data, depending on the complexity of the task to be accomplished.

When analyzing successive frames an accurate segmentation can be obtained in just one run of equation (5), starting from the previous frame segmentation, since the scene does not drastically vary in adjacent frames. This is accomplished by considering as ϕ^0 the steady-state level set function obtained in the first binarization of the previous frame. In this respect it is important to have a more accurate segmentation for the first frame; five runs proved to be enough for this purpose. Moreover the pair of interest (eyes or pupils) this time can be reliably identified among the segmented objects on the current frame, since the features vectors have minimal distance from the analogous pair of vectors of the previous frame. We remark that the efficiency of the segmentation procedure described in Section 2 allows us to obtain the optimal segmentation from frame to frame in just one run of equation (5) and it is exactly this property that makes the algorithm able to deal with the real time constraint.

In Fig. 6, eight frames of the first video sequence are displayed; the subject performs a wide head movement so that various eyes orientations are present; even though eye occlusion does not really happen, a very critical situation occurs when the subject is looking up. The algorithm was able to track the pupils in all the 300 frames and the error on the pupils position, compared to the positions estimate provided by IR illumination (as described in Zhu 2005) was characterized by an average of 1 pixel with a standard deviation of 0.8.

Experiment of Fig. 7 shows the robustness of the method with respect to eye occlusions; in this case eyes tracking was just performed. When one eye, or even both, is lost because of the occlusion, the algorithm keeps memory of the last position correctly identified; this is updated as soon as the occlusion is removed.

The algorithm has featured encouraging performances both in term of accuracy of the eyes or pupils identification and speed, this last being fundamental for real time applications.

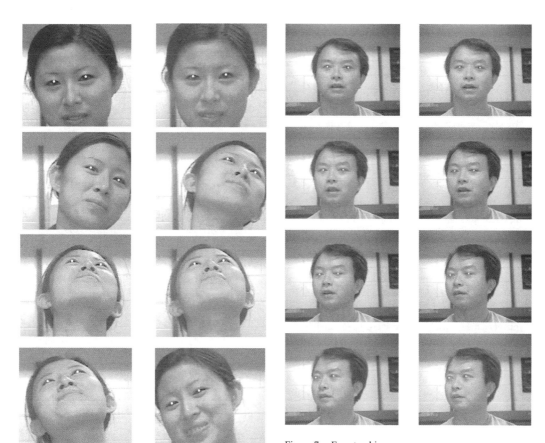

Figure 6.   Pupils tracking.

Figure 7.   Eyes tracking.

5   CONCLUSIONS

Eye tracking systems should have some key characteristics: non-invasive low cost equipment, extremely simple calibration and robustness with respect to changes in illumination conditions. The authors refer to a standard set up in which the subject is in front of a workstation endowed with a regular video camera; no specific source of illumination exploiting the retinal reflectance power is considered. An eye tracking system is a complex ensemble of units performing the ultimate tasks of identifying the point of the screen the subject is looking at. This paper is concerned with the signal data analysis unit that is supposed to track the subject eyes on the images captured by the camera. To this purpose a novel tracking procedure was designed, based on a discrete level set formulation of the optimal segmentation problem. This procedure is reliable: it enjoys the accuracy and robustness of the region based segmentation procedures, whereas high numerical efficiency is obtained due to the discrete formulation. These characteristics indeed allow us to analyze sequences of images in real time since the optimal segmentation of the current frame is obtained by a simple and fast update of the previous frame segmentation. The frame segmentation provides a simple image representation with all the elements of interest clearly separated one to another. For eye tracking the pupil position and shape can be accurately estimated also in critical conditions where the eyes are semi-closed or occluded. The method does not require any particular calibration procedure.

ACKNOWLEDGEMENT

The Authors are indebted to Prof. Qiang Ji, of the Electrical, Computer and System Engineering Department of Renssealer Polytechnic Institute, Troy, New York, who granted them the use of video data taken from Intelligent System Lab data base at http://www.ecse.rpi.edu/homepages/cvrl/database/database.html

REFERENCES

P. Majaranta, K.J. Raiha. "Tewnty Years of Eye Typing: System and Design Issues", in Eye tracking Research & Applications: Proceedings of the Symposium on ETRA, 15–22, New York, 2002.

D. Heckenberg. " Performance Evaluation of Vision-Based High DOF Human Movement Tracking: A Survey And Human Computer Interaction Perspective", in Proceedings of the Conference on Computer Vision and Pattern Recognition Workshop, 156–164, 2006.

L.H. Yu, M. Eizenman. "A New Methodology for Determining Point-of Gaze in Head-Mounted Eye Tracking Systems", IEEE Trans. on Biomedical Engineering, 10:51, 1765–1773, 2004.

D. Ishima, Y. Ebisawa. "Eye Tracking Based on Ultrasonic Position Measurement in Head Free Video-Based EYE-Gaze Detection", Proceedings of IEEE EMBS Asian-Pacific Conference on Biomedical Engineering, 258–259, 2003.

A. Glenstrup, T. Angell-Nielsen. "Eye Controlled Media, Present and Future State",technical report, University of Copenaghen, http://www.diku.dk/users/panic/eyegaze/, 1995.

A. Villanueva, R. Cabeza, S. Porta. "Eye tracking: Pupil orientation geometrical modeling", Image and Vision Computing, 24, 663–679, 2006.

Z. Zhu, Q. Ji. "Robust real-time eye detection and tracking under variable lighting conditions and various face orientations", Computer Vision and Image Understanding, 98, 124–154, 2005.

A. Yuille, P. Hallinan, D. Cohen. "Feature extection from faces using deformable templates", International Journal of Computer Vision, 8:2, 99–111, 1992.

K.M. Lam, H. Yan. "Locating and extecting the eye in the human face images", Pattern Recognition, 29, 791–779, 1996.

J. Huang, H. Wechsler. "Eye detection using optimal wavelet packets and radial basis functions", International Journal of Pattern Recognition and Artificial Intelligence, 13:7, 1009–1025, 1999.

W.M. Huang, R. Mariani. "Face detection and precise eyes location", Proceedings of the International Conference on Pattern Recognition, 2000.

T. Morris, P. Blenkhorn, F. Zaidi. "Blink detection for real-time eye tracking", Journal of Network and Computer Applications, 25, 129–143, 2002.

M.A. Turk. "Interactive-Time Vision: Face Recognition as a Visual Behaviour", MIT, Doctoral Thesis, 1992.

X. Xie, R. Sudhakar, H. Zhuang. "On improving eye feature extection using deformable templates", Pattern Recognition, 27, 791–799, 1994.

A. De Santis, D. Iacoviello. "Optimal segmentation of pupillometric images for estimating pupil shape parameters", Computer Methods and Programs in Biomedicine, special issue on Medical Image Segmentation, 84, 174–187, 2006.

A. De Santis, D. Iacoviello. "Discrete level set approach to image segmentation", Technical Report n. 13, http://www.dis.uniroma1.it/~iacoviel/Articoli/DeSantis_new_2006.pdf.

Computational Vision and Medical Image Processing – João Tavares & Natal Jorge (eds)
© 2008 Taylor & Francis Group, London, ISBN 978-0-415-45777-4

Multiscale analysis of short-term cardiorespiratory signals

L. Angelini
TIRES-Center of Innovative Technologies for Signal Detection and Processing, University of Bari, Italy
Dipartimento Interateneo di Fisica, Bari, Italy
Istituto Nazionale di Fisica Nucleare, Sezione di Bari, Italy

T.M. Creanza
Istituto di Studi sui Sistemi Intelligenti per l'Automazione, C.N.R., Bari, Italy

R. Maestri
Divisione di Cardiologia e Bioingegneria, Fondazione Salvatore Maugeri, IRCCS di Montescano (PV), Italy

D. Marinazzo & M. Pellicoro
TIRES-Center of Innovative Technologies for Signal Detection and Processing, University of Bari, Italy
Dipartimento Interateneo di Fisica, Bari, Italy
Istituto Nazionale di Fisica Nucleare, Sezione di Bari, Italy

G.D. Pinna
Divisione di Cardiologia e Bioingegneria, Fondazione Salvatore Maugeri, IRCCS di Montescano (PV), Italy

S. Stramaglia
TIRES-Center of Innovative Technologies for Signal Detection and Processing, University of Bari, Italy
Dipartimento Interateneo di Fisica, Bari, Italy
Istituto Nazionale di Fisica Nucleare, Sezione di Bari, Italy

S. Tupputi
Dipartimento Interateneo di Fisica, Bari, Italy
Istituto Nazionale di Fisica Nucleare, Sezione di Bari, Italy

ABSTRACT: We present the multi-scale entropy analysis of short-term physiological time series of simultaneously acquired samples of heart rate, blood pressure and lung volume, from healthy subjects and from subjects with Chronic Heart Failure. Evaluating the complexity of signals at the multiple time scales inherent in physiologic dynamics, we find new quantitative indicators which are statistically correlated with the pathology. As the multi-scale entropy analysis has been applied up to now to 24 hours electrocardiographic signals, these results on short-term recordings enlarge the applicability of the method. In the same spirit of the multi-scale entropy approach, we also propose a multi-scale approach, to evaluate interactions between time series, by performing a multivariate autoregressive modelling of the coarse grained time series. We then address the problem of classifying a subject as healthy or affected by Chronic Heart Failure on the basis of all the collected indicators.

1 INTRODUCTION

Physiological systems are ruled by mechanisms operating across multiple temporal scales. The underlying dynamics of such systems spans a great range of frequencies, therefore a complete analysis must include a multiple time-scale approach. A recently proposed approach, multiscale entropy analysis (MSE) (Costa et al. 2002), where the degree of complexity of time series is compared at varying temporal scale.

In this paper we show that MSE analysis may also be successfully applied to short term physiological recordings, still obtaining relevant information about the underlying mechanisms. In particular, we consider simultaneous recordings of electrocardiogram, respiration signal and arterial blood pressure, and discuss the ability to diagnose Chronic Heart Failure (CHF), a disease associated with major abnormalities of autonomic cardiovascular control.

Besides entropy-based methods, interactions between time series have been widely studied with linear predictive models (Korzeniewska et al. 2003). We thus implement the multiscale paradigm in this frame by considering a multiscale version of the classical multivariate autoregressive (AR) analysis of time series, to find scale-dependent patterns of interactions between the physiological time series here considered.

Once we have obtained the indicators described above at different time scales, we address the problem of optimal feature selection to use some of them as parameters for a classifier.

2 MODEL

We briefly recall the MSE method (Costa et al. 2002). Given a one-dimensional discrete time series, consecutive coarse grained time series, corresponding to scale factor τ, are constructed in the following way. First, the original time series is divided into non overlapping windows of length τ, where the length is defined in unit of samples; then, data points inside each window are averaged, so as to remove fluctuations with time scales smaller than τ. For scale one, the coarse grained time series is simply the original time series; the length of each coarse grained time series is equal to the length of the original time series divided by the scale factor τ. Finally an entropy measure S_E is calculated for each coarse grained time series and plotted as function of the scale factor τ. S_E coincides with the parameter $S_E(m,r)$, introduced by Pincus (Pincus 1991) and Richman and Moorman (Richman and Moorman 2000), termed *sample entropy*, and is related to the probability that sequences from the time series, which are close (within r) for m points, remain close at the subsequent data point.

Next, we quantify the interactions between the time series at different time scales, in the frame of linear predictive models. To do so, we implement a multiscale version of AR modeling of time series. For each scale factor τ, we denote $\mathbf{x} = (rri, sap, dap, ilv)$ the four-dimensional vector of the coarse grained time series of the lung volume (ilv), R-R interval (rri), systolic (sap) and diastolic (dap) blood pressure. At each scale, all coarse grained time series are normalized to have unit variance (see, e.g., (Kantz and Schreiber 1997)). A multivariate AR model of unity order is then fitted (by standard least squares minimization) to data:

$$\mathbf{x}(n) = C\,\mathbf{x}(n-1); \qquad (1)$$

C is a 4×4 matrix, depending on τ, whose element C_{ij} measure the causal influence of j–th time series on the i–th one. Note that, unlike the typical AR modeling and in the spirit of the multiscale approach, here the order of the AR model is kept fixed and small: $m = 1$. Indeed information from longer and longer time scales is processed, in the AR fitting, as τ is increased, and the variations with τ of the elements of matrix C describe interactions between time series as a function of the scale.

3 DATA

Our data are from 47 healthy volunteers (age: 53 ± 8 years, M/F: 40/7) and 275 patients with chronic heart failure (CHF) (age: 52 ± 9 years, Left Ventricular Ejection Fraction: $28 \pm 8\%$, New York Heart Association class: 2.1 ± 0.7, M/F: 234/41), caused mainly by ischemic or idiopathic dilated cardiomyopathy (48% and 44% respectively), consecutively referred to the Heart Failure Unit of the Scientific Institute of Montescano, S. Maugeri Foundation (Italy) for evaluation and treatment of advanced heart failure. Concerning the second group, cardiac death occurred in 54 (20%) of the patients during a 3-year follow-up. The cardiorespiratory data were collected both in conditions of spontaneous breathing, and in regime of paced breathing (Cooke et al. 1998; Rzeczinski et al. 2002). Paced breathing is a simple experimental procedure that allows a better standardization in the measure of spectral indexes of cardiovascular variability (see for example (Pinna et al. 2006) for a detailed discussion). After instrumentation and calibration were completed, the subjects, in supine position, carried out a session of familiarization with the paced breathing protocol. They were instructed to follow recorded instructions to breath in and out at a frequency of 0.25 Hz. After an initial trial, the subject were asked whether they felt comfortable with the paced breathing frequency, and an adjustment was made within $\pm 10\%$ of the target value. After signal stabilization, all the subjects were told to breath spontaneously, while they underwent a 10 min recording of ECG, lung volume (Respitrace Plus, Non-Invasive Monitoring Systems), and nonivasive arterial blood pressure (Finapres 2300, Ohmeda). The recordings were then repeated, again for 10 minutes, in the regime of paced breathing. R-R interval (resolution 1 ms), and systolic and diastolic arterial pressure values were obtained from the ECG and arterial pressure signals using custom made software (Maestri and Pinna 1998). The lung volume, R-R interval, systolic and diastolic pressure time series were re-sampled at a frequency of 2 Hz using a cubic spline interpolation.

4 RESULTS

4.1 *Multiscale entropy*

In figure 1 we depict the average S_E of *rri* time series of controls, patients and dead patients, in basal

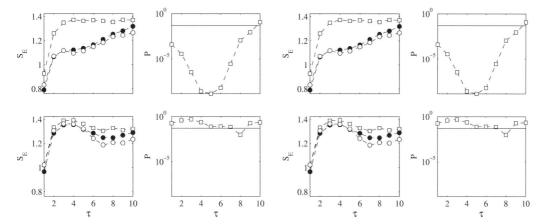

Figure 1. Sample entropy of *rri* time series plotted versus τ. Empty squares are the averages over the 47 healthy subjects, full circles are the averages over the 275 CHF patients, and empty circles are the averages over the 54 patients for whom cardiac death occurred. Top left: S_E in basal condition. Top right: the probability that basal S_E values from controls and patients were drawn from the same distribution, evaluated by non parametric test. Bottom left: S_E in paced breathing condition. Bottom right: the probability that paced breathing S_E values from controls and patients were drawn from the same distribution, evaluated by non parametric test.

Figure 2. Sample entropy of *sap* time series plotted versus τ. Empty squares are the averages over the 47 healthy subjects, full circles are the averages over the 275 CHF patients, and empty circles are the averages over the 54 patients for whom cardiac death occurred. Top left: S_E in basal condition. Top right: the probability that basal S_E values from controls and patients were drawn from the same distribution, evaluated by non parametric test. Bottom left: S_E in paced breathing condition. Bottom right: the probability that paced breathing S_E values from controls and patients were drawn from the same distribution, evaluated by non parametric test.

condition (high) and paced breathing (low). On the right we depict, as a function of the scale factor τ, the probability that *rri* entropy values from controls and patients were drawn from the same distribution, evaluated by non parametric Mann-Whitney test (Mann and Whitney 1947): the discrimination is excellent at intermediate τ's. It is worth mentioning that here we compare subjects from the same age group, so that the decrease in heart rate variability cannot be accounted to age rather than disease. In the case of paced breathing the three curves get closer and the discrimination, between patients and controls, is decreased: thus paced breathing, in the case of *rri* entropy, reduces differences between patients and controls. We note that, on average, patients for whom Cardiac Death occurred show further reduced S_E with respect to the remaining CHF patients, thus suggesting that the general *complexity–loss* theory also applies to the severity of the pathology. According to this theory, de-complexification of systems is a common feature of pathological conditions, as well as of aging; when physiologic systems become less complex, their information content is degraded. As a result, they are less adaptable and less able to cope with the exigencies of a constantly changing environment.

In figure 2 we depict S_E of systolic arterial pressure (*sap*) time series. We find that at low τ patients have higher entropy, whilst at large τ they have lower entropy than controls. The crossover occurs at $\tau = 3$ in

basal conditions, and $\tau \sim 6$ for paced breathing. The *complexity-loss* paradigm, hence, here holds only for large τ. This may be explained as an effect of respiration, whose influence becomes weaker as τ increases. This effect is more evident in conditions of paced breathing, due to synchronization (Rzeczinski, Janson, Balanov, and McClintock 2002; Schafer, Rosenblum, Kurths, and Abel 1998). Our results are consistent with those obtained in (Ancona , Maestri, Marinazzo, Nitti, Pellicoro, Pinna, and Stramaglia 2005) using a different approach and with $\tau = 1$. It is interesting to observe that curves corresponding to dead patients are farther, from the controls curve, than the average curve from all patients, up to $\tau = 7$; departure from the controls curve appears to be connected with the severity of the disease.

Now we turn to consider *ilv* time series, as depicted in figure 3. In the basal case, controls have higher entropy at small scales. On the other hand controls show lower entropy than patients at $\tau > 7$: patients pathologically display fluctuations of *ilv* at larger scales than healthy subjects. Under paced breathing, controls are characterized by reduced fluctuations at high τ; at $\tau = 4$, when the window size is half of the respiration period, controls show a local minimum of the entropy. These phenomena are not observed for patients, where paced breathing is less effective in regularizing the *ilv* time series. At this point, a comment is worth about the choice of parameters m and r for the

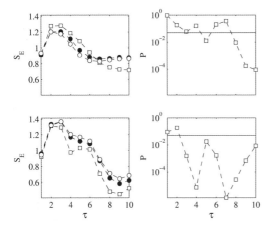

Figure 3. Sample entropy of *sap* time series plotted versus τ. Empty squares are the averages over the 47 healthy subjects, full circles are the averages over the 275 CHF patients, and empty circles are the averages over the 54 patients for whom cardiac death occurred. Top left: S_E in basal condition. Top right: the probability that basal S_E values from controls and patients were drawn from the same distribution, evaluated by non parametric test. Bottom left: S_E in paced breathing condition. Bottom right: the probability that paced breathing S_E values from controls and patients were drawn from the same distribution, evaluated by non parametric test.

calculation of sample entropy. The value of r determines the level of accepted noise; in the multiscale approach, m is kept fixed and small so that the influence of longer and longer time scales is probed as τ is increased. We verify that our results do not depend on the choice of m and r. Indeed, we find that as the value of r increases (m increases) the values of S_E, for both patients and controls, decrease. However the consistency of S_E values is preserved.

In order to show now that the dynamics of the system is actually being measured, this analysis is applied to surrogate data, obtained by random shuffling of the temporal ordering of data samples. We observed that any structure and any discrimination between the three classes of subject is lost in this case. Similar results are obtained with the other quantities in both regimes.

We also evaluate the diagnostic power of the indicators above described, by measuring the area under the Receiver-Operating-Characteristic (ROC) curve (Swets 1988), a well-established index of diagnostic accuracy; the maximum value of 1.0 corresponds to perfect assignment (unity sensitivity for all values of specificity) whereas a value of 0.5 arises from assignments to a class by pure chance. The results concerning the entropic indicators are reported in Table 1.

4.2 Multiscale AR analysis

Coming to the AR analysis, in Table 2 we list the causal relationships which we find to be the most

Table 1. Area under ROC curve for MSE analysis.

Series	τ	Area
rri, basal	5	0.7612
rri, paced breathing	8	0.6222
sap, basal	1	0.6884
sap, paced breathing	1	0.7388
dap, basal	1	0.7705
dap, paced breathing	1	0.7882
ilv, basal	9	0.6642
ilv, paced breathing	7	0.7258

Table 2. Relevant P-values for the separation between causal relationships in patients and control subjects.

Interaction	τ	P-value
$rri \rightarrow sap$, basal	1	1.17×10^{-09}
$rri \rightarrow sap$, paced breathing	1	5.76×10^{-11}
$dap \rightarrow rri$, basal	1	1.61×10^{-05}
	6	5.54×10^{-09}
$dap \rightarrow rri$, paced breathing	1	5.63×10^{-05}
	6	8.75×10^{-06}
$rri \rightarrow ilv$, basal	4	3.76×10^{-05}
$rri \rightarrow ilv$, paced breathing	4	2.09×10^{-05}
$ilv \rightarrow rri$, basal	2	8.13×10^{-06}
$rri \rightarrow rri$, basal	8	6.50×10^{-09}
$rri \rightarrow rri$, paced breathing	8	8.22×10^{-04}
$dap \rightarrow dap$, basal	1	1.61×10^{-05}
$dap \rightarrow dap$, paced breathing	1	5.63×10^{-05}
$sap \rightarrow dap$, basal	1	5.86×10^{-09}
$sap \rightarrow dap$, paced breathing	1	1.96×10^{-08}
	8	5.39×10^{-07}
$ilv \rightarrow sap$, paced breathing	4	5.32×10^{-04}

discriminating between patients and the healthy subjects (controls). Firstly we consider the interactions between heart rate and blood pressure. In physiological conditions heart rate and arterial pressure are likely to affect each other as a consequence of the simultaneous feedback baroreflex regulation from *sap-dap* to *rri* and feedforward mechanical influence from *rri* to *sap-dap* (Miyakawa et al. 1984).

The interactions between blood pressure and heart rate are weaker for patients with respect to controls.

Concerning the interaction of respiration with heart rate, originating the well known phenomenon of respiratory sinus arrhythmia, we find that the interaction $rri \rightarrow ilv$ is significantly stronger in controls than patients, under paced breathing and using $\tau = 4$. We also find that the interaction $ilv \rightarrow rri$ is negative and significantly stronger in patients, in basal conditions and at high frequencies ($\tau \leq 4$).

It is known that respiration interacts in an open loop way with arterial pressure, mainly through a mechanical mechanism. Our findings confirm it; indeed we find no significant $sap \rightarrow ilv$ interaction, but significant differences between patients and controls are found when the interaction $ilv \rightarrow sap$ is considered: controls show increased interaction w.r.t. patients.

Also for AR modeling we have applied the ROC analysis: for the two classes of controls and patients, in many cases we find fair discrimination (area between 0.7 and 0.8) or good discrimination (between 0.8 and 0.9). The best ROC area is 0.874 and obtained using the values of $dap \rightarrow dap$ interaction at $\tau = 1$ and in basal conditions. Excellent (i.e., between 0.9 and 1) discrimination performances are obtained considering pairs of indicators and using Fisher linear discriminant (FLD) analysis to find the best linear combination of the two indicators; for example combining basal $dap \rightarrow dap$ interaction at $\tau = 1$ (alone provides 0.874 ROC area) and basal $rri \rightarrow sap$ interaction at $\tau = 1$ (alone provides 0.778 ROC area), leads to a FLD with 0.941 ROC area.

4.3 Feature selection

We now address the problem of classifying the subjects as healthy or CHF patients on the basis of several measured quantities. Formally, given a set of n objects, characterized by p features, we must find a *prediction function* able to assign a label to each object as a function of the given features. Our variables are the entropies of the rri, sap, dap and ilv time series, together with the coefficients of the AR model, at different time scales $\tau = 1 \dots 10$, in two respiration regimes: in total 400 variables for 303 subjects. We want to find the subset of the q(out of p) most important features in order to separate the two classes. Our classifier is based on the Fisher's linear discriminant: we look for the equation of the hypersurface which separates our points in two groups, such that the difference between the averages of the two groups is maximized, and the variance within each group is minimized.

To better evaluate the goodness of a classifier when its dimensions are comparable with the dimension of the sample we used the regularized least squares method. We chose as regularization parameter the minimizer of the leave-one-out (LOO) error (Vapnik 1998). Using this selection criterion we applied three different techniques to reduce the number of subsets which have to be compared: the Forward Selection (FS) and the Backward Elimination (BE), and the Recursive Feature Elimination (RFE) (see (Guyon and Elisseeff 2003) for a complete review). The errors are reported in figure 4, compared with the curves which would be obtained using a random permutation of the variables. BE method is more effective for high values of $q(q > 100)$, while FS performs better for small $q(q < 50)$. In the intermediate region BFE seems to be

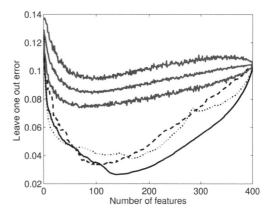

Figure 4. Leave-one-out error as a function of the dimension of the subsets individuated by three different selection procedures: FS(dotted line), RFE(full line), BE(dashed line). The random curve and the $\pm 95\%$ confidence intervals are also reported in grey.

Table 3. LOO errors.

Method	q_{min}	LOO error
RFE	96	0.0331 ± 0.0033
BE	137	0.0265 ± 0.0036
FS	168	0.0377 ± 0.0029

the best choice. For the sake of simplicity we suggest to choose the subset for which the minimum value of the leave-one-out error is attained at smallest $q(96)$, as long as the corresponding error is comparable with the minimum possible error, obtained with BE method.

5 CONCLUSIONS

We have successfully extended the MSE analysis to short term time series coming from a cardiovascular care center. Furthermore we have shown that the multiscale approach is useful also when applied to multivariate AR modeling. Both types of analysis have indeed proved useful in separating patients affected by Chronic Heart Failure from control subjects.

We have then addressed the problem of optimal feature selection in order to build a successful classifier.

REFERENCES

Ancona, N., R. Maestri, D. Marinazzo, L. Nitti, M. Pellicoro, G. D. Pinna, and S. Stramaglia (2005, Aug). Leave-one-out prediction error of systolic arterial pressure time series under paced breathing. *Physiol Meas* 26(4), 363–372. Clinical Trial.

Cooke, W.H., J. F. Cox, A. M. Diedrich, J. A. Taylor, L. A. Beightol, J. E. Ames, J. B. Hoag, H. Seidel, and D. L. Eckberg (1998, Feb). Controlled breathing protocols probe human autonomic cardiovascular rhythms. *Am J Physiol 274*(2 Pt 2), 709–718.

Costa, M., A. L. Goldberger, and C. K. Peng (2002, July). Multiscale Entropy Analysis of Complex Physiologic Time Series. *Phys. Rev. Lett. 89*(6), 068102.

Guyon, I. and A. Elisseeff (2003). An introduction to variable and feature selection. *Journal of Machine Learning Research 3*, 1157.

Kantz, H. and T. Schreiber (1997). *Nonlinear Time Series Analysis*. Cambridge, UK: Cambridge Univ. Press.

Korzeniewska, A., M. Manczak, M. Kaminski, K. J. Blinowska, and S. Kasicki (2003, May). Determination of information flow direction among brain structures by a modified directed transfer function (dDTF) method. *J Neurosci Methods 125*(1–2), 195–207.

Maestri, R. and G. D. Pinna (1998). Polyan: A computer program for polyparametric analysis of cardiorespiratory variability signals. *Computer Methods and Programs in Biomedicine 56*(1), 37–48.

Mann, H. B. and D. R. Whitney (1947). On a test of whether one of 2 random variables is stochastically larger than the other. *Annals of Mathematical Statistics* (18), 50–60.

Miyakawa, K., C. Polosa, and K. H. P (1984). *Mechanisms of blood pressure waves*. Berlin: Springer.

Pincus, S. M. (1991). Approximate entropy as a measure of system complexity. *Proc. Natl. Acad. Sci. USA 88*, 2297–2301.

Pinna, G. D., R. Maestri, M. T. La Rovere, E. Gobbi, and F. Fanfulla (2006). Effect of paced breathing on ventilatory and cardiovascular variability parameters during short-term investigations of autonomic function. *Am J Physiol Heart Circ Physiol 290*(1), H424–433.

Richman, J. S. and J. R. Moorman (2000, Jun). Physiological time-series analysis using approximate entropy and sample entropy. *Am J Physiol Heart Circ Physiol 278*(6), 2039–2049.

Rzeczinski, S., N. B. Janson, A. G. Balanov, and P. McClintock (2002, November). Regions of cardiorespiratory synchronization in humans under paced respiration. *Phys. Rev. E 66*(5), 051909.

Schafer, C., M. G. Rosenblum, J. Kurths, and H. H. Abel (1998, Mar). Heartbeat synchronized with ventilation. *Nature 392*(6673), 239–240. Letter.

Swets, J. A. (1988, June). Measuring the Accuracy of Diagnostic Systems. *240*, 1285.

Vapnik, V. N. (1998, September). *Statistical Learning Theory*. John_Wiley.

Computational Vision and Medical Image Processing – João Tavares & Natal Jorge (eds)
© 2008 Taylor & Francis Group, London, ISBN 978-0-415-45777-4

Human head models for the EEG forward problem

R. Rytsar & T. Pun
CUI - University of Geneva, Geneva, Switzerland

ABSTRACT: Solution of the forward problem using realistic head models is necessary for accurate EEG source analysis. For this purpose, realistic shapes of head tissues have been derived from a set of 2-D magnetic resonance images (MRI) by extracting surface boundaries for the major tissues, such as the scalp, the skull, the cerebrospinal fluid (CSF), the white matter and the gray matter. From boundary data a 3-D volume generic head model has been constructed and a mesh for an arbitrary complexity head shape has been generated for finite-element method (FEM) modelling. This paper first addresses the use of this realistic Finite Elements head model to solve the EEG forward problem. The influence of head model complexity on the electrical potentials values on the scalp is then examined with different human head models. It was shown that the structure of the anatomical surfaces of the CSF, the gray and the white matter could significantly influence the signal data calculated on the scalp.

1 INTRODUCTION

The electrical activity inside the brain consists of currents generated by biochemical sources at the cellular level. This activity can be measured by an electroencephalograph (EEG). Neurologists have been interested in the so-called EEG inverse problem that is the determination of the small active brain areas that significantly contribute to the generation of the electric field from the measured potentials on the scalp (Awada et al. 1997, Bertrand et al. 1991). Solving the EEG inverse problem requires an appropriate model of the corresponding forward problem (Awada et al. 1997, Bertrand et al. 1991, Rytsar & Pun 2006). In the forward calculations, the electric potentials are computed from given current source parameters, head geometry and conductivity properties. The human head, as a conductor, is often approximated by three or four spherical layers with different electrical conductivities representing the brain, the cerebrospinal fluid, the skull, and the scalp. It allows reducing the EEG forward problem to closed form analytic solution. In the case of a realistic head however, such solution does not exist and, therefore, the finite element method that describes the different parts of the head and their properties can be used.

Spherical head models are geometrically not sufficiently accurate for different applications (Hämäläinen & Sarvas 1989, Roth et al. 1997, Crouzeix et al. 1999, Yvert et al. 1997, Cuffin et al. 2001). Considering the complex shape of the head surface, a spherical surface approximates the shape of the head surface reasonably well in the occipital area while it is more or less inappropriate in the temporal and the frontal areas (Hämäläinen & Sarvas 1989). Several studies investigating the localization accuracy have estimated significant differences in the current source locations between patient-specific head models and spherical models (Roth et al. 1997, Crouzeix et al. 1999, Yvert et al. 1997). Moreover, the visualization of the spherical model results is not clinically very useful when magnetic resonance images are not available since there is not link to real anatomy.

Realistic models of the geometry can improve the accuracy of forward solutions and inverse source localizations. Head phantom studies show their high accuracy even with a simple three layers (brain, skull and scalp) model (Leachy et al. 1998). More recent studies indicate the necessity of highly heterogeneous models of the head for accurate simulations of scalp potentials (Ramon et al. 2006).

In this article a patient-specific 3-D head model has been constructed from 180 segmented coronal slices. The structure of anatomical surfaces of scalp, skull, CSF, gray and white matter is taken into account. This model has been applied to FEM computation of the scalp potentials arising from a source in the brain. We aim to provide a comparison of the results on the head models of different complexity: from a four-layer sphere to real head shape models with five tissues compartments. It allows examining the influence of different anatomical structures on the

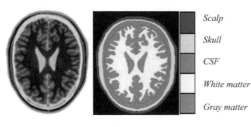

	Scalp
	Skull
	CSF
	White matter
	Gray matter

Figure 1. A single coronal slice and a segmented slice with five head compartments.

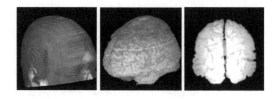

Figure 2. Examples of surfaces extracted from MR images: scalp, gray matter and white matter surfaces (from left to right).

scalp potentials. The accuracy of the surface potential numerical approximation has been validated on the spherical head model by comparison of the results with the analytical solution (Zhang 1995).

2 CONSTRUCTION OF THE REALISTIC FEM HEAD MODEL

High-resolution MRI allows to visualize neuroanatomical structures and can be used to extract morphological information about individual head shape (see Figure 1) for further FEM modelling. For this purpose, procedures such as segmenting the images and generating a mesh for an arbitrary complexity 3-D head shape are essential components.

The real head shape is represented as a set of 180 coronal slices from which the geometry of different tissues of the head can be derived. Automatically segmenting these images is difficult since different structures are often indistinguishable because of their low intensity in standard T1-weighted anatomical MRIs. Moreover, each structure does not have a unique gray scale value. Therefore, the segmentation of the head tissues has been performed interactively using the 3D Doctor software (3-D Doctor software). Figure 1 shows a single slice and a segmented slice with five head compartments: the scalp, the skull, the CSF, the gray and the white matter.

The head model with five compartments is obtained by replacing some tissues with the adjacent tissues (Ramon et al. 2006): soft bone was replaced with the hard bone in the skull, cerebellum was replaced with CSF, fat layer near to the scalp was replaced with the scalp, and eye sockets were replaced with soft tissues. According to the chosen model of the head all the slices were segmented. From boundary data the 3-D volume model was reconstructed using the volume-rendering. Figure 2 shows the surfaces of the segmented tissues: the scalp, the gray matter, the white matter.

The second essential prerequisite for FEM modelling is the construction of a mesh adequately representing the geometric and electric properties of the head volume conductor. The surface geometry of the head has been depicted by 14964 triangles and 7484

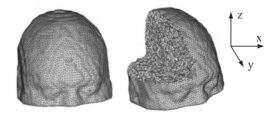

Figure 3. Surface-meshed and volume-meshed models.

nodes (see Figure 3). A mesh with first-order finite elements has been created using the HyperMesh software package (HyperMesh software).

The mesh quality can have considerable impact on the computational analysis, in terms of the accuracy of the solution and the time it takes to determine it. This aspect becomes important for further calculation of the electrical potentials on the scalp due to dipole sources inside the brain since the problem is ill conditioned. The mesh quality can be evaluated through the computation of the tetrahedral elements quality q:

$$q = f \frac{V^2}{(A+B+C+D)^3} \qquad (1)$$

where V denotes the volume of the tetrahedron; A, B, C and D are the areas of its faces and $f = 216\sqrt{3}$ is a normalizing coefficient ensuring that the quality of an equilateral tetrahedron equals 1. It is important to note that regular elements will allow a better convergence of the finite element solution. From this point of view we have analysed the quality of the "normal," "optimize meshing speed" and "interpolated" meshes proposed by the HyperMesh software. The latter was chosen for finite-element head modelling as the most regular. The average tetrahedral elements quality is greater than 0.6 as it presented by the histogram in Figure 4. It shows that the mesh is of the acceptable quality and uniformity.

We avoid the arduous work of 3D-construction of the complex tissues in the head by simplifying the modelling process in the following way. The head was meshed as a whole and then each finite-element unit was labelled: knowing the exact coordinates of all the mesh nodes and the kind of tissue at each of

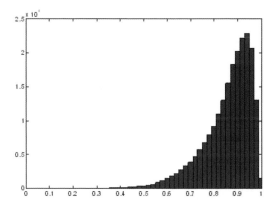

Figure 4. Mesh quality histogram (39575 nodes, 212439 tetrahedrons).

these points we can specify appropriate conductivity of each tiny volume. The following isotropic conductivities were assigned to the scalp (0.33 S/m), the skull (0.0042 S/m), the CSF (1.79 S/m), the gray matter (0.33 S/m) and the white matter (0.14 S/m) (Wolters et al. 2005).

The generated realistic FE head model includes 212439 tetrahedrons altogether, and after being labelled, 46040 tetrahedrons belong to the scalp, 37679 to the skull, 40531 to the cerebrospinal fluid, 41214 to the gray matter and 27270 to the white matter. The total number of the nodes is 39575.

3 THE EEG FORWARD PROBLEM

We have used this model for solving the EEG forward problem, which is defined as the potentials determination on the human head surface from a given configuration of the source in the brain, the geometry and the distribution of the electrical conductivity $\sigma(\vec{r})$ within the head. The source is assumed to be a dipole of amplitude $I_0 d$ in the direction $\hat{p}$. An ideal current dipole can be described as two-point sources of opposite polarity with an infinitely large current I_0 and an infinitely small separation $\vec{d}$. Mathematically, the current source density in the element volume (A/m^3) is defined by

$$\rho(\vec{r}) = \lim_{d \to 0} I_0 \left[\delta\left(\vec{r} - \vec{r}_o - \frac{\vec{d}}{2} \right) - \delta\left(\vec{r} - \vec{r}_o + \frac{\vec{d}}{2} \right) \right], \quad (2)$$

where $\delta(\cdot)$ denotes 3-D Dirac delta function. Theoretically, this forward problem is governed by the Poisson's equation for the electric potentials $U(\vec{r})$ in the head Ω:

$$\nabla \cdot (\sigma(\vec{r}) \nabla U(\vec{r})) = \rho(\vec{r}) \text{ in } \Omega. \quad (3)$$

At the outer boundary of the medium there exists a homogeneous Neumann boundary condition

$$\frac{\partial U(\vec{r})}{\partial n} = 0 \text{ on } \partial\Omega, \quad (4)$$

where n is the distance measured normally to the boundary.

The numerical solution of the forward problem is calculated at the nodes using the FEM. The interpolation over the volume elements is described by shape functions, which depend on the shape of the elements. In our case, the elements are tetrahedrons with four nodes, and the interpolation is linear. Thus, the numerical approximation $\tilde{U}$ of the solution U for a mesh consisting of N nodes is given by

$$\tilde{U} = \sum_{i=1}^{N} U_i H_i, \quad (5)$$

where U_i is the solution at node i and H_i is the shape function that describes the contribution of the value at node i. The approximated value at any point is a sum of the contributions of all nodal values.

Once the conducting volume is meshed with the tetrahedral elements, the potential values at the nodes U_i are then obtained by solving the standard FEM system of equations (Awada et al. 1997, Bertrand et al. 1991)

$$[K_{ji}][U_i] = [\rho_i], \quad (6)$$

where $[K_{ji}]$ is an $N \times N$ matrix that depends only on the shape functions and conductivities distribution and $[\rho_i]$ is an N-dimensional vector in which the source function is incorporated. The elements of the stiffness matrix $[K_{ji}]$ are defined as

$$[K_{ij}] = \int_{\Omega^{(e)}} \sigma(r)(\nabla H_j^{(e)}) \cdot \nabla H_i^{(e)} d\Omega^{(e)} \quad (7)$$

and the i-th component ρ_i of this vector is given by

$$\rho_i = \int_{\Omega^{(e)}} \rho(\vec{r}) H_i^{(e)}(\vec{r}) d\Omega^{(e)} =$$

$$= I_0 \lim_{d \to 0} \left[H_i^{(e)} \left(\vec{r}_0 + \frac{\vec{d}}{2} \right) - H_i^{(e)} \left(\vec{r}_0 - \frac{\vec{d}}{2} \right) \right] = \vec{M} \cdot \nabla H_i^2 \cdot \hat{p}, \quad (8)$$

where (e) signifies an element subdomain and $\vec{M} = I_0 \vec{d}$ is the dipole moment.

Since each tetrahedron directly interacts only with its immediate neighbours, the matrix is quite sparse; the conjugate gradient method is used for efficient solving of the system equation (6).

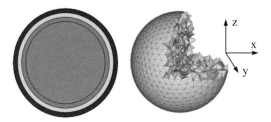

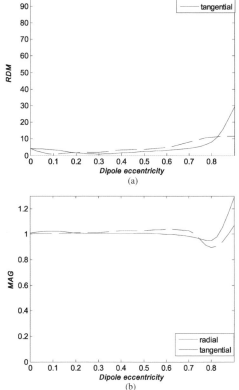

Figure 5. The isotropic four-layer spherical head model.

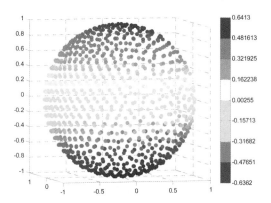

Figure 6. Electrical potentials on the surface of the four-layers sphere.

Figure 7. A plot of RDM (a) and MAG (b) of the surface potential due to the radial (solid line) and tangential dipole (dashed line).

4 RESULTS

Simulations were carried out to evaluate the influence of the head model complexity on the potentials values calculated on the scalp. The simplest proposed head model is four-layer sphere with diameters 0.84, 0.87, 0.92, and 1.0 m delimiting domain of conductivities 0.33, 1.79, 0.0042, 0.33 $\Omega^{-1}m^{-1}$, representing the brain, CSF, skull and scalp respectively (see Figure 5). The head surface was approximated by 1685 triangles. The volume mesh has been built up with 65699 nodes and has 374797 first-order tetrahedral elements. There are 292052, 14946, 22515 and 45284 elements in the brain, CSF, skull and scalp volumes respectively.

The source is a z-oriented dipole placed in the centre of the sphere. Refer to Figure 5 for the coordinate orientation. The electrical potentials were calculated at 844 nodes on the surface. Their values are distributed symmetrically according to the symmetry of the head model geometry and are indicated on Figure 6.

Although the four-layer sphere is quite coarse from a modeling point of view, most of the conclusions about modeling accuracy can be demonstrated with this simple model (Awada et al. 1997, Bertrand et al. 1991, Rytsar & Pun 2006). The relative difference measure (RDM) and the magnitude factor (MAG) were used

to measure the similarity between the analytic and numeric solutions:

$$RDM = \sqrt{\frac{\sum_{i=1}^{N_b}\left(U_i^A - U_i^{FEM}\right)^2}{\sum_{i=1}^{N_b}\left(U_i^A\right)^2}}, \qquad (9)$$

$$MAG = \sqrt{\frac{\sum_{i=1}^{N_b}\left(U_i^{FEM}\right)^2}{\sum_{i=1}^{N_b}\left(U_i^A\right)^2}}, \qquad (10)$$

where N_b is a number of nodes on the boundary, U_i^A is a exact potential at boundary node i, and U_i^{FEM} is the numerical solution at boundary node i.

In order to evaluate accuracy of the numerical results the dipole is placed along the z-axis with the different eccentricities pointing in the radial and tangential directions. Figure 7 shows the RDM and MAG for different source location. The RDM increases with

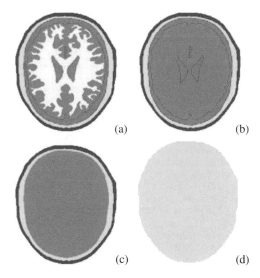

(a)   (b)

(c)   (d)

Figure 8. The human head models of different complexities: a) the full model of the head includes structures of the scalp, the skull, the CSF, the white matter and the gray matter; b) full model without distinction between the gray and the white matter; c) full model without distinction between the gray matter, the white matter and the CSF; d) the homogeneous head model.

source eccentricity and is lower than 5% for eccentricities less than 0.7 ($N = 65699$). However, RDM increases exponentially while eccentricity is greater 0.7. RDM reaches 17% at the point where eccentricity equal to 0.85. The MAG remains very close to 1 for eccentricities less than 0.7.

In the case of an arbitrarily shaped volume conductor the accuracy of the potential calculations cannot be easily verified. The obvious way to proceed would be to increase the density of the tetrahedrons gradually and to show that obtained result approaches some limiting value. However, this method can only verify how the FEM works.

The influence of the choice of the human head model was investigated with four head models of different complexities ranging from the full model to the less-tissue amount type model. The scalp potentials due to z-oriented dipole source in the brain were calculated for all models.

The full head model takes into account the structure of five different tissues of the head: the scalp, the skull, the CSF, the white and gray matters (see Figure 8a). This model was used as a reference model. The scalp potentials distribution and their magnitude scale are shown in Figure 9a for this model.

The model composed of four tissue-types in comparison with the full model was obtained by the replacing the white matter with the gray matter (see Figure 8b). This assumption does not significantly

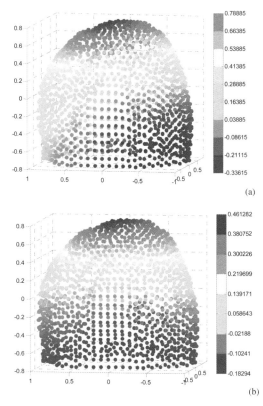

(a)

(b)

Figure 9. Electrical potentials a) on the scalp of the reference head model; b) on the surface of the homogeneous head shape medium.

influences the scalp potentials values in comparison with the corresponding results obtained with the full model. These differences are shown in Figure 10a. The differences in negative and positive peak values are 0.0103 and 0.0137 V, respectively.

The next proposed model was obtained from the previous one by the elimination of the difference between the brain matter and the CSF (see Figure 8c). The distribution of the potentials on the scalp is more uniform in comparison with the previous models. Moreover, there are noticeable changes in the scalp potentials over a large portion of head surface in comparison with the results of the full model. These differences are more visible in the scalp potential differences plot given in Figure 10b.

The differences in negative and positive peak values are −1.0532 and 0.5382 V, which are significant. This is expected result because the CSF has much higher conductivity than adjacent regions. This leads to a large gradient of potentials in the CSF. Once again this highlights the important role of the CSF in redistributing the volume currents and thereby modifying the scalp potentials (Ramon et al. 2004).

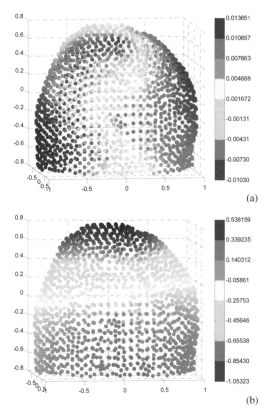

(a)

(b)

Figure 10. Differences in scalp potentials between the full head model and a) the head model without distinction between the gray and the white matter; b) the head model without distinction between the CSF, the gray matter and the white matter.

In the case of homogeneous medium the boundary between all tissues inside the head are removed and gradient of the conductivity is neglected (see Figure 8d). Electrical potentials on the surface quite significantly differ from the corresponding values calculated on the scalp of the full head model (see Figure 9b), wherein the conductivity varies over the whole medium. Their values only reflect the influence of the head shape.

5 CONCLUSIONS

What tissues can be neglected in the process of constructing a head model for solving the EEG forward problem using FEM? This article answers this question by evaluating the influence of each tissue on the calculated potential values. The simulation results indicate that the complexity of head model impacts on the scalp potentials. The structures of a high conductivity contrast, such as the CSF and the skull, play an important role in modifying of the potential values. In order to obtain an accurate EEG forward problem solution one needs to have highly heterogeneous model of the head.

REFERENCES

3-D Doctor software: http://www.ablesw.com/3d-doctor/
Awada K.A., Jackson D.R., Williams J.T., Wilton D.R., Baumann S.B. & Papanicolaou A.C. 1997. Computational aspects of finite element modeling in EEG source localization. *IEEE Trans. Biomed. Eng.* 44 (8): 736-752.
Bertrand O., Thevenet M. & Perrin F. 1991. 3-D finite element method in brain electrical activity studies. In J. Nenonen, H.-M. Rajala & T. Katila (eds), *In Biomagnetic localization and 3D modelling. Report TKK-F-A689, Helsinki University of Technology, Dept. of Technical Physics, Lab. of Biomed. Engr. Espoo*, Finland.
Crouzeix A., Yvert B., Bertrand O., & Pernier J. 1999. An evaluation of dipole reconstruction accuracy with spherical and realistic head models in MEG. *Clin. Neurophysiol.* 110 (12): 2176–2188.
Cuffin B., Schomer D., Ives J. & Blume H. 2001. Experimental tests of EEG source localization accuracy in realistically shaped head model. *Clin. Neurophysiol.* 112: 2288–2292.
Hämäläinen M. & Sarvas J. 1989. Realistic conductivity geometry model of the human head for interpretation of neuromagnetic data. *IEEE Trans. Biomed. Eng.* 36 (2): 165–171.
HyperMesh software: http://www.altair.com/
Leachy R.M., Mosher J.C., Spencer M.E., Huang M.X. & Lewine J.D. 1998. A study of dipole localization accuracy for MEG and EEG using a human skull phantom. *Electroencephalogr. Clin. Neurophysiol.* 107 (5): 159–73.
Ramon C., Schimpf P., Haueisen J., Holmes M. & Ishimaru A. 2004. Role of Soft Bone, CSF and Gray Matter in EEG simulations. *Brain Topography* 16 (4): 245–248.
Ramon C., Schimpf P. & Haueisen J. 2006. Influence of head models on EEG simulations and inverse source localizations. *Biomed Eng Online* 5 (1).
Roth B., Ko D., von Albertini-Carletti I., Scaffidi D., & Sato S. 1997. Dipole localization in patients with epilepsy using the realistically shaped head model. *Electroencephalogr. Clin. Neurophysiol.* 102 (3): 159–166.
Rytsar R. & Pun T. 2006. The forward EEG solutions: finite element modelling of the dipole source. In *Proc. of European BioAlpine Convention, Grenoble, France.*
Wolters C.H., Anwander A., Tricoche X., Lew S. & Johnson C.R. 2005. Influence of Local and Remote White Matter Conductivity Anisotropy for a Thalamic Source on EEG/MEG Field and Return Current Computation. *International Journal of Bioelectromagnetism* 7 (1): 203–206.
Yvert B., Bertrand O., Thevenet M., Echallier J., & Pernier J. 1997. A systematic evaluation of the spherical model accuracy in EEG dipole localization. *Electroencephalogr. Clin. Neurophysiol.* 102(5): 452–459.
Zhang Z. 1995. A Fast Method to Compute Surface Potentials Generated by Dipoles within Multilayer Anisotropic Spheres. *Phys. Med. Biol.* 40: 335–349.

Computational Vision and Medical Image Processing – João Tavares & Natal Jorge (eds)
© 2008 Taylor & Francis Group, London, ISBN 978-0-415-45777-4

A system for ploidy analysis of malignant nuclei

S. Isotani
Institute of Physics of University of São Paulo

Y.I. Sakai
Cytopatology section, Pathology Division, Adolfo Lutz Institute

A.T. Sakai
Department of Urology, Escola Paulista de Medicina, Federal University of São Paulo

O.M. Capeli
Centro Universitário FIEO

A.R.P.L. de Albuquerque
Paulista University

ABSTRACT: This paper describes SPCIM, a system developed for the processing of images from microscope. The main functions of this system are Acquisition of Images and Dimension Calibration, Determination of Images Background, Calibration of DNA, Histogram of Calibration of DNA, Analysis, Histogram of Analysis and Visualization of Images. These functions make use of the modules Image Acquisition, Dimension Calibration, Background Determination, Nuclei Demarcation, Histogram and Database.

1   INTRODUCTION

The computerized nuclear morphometry as an objective method for characterizing populations of human cancer cells since the pioneering study of Stenkvist et al. (1978) was object of several works (Zajdela et al. 1979, Crocker et al. 1982, Crocker et al. 1983, Smeulders & Dorst 1985, Tosi et al. 1986, Young 1988, Russ 1989, Bilbbo 1997, Underwood 2001, Zheng 2004).

The processing of images from optical microscopy, has several commercial systems at the market, but not always assists the specific needs of the research. Especially, in Brazil, the systems of processing of microscopic images need to be modest without losing its functionality.

The SPCIM (System of Processing of Microscopic Images) is a system developed for the study of the morphometry and ploidy of human cancerous cells that can be used in modest microcomputers. It was applied in the analysis of fine needle aspiration specimens (Maeda et al. 1998) and morphometry evaluation of nucleolar organizer regions in cervical intraepithelial neoplasia (Sakai et al. 2001), images obtained by video (Marques et al. 2001) and tomography of X-rays (Marques et al 2007). Also SPCIM was applied in the

quality control system of the Papanicolaou exam of the Adolfo Lutz Institute (Sakai et al. 2004a), the Brazilian reference laboratory, because the exam depends on the quality of the laboratories and training of the human resources (Mody et al. 2000, Sankaranarayanan et al. 2001).

We report in the present work, a general description of the modules and functions of SPCIM.

2   THE STRUCTURE OF SPCIM

The SPCIM is managed through a browser that allows access their main functions. The main functions are accessible through a control box that has the following options: Acquisition of Images and Dimension Calibration, Determination of Background, Calibration of DNA, Histogram of Calibration of DNA, Analysis of Images, Histogram of Analysis and Visualization of Images. The control box is shown in Figure 1. The main browser of SPCIM was built using Visual Basic language.

The function of Acquisition of Images and Dimension Calibration is used to acquire images of the slide for analysis and for dimensional calibration. This function is begun attributing a name to the

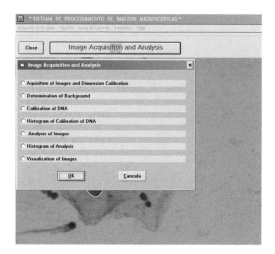

Figure 1. The main browser of SPCIM.

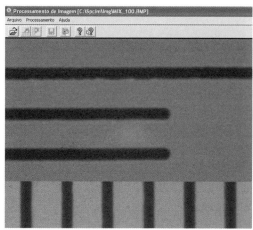

Figure 2. Acquired image for dimension calibration using a calibration slide.

slide from where the images will be collected. The collected images of this slide receive names among *******1.bmp until ******99.bmp. The numbering is sequential following the order of the slides. Then the attribution of the name of the image to slides is univocous. For instance, there is only an image 00011638.bmp and it corresponds to slide TGF-01. The image acquisition is made using a module of Image Acquisition written in Visual C language. The association of the name of each image with their slide and this with the user is filed in a database managed using the module Database.

To assure that the absolute values of length and area are obtained with safety, the Dimension Calibration is executed simultaneously with the Image Acquisition. Figure 2 shows the interface of the module Dimension Calibration.

The slides containing samples of tissues are stained to emphasize the images of the nuclei and cytoplasm. The function Determination of Background is used to determine the absorbance, the difference between the optical density of the glass and of tissue.

The function Analysis of Images makes use of the module Nuclei Demarcation in the determination of the optical density, perimeter and area of the cell nuclei. The absorbance is calculated discounting from the optical density of the sample the optical density of the background. Figure 3 shows the result of analysis of Nuclei Demarcation.

The function of calibration of DNA uses the module Nuclei Demarcation for the determination of the mean absorbance of reference nuclei. Since the amount of DNA of the reference nuclei is 2 DNA, the mean value of the absorbance of these nuclei is used in the determination of the amount of DNA of the analyzed nuclei.

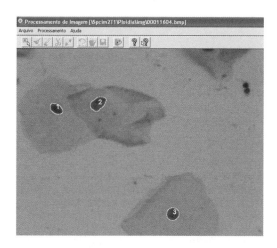

Figure 3. Interface of the module Nuclei Demarcation after segmentation, but before saving the results of analysis. The image is from slide TGF-01, image 00011604.bmp of a smear of healthy reproductive women.

The functions Histogram of Calibration of DNA and Histogram of Analysis show the distribution of values of absorbance of the analyzed nuclei using the module Histogram. Figure 4 shows the histogram of analysis of a slide.

3 THE STRUCTURE OF THE MODULES

The algorithms are in its majority of public domain (Gonzalez & Woods 2001, Russ 1995, Rimmer 1992) and some of them were adapted to improve the operation of SPCIM.

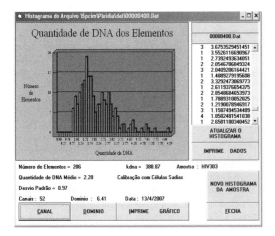

Figure 4. Histogram of analysis showing DNA ploidy. The analysis is from cervical smears slides of HIV-positive patients slide HIV303.

The Image Acquisition module was built using the Visual C language. The access to VideoBlaster SE 100 was done using the files supplied by the manufacturer. The file is saved in 256 colors bmp files. The images were acquired using a microscope Hund, with camera CCD of Sansung. The time of acquisition of 98 images was approximately of 30 minutes.

The module of Dimension Calibration was projected to acquire or to read image files of slides of dimension calibration. Both tools of reading, horizontal and vertical, measure the number of pixels among two defined points through click and drag of the mouse. This module was built using Visual C language.

The module of Background Determination receives the address of the image to be treated, open it and allow the user to choose the area for determination of background. This module was built using Visual C.

The images of the cellular tissues possess background in general clear, with well defined dark regions, associated to the cellular nuclei. This was used to build the algorithms of segmentation of the module Nuclei Demarcation which was presented a previous report [Isotani et al. 2007]. This module was built using Visual C language. The module Nuclei Demarcation includes cut and join algorithms that are of public domain, and algorithms for determination of dark regions, segmentation, area threshold, counting of regions, borderline of regions and validation of regions that were adapted in this work to optimize the functionality of SPCIM. Figure 5 shows some tools of the module Nuclei Demarcation. The time of analysis of 98 images using the module Nuclei Demarcation was about 30 minutes.

The segmentation algorithm generates the segmentation parameter analyzing the intensities. Therefore the segmentation will always give the same results for

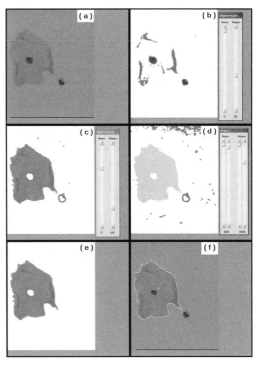

Figure 5. Application of Nuclei Demarcation module. Figure 5a show an original image, Figure 5b the result of segmentation, Figure 5c the result of application of control bars using the minimum and maximum to isolate the cell cytoplasm, Figure 5d the result of application of area threshold bars, Figure 5e the result of application of regions borderline algorithm and Figure 5f the final result. The image is from slide TGF-01 of a smear of healthy reproductive women.

the values of the absorbance, area and perimeter. However the image segmented like this cannot represent the image of the nucleus correctly because of variations in the intensity of coloration of these nuclei. These variations in the intensity of the color are resulted of factors as the collecting, transport, fixation, coloring, temperature and time of coloration of the slide. Also the segmentation may fail depending on the contrast and noise in the image acquisition.

To correct the segmentation in these cases, and to assist other needs we introduced in the module Nuclei Demarcation two control bars for alteration of the segmentation parameter, one for minimum and another for maximum. Figure 5a show an original image, Figure 5b the result of segmentation, Figure 5c the result of application of control bars using the minimum and maximum to isolate the cell cytoplasm, Figure 5d the result of application of area threshold bars, Figure 5e the result of application of regions borderline algorithm and Figure 5f the final result.

Together with the dyeing of the nuclei the cytoplasm can be colored also. In these cases the borderline between the cytoplasm and the nucleus can be difficult to distinguish with accuracy. For this reason it can have differences in the segmentation done by different people and by these same people in different moments. A deepened study of the effect of simultaneous dyeing of nuclei and cytoplasm is in course. A preliminary result using ten slides of a smear of healthy reproductive women show that the differences are smaller than 5% in the measures of absorbance, area and perimeter.

The module Histogram was built using public algorithms in Visual Basic language. The number of channels is determined in a way to obtain better symmetry of the graph or can be chosen. Also in this module the histogram of the sum of the analysis of all images of a slide can be shown as well as the analysis of each image. This module was built using Visual Basic language.

Each user, when being admitted in the system after informing the password, will just visualize the exams that it accomplished, which will be identified for a code associated to the analyzed slide. We used Structured Query Language to build the module Database, with three filing shells: user, report and slide. The user's shell contains the name and each user's password. The password is encrypted with the objective to maintain the privacy of the data. The report shell contains data about the analysis with the specifications of the user. The slide shell contains data such as the number of images collected from the slide, the images that were analyzed and data of calibrations.

4 CONCLUSIONS

We described in this work the structure of SPCIM, a system of computerized processing of microscopic images for the study of developed for the study of the morphometry and ploidy of human cancerous cells.

The main functions are accessed through a control box. These functions are: Acquisition of Images and Dimension Calibration, Determination of Background, Calibration of DNA, Histogram of Calibration of DNA, Analysis of Images, Histogram of Analysis and Visualization of Images.

The functions make use of modules that are tools for the accomplishment of specific tasks as: Image Acquisition, Dimension Calibration, Background Determination, Nuclei Demarcation, Histogram and Database.

The SPCIM was applied in the study of fine needle aspiration specimens, cervical intraepithelial neoplasia, images obtained by video and tomography of X-rays, and in the quality control system of the Papanicolaou exam.

In these applications the functions of SPCIM showed good manageability and robustness. Besides,

the analysis of the images was made with larger speed and precision that in manual processes.

The analysis of the performance of SPCIM in the studies described above showed that this system fits the demands of good performance in low cost equipments.

We verified that segmented image can represent the image of the nucleus incorrectly because of variations in the intensity of coloration of these nuclei and contrast and noise in the image acquisition. In this field, color attributes serve to do the analysis of cell images (Bonventi Jr et al. 1994, Underwood 2001). At first place a considerable part of processing can be reduced removing the background. With this objective we are implementing the method of image background filtering using elliptical color coordinates (Bonventi Jr. et al. 1994). After the removal of the background it can still remain little or no background and cytoplasms that are colored differently of the nuclei. Using the images treated like this, we are implementing a method using standard cell nuclei images in the evaluation of correlation functions (Brasil Fo et al. 1994).

5 ACKNOWLEDGEMENTS

This work was supported by the FINEP (Financiadora de Estudos e Projetos), CNPq (Conselho Nacional de Desenvolvimento Científico e Tecnológico), FAPESP (Fundação de Amparo à Pesquisa do Estado de São Paulo) and PSF (Pesquisadores Sem Fronteira). We would like to thank Venâncio Avancini Ferreira Alves for his valuable participation in the idealization of the present work.

REFERENCES

Bilbbo, M. 1997. *Comprehensive Cytophatology*, 2nd ed. Philadelphia, Saunders Company, Chap. 38, 971–988.

Brasil Fo, N. & de Albuquerque, A.R.P.L. & Isotani, S. 1994. A Method for Image Segmentation of Colored Images Using Correlation Function Maximization. In Ohzu, H. & Komatsu, S.: *Optical Methods in Biomedical and Environmental Sciences*: 71–74. Amsterdam: Elsevier.

Bonventi Jr, W. & de Albuquerque, A.R.P.L. & Isotani, S. 1994. A Method for Image Background Filtering of Colored Images Using Elliptical Coordinates in Ohzu, H. & Komatsu, S.: *Optical Methods in Biomedical and Environmental Sciences*: 75–78. Amsterdam: Elsevier.

Crocker, J. &, Jones, E.L. & Curran, R.C. 1982. Study of nuclear diameters in non-Hodgkin's lymphomas. *Journal of Clinical Pathology*, 35: 954–958.

Crocker, J. &, Jones, E.L. & Curran, R.C. 1983. A comparative study of nuclear form factor, area and diameter in non-hodgkin's lymphomas and reactive lymph nodes. *Journal of Clinical Pathology*, 36: 298–302.

Gonzalez, C.R. & Woods, R.E. 2001. *Digital Image Processing*, 2nd ed., Prentice Hall.

Isotani, S. & Brasil Fo, N. & Capeli, O.M. & Sakai, Y.I. & de Albuquerque, A.R.P.L. 2007. A system for the determination of image density and morphometry developed for the analysis of malignant nuclei. In *Proceedings of 20th IEEE International Symposium on Computer-Based Medical Systems, Maribor, Slovenia*.

Maeda, M.Y.S. & Sakai, Y.I. & Sakai, A.T. & Isotani, S. & Srough, M. 1998. Prostatic Cancer: DNA Ploidy, Karyometry and Gleason Score on Fine Needle Aspiration Specimens, *Acta Cytologica, The Journal of Clinical Cytology and Cytopathology*, 42: 528.

Marques, S.R. & Bonald, L.V. & DeAngelis, M.A. & Isotani, S. & Smith, R.L. 2001. The External opening of the cochlear aqueduct in infants and adults. *Oto-Rhino-Laryngologia Nova*, 11: 298–301.

Marques, S.R. & Smith, R.L. & Isotani, S. & Alonso, L.G. & Anadao, C.A. & Prates, J.C. & Lederman, H.M. 2007. Morphological analysis of the vestibular aqueduct by computerized tomography images. *European Journal of Radiology*, 61: 79–83.

Mody, D.R. & Davey, D.D. & Branca, M. 2000. Quality assurance and risk reduction guidelines. *Acta Cytologica*, 44(4): 496–507.

Rimmer, S. 1992. *Supercharged Bitmapped Graphics*, Windcrest/McGraw-Hill.

Russ, J.C. 1989. Automatic methods for the measurement of curvature of lines, features and feature alignment in images. *Journal of Computer Assisted Microscopy*, 1: 39–77.

Russ, J.C. 1995. *The Image Processing Handbook*, CRC Press.

Sakai, Y.I. & Sakai, A.T. & Isotani, S. & Cavaliere, M.J. & Almeida, L.V. & Calore, E.E. 2001. Morphometric evaluation of nucleolar organizer regions in cervical intraepithelial neoplasia. *Pathology Research and Practice*, 197: 189–192.

Sakai, Y.I. & Sakai, A.T. & Isotani, S. & Pereira, S.M.M. & Yamamoto, L.S.U. & Maeda, M.Y.S. & Luvizotto, H.B. & Veiga, E. 2004a. Evaluation of dyeing capacity of component of papanicolaou technique by computer image analysis system. In *Proceedings of XVth International Congress of Cytology, Santiago de Chile*, 92.

Sakai, Y.I. & Sakai, A.T. & Isotani, S. & Calore, E.E. & Cavaliere, M.J. & Maeda, M.Y.S. 2004b. Morphometric nuclear analysis in cervical intraepithelial neoplasia of HIV-positive women. In *Proceedings of XVth International Congress of Cytology, Santiago de Chile*, 91.

Sankaranarayanan, R. & Buduck, A.M. & Rajkumar, R. 2001. Effective screening programs for cervical cancer in low and middle-income developing countries. *Bull World Health Organ*, 79: 954–62.

Smeulders, A.W.M. & Dorst, L. 1985, Measurement issues in morphometry. *Analytical and Quantitative Cytology and Histology*. **7**: 242–249.

Stenkvist, B. & Westman-Naeser, S. & Holmquist, J. & Nordin, B. & Bengtsson, E. & Vegelius, J. 1978. Computerized nuclear morphometry as an objective method for characterizing human cancer cell populations. *Cancer Research*, 38: 4688–4697.

Tosi, P. & Luzi, P. & Baak, J.A.P. & Miracco, C. & Santopietro, R. & Vindigni, C. & Mattei, F.M. & Acconcia, A. & Massai, M.R. 1986. Nuclear morphometry as an important prognostic factor in stage 1 renal cell carcinoma. *Cancer*, 58: 2512–2518.

Underwood, R.A. & Gibran, N.S. & Muffley, L.A. & Usui, M.L. & Olerud, J.E. 2001. Color subtractive – Computer assisted image analysis for quantification of cutaneous nerves, *Journal of Histochemistry and Cytochemistry*, 49(10): 1285–1291.

Young, I.T. 1988. Sampling density and quantitative microscopy. *Analytical and Quantitative Cytology and Histology*. 10: 269–275.

Zajdela, A & De Lariva, L.S. & Ghossein, N.A. 1979. The relation of prognosis to the nuclear diameter of brest cancer cells obtained by cytologic aspiration. *Acta Cytological*, 23(1): 75–80.

Zheng, Q. & Milthorpe, B.K. & Jones, A.S. 2004. Direct neural network application for automated cell recognition, *Cytometry*, 57A: 1–9.

Computational Vision and Medical Image Processing – João Tavares & Natal Jorge (eds)
© 2008 Taylor & Francis Group, London, ISBN 978-0-415-45777-4

Microscale flow dynamics of red blood cells in a circular microchannel

R. Lima
Dept. Bioeng. & Robotics, Grad. Sch. Eng., Tohoku Univ., Aoba, Sendai, Japan
Dept. Mechanical Tech., ESTiG, Braganca Polyt., C. Sta. Apolonia, Braganca, Portugal.

M. Nakamura
Dept. Mechanical Science and Bioeng., Grad. Sch. Eng., Osaka Univ., Toyonaka, Osaka, Japan

T. Ishikawa
Dept. Bioeng. & Robotics, Grad. Sch. Eng., Tohoku Univ., Aoba, Sendai, Japan

S. Tanaka
Dept. Nanomechanics, Grad. Sch. Eng., Tohoku Univ., Aoba, Sendai, Japan

M. Takeda
Dept. Bioeng. & Robotics, Grad. Sch. Eng., Tohoku Univ., Aoba, Sendai, Japan
Div. Surgical Oncology, Grad. Sch. Medicine, Tohoku Univ., Seiryo-machi, Aoba-ku, Sendai, Japan

Y. Imai & K. Tsubota
Dept. Bioeng. & Robotics, Grad. Sch. Eng., Tohoku Univ., Aoba, Sendai, Japan

S. Wada
Dept. Mechanical Science and Bioeng., Grad. Sch. Eng., Osaka Univ., Toyonaka, Osaka, Japan

T. Yamaguchi
Dept. Bioeng. & Robotics, Grad. Sch. Eng., Tohoku Univ., Aoba, Sendai, Japan

ABSTRACT: The blood flow dynamics in microcirculation depends strongly on the motion, deformation and interaction of RBCs within the microvessel. This paper presents the application of a confocal micro-PTV system to track RBCs through a circular polydimethysiloxane (PDMS) microchannel. This technique, consists of a spinning disk confocal microscope, high speed camera and a diode-pumped solid state (DPSS) laser combined with a single particle tracking (SPT) method. By using this system detailed motions of individual RBCs were measured at a microscale level. Our results showed that this technique can provide detailed information about microscale disturbance effects caused by RBCs in flowing blood.

1 INTRODUCTION

The blood is composed with approximately half volume of red blood cells (RBCs) which is believed to strongly influence its flow properties. Blood flow in microvessels depends strongly on the motion, deformation and interaction of RBCs. Several studies on both individual and concentrated RBCs have already been performed in the past [1, 2]. However, all studies used conventional microscopes and also ghost cells to obtain visible trace RBCs through the microchannel. Recently, considerable progress in the development of confocal microscopy and consequent advantages

of this microscope over the conventional microscopes have led to a new technique known as confocal micro-PIV [3, 4]. This technique combines the conventional PIV system with a spinning disk confocal microscope (SDCM). Due to its outstanding spatial filtering technique together with the multiple point light illumination system, this technique has the ability to obtain in-focus images with optical thickness less than 1 μm.

The main purpose of this paper is to examine the potential of the confocal micro-PTV system to measure individual RBCs at different haematocrits (Hct) through a 75 μm circular PDMS microchannel. Moreover we would like to compare our results with a large-scale simulation technique in order to obtain more detailed insights about the blood rhelogical properties at cellular level.

2 MATERIALS AND METHODS

2.1 Working fluids and microchannel

Three working fluids were used in this study: dextran 40 (Dx40) containing about 3%(3Hct) 14% (14Hct) and 37% (37Hct) of human red blood cells (RBCs). The blood was collected from a healthy adult volunteer, where ethylenediaminetetraacetic acid (EDTA) was added to prevent coagulation. The RBCs were separated from the bulk blood by centrifugation (1500 RPM for 5 minutes) and aspiration of the plasma and buffy coat and then washed twice with physiological saline (PS). The washed RBCs were labeled with a fluorescent cell tracker (CM-Dil, C-7000, Molecular Probes) and then diluted with Dx40 to make up the required RBCs concentration by volume. All blood samples were stored hermetical at 4°C until the experiment was performed at controlled temperature of about 37°C.

The microchannel used in this study was a PDMS circular microchannel (75 μm in diameter) fabricated by a wire casting technique [5].

2.2 Confocal micro-PTV experimental set-up

The confocal micro-PIV system used in our experiment consists of an inverted microscope (IX71, Olympus, Japan) combined with a confocal scanning unit (CSU22, Yokogawa) and a diode-pumped solid state (DPSS) laser (Laser Quantum Ltd) with an excitation wavelength of 532 nm. Moreover, a high-speed camera (Phantom v7.1) was connected into the outlet port of the CSU22 (see Figure 1). The PDMS microchannel was placed on the stage of the inverted microscope

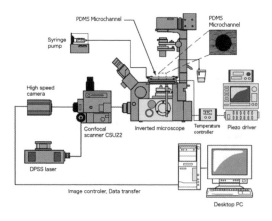

Figure 1. Experimental set-up.

where the flow rate of the working fluids was kept constant (Re = 0.004) by means of a syringe pump (KD Scientific Inc.). A thermo plate controller (Tokai Hit) was set to 37°C. All the confocal images were captured in the middle of the microchannels with a resolution of 640 × 480 pixels, 12-bit grayscale, at a rate of 100 frames/s with an exposure time of 9.4 ms. The recorded images were transfered to the computer and then evaluated in Image J (NIH) [6] by using a manual tracking MTrackJ [7] plugin. As a result it was possible to track single RBCs through the middle plane of the microchannel.

2.3 RBC radial displacement

The radial displacements (ΔR) of the tracked RBCs were determined by using a cumulative radial displacement, given by:

$$\Delta R = \sum_{i=0}^{n} |R_0 - R_i| \tag{1}$$

where R_0 is the initial radial position and R_i is the cumulative radial displacement for a defined time interval.

2.4 Flow model of multiple RBCs

A simulation method for multiple RBCs was proposed for understanding the rheological properties of blood from a viewpoint of multiscale mechanics. Assuming that macroscopic flow field is not affected by each RBC motion, macroscopic flow field was determined by theoretical/numerical analysis. The momentum and viscous fluid forces acting on RBC were evaluated from the difference in the velocities between the RBC and the prescribed flow field. Moreover, the mechanical interaction among the multiple RBCs was expressed by an attraction-repulsive potential function assigned at each nodal point on the RBC membrane [8].

Very recently the elastic RBC flow model [8, 9] was successfully extended to a three-dimensional large scale computer simulation by using parallel computation (512 processors). As result it was possible to analyse the flow behaviour of RBCs in detail [9].

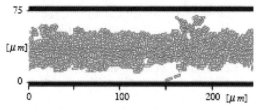

Figure 2. Simulation of RBCs flowing in a 75 μm microchannel [8].

3 RESULTS AND DISCUSSION

Figure 3 shows images with both RBCs (halogen illumination) and labeled RBCs (laser-emitted light) at different Hcts.

Figure 4 shows the RBC paths at the middle plane with Hct up to 37%, whereas Figure 5 shows the radial displacement (ΔR).

Our preliminary results suggest that the RBC paths are strongly dependent on the Hct and as a result the radial RBC displacement increases with the heamatocrict. Moreover, our results also indicate that the interactions of RBCs are more predominant around the plasma layer. The present work demonstrates that the proposed confocal micro-PTV system can measure the motion of labeled RBCs at different Hcts and consequently provide detailed information about microscale disturbance effects caused by RBCs in flowing blood.

The three-dimensional elastic RBC flow model reproduced realistic RBC flow behaviour such as tank tread and tumbling motion, and also axial migration, which are often observed *in vivo* microvessels. Some preliminary results on multiple RBC behavior in a Poiseuille flow are shown in Figure 6.

By comparing the results from this numerical model with the experimental data, it is possible to observe that in both cases the RBCs radial displacement tend to increase as we move a way from the centre of the microchannel. An ongoing study to compare in more detail the present experimental results with the three-dimensional elastic RBC flow model is currently under way.

(a)

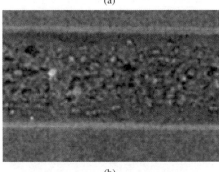

(b)

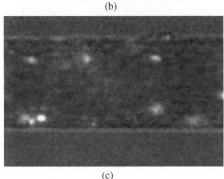

(c)

Figure 3.   Both normal and labeled RBCs (bright spots) with (a) 3% Hct, (b) 14% Hct, (c) 37% Hct (20×, 1.6 zoom).

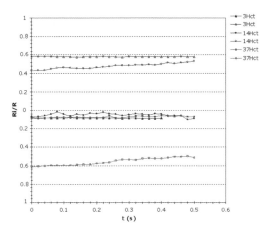

Figure 4.   RBCs streamlines at several haematocrits: 3% Hct, 14% Hct, 37% Hct.

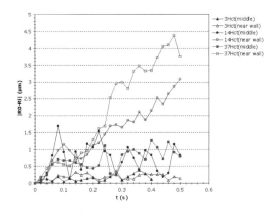

Figure 5.   Radial displacement (ΔR) of labeled RBCs at several haematocrits: 3% Hct, 14% Hct, 37% Hct.

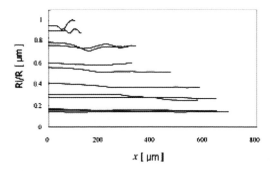

$x \, [\, \mu m \,]$

Figure 6.   RBCs paths by using three-dimensional elastic RBC flow model for 15.7% Hct in straight blood vessel with 106 diameter [9].

REFERENCES

Goldsmith, H., 1971. Red cell motions and wall interactions in tube flow, *Federation Proceedings*, 30 (5). 1578–1588.

Goldsmith, H. and Marlow J., 1979. Flow behavior of erythrocytes. II. Particles motions in concentrated suspensions of ghost cells, *Journal of Colloid and Interface Science*, 71(2): 383–407.

Lima, R., et al., 2006. Confocal micro-PIV measurements of three dimensional profiles of cell suspension flow in a square microchannel, *Meas. Sci. Tech.*, (17): 797–808.

Lima, R., et al., 2007. *In vitro* confocal micro-PIV measurements of blood flow in a square microchannel: the effect of the haematocrit on instantaneous velocity profiles, *J. Biomech.*, (in press).

Lima, R., 2007. Analysis of the blood flow behavior through microchannels by a confocal mico-PIV/PTV system, PhD Thesis, Dep. of Bioengineering and Robotics, Tohoku University, Japan.

Abramoff, M., Magelhaes, P., and Ram, S., 2004. Image Processing with ImageJ, *Biophotonics International*, 11(7): 36–42.

Meijering, E., Smal, I., and Danuser, G., 2006. Tracking in Molecular Bioimaging, *IEEE Signal Processing Magazine*, 23 (3): 46–53.

Tsubota, K. et al., 2006. A particle method for blood flow simulation, *J. Earth Sim.*, (5): pp. 2–7.

Kitagawa, Y., 2006. Large scale computer simulation of the flow of elastic red blood cells using parallel computation, Master Thesis, Dep. of Bioengineering and Robotics, Tohoku University, Japan (in Japanese).

Computational Vision and Medical Image Processing – João Tavares & Natal Jorge (eds)
© 2008 Taylor & Francis Group, London, ISBN 978-0-415-45777-4

Movement detection in 2-D images with variable dimension cells

Frederico Lapa Grilo
CEM, ESTSetubal/IPS, CESET

João M.G. Figueiredo
CEM-IDMEC, Universidade de Évora – Eng. Mecatrónica

O.P. Dias
ESTSetubal/IPS, CESET, INESC

T.G. Amaral
ESTSetubal/IPS, CESET, ISR – Coimbra

ABSTRACT: This paper proposes a movement searching methodology based on 2-D Images, with variable dimension of rectangular cells. The cell dimension is defined automatically based on a statistical adaptive method, which accounts for previous searching results. The main contribution of the present paper in relation to usual image movement searching methods is the self adapting methodology that reduces significantly the computational work in comparison with fixed-dimension image working cells.

1 INTRODUCTION

Artificial vision is a robust tool for pattern recognition with major application fields in industry, building automation and more recently in medical diagnosis. Geometry restoration and visual servoing are domains where artificial vision gives important contributions. Works from Nelson (2003) and Hlou & Lichioui & Guennoun (2003) are good examples of vision potential.

This paper deals with movement detection in consecutive time frames.

Among the different movement detection methods there are two main basic classes: *feature based methods and optical flow based methods*, (Alexandre & Campilho, 1998). In this paper two main processing methods had been used: **inter-frame differencing** and **reference frame differencing**. The performance of both methods is drastically influenced by the pixel area where the method is applied and the subsequent use of the obtained results. These processing results are usually used in real-time monitoring (Fathy & Siyal, 1995), and video compression applications (Jain, 1981), (Mitchell & Pennebaker & Fogg & LeGall, 1996).

In the **inter-frame differencing method** (Jain & Kasturi & Schunck, 1995) it is defined for each pixel (x,y) between the frame obtained at instant

t_1 (t_1-frame) and the frame obtained at instant t_2 (t_2-frame):

$$Dp(x,y) = \begin{cases} 1, & |F(x,y,t_1) - F(x,y,t_2)| \geq \tau \\ 0, & otherwise \end{cases} \quad (1)$$

where $Dp(x,y)$ is the difference in the intensity level of the pixel (x,y) and τ represents the threshold in number of different pixels.

In the **reference frame differencing method** (Jain & Kasturi & Schunck, 1995) it is defined for each pixel (x,y) between the reference frame F_r, and the t_n-frame:

$$Dp(x,y) = \begin{cases} 1, & |F_r(x,y,t_r) - F(x,y,t_n)| \geq \tau \\ 0, & otherwise \end{cases} \quad (2)$$

where t_r and t_n represent the instants where the frames F_r and F_n were acquired, respectively.

The image processing method proposed in this paper, changes the cell working area, taking into account the detected movement amplitude (given by the number of different pixels). The reducing of the cell searching area improves the method processing speed; it should be enhanced that the reducing criteria for the cell working area, is itself adaptable, taking into account the statistical evaluation of previous processed images, the method efficiency increases drastically.

The adaptive algorithms are largely used in digital signal processing applications such as noise cancellation systems, digital filters and data transmission.

Usually these adaptive algorithms use an adaptive function that varies with a specific system characteristic that describes the system evolution regarding the characteristics that are being controlled. This capacity is implemented through rated weights that are defined according to a statistical evaluation of previous processed results.

Referring to the FIR digital filter, with dimension n, and following the usual LMS algorithm (Haykin, 1991) we get:

$$y(i) = k_{0_i} x(i) + k_{1_i} x(i-1) + ... + k_{n_i} x(i-n) \qquad (3)$$

where $y(i)$ represents the filter response to the input $x(i)$ at the instance i, and k_i's are the weighting factors.

Considering $d(i)$ the desired filter response at instant i, the error can be computed by:

$$e(i) = d(i) - y(i) \qquad (4)$$

The changed weighting factors to be used in the next iteration $(i+1)$ are computed according to the adaptive rule:

$$k_{n_{i+1}} = k_{n_i} + \mu.e(i)X(i); \quad n \in N_0; \quad \mu \in [0,1] \qquad (5)$$

where μ is an adaptive learning constant and $X(i)$ is given by (6).

$$X(i) = [x(i)x(i-1)..x(i-n)]^t \qquad (6)$$

The proposed method in this paper is slightly different from the above described method in a way that in the actualization of the weighting factors k_i the prior computed error $e(i)$ is substituted by the movement detection, and the obtained output is the new dimension of the image working cell, as it will be explained in the next section.

2 IMAGE PROCESSING ALGORITHM

2.1 Algorithm description

The proposed movement detection algorithm is initiated with a maximal working cell dimension, equal to the image resolution $P \times Q$.

This working cell is first divided into four searching sections with dimension $P/2 \times Q/2$ as shown in fig. 1.

According to the first division of the original cell into four searching cells we use the inter-frame differencing searching methodology, applied along four lines, as shown in fig. 2.

Figure 1.   Working cells with the dimension $P/2 \times Q/2$.

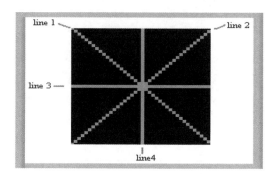

Figure 2.   Detection lines 1–4.

When a movement between two consecutive time frames is detected the correspondent cell is further divided into: $P/a_i \times Q/a_i$. The parameter a_i is the division coefficient of the cell and it is computed by (7).

Depending on the adapting n-dimension function, the computation of the a_i factors is related with the amplitude of the detected movement between two consecutive time frames:

$$a_i = k_i \times a_{i-1} + k_{i-1} \times a_{i-2} + ... + k_{i-n} \times a_{i-n-1} \qquad (7)$$

where

$$k_i = D_i \times \alpha; \qquad (8)$$

The parameter D_i represents the number of cell pixels at instance i, referring the cell where the movement was detected, and α is an adapting factor that defines the evolution of the learning algorithm. The higher the α value, the faster the algorithm becomes.

2.2 Algorithm reduction

In order to reduce the computation effort due to the adapting algorithm presented in (7) and therefore improving the method sample frequency, a possible simplification can be obtained according to (for dimension $n = 1$):

$$k_i = \begin{cases} 2, Di \geq \tau_p \\ 1/2, otherwise \end{cases} \qquad (9)$$

where a_i is computed by (10).

$$a_i = k_i \times a_{i-1} \qquad (10)$$

This simplified algorithm is very effective in terms of computational effort as the binary operations; multiplying or dividing by 2, is very simple as it implies only a bit shift, to the right or to the left, respectively.

3 SYSTEM IMPLEMENTATION

To validate the proposed methodology, several tests were accomplished in the National Instruments Lab-View working environment, based on the acquired data, through a webcam connected to a USB port.

Two software programs were developed: one to accomplish tests on image acquired by the webcam, in real time, and the other to perform tests on AVI video format.

In the two software applications the acquired images were converted into the IMAQ format of Lab-View. The format IMAQ can be used in the vision functions of the LabView library, although it can be converted for other standard image formats. In the work described in this paper, the IMAQ files were converted into a 24 bits RGB matrix (8 bits for each colour component).

For some processes the matrix RGB is converted into an 8 bits matrix of grey levels. In the present work the *inter-frame differencing method* functions for movement detection, as well as the adapting functions, affect directly the matrix elements. In this paper we didn't use predefined functions from libraries.

To accomplish the validation tests of the proposed methodology, it was developed a user interface, in the LabView environment, that holds two areas, whose functions are described below.

Area 1 (fig. 3) – In this area four views are presented. **The first view** presents the coloured video or images, collected in real time; **the second view** shows the same image in levels of grey; **the third view** identifies all points in the image where some movement was detected, through the *inter-frame differencing* method; and finally, **the fourth view** illustrates the image with the cells, achieved with the adaptive algorithm, which illustrates the evolution of the cell dimensions.

Area 2 (fig. 4) – The following graphs are shown: **(i)** the graph with the accumulated number of tested pixels (with and without the algorithm); **(ii)** the graph

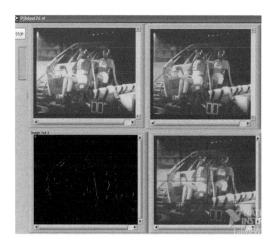

Figure 3. Area 1.

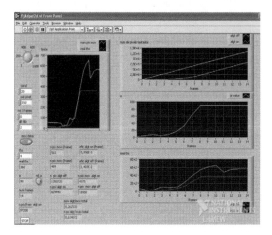

Figure 4. Area 2.

with the factor a_i and **(iii)** the graph with the variation of the dynamic threshold, used by the algorithm, in each frame. In this area it is also presented the system performance evaluation. The meaning of each parameter used to evaluate the system performance is shown in section 4.

The software developed to implement the proposed methodology, allows also the generated data output to be delivered as Excel format files, and as tables or graphs that can be included in *html* pages.

4 PERFORMED TESTS

The accomplished tests were driven for the worst case. Therefore the option of using a video with 157 frames in slow motion was taken, because in this situation,

59

n pix algth off	n pix mov algth off
15147360	42375
n pixl algth on	n pix mov algth on
6320090	30756

n px mov algth on / n px mov algth off
0,725805

n px tested algth on / n px tested algth off
0,417241

Figure 5.   Test results.

there is always movement between two consecutive frames.

It was noticed that the threshold τ_p, used in the adapting algorithm (9) should be dynamic, because with the increase of the number of cells, resulting from image partitions, it implies an increase in the number of movement points detected between two consecutive time frames. The new threshold τ_{pi} has to be adapted to the increasing number of cells. The dynamic threshold is therefore given by:

$$\tau_{pi} = a_{i-1} \times \tau_p \tag{11}$$

Also the expression (9) can be written as:

$$k_i = \begin{cases} 2, Di \geq \tau_{pi} \\ 1/2, otherwise \end{cases} \tag{12}$$

The parameters selected to evaluate the system performance are the following ones:

- Accumulated number of tested pixels **without** the algorithm (*n pix algth off* – fig. 5);
- Accumulated number of tested pixels **with** the algorithm (*n pix algth on* – fig. 5);
- Ratio between the accumulated number of tested pixels **with** and **without** algorithm (*n pix algth on/n pix algth off* – fig. 5);
- Number of pixels with movement detected **without** algorithm (*n pix mov algth off* – fig. 5);
- Number of pixels with movement, detected **with** algorithm (*n pix mov algth on* – fig. 5);
- Ratio between the number of pixels with movement, detected **with** and **without** algorithm (*n pix mov algth on / n pix mov algth off* – fig. 5).

The parameter values obtained in the performed tests are presented in figure 5.

As we can see, in figure 6, with the test progression the value of the factor a_i changes, varying the dimension and the number of the cells. From frame 1 to frame 6 we assist to a change in cells due to the action of the

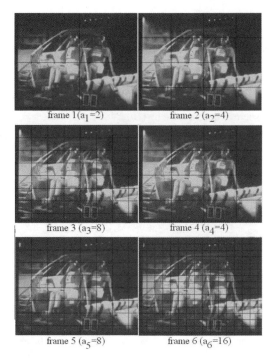

frame 1 (a_1=2)   frame 2 (a_2=4)

frame 3 (a_3=8)   frame 4 (a_4=4)

frame 5 (a_5=8)   frame 6 (a_6=16)

Figure 6.   Test frames from 1 to 6.

adapting algorithm (12). This change is the function of the movement quantity detected in the cells. The movement quantity is measured by the number of changed pixels, from one frame to the next.

The tests done have shown that the accumulated number of tested pixels, with the adapting algorithm, reduces the growing trend relatively to the tests accomplished without the adapting algorithm, as it is illustrated in fig. 7. In fact the difference of 8.8×10^6 pixels is obtained when we compare the tests done, with and without the adapting algorithm (fig. 7). The ratio among the number of tested pixels with, and without, the adapting algorithm is 0.41721 (fig. 5).

In figure 8, the obtained results with the adapting algorithm active are presented. This figure presents the dynamic behaviour of the factor a_i and the correspondent number of tested pixels. Analysing these results it can be concluded that: **a)** the number of tested pixels, from frame1 to frame10, has a constant growth until it reaches the value of 90; **b)** the value of 90 stays constant until frame55; **c)** between frames 55 and 62 the number of tested pixels varies from 45 to 90; **d)** frame 63 presents the value 23; **e)** between frames 64 and 72 the number of tested pixels varies from 11 to 23; **f)** in frame 74 the tested pixels are 45; **g)** between frames 74 and 78 the number of tested pixels varies from 45 to 90; **h)** between frames 79 and 94, the number of tested pixels varies from 11 to 23; **i)** the number of tested pixels are 2 in both frames 95 and 96; one wood **j)**

60

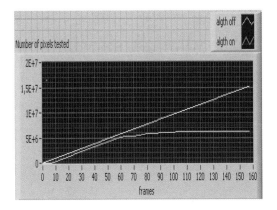

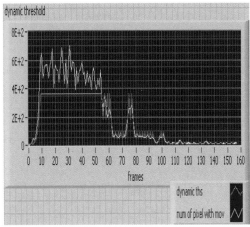

Figure 7. Number of pixels tested with and without the adaptation algorithm.

Figure 9. Action of the dynamic update threshold τ_{pi}.

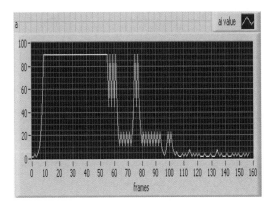

Figure 8. Value of a_i factor.

between frames 97 and 102, the number of tested pixels varies from 11 to 23; **k)** between frames 103 and 106 the number of tested pixels varies from 3 to 6; and finally **l)** after frame 107 the number of tested pixels varies from 2 to 4. This analysis, based on the information illustrated in the figure 8, characterizes the complete dynamic behaviour of the adapting algorithm proposed in this paper.

Figure 9 shows that the dynamic threshold τ_{pi} adapting feature, avoids the saturation of the number of tested pixels. In this way, any information concerning moving characteristics in the analysed frames can be kept.

5 CONCLUSIONS

The tests performed with the adapting algorithm proposed in this paper, show that, relatively to the results obtained without the proposed algorithm, the number of tested pixels is reduced in about 60%, without losing information concerning movement detection.

For the case study presented in this paper (string of 157 frames), it is shown a reduction of 8.8×10^6 tested pixels. In fact, the proposed algorithm presents a greater efficiency in relation to movement detection, what is achieved through the reduction of the tested pixels and, at the same time, increasing the probability to identify movement between consecutive time frames.

Future works will focus on the application of the current algorithm, in an independent way, to each cell, as well as with the help of other methodologies, in the context of genetic algorithms and neural networks. The team, already accomplished some works based on these approaches, with promising results.

REFERENCES

Alexandre, L.; Campilho, A. (1998); *A 2D Image Motion Detection Method Using a Stationary Camera*, REC-PAD98, 1998.

Fathy, M.; Siyal, M. (1995); An image detection technique based on morphological edge detection and background differencing for real-time traffic analysis, Pattern Recognition Letters, 16,1995.

Haykin, S. (1991); *Adaptive Filter Theory.* Englewood Cliffs, NJ: Prentice-Hall, 1991.

Hlou, L.; Lichioui, A.; Guennoun, Z. (2003); *Degraded 3D-Objects Restoration and their Envelope Extraction*, Intl. Journal of Robotics and Automation, Vol. 18, N. 2, 2003.

Jain, A. (1981); *Image data compression: A review*, Proceeding of IEEE, Vol. 69, No 3, 1981.

Jain, R.; Kasturi, R.; Schunck, B. (1995); *Machine Vision,* McGraw Hill, 1995.

Mitchell, J.; Pennebaker, W.; Fogg, C.; LeGall, D. (1996); *MPEG video compression standard*, Chapman & Hall, 1996.

Nelson, B. (2003); *A Distributed Framework for Visually Servoed Manipulation using an Active Camera*, Intl. Journal of Robotics and Automation, Vol. 18, N. 2, 2003.

Computational Vision and Medical Image Processing – João Tavares & Natal Jorge (eds)
© 2008 Taylor & Francis Group, London, ISBN 978-0-415-45777-4

Behaviour of the medial gastrocnemius and soleus muscles during human drop jumps

F. Sousa & J.P. Vilas-Boas
Laboratory of Biomechanics, Faculty of Sports, University of Porto, Porto, Portugal

M. Ishikawa & P.V. Komi
Neuromuscular Research Center, Department of Biology of Physical Activity, University of Jyväskylä, Jyväskylä, Finland

ABSTRACT: The aim of this study was to examine how the fascicle–tendon interaction takes place in the medial *gastrocnemius* (MG) and *soleus* (SOL) muscles during drop jumps (DJ) performed from different drop heights (DH). Eight subjects performed unilateral DJ on a sledge apparatus from different DH with maximal rebounds. Reaction forces and electromyography, together with ultrasonography of MG and SOL were collected. Results showed that the fascicles of the MG and SOL behaved differently during contact phase. The SOL fascicles lengthened prior to shortening in all conditions. The MG fascicles primarily shortened in the lower DH condition. Thereafter, in the higher DH conditions, the MG fascicles behaved isometrically or were lengthened during the braking phase. Results suggest that the fascicles of synergistic muscles can behave differently during DJ, and that there may be specific length change patterns of the MG fascicles with increasing DH, not observed in the SOL fascicles.

The recent advances in real-time ultrasonography allowed the assessment of superficially located soft-tissue structures like fascicles and tendons, during movements of the major leg extensor muscles. This technique has provided a none invasive method for studying the muscle architecture (length and pennation angle of fascicles) and elongation of tendinous structures of human muscles *in vivo* (Fukashiro et al. 1995; Fukunaga et al. 1997; Fukunaga et al. 2001; Kawakami et al. 1998). The function of the fascicles in the *triceps surae* muscle has received special attention during human locomotion from both neuro-physiological and biomechanical points of view. These muscles are known to have either mono- (*soleus* muscle) or bi-articular (*gastrocnemius* muscle) function and they may consequently show different functional behaviour. This question can be answered by performing real-time ultrasonographic scanning of the fascicle and tendon behaviour during human movements, provided that the ultrasound scanning frequency is high enough to capture the fast movement (Ishikawa et al., 2005a; Ishikawa et al., 2005b). A previous study done during human walking show that fascicles of the synergistic MG and SOL behaved differently, nevertheless it is believed that this low impact condition cannot be used to generalize the fascicle behaviour of these muscles, as slow speed walking clearly differs from other forms of movement. Consequently, there is

a need for a more detailed understanding of the muscle specific fascicle behaviour during higher Achilles tendon loading conditions of human Stretch-Shortening Cycle (SSC) movements. In natural and complex human locomotion movements such as running and jumping, muscle activity undergoes a SSC, in which the muscle-tendon unit (MTU) is firstly stretched and immediately followed by a shortening contraction (Norman and Komi, 1979). The objectives of the present study were to investigate the fascicle-tendon interaction in a synergistic pair of muscles (MG and SOL) during DJs of different impact loading conditions. We hypothesized that in this high intensity exercises there is a critical level of stretch load (dropping height), beyond which the fascicles loose their ability to tolerate the imposed load. Additionally we believed that the fascicle behaviour in the bi-articular MG muscle is probably muscle specific and task dependent, not necessarily following the SSC behaviour.

1   METHODS

1.1   *Subjects and experimental procedure*

Eight physically active subjects: age 26.6 (SD 2.8) yr, height 170.1 (6.8) cm and body mass 61.9 (9.9) kg,

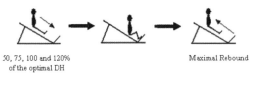

50, 75, 100 and 120%   Maximal Rebound
   of the optimal DH

Figure 1. Schematic presentation of the experimental protocol.

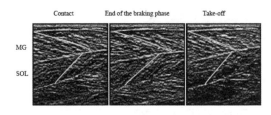

Figure 2. A typical example of the sequence of longitudinal ultrasonic images of the MG and SOL muscles during the drop jumps. Each longitudinal ultrasonic image was recorded at 96 Hz.

who were familiarized with the sledge DJ exercises, participated voluntarily in this study after giving informed consent of the procedures and risks. The study was approved by the Ethics Committee of the University of Jyväskylä. After determining the lowest position of the sledge seat at rest with knee and ankle angles of 50 and 85 degrees, respectively (0 degree is extended position), the subjects performed several unilateral maximum DJs from different DH on the sledge apparatus (Horita et al., 2003; Ishikawa et al., 2003; Ishikawa and Komi, 2004; Kyrolainen and Komi, 1995) in order to decide the individual optimal DH for each subject. This was determined by dropping the subject from different heights to find the greatest rebound height of the center of gravity (Komi and Bosco, 1978). Thereafter, subjects performed the DJ with maximal rebounds from the four individually predetermined DHs: 50, 75, 100, and 120% of the optimal DH (DJ1, DJ2, DJ3 and DJ4, respectively), in a random order. The inclination of the sledge apparatus was set 43 degrees from the horizontal position. During the jumping tasks, the lowest sledge seat position (see above) and maximal jumping height were confirmed by monitoring signals of the position sensor attached to the sledge seat during trials. Three successful trials were required for each condition. Rest between conditions was kept as 5 minutes.

Reaction forces (F_z; parallel to the movement plane of the sledge seat), sledge displacement, velocity of sledge displacement and electromyographic (EMG) activity of the MG and SOL muscles of the right leg were stored simultaneously on a personal computer via an analog-to-digital converter (sampling rate 2 kHz; Power 1401, Cambridge Electronics Design Ltd, England). Surface bipolar EMG electrodes were used to record the MG and SOL muscle activities (Ag/AgCL miniature surface bipolar electrodes, inter-electrode distance of 20 mm; Beckman skin electrode 650437, USA). To place the EMG electrodes on the muscle belly, the B-mode ultrasound images were used to precisely locate the midbellies of the muscle exactly and individually. EMG signals were amplified (input impedance 25 MΩ, common mode rejection ratio > 90 dB) and then sent telemetrically to the analog-to-digital converter. The skin was lightly treated with sandpaper to secure an inter-electrode resistance value below 5 kΩ. In all DJs, the subjects were video-recorded with a high speed video camera

(200 fps; Peak Performance Inc, USA) from the right side, perpendicular to the plane of motion, to calculate the joint angles of the lower limb (knee and ankle). Reflective markers were placed on the centre of rotation of the shoulder, *trochanter major*, centre of rotation of the knee, lateral *malleolus*, heel and fifth metatarsal head, and then digitized automatically using Peak Motus software (Peak Performance Inc, USA). The transformed coordinates were digitally filtered with a Butterworth fourth-order zero–lag low-pass filter (cut-off frequency: 8 Hz). Longitudinal images of the MG and SOL muscles of the right leg were obtained during movement using a real-time B-mode ultrasound apparatus (SSD-5500, Aloka, Japan). An adhesive silicon pad was placed between the skin and the ultrasound probe to avoid movement of the ultrasound transducer. The visibility of the echoes from the fascicle interspaces during movement was confirmed and the probe was fixed securely with a special support device made of polystyrene. The ultrasonographic apparatus was used to measure two-dimensional fascicle length in the MG and SOL muscles during the DJ (96 images·s^{-1}, 6 cm linear array probes with scanning frequency 10 MHz, Aloka, JAPAN) (Ishikawa et al., 2005a; Ishikawa et al., 2005b). The width and depth (thickness) of the scanning images were 5.91 (330 pixels) and 5.38 (248 pixels) cm, respectively. The superior and inferior *aponeurosis* and the MG and SOL fascicles were identified and digitized in each image (Fukunaga et al., 2001; Ishikawa et al., 2005a; Ishikawa et al., 2005b; Kawakami et al., 2002). For each subject, the entire fascicle length of the MG and SOL muscles were estimated by trigonometry (Finni et al., 2001; Ishikawa et al. 2003) in case that the entire fascicle length of the muscle could not be visualized throughout the contact phase of the jumps.

The reliability of the ultrasound method of the fascicle length calculation has been established elsewhere with the coefficient of variation between 0–6% (Fukunaga et al., 1997; Ishikawa et al., 2006; Kawakami et al., 1998; Kawakami et al., 2002; Kurokawa et al., 2001). An electronic pulse was used to synchronize the EMG, kinetic, kinematic and ultrasonographic data.

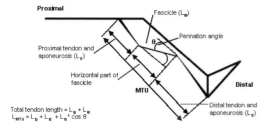

Figure 3. Schematic model of the triceps surae muscle group. The method requires that the total MTU length is kinematically continuously recorded, during the drop jumps.

1.2 Analysis

The model of Hawkins & Hull (Hawkins and Hull, 1990) was used to estimate MTU length (L_{MTU}) changes in MG and SOL muscles from the joint angular data. The fascicle length data acquired at 96 Hz were interpolated to 100 Hz. After the L_{MTU} of MG and SOL data were re-sampled at 100 Hz, the length changes in tendinous tissue (TT) were calculated by subtracting the horizontal length component of the identified MG and SOL fascicle from the L_{MTU} (Fukunaga et al., 2001; Ishikawa et al., 2005a; Ishikawa et al., 2005b; Kurokawa et al., 2001).

$$L_{TT} = L_{MTU} - L_{fa} \cdot \cos \alpha,$$

where L_{TT} is the TT length, L_{MTU} is the muscle-tendon unit length, L_{fa} is the fascicle length and α is the angle between the fascicle line and the *aponeurosis* (pennation angle) (Figure 3). The estimated TT strain was calculated from the length changes of TT from the point of contact to the point of the peak TT length divided by the TT length at the contact moment.

EMG signals were full-wave rectified and low-pass filtered at 75 Hz (Butterworth type fourth-order low-pass digital filter). The EMG signals were integrated and then averaged (aEMG) individually and separately for the following three phases during the ground contact of DJ: pre-activation, braking and push-off phases. The pre-activation phase was defined as the 100ms period preceding the ground contact. The transition from the braking phase to the push-off phase was determined while the sledge was at its lowest position.

1.3 Statistics

Values are presented as means and standard deviations (SD) unless otherwise stated. To analyse the differences between DH conditions for the sledge speed and Fz, a repeated measures one-way Analysis Of Variance (ANOVA) was used, with a post hoc least significant difference multiple comparison. A multivariable ANOVA was used to access the difference in the EMG and length data as well as interactions of

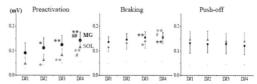

Figure 4. Averaged EMG values ($+/-$SD) for MG and SOL muscles in preactivation, braking and push-off phases for all dropping conditions (DJ1, DJ2, DJ3 and DJ4). Significantly different from DJ1 at * $P < 0.05$ and ** $P < 0.01$. respectively. Significantly different from DJ2 at # $P < 0.05$ and ## $P < 0.01$.

the DH intensity with each muscle, and of the DH intensity with each phase. If interactions were present, a post hoc Tukey test was used to test the difference between them. Spearman's rank correlation coefficient for polynomial regression analysis of variables was used to calculate the statistical significance of the relationship between the dropping sledge speed and the fascicle stretch length. The probability level accepted for statistical significance was P < 0.05.

2 RESULTS

The SSC of muscle function can be divided into three important parts: the preactivation, braking and push-off phases (Komi, 2000). In this way the recorded EMG signals were analysed for these phases during the four DH conditions (DJ1, DJ2, DJ3 and DJ4) performed with maximal rebounds. These results are shown in Figure 4. When a significant interaction between a muscle and a particular phase (P < 0.01) was found, the difference of the DH intensity was examined for each muscle. The most obvious finding was the significant increase of preactivation of MG and SOL muscles as a function of the DH. In addition, in the braking phase, the DJ3 condition showed higher aEMG values of both muscles as compared to DJ1 (Fig 2). When DJ3 was compared with DJ4, the SOL aEMG continued to increase in the braking phase. This was not the case for the MG muscle, although its aEMG at DJ4 was still higher than at DJ1 (P < 0.05). In the inter-muscular comparison, significant differences were observed between MG and SOL muscles from DJ1 to DJ3, but this significance disappeared at DJ4. The push-off phase EMG activities showed similar amplitudes, independent of the DH conditions in the evaluated muscles.

Figure 5 was constructed to give a representative example of the changes in the examined parameters during the contact phase (DJ1, DJ2, DJ3 and DJ4). The curves were averaged from three repetitions, with the first contact on the force plate used as a trigger point for averaging. As shown in this figure and also verified for the entire sample group, the MG and SOL

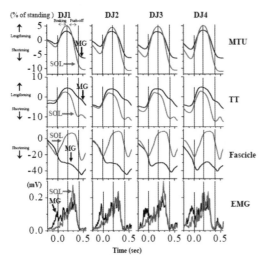

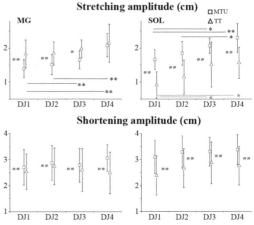

Figure 5. Representative example of the time course data of the lengths of the medial *gastrocnemius* (MG) and *soleus* (SOL) muscle-tendon units (MTU), tendinous tissue (TT) and fascicle, together with the EMG activity of the MG and SOL muscles. 0.0 in the x-axis shows the contact moment of the DJ. "% of standing" in the y-axis for the first 3 rows show the relative changes from the length in the upright position. Vertical dotted lines show the points of the contact, end of the braking phase and take-off.

muscles both exhibited similar MTU behaviour across all jumping conditions. In all DJs, the MTU of the two muscles demonstrated a typical SSC behaviour: stretch prior to shortening during contact with the force plate. The same was also observed for the behaviour of the TT in both muscles.

Figure 6, shows the amplitudes of the respective stretch and shortening of the MTU and TT in the braking and push-off phases. The stretch amplitudes of the two muscles increased as a function of DH, and the amplitudes of shortening remained constant within each muscle, across all conditions.

The situation became more complicated, however, when fascicle length changes were compared. Different fascicle length behaviour was observed between the two muscles when the DH was increased. Until the point of contact, the fascicles of both muscles shortened, but thereafter the patterns became different. After the initial contact, the SOL fascicles began lengthening before shortening in all subjects (Table 1), following the SSC concept. The MG fascicles, however, continued shortening during the early braking phase in all DH conditions in all subjects (Figure 5, Table 1). Thereafter, the MG fascicles behaved differently during the late braking phase depending on DH condition. In DJ1 and DJ2, the MG fascicles further shorten or remained the same length during the late braking phase. In DJ4, which represents very

Figure 6. Amplitudes of stretch (*top*) and shortening (*bottom*) in the braking and push-off phases for MTU and TT of the MG and SOL muscles, for all dropping conditions (DJ1, DJ2, DJ3 and DJ4). Significantly different between conditions at $*P < 0.05$ and $** P < 0.01$. Significantly different between MTU and TT at $\#P < 0.05$ and $\#\# P < 0.01$.

Table 1. The fascicle length (cm) at the contact, end of the braking phase and take-off moments during the contact of drop jumps.

	Contact	End of braking phase	Take-off
MG			
DJ1	3.76 (0.39)⌐	3.54 (0.27) *	3.27 (0.60) **, $
DJ2	3.62 (0.40)	3.51 (0.27) *	3.17 (0.40) **, $
DJ3	73.57 (0.41)	3.48 (0.27)	3.21 (0.42) *, $
DJ4	3.25 (0.31)⌐	3.38 (0.27) *	3.17 (0.26) $
SOL			
DJ1	4.41 (0.99)	5.04 (0.96) *	4.29 (0.96) $
DJ2	4.35 (1.05)	5.05 (0.96) *	4.44 (0.95) $
DJ3	4.35 (1.06)	5.02 (0.94) *	4.43 (1.01) $
DJ4	4.34 (1.04)	4.97 (0.92) *	4.52 (0.84) $

high stretch loads upon impact, a sudden stretch of the MG fascicles after 30–60 ms of the contact were observed in all subjects (also in DJ3 in some subjects). In the subsequent push-off phase, the MG fascicles shortened again in all subjects (Figure 5, Table 1).

3 DISCUSSION

Previous investigations have suggested that the patterns of fascicle length changes during SSC exercises are muscle and intensity (DH and rebound effort) specific (Ishikawa et al., 2003; Ishikawa and Komi,

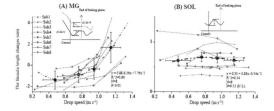

Figure 7. The slope of MG (A) and SOL (B) fascicle length changes vs. the peak dropping speed of the sledge displacements in the braking phase (n = 8). The solid circles and bars show the averaged data and standard errors. *Inset:* how the fascicle length changes were calculated during the braking phase (see also Ishikawa et al., 2005b). NS, not significant.

2004; Ishikawa et al., 2005b). The present work represents the first attempt to compare the fascicle behavior (with ultrasonography) in synergistic muscles (MG and SOL) during different SSC exercises, including very high impact load (DH) conditions. The present results thus emphasize that the MG fascicle behavior varies during the braking phase of DJ depending on DH and the fascicles of MG and SOL muscles do not show the same behavior during the braking phase of DJ. This is despite the fact that the EMG activation of both muscles trended to increase similarly from DJ1 to DJ3 (Fig 4). These results thus clearly indicate the existence of muscle-specific fascicle behavior (MG and SOL) that is also dependent on the DH. In order to examine the influence of the DH on fascicle length changes during the braking phase, Figure 7 plots these attributes against each other for this particular phase. The length change patterns tend to have an opposite quadratic relationship relative to the prestretch intensity (dropping speed). In the MG muscles, there is perhaps a critical level of the stretch load (DH), beyond which the MG fascicles lose their ability to tolerate the imposed load (Fig. 7A). It is interesting to observe that the result shown in Figure 7A is almost identical to that presented in a previous report (Ishikawa et al., 2005b), which used a different SSC exercise to the present study. These results also suggest that the fascicle behavior of MG and SOL during DJ and again suggest that there may be specific length change patterns of the MG and SOL fascicles, depending on the DH.

The observed difference in the fascicle behavior between the two muscles is interesting from several points of view. One may first think that the result can be explained from the pure anatomical differences between the muscles. The MG muscle is a bi-articular muscle and has unique functions in saving and transferring energy and power flow from one joint to another during human locomotion. The results presented in Figure 5 and 6 also highlight the differences between mono- and bi-articular muscles, in which the stretched MTU length was significantly shorter in MG than in

SOL muscle for all DH conditions. At the fascicle level, the stretched fascicle length was also smaller in MG than in SOL (Figure 5, Table 1).

In the present study, the SOL muscle behaved differently from MG in several ways. In the preactivation and braking phases, muscle activity tented to be higher in DJ4 than DJ3 in the SOL muscle but not in MG (Figure 4). At the end of the braking phase, the correspondingly SOL fascicle length was shorter in DJ4 than in DJ3 (only significant difference between DJ2 and DJ4, Table 1). This suggests that the SOL muscle is still able to function "normally" without any additional rapid fascicle length changes (MG).

The observed intermuscular differences in fascicle behavior are very important findings of the present study. The results suggest that the whole concept of the SSC of muscle function should be critically assessed, especially regarding the way in which fasciles interact with TT during normal locomotion.

4 CONCLUSIONS

The real-time ultrasonography method provided a none invasive method for studying the muscle architecture in dynamic situations, allowing the assessment of fascicles and tendons behaviour during exercises.

The results of the present study highlight different fascicle behaviour in the synergistic muscles, MG and SOL during DJ exercises, and that there may be specific length change patterns of the MG fascicles with increasing DH, not observed in the SOL fascicles.

REFERENCES

Finni, T.; Ikegawa, S.; Lepola, V. and Komi, P. (2001). *In vivo* Behaviour of Vastus Lateralis Muscle During Dynamic Performance. *Eur J Sports Sci* (Online) 1. http://www.humankinetics.com/ejss
Fukashiro, S.; Itoh, M.; Ichinose, Y.; Kawakami, Y. & Fukunaga, T. (1995). Ultrasonography gives directly but noninvasively elastic characteristic of human tendon in vivo. Eur J Appl Physiol Occup Physiol 71:555–557.
Fukunaga, T.; Ichinose, Y.; Ito, M.; Kawakami, Y. and Fukashiro, S. (1997). Determination of fascicle length and pennation in a contracting human muscle in vivo. J Appl Physiol 82: 354–358.
Fukunaga, T.; Kubo, K.; Kawakami, Y.; Fukashiro, S.; Kanehisa, H. and Maganaris, C. N. (2001). *In vivo* behaviour of human muscle tendon during walking. Proc R Soc Lond B Biol Sci 268: 229–233.
Hawkins, D. and Hull, M. L. (1990). A method for determining lower extremity muscle-tendon lengths during flexion/extension movements. J Biomech 23: 487–494.
Horita, T.; Komi, V.; Hamalainen, I. and Avela, J. (2003). Exhausting stretch-shortening cycle (SSC) exercise causes greater impairment in SSC performance than in pure concentric performance. Eur J Appl Physiol 88: 527–534.

Ishikawa, M. and Komi, P. V. (2004). Effects of different dropping intensities on fascicle and tendinous tissue behavior during stretch-shortening cycle exercise. J Appl Physiol 96: 848–852.

Ishikawa, M.; Finni, T. and Komi, P. V. (2003). Behaviour of vastus lateralis muscle-tendon during high intensity SSC exercises *in vivo*. Acta Physiol Scand 178: 205–213.

Ishikawa, M.; Komi, P. V.; Grey, M. J.; Lepola, V. and Bruggemann, G. P. (2005a). Muscle-tendon interaction and elastic energy usage in human walking. J Appl Physiol 99: 603–608.

Ishikawa, M.; Niemela, E. and Komi, P. V. (2005b). Interaction between fascicle and tendinous tissues in short-contact stretch-shortening cycle exercise with varying eccentric intensities. J Appl Physiol 99: 217–223.

Ishikawa, M.; Pakaslahti, J. and Komi, P. V. (2006). Medial gastrocnemius muscle behavior during human running and walking. Gait Posture (in press)

Kawakami, Y.; Ichinose, Y. and Fukunaga, T. (1998). Architectural and functional features of human triceps surae muscles during contraction. J Appl Physiol 85: 398–404.

Kawakami, Y.; Muraoka, T.; Ito, S.; Kanehisa, H. and Fukunaga, T. (2002). In vivo muscle fibre behaviour during counter-movement exercise in humans reveals a significant role for tendon elasticity. J Physiol 540: 635–646.

Komi, P. V. and Bosco, C. (1978). Utilization of stored elastic energy in leg extensor muscles by men and women. Med Sci Sports 10: 261–265.

Komi, P. V. (2000). Stretch-shortening cycle: a powerful model to study normal and fatigued muscle. J. Biomech 33:1197–1206.

Kurokawa, S.; Fukunaga, T. and Fukashiro, S. (2001). Behavior of fascicles and tendinous structures of human gastrocnemius during vertical jumping. J Appl Physiol 90: 1349–1358.

Kyrolainen, H. and Komi, P. V. (1995). Differences in mechanical efficiency between power- and endurance-trained athletes while jumping. Eur J Appl Physiol Occup Physiol 70: 36–44.

Norman, R. W. and Komi, P. V. (1979). Electromechanical delay in skeletal muscle under normal movement conditions. Acta Physiol Scand 106: 241–248.

Computational Vision and Medical Image Processing – João Tavares & Natal Jorge (eds)
© 2008 Taylor & Francis Group, London, ISBN 978-0-415-45777-4

Fast and straightforward method to estimate sub-pixel level displacement field

A. Sousa, J. Morais, V. Filipe & A. Jesus
CETAV/Universidade de Trás-os-Montes e Alto Douro, Vila Real, Portugal

M.A.P. Vaz
Faculdade de Engenharia da Universidade do Porto, Porto, Portugal

ABSTRACT: This paper presents a fast and straightforward method to assess displacement fields on the surface of a planar specimen with sub-pixel resolution. First, pixel level displacements are obtained by cross correlation (CC) of two consecutive images, used as the image similarity measure. Next, sub-pixel displacement information is calculated from the CC coefficients matrix using a similarity interpolation method. Finally, sub-pixel estimation error cancellation is performed. Several experiments were conducted to validate the proposed method effectiveness, applied to chestnuts compression test for quality control purposes.

1 INTRODUCTION

Displacement estimation between images can be very important in many fields of digital image processing like motion estimation (Aggarwal & Nandhakumar 1988), image measurement (West & Clarke 1990) or image registration (Tian & Huhns 1986). In the last decade, many techniques used in these fields are assuming greater importance on the evaluation of materials mechanical behavior (Corr et al. 2006). Some of the advantages of using such techniques over the strain gauges or moiré methods (Corr et al. 2006, Lin 2001), is their non-evasive nature. With this method no surface preparation is needed and no mass is added to the studied objects. This way the method is insensitive to vibration (Yaofeng et al. 2005) and enables an easier extraction of surface displacement or strain fields.

Pixel level displacements between pairs of images are obtained searching the minimum or maximum likelihood from the similarity or dissimilarity of images regions. Either, the sum of absolute differences (SAD) (Hill et al. 2006), sum of squared differences (SSD) or cross correlation (CC) are usually used to evaluate the matching between regions. In the method here proposed the normalized cross correlation was selected.

Digital image correlation (DIC) algorithms can be divided into two main groups (Yaofeng et al. 2005)

according to the image analysis, which can be performed in the spatial domain or the spectral domain (Balci & Foroosh 2006). DIC in the spatial domain was first proposed by Sutton and coworkers (Chu et al. 1985). They used coarse-fine search to find the accurate displacements in pixel level, and the sub-pixel accuracy was achieved by combining the intensity pattern of an image with bilinear, polynomial or bi-cubic spline. A decade after the presentation of DIC in spatial domain, Chen and coworker (Chen & Chiang 1992) began to perform digital image correlation in the spectral domain. In this technique, known by digital speckle displacement measurement (DSDM), the sub-pixel accuracy was achieved by biparabolically fitting the correlation coefficients around the peak position.

In our proposed method the cross correlation coefficients are used to measure the similarity between images in the spatial domain. Greater resolution, can be achieved when sub-pixel information is calculated increasing the accuracy of the displacement field characterization. There are a wide variety of algorithms performing this calculation (Aggarwal & Nandhakumar 1988), although they can be classified into four categories (Tian & Huhns 1986): image interpolation, similarity interpolation, gradient-based and phase-correlation method. In the method here described the sub-pixel information is calculated from the cross correlation coefficients matrix. This improvement is obtained from pixel level

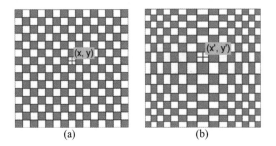

(a)   (b)

Figure 1. Reference (a) and deformed (b) sub-image mapped using correlation.

displacement data, using a similarity interpolation technique.

In this paper the pixel level displacement field calculation is described in section 2. An easy way to obtain the similarity measure and speedup the calculation is also presented along with some performance evaluation tests. Section 3 proposes a method to estimate sub-pixel information from cross correlation matrix using similarity interpolation technique and sub-pixel estimation error cancellation. Section 4 shows experimental results using the proposed method on displacement field calculation. The results shown were obtained over the plain surface of chestnuts samples under uniaxial compression loading.

2 SIMILARITY MEASURES

The present method computes normalized cross correlation in the spatial domain. Although it is well known that cross correlation can be efficiently implemented in the frequency domain, the normalized form of cross correlation, preferred for feature matching applications, does not have a simple frequency domain expression (Lewis, Haralick & Shapiro 1992).

2.1 Correlation method

This method is applied over a pair of images obtained before (reference) and after surface deformation (deformed). The reference image is divided in a set of sub-images and a normalized cross correlation is applied to obtain the displacement of each sub-image center point. To perform this task each sub-image in the reference recording is used to map another sub-image on the deformed image (Figure 1). The matching of a sub-image is found by maximizing the normalized cross correlation coefficient between intensity patterns of two sub-images.

The use of cross-correlation for sub-image matching is motivated by the distance measure (squared Euclidean distance):

$$d_{f,t}^2(u,v) = \sum_{x,y}[f(x,y) - t(x-u, y-v)]^2 \qquad (1)$$

(where f is the image function and the sum is over x, y under the window containing the feature t positioned at u, v)

In the expansion of d^2

$$d_{f,t}^2(u,v) = \sum_{x,y}[f^2(x,y) - 2f(x,y)t(x-u, y-v) \qquad (2)$$
$$+ t^2(x-u, y-v)]$$

the term $\sum t^2(x-u, y-v)$ is constant. If the term $\sum f^2(x,y)$ is approximately constant then the remaining cross-correlation term,

$$c(u,v) = \sum_{x,y} f(x,y)t(x-u, y-v) \qquad (3)$$

is a measure of the similarity between the image f and the feature t. Among others disadvantages in using $c(u, v)$ for template matching it can be noticed that the range of $c(u, v)$ is dependent on the size of the feature and is not invariant to changes in image amplitude, such as those caused by changing lighting conditions across the image sequence.

The *correlation coefficient* overcomes these difficulties by normalizing the image and feature vectors to unit length, yielding a cosine-like correlation coefficient

$$\gamma(u,v) = \frac{\sum_{x,y}[f(x,y) - \overline{f}_{u,v}][t(x-u, y-v) - \overline{t}]}{\{\sum_{x,y}[f(x,y) - \overline{f}_{u,v}]^2 \sum_{x,y}[t(x-u, y-v) - \overline{t}]^2\}^{0.5}} \qquad (4)$$

where $\overline{t}$ is the mean of the feature and $\overline{f}_{u,v}$ is the mean of $f(x,y)$ in the region under the feature. $\gamma(u, v)$ is referred as the *normalized cross-correlation* (Lewis, Haralick & Shapiro 1992).

2.2 Searching optimization

Let us consider a reference and a deformed image divided into M lines by N columns sub-images as shown in Figure 1. Each sub-image mapping generates a correlation coefficient matrix (CCM), where each element represents the correlation between sub-images obtained from sliding the reference sub-image (T) over deformed sub-image (S) by a maximum offset (O) in each shifted position (P). The displacement between sub-images is obtained maximizing the CCM. A typical CCM is shown in Figure 2.

The calculation of CCM can be a very time consuming process once it needs to correlate T with S in every shifted position P. If we assume that the analyzed specimen has a uniform displacement field a connection between sub-image displacements could be established. This information enables the reduction of the maximum offset needed for the search and consequently the sub-image search area. In this work

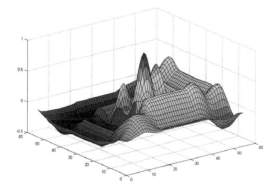

Figure 2. Typical CCM showing correlation coefficient peak.

a new approach to optimize sub-image mapping performance is proposed. In this method two easy steps should be taken to achieve this goal:

1. Calculate *CCM* elements:

   – In every shifted position P within offset O, if sub-image column = 1;
   – In shifted position $P = 1, 2, 3$ and 4 in most cases, otherwise.

2. Shift next S center with displacement values obtained from current sub-image.

This optimization performs faster sub-image mapping than usual procedure (Table 1), which may prove useful in real-time applications such as quality process control. The proposed method assumes non occluded surfaces which occur in most cases when assessing displacement fields on the surface of planar specimens.

2.3 *Performance evaluation*

Synthetic images were used to establish a comparison between common and proposed search method for a given input displacement. Image function is shown in Equation 5. Each synthetic image consists of a horizontal and vertical sinusoidal intensity sweep pattern. The normalized intensity at the position (u, v) is

$$I(u, v) = \frac{1}{2} + \frac{1}{4}\left(\cos\left(\frac{\pi u^2}{R}\right) + \cos\left(\frac{\pi v^2}{R}\right)\right) \quad (5)$$

where R is the position at which the spatial frequency becomes 1. $R = 1000$ is used in this case. Figure 3 shows the resulting synthetic image. Two images are generated $I_n = (u - n, v - n), n = 0, 1$ where I_0 is used as the reference image and I_1 as the deformed image, both with 640×480 pixels.

In Table 1 a comparison of processing speed is established between common and the new proposed method. The tests were computed with synthetic

Table 1. Comparison between common and proposed sub-image mapping optimization method.

Sub-image size (pixels)		Processed positions	Average processing time (secs)		Performance improvement
T	S	M × N	common	proposed	%
21 × 21	25 × 25	19 × 25	1.5266	1.4876	2.55
21 × 21	31 × 31	15 × 20	1.1968	0.9579	19.96
21 × 21	41 × 41	11 × 15	0.9139	0.5216	42.93
21 × 21	51 × 51	09 × 12	0.8259	0.3628	56.07
21 × 21	61 × 61	07 × 10	0.7156	0.2686	62.47

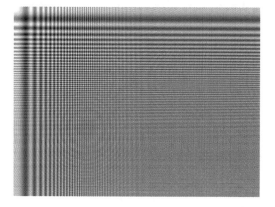

Figure 3. Synthetic image.

images (Figure 3) in a 2.53 GHz P4 CPU with 512 Mb RAM. The proposed sub-image mapping optimization method consistently improves performance, when compared with the common method every time the search window size (S) is increased.

3 SUB-PIXEL ESTIMATION

After finding the pixel level displacement of the two images, sub-pixel estimation is applied to obtain greater resolution. Let us consider $R_{CC}(s)$ as the *CCM* similarity value peak and its four neighboring values as illustrated in Figure 4.

3.1 *Proposed method*

The sub-pixel displacement position can be estimated using the similarity values. Some symmetric functions are generally used for fitting over three similarity values to find the sub-pixel displacement:

1. First order symmetric fitting function estimates the sub-pixel position as the intersection point of two lines; one line passes $(-1, R(-1))$ and $(0, R(0))$,

71

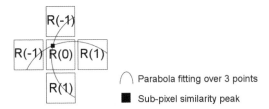

Parabola fitting over 3 points

■ Sub-pixel similarity peak

Figure 4. Cross correlation matrix with pixel level similarity peak value at $R(0)$.

while the other line passes $(1, R(1))$ with the inverse-sign gradient;

2. Parabola fitting (Shimizu & Okutomi 2006), used in our method, estimates the sub-pixel position as the centerline of the fitted parabola (Figure 4). This way the sub-pixel estimation d is:

$$d = \frac{R(-1) - R(1)}{2R(-1) - 4R(0) + 2R(1)} \quad (6)$$

Three pixels are required as references for each axis. It is then chosen the one observed pixel that best matches to the center of the reference pixels, at a given sub-pixel position, based on considering the similarity matrix $R_{CC}(s)$ in pixel unit. $R(0)$ is the similarity peak value (Figures 2, 4). The displacement of the observed pixel against the center of the reference pixels is limited in -0.5 to $+0.5$ pixel.

3.2 Estimation error cancellation

Sub-pixel displacements estimation includes a systematic error called "pixel-locking". This error is such that estimated displacements d are greater than true values when $d < 0$, and less than true values when $d > 0$. This causes the histogram of the estimated values d to have peaks at integer pixel locations.

In our approach this systematic error is canceled using a method firstly introduced by Shimizu and coworker (Shimizu & Okutomi 2001), called Estimation Error Cancellation (EEC). The EEC method uses de following features of the sub-pixel estimation error (Shimizu & Okutomi 2006):

1. The error is periodic with its period equal to the pixel interval;
2. The error magnitude is symmetric about the point at $d = 0$;
3. It is also nearly symmetric with respect to $d = 0.25$ for the range $[0, 0.5]$, and with respect to $d = -0.25$ for the range $[-0.5, 0]$.

A half-pixel shifted image is generated from averaging the original image and one pixel shifted image. Next the method uses a half pixel shifted sub-pixel estimation matrix obtained from using image matching between the half-pixel shifted image and the reference

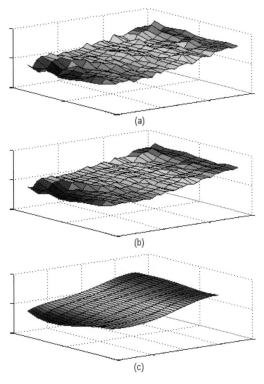

Figure 5. Typicall (a) pixel-locking effect, (b) its cancellation using EEC method and (c) 4th degree polynomial smoothing.

image. This matrix has a negative phase versus the original, allowing this way the estimation error cancellation which results in a more precise 2D sub-pixel displacement estimation.

First, the conventional sub-pixel estimation using the original images is executed. Second, one of the interpolated images is generated by checking the sign of the sub-pixel estimation result d. Third, further sub-pixel estimation is performed. Finally, the estimation errors are cancelled (Shimizu & Okutomi 2001). Figure 5 illustrates the typical pixel-locking effect contained in common displacement field estimation, its cancellation using EEC method and 4th degree polynomial smoothing.

4 RESULTS

Several experiments were conducted to illustrate the effectiveness of the proposed method. Its application for quality control is illustrated with a real chestnuts specimen image from uniaxial compression tests (Figure 6). In this experiment optimization performs about 1/3rd faster sub-image mapping (1.95 secs) than

Figure 6. Real chestnut specimen reference image (1100 × 351 pixels).

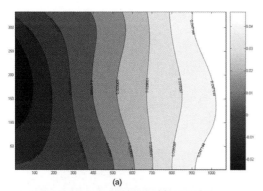

(a)

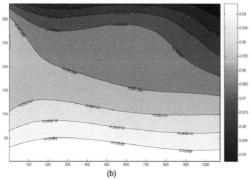

(b)

Figure 7. Displacement field of chestnut specimen surface: (a) u field, (b) v field.

usual procedure (2.94 secs) with the following search optimization parameters: $P = 4$ and $O = 9$ with T [19 × 19 pixels] and S [37 × 37 pixels].

Figure 7 shows the sub-pixel displacement field estimation obtained from applying the new method to a real chestnuts specimen. Two images were extracted from the image sequence:

1. Reference image – obtained before deformation and illustrated in Figure 6;
2. Deformed image – obtained after deformation.

The parallel horizontal u field and vertical v field contour bands would indicate a homogeneous and simple compression state. In the present case non aligned

bands indicate specimen undergoes a complex deformation during loading process. This effect could be due to friction between the loading platens and the specimen or due to the heterogeneous structure of chestnut.

REFERENCES

Aggarwal, J.K. & Nandhakumar, N. 1988. On the Computation of Motion from Sequences of Images – a Review, *Proceedings of the IEEE* 76(8): 917–935.

Balci, M. & Foroosh, H. 2006. Subpixel Estimation of Shifts Directly in the Fourier Domain, *IEEE Transactions on Image Processing* 15(7): 1965–1972.

Chen, D.J. & Chiang, F.P. 1992. Optimal Sampling and Range of Measurement in Displacement-Only Laser-Speckle Correlation, *Experimental Mechanics* 32: 145–153.

Chu, T.C., Ranson, W.F., Sutton, M.A. & Peters, W.H. 1985. Applications of Digital Image Correlation Techniques to Experimental Mechanics, *Experimental Mechanics* 25: 232–244.

Corr, D.J., Accardi, M., Graham-Brady, L. & Shah, S. 2006. Digital image correlation analysis of interfacial debonding properties and fracture behavior in concrete, *Engineering Fracture Mechanics* 74(1–2): 109–121.

Haralick, R.M. & Shapiro, L.G. 1992. Computer and Robot Vision, *Addison-Wesley* II: 316–317.

Hill, P.R., Chiew, T.K., Bull, D.R. & Canagarajah, C.N. 2006. Interpolation Free Subpixel Accuracy Motion Estimation, *IEEE Transactions on Circuits and Systems for Video Technology* 16(12): 1519–1526.

Lewis, J.P. 1995. Fast Normalized Cross-Correlation, In Vision Interface, *Canadian Image Processing and Pattern Recognition Society*: 120–123.

Lin, S.T. 2001. A New Moiré Interferometer for Measuring In-plane Displacement, *Experimental Mechanics Journal* 41(2): 140–143.

Shimizu, M. & Okutomi, M. 2001. Precise sub-pixel estimation on area-based matching, *Proceedings of the 8th IEEE International Conference on Computer Vision, ICCV'01* 1: 90–97.

Shimizu, M. & Okutomi, M. 2006. Multi-Parameter Simultaneous Estimation on Area-Based Matching, *International Journal of Computer Vision* 67(3): 327–342.

Tian, Q. & Huhns, M.N. 1986. Algorithms for Subpixel Registration, *Computer Vision, Graphics, and Image Processing* 35: 220–223.

Walker, C.A. 1994. A historical review of moiré interferometry, *Experimental Mechanics* 34: 281–299.

West, G.A.W. & Clarke, T.A. 1990. A Survey and Examination of Subpixel Measurement Techniques, *Proceedings of the SPIE: Close-Range Photogrammetry Meets Machine Vision* 1395: 456–463.

Yaofeng, S., Meng T.Y., Pang, J.H.L. & Fei, S. 2005. Digital Image Correlation and its Applications in Electronics Packaging, *Proceedings of the 7th Electronic Packaging Technology Conference (EPTC2005)* 1: 129–134.

Computational Vision and Medical Image Processing – João Tavares & Natal Jorge (eds)
© 2008 Taylor & Francis Group, London, ISBN 978-0-415-45777-4

Acquisition electronic system for visualization, registering and filtering of ECG signals

Ricardo E. Martines, Rosa C. David, Juan M. Vilardy, Cesar O. Torres & Lorenzo Mattos

Optics and Computer Science Laboratory, University Popular of the Cesar, Valledupar, Cesar, Colombia

ABSTRACT: A data acquisition system of electrocardiographic signals (ECG) is presented, using three surface electrodes (transducers) that convert the heart signal in bioelectric potentials, and the design of an acquisition card (bipolar electrocardiograph) where we take the bipolar derivations DI, DII and DIII, considering the standards of design for this type of circuits, norm: IEC 60601-1/IEC 60601-25, which were adopted in Colombia by the ICONTEC (NTC IEC 60601/NTCIEC 60601-25). The electrocardiographic signal processing was done using the Discrete Wavelet Transform for the filtering with an interactive graphical user interface (GUI) implemented in Matlab®v7.1 using GUIDE (Graphical User Interface Development Enviroment). This electronic system of low cost can be used as a essential support for authorized personal to diagnose cardiac pathologies.

1 INTRODUCTION

The heart has the property of the formation of an automatic impulse and a rythmical contraction, the electrical impulse takes place in the system of conduction of the heart. The excitation of muscular fibers of all the myocardium result in the contraction cardiac, the production and conduction of these electrical impulses originates small currents that propagate to all the body. The electrocardiographic signals are functions that represent the changes that happen in the electrical potentials of the heart in the dominion of the time, in where the voltage is indicated on the vertical axis, and the time on the horizontal axis. (Fig. 1) (Matiz et al., 1999). The wave P is the deflection produced by the auricular depolarization, the complex QRS represents the time of ventricular depolarization, the wave T is the ventricular fast repolarización and the wave U is a deflection of low voltage, normally is positive (Matiz et al. 1999).

A data acquisition system is a tool for the signal processing as much in the dominion of the time like in the dominion of the frequency. The signal processing is used in many branches of the science, as it is the medicine and engineering. Some diseases are related to alterations of the waves of the ECG, one of these is the arrhythmia that is an alteration of the frequency and heart rate, or a modification of the duration, position or forms of the waves that compose to the cycle of work of the heart.

The Wavelet transform is tool mathematical (well-known like "the mathematical microscope") which it

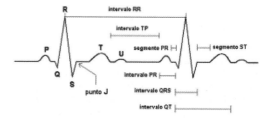

Figure 1. Waves of the ECG.

has like antecedent to the Fourier transform, and comes from the iteration of banks of filters. It's basic characteristics is the multi-resolution which allows the scale change, constituting an form suitable for the analysis of nonstationary signals, mainly those that present fast transitions. The discrete Wavelet transform analyzes the signal to different frequency bands with different resolutions when it's breaking up in approximations and details, use for it two sets of functions: scale functions and wavelet functions respectively. A description of the stages developed in this work is done next, describing the acquisition electronic system and the tests made with MATLAB®.

2 ACQUISITION STAGE

The electrocardiographic signals are taken directly from the patient, or a patient simulator. In this stage

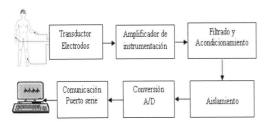

Figure 2. Block diagram of the acquisition electronic system.

Figure 3. Designed electrocardiograph (CARDIOBIP).

we directly captured the signal of the patient through transducers, in this case surface electrodes type clamp, which change this signal in bioeléctricos potentials. In order to connect the electrodes it's due to consider the bipolar derivations or conventional. The bioelectric potentials are of the order of the millivolts, which are amplified using the instrumentation amplifier (Carr et al., 2001). The electrocardiographic signals are of low frecuency, 0.05 Hz–100 Hz, and thus this signal can brings some type of noise signals (Cromwell et al. 1995); by this it was necessary to filter the ECG signal with low-pass and notch filter (Coughlin et al. 1993). After of filtering the signal, it must be prepared and isolated to digitize using a PIC16F876 microcontroller, which makes analogous conversion to digital by successives approximations and sends the signal through the serial communication protocol RS232 by means of the USART using integrated circuit MAX232. This serial communication is controlled by the interactive graphical user interface which was made in Matlab® V.7.1 using GUIDE (Graphical User Interface Development Enviroment). In figures 2 and 3, it's the block diagram of the acquisition stage and the hardware of the electrocardiographic acquisition system, respectively.

3 SIGNAL PROCESSING WITH MATLAB®

3.1 ECG acquisition: design and development

This part of the application was designed and made by means of the interface of the serial port taken from the basic tools Matlab (MathWorks[1] et al., 2005), for this

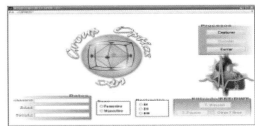

Figure 4. Interactive graphical user interface.

remembers the serial communication with the acquisition card designed and all the parameters that the ECG signal require for their later storage and processing. For the acquisition of the derivations, DI, DII and DII of the ECG signal, it synchronizes the created graphical application in Matlab with sent of digitized data of the PIC and the MAX232, for this the graphical interface that controls the taking of derivations of the ECG signal, is the following:

Before coming to capture ECG signals, it is necessary to fill the important data of the patient to who is going away to take the samples: Name, Age, identification number and Sex, these data are taken to fill the file of the patient and to store them in the hard disk of the PC to use. As soon as the data of the patient are filled and the derivation is selected to take, when pressing the *Capture* button, appears the signal that is taken from the patient by means of the data acquisition card, this signal constantly is updated in the graphical interface, and at the moment that the user of the application decides to acquire the definitive ECG signal for the selected derivation, it is must press the *Save* button, soon when finishing this process, the ECG signal acquired is stopped and it is asked to the user of the application if you wish to save the derivation that has captured (this by a dialog box, which it indicates the step for saving ECG signal). In moment that the user save the first captured derivation, it creates a folder with name of the identification number of the patient within *Electrocardiograms* folder, this folder is in the root directory of the application, the derivation taken is stored within folder created in a file with extension *.txt and name depends of the derivation captured, then if the captured derivation is DI, the name of the file is: *DI.txt* and in the same way for the other two derivations. The process to capture the remaining derivations is equal to the described previously in this section. The following figures show the three captured derivations of the patient.

Finally if signals ECG already were captured and previously stored by the application created in this project, these can be loaded by the acquisition ECG signals interface using the menu *File → Load*, which will show a dialog box, where is possible to search

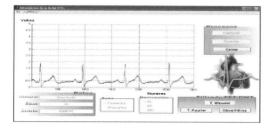

Figure 5.   DI derivation.

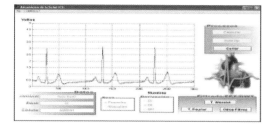

Figure 6.   DII derivation.

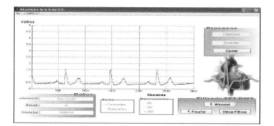

Figure 7.   DIII derivation.

the derivation to load, so that the import of ECG signals towards the interface is successful, it is necessary the following: The file must have extension *.txt, it content must be in the format given by the application and the three derivations must be located in the same directory, since at the time of loading a derivation, the application automatically loads the two rest and if these are not, it will display an error message and ECG signals will not be able to be loaded.

3.2   *ECG filtering and multiresolution: design and development*

In this point of the application the design and development were made mainly using toolboxes of Matlab: *Wavelet Toolbox (MathWorks[2] et al., 2005), Signal Procesing Toolbox (MathWorks[3] et al., 2005) and GUIDE (Graphical Users Interfaces Development Enviroment) (MathWorks[4] et al., 2005)*, mainly considered the traditional filtering techniques and the filtering of noise with the discrete Wavelet transform

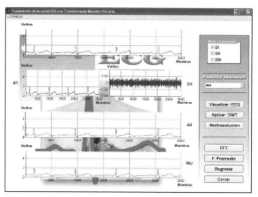

Figure 8.   DI filtering with DWT, Family db4.

and the multiresolution that offers the **DWT** when treating ECG signal.

3.2.1   *ECG filtering with discrete wavelet transform (DWT)*

The new signal processing tools as the DWT, offer a good scene for filtering techniques, specially for noise filtering, when separating the ECG signal in their low (Approximations) and high components frequencies (Details) and then carrying out an adapted reconstruction of the signal, conserving the important approximations, because that signal ECG this composing mainly of low frequencies. The graphical interface that makes the noise filtering with DWT using in the reconstruction of the signal only the importants approximations and not it consider the details, as well as for the reconstruction of the signal using thresholds for the details. In this GUI the visualization of the three derivations by means of the *Derivation* selected at the moment that it pressing the *Visualize: ECG* button, to be able to filter some of the selected derivation, first it due to select the *Wavelet Family* in the popup menu of this GUI (the families used in this application are: **bior3.7, db3, db4 and symlet4**), when the family wavelet is selected to make the noise filtering it due to press the *Apply: DWT* button, the results of this filtering are in the following figure: *A1 and D1,* Approximation and Detail at the level of ECG signal respectively, *A3* reconstructed Approximation of ECG signal at level three, that is to say, filtering signal (with only the levels one, two and three approximation) and *RU* ECG Signal filtering or reconstructed using uniform thresholds for all the details of the three different decomposition levels.

3.2.2   *ECG multiresolution (DWT)*

Within the possibilities of signal processing that offers the DWT is the signals multiresolution, in which it is possible to be observed the interest signal in different scales from resolution by the number of samples

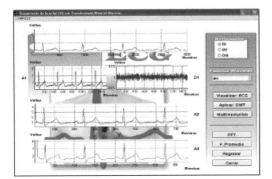

Figure 9.   DI multiresolution with DWT, Family db4.

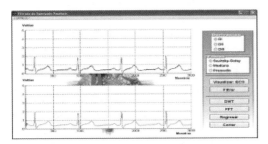

Figure 10.   DI filtering with Savitzky-Golay filter.

or time and amplitude, for to observe more in detail the different characteristics of the signal to analyze so much in the time as in amplitude, this characteristic it is important so that it allows us to emphasize events or particularitities of the ECG signal that can be small. Like the filtering operation with DWT, to make the multiresolution application of the ECG signal it due to select the **Wavelet family**, and finally to press the **Multiresolution** button, the outputs signals for this procedure are: **A1 and D1,** Approximation and Detail at the level one with half of samples of the ECG signal respectively, **A2** Approximation reconstructed of the ECG signal at level two, that is to say, the filtering signal with a quarter of samples of the signal ECG and **A3** ECG Signal filtering or reconstructed at level three with a eighth of samples of the signal ECG (with only the levels one, two and three approximation). The results of the multiresolution with DWT, family db4 are shown in the above figures:

3.2.3   ECG filtering Low-Pass
Analyzing the content in frequencies of ECG signal, we can say that the present noise in this signal can be associate to high frequency components and therefore to eliminate this noise we can apply low-pass filters: Median, average and/or Savitzky-Golay (De La Cruz et al., 2004). In this GUI the results of the application low-pass filters previously mentioned appear, for the

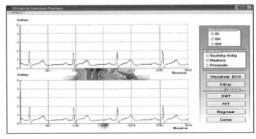

Figure 11.   DI filtering with median filter.

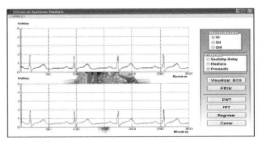

Figure 12.   DI filtering with average filter.

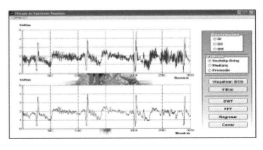

Figure   13.   Noisy   derivation   (DI)   filtering   with Savitzky-Golay filter.

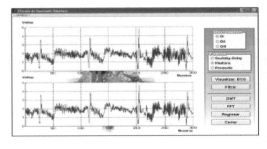

Figure 14.   Noisy derivation (DI) filtering with Median filter.

selected derivation. In the following figures we have the graphical results of the filtering low-pass.

The results of the low-pass filters application are moderate, since within the frequency components that

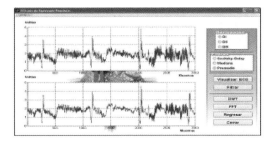

Figure 15. Noisy derivation (DI) filtering with Median filter.

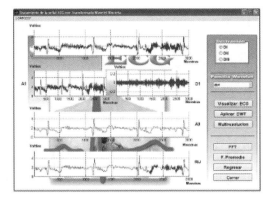

Figure 16. Noisy derivation (DI) filtering with DWT, Family db4.

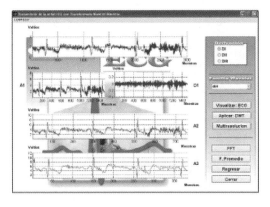

Figure 17. Multiresolution of noisy DI with DWT, Family db4.

it is preserve after the filtering of the low frequencies, are noisy components due to the movement of the patient at the moment that takes the sample from the derivation among other possibilities of noise.

A graphical example for the filtering of a very noisy derivation (DI) using filters low-pass and DWT, it is observed in the following graphs:

When analyzing the results of the figures 13 to 16, by simple observation it note that the filtering of noise

using DWT is superior to the filtering in the frequency domain and low-pass. In order to conclude the example of noisy derivation (DI), the results of the applied multiresolution to this derivation it is shown in the figure 17.

4 CONCLUSIONS

By means of our acquisition electronic system, the heart signal was registered and simultaneously was captured and acquired through Matlab v7.1 in an interactive graphical user interface (GUI). The acquisition was made using electrodes connected to the patient and an initial circuit where an instrumentation amplifier was used; the analogous filtering of the signal was implemented due to the originating noise of the breathing movement of the patient and other noises of the outside, like the electrical public net and the cables of he himself circuit. The digital treatment of the signal was based on the acquisition and the application of digital filters, with the application of an algorithm that uses the discrete Wavelet transform to emphasized the different waves of an ECG and better results were obtained for the signal ECG filtering in comparison with the conventional filtering low-pass. Although many techniques exist to carry out studies to analogous signals how the heart signal, the contribution that is made with this acquisition electronic system can be very useful in centers of clinical practice (hospital) and investigation.

REFERENCES

Matiz H, & Gutiérrez O. 1999. Gutiérrez, Electrocardiografía básica, del trazado al paciente. In Kimpres (eds). Bogotá.

Carr J, & Brown J. 2001. Introduction to biomedical equipment technology. In Prentice Hall (eds).

Cromwell L, & Weibell F. 1995.Instrumentación y medidas biomédicas. In Marcombo (eds), Barcelona (ESPAÑA),

Coughlin R, & Driscoll F. 1993. Amplificadores operacionales y circuitos integrados lineales. In Prentice Hall (eds).

De La Cruz, J. & Pajares G. 2004. Visón por Computadora. In Alfaomega (eds), 65 – 179. Colombia.

MathWorks[1]. "MATLAB External Interfaces". http://www. mathworks.com/access/helpdesk/help/pdf_doc/ matlab/ apiext.pdf, September 2005.

MathWorks[2]. "MATLAB Wavelet Toolbox help". http://www. mathworks.com/access/helpdesk/help/pdf_doc/ wavelet / wavelet_ug.pdf, September 2005.

MathWorks[3]. "MATLAB Signal Processing Toolbox help". http://www.mathworks .com /access/helpdesk/help /pdf_ doc/ signal/ signal_tb.pdf, September 2005.

MathWorks[4]. "MATLAB Creating Graphical User Interfaces". http://www . mathworks . com /access /helpdesk /help /pdf_ doc/ matlab/ buildgui.pdf, September 2005.

Computational Vision and Medical Image Processing – João Tavares & Natal Jorge (eds)
© 2008 Taylor & Francis Group, London, ISBN 978-0-415-45777-4

Macerals detection and quantization in coal using optical methods

Jose C. Peña, Yamelys Navarro & Cesar O. Torres

Optics and Computer Science Laboratory, University Popular of the Cesar, Valledupar, Cesar, Colombia

ABSTRACT: A new image analysis technique has been developed which allows maceral analysis of coal to be carried out. The technique is able to separate the liptinite, vitrinite and inertinite component from the background resin by using quantization method, this method identify the five most important colors in each image of each maceral. The colors images identified are introduced in the VanderLugt correlator to be compare with images coal's microstructures without characterize and show the correlation's peaks.

1 INTRODUCTION

Maceral analysis can be carried out using manual and automated techniques, although both have drawbacks. Manual analysis is time consuming and unavoidably subjective, especially when comparing results of different analysts. Automated analysis is one possible alternative to manual analysis but is not without its difficulties, especially when resolving the various macerals from each other and from the support media used to suspend the coal in a polished block.

The background resin, and there are many types that are in use at present, will normally appear as a black background during oil immersion microscopy. Image analysis systems can carry out complex shape analysis on any given phase but only if the phase can be detected. Pattern recognition work on images that have failed to identify the coal phase in an image can produce poor result.

There have been several published techniques for carrying out maceral analysis that are able to distinguish macerals from background resin. All of these techniques have been shown to generate acceptable results compared with manual data. However, none of these techniques are without drawbacks. Edge effects and internal reflection can occur with colored resins, particle grinding can reduce the effectiveness of morphological techniques and changing between blue and white light sources doubles the analysis time.

A new technique has been developed which allows liptinite, vitrinite and inertinite to be distinguished from the resin background. Whilst carrying out the analysis, sufficient data is collected to allow the maceral composition to be evaluated. [1]

2 METHODOLOGY

2.1 *Image processing*

With breathtaking pace, computers are becoming more powerful and at the same time less expensive, so that widespread applications for digital image processing emerge. In this way, image processing is becoming a tremendous tool to analyze image data in all areas of natural science. For more scientists digital image processing will be the key to study complex scientific problems they could not have dreamed to tackle only a few years ago. A door is opening for new interdisciplinary cooperations merging computer science with the corresponding research areas [2].

2.1.1 *Images processing with color quantization*

Segmentation is the partition of a digital image into multiple regions (sets of pixels), according to some criterion. The goal of segmentation is typically to locate certain objects of interest which may be depicted in the image. Segmentation could therefore be seen as a computer vision problem [2].

Quantization is the process of approximating a continuous range of values (or a very large set of possible discrete values) by a relatively-small set of discrete symbols or integer values. More specifically, a signal can be multi-dimensional and quantization need not be applied to all dimensions. A discrete signal need not necessarily be quantized (a pedantic point, but true nonetheless and can be a point of confusion).

For this process has been developed and carried out, in the Matlab platform, an algorithm which does it these operations: image capture, show the histogram to every one of RGB components, segmentation, quantization and obtaining of Image predominant colors percentage.

2.2 Correlation for Fourier Transform

2.2.1 Correlation output peak intensity
The correlation output peak intensity (COPI) refers to the maximum value of the correlation output plane that is ideally located at $(x, y) = (0, 0)$. COPI is mathematically defined as $COPI = \max_{x,y}\{|C(x,y)|^2\}$ where C(x, y) is the output correlation value at position (x, y). A filter that can generate a high COPI value has good performance and better detectability [3].

2.2.2 Peak sharpness
To provide the best detectability and performance, a filter must yield a sharp correlation peak as well as high COPI value. The sharpness of the correlation peak can be measured with the peak-to-correlation energy (PCE) measure, which is desired to be as large as possible.

3 MATHEMATICAL FORMULATION

3.1 Fourier transform

The Fourier Transform is amply used in the image domain since codify the information in the image (real space) to spatial frequency space (reciprocal space), in him can do it easy operation like filtered, gradients and contrast improvement, inter other. By means of Fourier Transform is possible correlate two functions.

The correlation operation is defined like:

$$C(\xi,\eta) = \int\limits_{-\infty}^{\infty}\int\limits_{-\infty}^{\infty} U_0(\xi_0,\eta_0)H_0^*(\xi_0 - \xi,\eta_0 - \eta)d\xi_0 d\eta_0 \quad (1)$$

The Fourier transforms of images are multiplied one each other before to conjugate a one of them. The inverse Fourier Transform of this product is the correlation between functions. The diagram of fig. 1 is based in one property of Fourier Transform this guarantee her continuity. The principal macerals have color, shape and structure is permitted achieve a frequency

correlation of each coal sample to study. Use the Standard Fourier Transform to correlate the colors images, identified with the quantization method, with the image to identify obtaining the correlations peaks and the similitude percentage with each maceral. [4]

4 IMPLEMENTATION

4.1 Design and development of the investigation and software

Several computer simulations were carried out to demonstrate briefly the performance of the new transformation. The first step is do it the image study supplied for "Instituto Nacional del Carbón" INCAR (Spain), to quantizate for minimal variance to the all image that has been characterized in order to obtain a ocurrence measure of every one of color quantized in percentage with respect to the pixels total's sample, in this manner is determined what colors identify to each sample. A software has been developed in Matlab 7.01 platform to the realization for the last referred; for this process is used a Neural Network, this is trained with the colors percentages.

Having the colors data that identify to each maceral will develop a new software to take the images (100.000 for each sample approx.) and apply the correlation for Fourier Transform with the colors images previously identify in the last process to determine as much as percentage of each maceral have the image, to finish find the general percentage and conclude which maceral prevail.

5 RESULTS

5.1 Vitrinite's samples

The results obtained after analyzed this maceral that are derived from the cell wall material (woody tissue) of plants, which are chemically composed of the polymers, cellulose and lignin [5].

5.2 Liptinite's samples

The results are based in the Liptinite maceral compose; this maceral is derived from the waxy and resinous

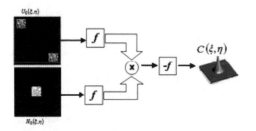

Figure 1. Flow diagram of VanderLugt correlator.

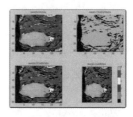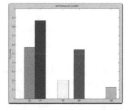

Figure 2. Vitrinite image processed. This figure show a Vitrinite image previously analyzed.

parts of plants such as spores, cuticles, and resins, which are resistant to weathering and digenesis [5]. The presence of green and black colors is stronger in this structures type.

Table 1.   Vitrinite color percentages quantized.

Colors	282	415	700	899	1299
Colors Quantity	11327	17156	3999	10786	2500
Pixels Percentages	17.283	26.178	6.102	16.45	3.814

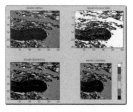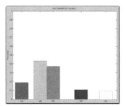

Figure 3.   Liptinite Image processed. This figure show a Liptinite image previously analyzed.

Table 2.   Liptinite color percentages quantized.

Colors	214	446	608	955	1261
Colors Quantity	5646	12966	11046	3240	2769
Pixels Percentages	8.615	19.784	16.854	4.943	4.2252

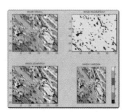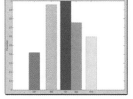

Figure 4.   Inertinite Image processed. This figure show an Inertinite image previously analyzed.

Table 3.   Inertinite color percentages quantized.

Colors	357	555	721	843	1015
Colors Quantity	6544	14907	15597	11787	9259
Pixels Percentages	9.985	22.746	23.799	17.985	14.128

5.3   *Inertinite's samples*

The inertinite macerals are derived from plant material that has been strongly altered and degraded in the peat stage of coal formation [5], for this the colors to prevail are yellow and grey.

6   CONCLUSIONS

The new method implement the color as in images processing, this give place to powerful descriptor that simplify the identification and extraction of the maceral sample.

The result reflect the important that have the quantization's method implemented, by means of himself, get establish the quantity colors and pixels' percent present in macerals sample.

The technique described in the paper represents a new means of carrying out automated maceral analysis. The technique is possibly more practical than other methods that involve colors images. The investigation purports apply a new automated and efficient method.

ACKNOWLEDGMENT

The authors gratefully acknowledge lent aid from the Instituto Nacional del Carbón (INCAR – Spain), CSIC (Spain), in special to the petrography group and the Dr. Angeles Borrego who has been ready to collaborate for carry out this project.

REFERENCES

[1] Edward Lester, David Watts, Michael Cloke. A novel automated image analysis method for maceral analysis. Fuel 81 (2002) 2209–2217.
[2] Jähne, B. "Digital Image Processing" Ed. Springer – Verlag. 1993.
[3] Lamia S.Jamal – Aldin, Rupert C.D. Young, and chris R. Chatwin. Application of nonlinearity to wavelet-transformed images to improve correlation filter performance. Appleid Optics vol 36. 1997.
[4] Goodman, J, "Introduction to Fourier Optics" Ed. Mc Graw Hill, pp 218–243. Estados Unidos. 1996.
[5] mccoy.lib.siu.edu/projects/crelling2/atlas/macerals/mactut. html

Computational Vision and Medical Image Processing – João Tavares & Natal Jorge (eds)
© 2008 Taylor & Francis Group, London, ISBN 978-0-415-45777-4

Medical interface for echographic free-hand images

J.B. Santos & D. Celorico

Department of Electrical Engineering and Computers, Institute of Science and Materials Engineering,
University of Coimbra, polo II, Coimbra, Portugal

J. Varandas & J. Dias

Department of Electrical Engineering and Computers, Institute of Systems and Robotics,
University of Coimbra, polo II, Coimbra, Portugal

ABSTRACT: This paper deals with the development of a medical imaging interface which primary goal is the 3D reconstruction of human organs from 2D ultrasound images. This particular tool was developed to work with a conventional ultrasound probe having a position sensor attached to it. Important requirements like acquisition, calibration, segmentation, interpolation and visualisation for 3D reconstruction are provided by this friendly interface. This tool was developed in C++ language integrating some Matlab executable files and the Visualization Toolkit (VTK) that is free software oriented for 3D graphical computation, visualization and image processing. In this work, we concentrated our efforts in the procedures for the volumetric reconstruction of the Left Ventricle (LV). Special attention is dedicated to the segmentation of the LV contours due to its importance to the 3D volumetric reconstruction. This is not an easy task when the 2D images are noisy and the LV boundaries are not clear as is the case verified this work.

Keywords: Medical images, volumetric images, ultrasound.

1 INTRODUCTION

Echocardiography is a valuable non-invasive tool for imaging the heart and surrounding intrathoracic structures and is an essential way to diagnosis heart diseases. Concerning to the LV, the endocardial and epicardial boundaries are useful quantitative measures for various cardiac functions such as pressure-volume ratio, ejection fraction and cardiac wall motion. To quantify these measures the contours of the LV should be extracted. The extraction of the LV boundaries by image processing techniques is a hard task inherent to the quality of the 2D images, namely due to their low spatial resolution, high level of speckle noise and myocardial boundaries poorly defined. Of course that some of these problems are more or less pronounced according to the quality of the used equipment.

Manual tracing of these borders requires an expert and besides, it is a time consuming and labour intensive task, when lots of images need to be analysed.

Automatic boundary extraction from echocardiographic images thus, appears as a clinical important need to produce most effective and reliable results.

In spite of many researchers have attempted to identify the LV boundaries on 2-DE images, automatically or semi-automatically [Melton 1983, Chu 1988, Detmer, 1990, Setarehdan 1998], this goal still is a challenge.

The LV contour extraction is based on the radial search-based approach where the intensity profiles along a set of equiangular radial lines emanating from the centre point of the LV are analysed. To improve the robustness of the detection, an additional and complementary approach making use of parallel lines traced along with the small dimension of the LV will also be implemented.

The volumetric reconstruction is based on filling an isotropic structure of voxels. Since a point can be located between voxels or a voxel can have more than one or none points, interpolation techniques are required. This tool allows the user to select three distinct interpolation processes: closest neighbour interpolation, Gaussian Splatter method and Shepard method (inverse method of weighed distance). The last section of the paper deals with the application of 3D images visualisations techniques. The user has three options: ray casting, isosurfaces and slicing.

2 B-SCAN IMAGE ACQUISITION

The 2D echographic images were acquired from an ultrasonic equipment from SonoSite (model: SonoSite Titan) with a conventional probe and a position sensor attached to it. This sensor works as the receiver of an electromagnetic position sensor (MiniBIRD_Model 500). The images are visualised in real time, allowing the user to select them at the same time as he performs the scanning. Figure 1 illustrates the front page of the application and Figure 2 the acquisition process showing a 2D image of the heart called apical long-axis four-chamber view.

When the user selects an image (B-scan) its position and orientation is registered using the position sensor. This information is used to determine the positions and orientations of the B-scans with respect to the fixed transmitter. The following procedure consists of the calibration of the system. This is an important requirement for the freehand images collection.

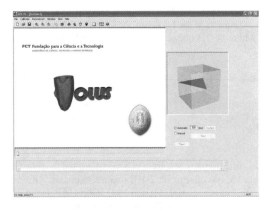

Figure 1. First page of the user interface.

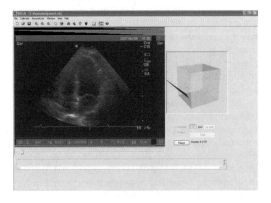

Figure 2. A 2D image of the heart: apical long-axis four-chamber view.

3 CALIBRATION PROCEDURE

Calibration implies the calculation of the position and orientation of the 2D images relating to the sensor attached to the probe. The results of calibration (six parameters: three for position and three for orientation) are used to calculate the correct position of each image pixel in a three-dimensional space, allowing the volume reconstruction in an accurate way. This tool allows the calibration process being accomplished in two different ways: manual and automatic. In general, the calibration problem can be formulated by the following equation [Varandas 2004]:

$$m = {}^{M}T_{O}.{}^{O}T_{P}.{}^{P}T_{I}.q \qquad (1)$$

with

$$q = \begin{bmatrix} s_x.u \\ s_y.v \\ 0 \\ 1 \end{bmatrix}$$

where the matrix ${}^{P}T_{I}$ denotes the transformation from B-scan image coordinates frame {I} to the position sensor on the ultrasound probe with coordinates frame {P} (see ref. (Varandas 2004) for details).

The transformation ${}^{O}T_{P}$ describes the relation between the position of the receiving sensor and the coordinates of the transmitter. The transformation ${}^{M}T_{O}$ represents the relation from the transmitter coordinates to the coordinate frame on the scanned volume. The variables u and v are the column and row indices of the pixel in the B-scan image, and s_x and s_y are the corresponding scaling factors (mm/pixel).

The calibration consists of the estimation of matrix ${}^{P}T_{I}$, matrix ${}^{M}T_{O}$, and the coefficients s_x and s_y, according to Equation 1. From the 14 unknown parameters only eleven will be identified for which, there are many possible methods. The used system device is similar to the Cambridge phantom [Varandas 2004].

The technique is based on the scanning of the bottom of a container with water where each acquired B-scan image has one line that represents the intersection of the ultrasound beam with the bottom of the container.

The automatic procedure, use all acquired images in the parameters calculation, which could lead to a calibration less effective when the images present poor quality. On the contraire, the manual calibration could give best parameter values, since all bad images can be rejected. However, it is a very time consuming process when compared with the automatic one.

4   IMAGES SEGMENTATION

For volumetric reconstruction of the ventricle cavity it is first necessary to extract its boundaries. In this work, we apply a method that is a combination of radial and parallel search-based edge detection. Before proceeding with its application, some image pre-processing is necessary to reduce the noise and enhance the edges of the ventricular boundary. Also the creation of a region of interest (ROI) will reduce the processing time. The ROI characterisation is made after a preview location of the cavity centre point. The results provided in this work were based on a manual centre point extraction. Considering that a reasonable number of images for volumetric reconstruction is about 25, this task is not too time consuming and, as an advantage, allows a much more precise reconstruction compared to an automatic selection of the centre point. After the location of the centre point, the image is divided in two sections (left and right) both referenced to an axis passing by the central point as illustrated in Figure 3. These two sections are then divided in small rectangular blocks. For each block a threshold is obtained using the Otsu's method [Otsu 1979]. Formulated as discriminant analysis, the method separates pixels into two classes (objects and background), by a threshold. The criterion function involves between-classes variance (σ_b^2) to the total variance (σ_t^2), and is defined as:

$$\eta = \frac{\sigma_b^2}{\sigma_t^2} \qquad (2)$$

All possible thresholds, through each selected block are evaluated in this way, and the intensity that maximizes this function is the optimal threshold.

At the end of that operation the LV boundaries were clearly more pronounced, however, the contours were not totally closed and additionally the cavities presented some noise. Some improvement was achieved by using morphological operations like dilation and erosion. The dilation operation adds pixels to the boundaries of objects in an image, while erosion removes pixels from object boundaries. The number of pixels added or removed depends on the size and shape of the structuring element used to process the image. Based on this morphological operations it was performed the image closure followed by the image opening with a square as a structuring element, thus, allowing that the location of the boundaries remain approximately at same positions (Fig. 4).

As illustrated in Figure 4, it is obvious that the extraction of boundaries could still be erroneous due to some remaining noise observed into the cavity. To guaranty an unambiguous identification of LV boundaries by the combined method of radial and parallel search-based edge detection, an additional path in the algorithm was introduced.

That consisted of splitting the cavity in four quadrants and performing edge detection through each a quadrant line. This process, associated with the application of a function of boundary tracing, permitted to evaluate the LV contour though neither totally closed nor smoothed. This goal was reached by the radial and parallel edge detection approaching above mentioned. In the radial approach an ellipse, centred at the centre point of the ventricle, was used and the intensity profile extracted along its radials. This results in a high density of points in the upper and lower region of the ventricular cavity and much less points in the left and right sides. To reduce the uncertainty in the LV walls detection a parallel approach was used, like the one shown in Figure 3. Here the intensity profiles were extracted along with the parallel lines from the central axis to the right and left sides.

The next step consisted of the elimination of all points outside of the cavity region. Then we proceeded with the smoothing of the extracted LV edge points followed by the application of a cubic spline interpolation, as illustrated in Figure 5.

It is important to mention that the processed images were very noisy. Most of them presented ill-defined boundaries making the reconstruction process very laborious, as was demonstrated before. Of course that

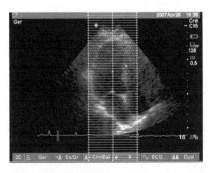

Figure 3.   ROI selection.

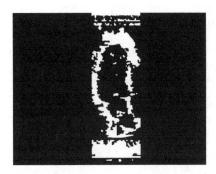

Figure 4.   ROI after morphological operations: erosion and dilation.

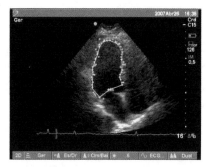

Figure 5. Extracted boundary points and contour tracing.

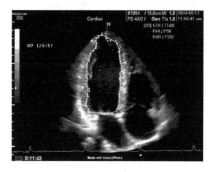

Figure 6. Application of the technique to higher quality images.

with images like the one shown in Figure 6, obtained from a higher resolution echographic machine, procedures are easier, essentially due to the fact that LV boundaries present a very good contrast when compared with the cavity region.

The LV tracing points resulting from the segmentation procedure, for each image, can be used by the application to create a mask. In the reconstruction path, the masks are very useful to produce a model of the left ventricular cavity.

5 INTERPOLATION

The volume reconstruction is based on filling an isotropic structure of voxels (volume unit) with each voxel having an associated luminance level. To determine this level it is necessary to identify the regions of interest in the two-dimensional images (B-scans) and localize them into the space. With the use of the information regarding to the coordinate position and orientation of each pixel in the image it is possible to determine its position in the 3D space. However a point can be located between voxels or a voxel can have more than one or none points, so it becomes necessary to use interpolation techniques. These techniques rely

on processes of constructing new data points with a specific attribute from a discrete set of known data points in the same area or region.

The developed application allows the user to select three distinct interpolation processes: closest neighbour interpolation, Gaussian Splatter method and Shepard method (inverse method of weighed distance). The interpolation method of the closest neighbour (implemented using vtkSurfaceReconstructionFilter) is based on the volume construction by identification of surfaces inside it [Hoppe 1994]. This interpolation surface corresponds to places were a function (that depends on the distance of points to the limit of the object) is equal to a constant (f(x, y, z) = c).

The Gaussian Splatter method (implemented using vtkGaussianSplatter) is based on the Gaussian distribution function. This interpolation technique uses the point data in a voxel structure and interpolates these point attribute values with the voxels according to the Gaussian distribution. The uniform Gaussian distribution centred at a point p_i can be cast in the form [Shroeder & al 2004]:

$$G(x,y,z) = s\, e^{-f(r/R)^2} \tag{3}$$

where s = a scale factor; f exponent scale factor; $r = ||p\text{-}p_i||$ is the distance between any point and the Gaussian centre point; and R is the radius of influence of the Gaussian.

The Shepard method (implemented using vtkShepardMethod) is a weighted inverse distance interpolation technique, which attributes a weight to each point. For each voxel, it is calculated the distance to every point and the weight attributed is dependent on that distance, thus, points that are closer have more weight decreasing then exponentially with the increased distance. Mathematically, given a set of points $p_i = (x_i, y_i, z_i)$ and the function values $F_i(p_i)$, the new points can be interpolated using the equation [Shroeder & al 2004]:

$$F(p) = \frac{\sum_{i=1}^{n} \dfrac{F_i}{|p - p_i|^2}}{\sum_{i=1}^{n} \dfrac{1}{|p - p_i|^2}} \tag{4}$$

6 VISUALISATION

The visualisation of 3D images can be made by volume rendering, slicing or surface rendering. This tool allows the technician to visualise the voxels structure by implementing one of the following three options: ray casting, isosurfaces and slicing.

6.1 Volume rendering

In order to visualize a volume, the ray casting method which is based on the Blinn/Kajiya model [Watt & Watt 1992] is a technique widely used. It permits visualizing sampled functions of three spatial dimensions by computing 2D projections of a colored semitransparent volume. It basically consists in casting a ray from a given position with a given direction, tracing rays from the view point into the viewing volume. In this model we have a volume which has a density $D(x,y,z)$, penetrated by a ray R. At each point along the ray there is an illumination $I(x,y,z)$ that comes from the light source and reaches the point (x,y,z). The intensity scattered along the ray depends on the intensity value on a reflection function or phase function P and on the local density $D(x,y,z)$. Along the ray the density function and the illumination from the source are parameterized as:

$$D(x(t), y(t), z(t)) = D(t) \qquad (5)$$

$$I(x(t), y(t), z(t)) = I(t) \qquad (6)$$

The illumination scattered along R from a point distance t along the ray is given by:

$$I(t)\, D(t)\, P\,(\cos \theta) \qquad (7)$$

where θ = angle between the ray and the light vector. The attenuation due to the density function along a ray can be calculated as:

$$\exp\left(-\tau \int_{t_1}^{t_2} D(s)\,ds\right) \qquad (8)$$

where τ = proportionality constant between density and attenuation. The intensity of the light that arrives to the view point along the direction of R due to all the elements along the ray is given by:

$$B = \int_{t_1}^{t_2}\left(\exp\left(-\tau \int_{t_1}^{t} D(s)ds\right)\right)(I(t)D(t)P(\cos\theta))dt \qquad (9)$$

The attenuation is represented in the view plane (2D image) using a correspondent colour. Thus each a pixel in the image plane has an associated colour according to the interaction of the ray with the volume.

Figure 7 shows a 3D image of the LV based on this method. The apparent no uniformity of the volume is a consequence of the free-hand technique.

6.2 Isosurface

Isosurfaces show 3-D scalar fields by drawing surfaces that represent points of constant values. The

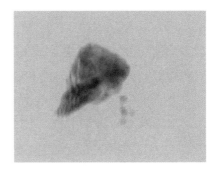

Figure 7.   Volume visualisation of the LV.

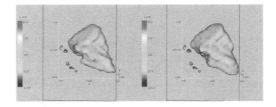

Figure 8.   Isosurface visualisation using Gaussian interpolation with automatic (left) and manual (right) segmentation.

isosurfaces representation uses the Marching Cubes algorithm [William & Cline 1987] (implemented using vtkMarchingCubes) that is a divide-and-conquer approach to locate the surface in a logical cube created from eight pixels. The values of the eight corners of the cube are used to determine how the surface intersects this cube. To find the surface intersection in the cube, it is assigned a value "1" to the cube's vertex if the data value at that vertex exceeds (or equals) the value of the surface to be constructed. These vertices are considered inside the surface. Cube vertices with values below the surface value receive a "zero" and are considered outside the surface. The surface intersects those cube edges where one vertex is outside the surface and the other is inside the surface. With this assumption, the topology of the surface within the cube is determined. In 3D space there are 256 different situations for the marching cubes representation but using rotations and symmetries they can be generalized in 15 families. Figure 8 shows an isosurface for the left ventricle. The left one is based on the automatic segmentation explained before and the right one is the result of a manual segmentation.

This application offers the possibility for the user to modify the isosurface level using a simple cursor. The visualized data is automatically updated and the operator can also obtain the volume value of the rendered object. The calculation of this value (implemented using vtkMassProperties) is based on the discrete form of the divergence theorem [Alyassin 1994].

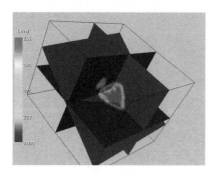

Figure 9. Visualisation of the three plane cuts: along the three major axes (x,y,z) and an arbitrary plane.

6.3 *Slicing*

Slicing allows the user to select different volume perspectives of the organ being studied. In practice this application allows us to perform cuts in the 3D volume. The user can automatically see the cuts in the three principal axes (x, y, and z) and it also provides the option of cutting an arbitrary plane. This arbitrary plane is selected using the cursor to change the plane orientation. This is important when particular region of the volume need to be analysed. All combinations of the four different cuts are permitted. These cuts are 2D images that have associated to each pixel a colour representation of the voxel intensity that is cut by the plane. Figure 9 illustrates cuts along the three major axes and also along an arbitrary plane.

7 CONCLUSIONS

The developed medical interface for echographic images presents as an important and valuable tool for a variety of clinical applications. Technicians are able to acquire and select the most important B-scan images in real-time, manipulate them and make 3D reconstruction in an acceptable time. Almost everything related with the volumetric reconstruction of freehand 2D images can be done with this tool. This tool is dynamic being in continuous expansion for additional functions.

It was demonstrated that the most difficult task associated with this application is the images segmentation. The used technique, combining the radial and parallel search-based edge detection, has proved to be effective in the boundaries extraction even for the noisy and low quality images analysed. As also demonstrated, if images present good quality the extracted contours are significantly improved.

This application was developed with the help of technicians giving valuable and important suggestions in its construction.

ACKNOWLEDGEMENTS

This project is funded by the Knowledge Society Operational Programme (POS_C) and by FEDER.

REFERENCES

Alyassin, A.M. & al. 1994. Evaluation of new algorithms for the interactive measurement of surface area and volume, Med Phys 21(6).

Chu, C. H. & al. 1988. Detecting left-ventricular endocardial and epicardial boundaries by digital two-dimensional echocardiography, vol. 7, pp. 81–90.

Detmer, P. R. & al. 1990. Matched filter identification of left-ventricular endocardial borders in transesophageal echocardiograms. IEEE Trans. Med. Imag., vol. 9, pp. 396–404.

Hoppe, H. 1994. PhD Thesis, Dept. of Computer Science and Engineering, University of Washington, June. IEEE Trans. Med. Imag., vol. 7, pp. 81–90.

Melton, H. E. & al. 1983. Automatic real-time endocardial edge detection in two-dimensional echocardiography. Ultrason. Imag., vol. 5, pp. 300–307.

Otsu, N. 1979, "A Threshold Selection Method from Gray Level Histogram," IEEE Trans. Systems, Man., and Cybernetics, vol. SMC-8, pp. 62–66.

Setarehdan, S. K. & Soraghan, J. J. 1998. "Cardiac Left Ventricular Volume Changes Assessment by Long Axis Echocardiographical Image Processing", IEE Proceedings-Vision, Image and Signal Processing, Vol. 145, n° 3, pp. 203–212.

Shroeder, W. & al. 2004. The Visualization Toolkit An Object-Oriented Approach To 3D Graphics Kitware, Inc (3th Edition).

Varandas, J. & al. 2004. VOLUS–A visualization system for 3D ultrasound data. Ultrasonics 42, 689–694.

Watt, A. & Watt, M. 1992. Advanced Animation and Rendering Techniques: Theory and Practice, Addison-Wesley, Reading, Massachusetts.

Computational Vision and Medical Image Processing – João Tavares & Natal Jorge (eds)
© 2008 Taylor & Francis Group, London, ISBN 978-0-415-45777-4

Numerical experiments for segmenting medical images using level sets

A. Araújo
CMUC, Department of Mathematics, University of Coimbra, Portugal

D.M.G. Comissiong
Department of Mathematics and Computer Science, University of the West Indies, Trinidad and Tobago

G. Stadler
Institute for Computational Engineering and Sciences, The University of Texas at Austin, USA

ABSTRACT: Image segmentation is the process by which objects are separated from background information. Structural segmentation from 2D and 3D images is an important step in the analysis of medical image data. In this work, we utilize level set algorithms and active contours without edges to segment two and three-dimensional image data. Besides synthetical data, we also use magnetic resonance images of the human brain provided by the Institute of Biomedical Research in Light and Images of the University of Coimbra (IBILI).

1 INTRODUCTION

Image segmentation refers to the process of separating objects in images from background information. Among others, it of great interest for medical practitioners, for example for surgery planning or tumor treatment. In this context, often magnetic resonance (MR), computer tomography (CT), position emission tomography (PET) or single photon emission computed tomography (SPECT) images are used.

When segmenting a medical image, the actual surfaces of identified objects are often blurred as a result of background noise generated by signal interference with surrounding tissues. Thus, efficient computational segmentation algorithms are required to accurately detect the outline of the regions of interest.

The level set method for capturing moving fronts was introduced by Osher and Sethian in 1988 (OS88). This Eulerian-type method utilizes a fixed mesh of grid points, and matches the evolving interface implicitly with the zero-level contour of a signed-distance function. The resulting initial value partial differential equation is essentially a Hamilton-Jacobi equation. It is then possible to utilize entropy-satisfying schemes developed for the numerical solution of hyperbolic conservation laws.

The success of level set methods is due to their efficiency and flexibility to deal with geometrical objects such as free boundaries and surfaces. Among its advantages, we mention the following: (i) Curvatures and normals are easily evaluated; (ii) Topological changes occur naturally; (iii) The technique is easily extended to three (or even more) dimensions. Moreover, level set theory is to a large amount based on the theory of partial differential equations and benefits from many of the results and techniques developed for such systems. A good introduction to the level set method and its use in image segmentation can be found in the texts by Osher and Fedkiw (OF03) and Sethian (Set99).

In this work we explain the basic ideas behind level set-based image segmentation and report on some numerical tests for both 2D and 3D data. For our purposes, we have chosen to use active contour methods without edges.

2 ACTIVE CONTOURS

The main idea behind algorithms for detecting active contours in image segmentation is easily explained. The user provides an initial guess for the contour. This guess is subsequently moved by image-driven forces to the boundaries of the desired objects. In such models, internal and external forces are considered. The internal forces defined within the curve are designed to keep the model smooth during the deformation process. External forces computed from the underlying image data are defined to move the model toward an object boundary or other desired features within the image.

2.1 *The classical models*

Let Ω be a bounded open set of $\mathbb{R}^2$, with $\partial\Omega$ its boundary. Let $u_0 : \overline{\Omega} \longrightarrow \mathbb{R}$ be a given image and $C : [0, 1] \longrightarrow \mathbb{R}^2$ be a parameterized curve.

The main method used here is, that the curve C is evolved towards the object boundary under a force, until it stops at the boundary. To stop the process, an edge-detector is used, depending on the gradient of u_0.

The use of the level set method, and in particular the motion by mean curvature of Osher and Sethian (OS88) is in this context specially justified due to its ability to naturally execute topological changes. With this method, the curve is represented implicitly via a Lipschitz function ϕ by the zero level set

$$C = \{(x,y) \in \Omega : \phi(x,y) = 0\},$$

and the evolution of C is given by the zero-level curve at time t of the function $\phi(x,y,t)$.

If we want the curve C to evolve in normal direction with speed a, we need to solve the partial differential equation

$$\frac{\partial \phi}{\partial t} = a|\nabla \phi|, \qquad \phi(x,y,0) = \phi_0(x,y),$$

with suitable boundary conditions. Here, the initial contour is given by the points (x,y) that satisfy $\phi_0(x,y) = 0$. Using motion by mean curvature, we have $a = k$, where k is the curvature of the level curve passing through (x,y), i.e.,

$$k = \operatorname{div}\left(\frac{\nabla \phi(x,y)}{|\nabla \phi(x,y)|} \right).$$

An example of a geometric active contour model based on the mean curvature motion is the variational formulation of Zhao et al. (ZCMO96)

$$\frac{\partial \phi}{\partial t} = |\nabla \phi| g(\nabla u_0) k + \nabla g(\nabla u_0)^T \nabla \phi,$$

where $g(\nabla u_0)$ is the edge function

$$g(\nabla u_0(x,y)) = \frac{1}{1 + |\nabla G_\sigma(x,y) \star u_0(x,y)|^2}, \qquad (1)$$

with $G_\sigma \star u_0$ is a smoother version of u_0. The first term of the corresponds to the motion in normal direction with velocity $g(\nabla u_0)k$ and the second one is a convective term in the direction of $\nabla g(\nabla u_0)$. Note that, if $g(\nabla u_0)$ vanishes, motion in normal direction stops and for constant $g(\nabla u_0)$, the convection term vanishes.

In practice, the edge-detector is never zero on the edges and therefore the evolving curve may not stop after having reached the desired contour. In (CV01), Chan and Vese propose a different active contour model without edges by using an edge-detector function that depends on the gradient ∇u_0. The stopping term is based on the so-called Munford-Shah segmentation technique.

2.2 Active contours without edges

We will now give a brief description of the algorithm presented in (CV01).

Let us consider an open subset $\omega \subset \Omega$, such that $C = \partial \omega$. We will use the notation inside $(C) := \omega$ and outside $(C) := \Omega \setminus \omega$. Assume that the image u_0 is formed by two regions of approximately piecewise-constant intensities of distinct values u_0^i (the object to be detected) and u_0^o. Let us denote by C_0 the boundary of u_0^i. Then we have

$$u_0 \approx \begin{cases} u_0^i, & \text{inside } (C_0), \\ u_0^o, & \text{outside } (C_0). \end{cases}$$

Let us now consider the following "fitting term"

$$F_1(C) + F_2(C) = \int_{\text{inside }(C)} |u_0(x,y) - c_1|^2 dx\, dy$$

$$+ \int_{\text{outside }(C)} |u_0(x,y) - c_2|^2 dx\, dy,$$

where C is any curve and c_1, c_2 depend on C and are the averages of u_0 inside C and outside C, respectively. Obviously, C_0 – boundary of the object – is the minimizer of the fitting term, i.e.,

$$\inf_C (F_1(C) + F_2(C)) \approx 0 \approx F_1(C_0) + F_2(C_0).$$

Our goal is to minimize the above fitting term plus some regularizing terms such as the length of the curve C and/or the area of the region inside C. Let us introduce the energy functional $F(c_1, c_2, C)$, defined by

$$F(c_1, c_2, C) = \mu \operatorname{Length}(C) + \eta \operatorname{Area}(C)$$

$$+ \lambda_1 F_1(C) + \lambda_2 F_2(C), \qquad (2)$$

where $\mu \geq 0$, $\eta \geq 0$, $\lambda_i > 0$, $i = 1, 2$, are fixed parameters. Then, we are interested in solving the minimization problem

$$\min_{c_1, c_2, C} F(c_1, c_2, C).$$

We now rewrite the original model (2) in the level set formulation. Let the evolving curve C be

$$C = \partial \omega = \{(x,y) \in \Omega : \phi(x,y) = 0\},$$

assuming that

$$\begin{cases} \text{inside } (C) = \omega = \{(x,y) \in \Omega : \phi(x,y) > 0\}, \\ \text{outside } (C) = \Omega \setminus \omega = \{(x,y) \in \Omega : \phi(x,y) < 0\}, \end{cases}$$

where ω is an open subset of Ω.

For the level set formulation of our active contour model, we replace C by the unknown variable ϕ. Using the Heaviside function

$$H(z) = \begin{cases} 1, & \text{if } z \geq 0, \\ 0, & \text{if } z < 0, \end{cases}$$

and the Dirac delta function $\delta(z) = \frac{d}{dz}H(z)$ (in the sense of distributions), we can rewrite the energy functional in the following way:

$$F(c_1, c_2, C) = \mu \int_\Omega \delta_0(\phi(x,y))|\nabla\phi(x,y)|dx dy$$

$$+\eta \int_\Omega H(\phi(x,y))dx dy$$

$$+\lambda_1 \int_\Omega |u_0(x,y) - c_1|^2 H(\phi(x,y))dx\, dy$$

$$+\lambda_2 \int_\Omega |u_0(x,y) - c_1|^2 (1 - H(\phi(x,y)))dx\, dy.$$

Let now

$$u = \begin{cases} \text{average } (u_0) \text{ inside } C, \\ \text{average } (u_0) \text{ outside } C. \end{cases}$$

Then, for $(x,y) \in \overline{\Omega}$,

$$u(x,y) = c_1 H(\phi(x,y)) + c_2(1 - H(\phi(x,y))).$$

Fixing ϕ and minimizing $F(c_1, c_2, C)$ with respect to c_1 and c_2, it is easy to express these constant functions of ϕ by

$$c_1(\phi) = \frac{\int_\Omega u_0(x,y)H(\phi(x,y))dx\, dy}{\int_\Omega H(\phi(x,y))dx\, dy},$$

if $\int_\Omega H(\phi(x,y))dx\, dy > 0$, and

$$c_2(\phi) = \frac{\int_\Omega u_0(x,y)(1 - H(\phi(x,y)))dx\, dy}{\int_\Omega (1 - H(\phi(x,y)))dx\, dy},$$

if $\int_\Omega (1 - H(\phi(x,y)))dx\, dy > 0$. In the degenerate case

$$\int_\Omega H(\phi(x,y))dx\, dy = \int_\Omega (1 - H(\phi(x,y)))dx\, dy = 0,$$

we obtain

$$\begin{cases} c_1(\phi) = \text{average } (u_0) \text{ in } \{\phi \geq 0\}, \\ c_2(\phi) = \text{average } (u_0) \text{ in } \{\phi < 0\}. \end{cases}$$

In order to compute the Euler-Lagrange equations, we use the variational level set approach and we arrive at

$$\frac{\partial\phi}{\partial t} = |\nabla\phi| \left[\mu k - \eta - \lambda_1(u_0 - c_1^2 + \lambda_2(u_0 - c_2)^2 \right],$$

in $(0, +\infty) \times \Omega$, and $\phi(x,y,0) = \phi_0(x,y)$ in Ω, where k is the curvature.

3 NUMERICAL EXPERIMENTS

In this section, we report on our numerical testing for the active contour algorithm without edges. By means of Example 1, basic properties of the algorithm are explained. The next example is a two dimensional medical image, in which we intend to segment parts of the spine. Finally, we apply the method to three-dimensional magnetic resonance data. Our numerical implementation is based on the MATLAB level set toolbox (MT05).

3.1 Example 1 (CMUC logo)

For our first test, we use the image shown in Figure 1. Note that the image in Figure 1 is not a binary image (i.e., it does not only contain black and white pixels), since the letters are drawn in different gray values. This image have a resolution of 120×120 pixels. We use the working domain $\Omega = [-1, 1] \times [-1, 1]$, the parameters $\mu = 2$, $\lambda_1 = \lambda_2 = 1$ and the final time $t_f = 0.06$. This implies a mesh with mesh size $h = 1/60$ for the space discretization. We remark that, in (CV01) the scaling $h = 1$ is used, which results in $\Omega = [-60, 60] \times [-60, 60]$ and in different scalings for the parameters $\mu, \lambda_1, \lambda_2$ and for Ω.

In Figure 2, the initial and final contours as well as seven intermediate steps of the time evolution for segmenting the CMUC logo are shown. As initial contour a circle is chosen, and as the time increases, this contour approaches the contour of the original image.

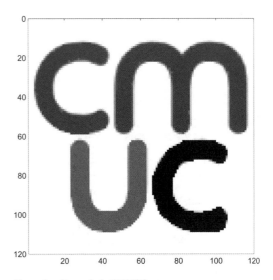

Figure 1. Example 1: CMUC logo.

93

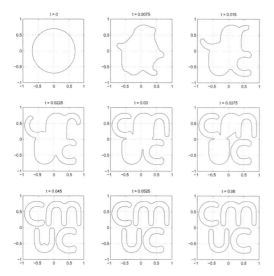

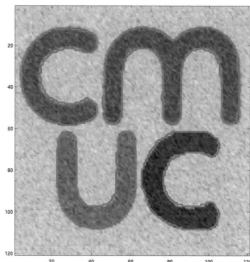

Figure 2. Example 1: Time evolution of level sets for the image shown on Figure 1.

Figure 3. Example 1: Segmentation obtained for a noisy image.

Note that the topological changes needed to segment the different letters do not represent any problem for the algorithm.

Another attractive feature of the active contour algorithm without edges is that it also works in the presence of noise. This is essential when dealing with medical images, in which a certain amount of noise is unavoidable due to background interferences or other perturbations. Our result obtained for a noisy image, obtained from the image in Figure 1 by the introduction of 25% of white noise, can be seen in Figure 3. While noise presents a serious problem for segmentation algorithms that use image gradient information, the active contour method is stable with respect to noise in the image data and robustly detects the image contours.

3.2 Example 2 (2D medical image)

In the second example, we attempt the segmentation of bones from the image shown on the left of Figure 4. We use the image domain $\Omega = [-1, 1] \times [-1, 1]$ and, for all tests, keep the parameters $\lambda_1 = \lambda_2 = 1$, $\nu = 0$ and the final time $t_f = 0.5$ fixed. We study the influence of μ, i.e., the weight for the term corresponding to motion in normal direction.

In the top of Figure 4 we show the evolution of the zero level set for $\mu = 0.1$. In the top plot of Figure 5, the resulting contour for this value of μ can be seen. In the bottom plot of Figure 5, the contour obtained with a larger values of μ, namely $\mu = 0.5$, is shown. Obviously, the parameter μ allows to control the sensitivity of the contour in capturing details. Obviously, larger

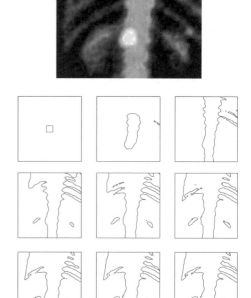

Figure 4. Example 2: Medical image (top) and level set evolution (bottom).

94

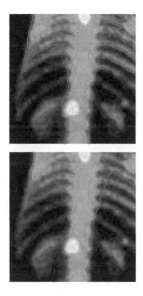

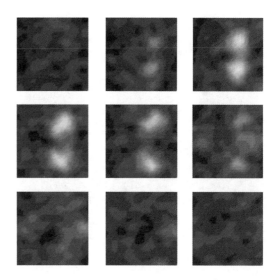

Figure 6.   Example 3: Slices of magnetic resonance image.

Figure 5.   Example 2: Contour found using $\mu = 0.1$ (top) and $\mu = 0.5$ (bottom).

values of μ lead to smoother contours; these might be more stable with respect to noise in the image, but they possibly do not have the ability to find small structures.

3.3 Segmenting brain regions in 3D

The data for this problem is provided by the Institute of Biomedical Research in Light and Images of the University of Coimbra (IBILI). The idea is to identify the form of the putamen and caudate which are structures in the brain. The putamen is a portion of the basal ganglia. The basal ganglia system was associated with motor functions, as lesions of these areas would often result in disordered movement in humans (Chorea, athetosis, Parkinson's disease). If the SPECT/PET scan shows decreased dopamine activity in the basal ganglia this aids in diagnosing Parkinson's disease.

The original images were obtained by SPECT scan with a radiopharmaceutical used in nuclear medicine for the differential diagnosis of Parkinsonian syndromes versus essential tremor: the DaTSCAN™ (Ioflupane, 123-I FP-CIT). The drug is administered to the patients via slow intravenous injection and a SPECT image is performed 3–6 hours after the injection. After its administration to the patient, the concentration of radiopharmaceutical is measured using a gamma-camera. SPECT imaging with DaTSCAN™ therefore produces images of the brain structures that are involved in the pathophysiology of parkinsonian syndromes. The procedure usually takes between 20

and 45 minutes depending on the type of gamma camera used.

We use a part of size $41 \times 41 \times 41$ of the original image. As before, the values of the data are between 0 and 255 and correspond to the gray values in the magnetic resonance image. We refer to Figure 6, where 9 two-dimensional slices of this data are shown. For our segmentation algorithm, we use the parameters $\lambda_1 = \lambda_2 = 1$, $\nu = 0$, $\mu = 0.05$ and the final time $t_f = 0.8$. As domain we choose $\Omega = [-1, 1]^3$ and as initial surface the cube $[-0.2, 0.2]^3$ is chosen. In our tests, we also used other initializations and obtained the same segmentation. However, if the initial zero level set was too small, it vanishes after the first steps of the iteration. The zero level sets for $\mu = 0.05$ at times $t = 0.0, 0.1, \ldots, 0.8$ are shown in Figure 7. We compared the results for various values of μ and observed a similar behavior as for the two-dimensional examples. For $\mu = 0.001$, the solution captures more details of the data and the resulting surface is relatively rough (see the bottom plot in Figure 8). For $\mu = 0.05$, the zero level set upon convergence is slightly smaller and smoother (see the top plot in Figure 8). For $\mu \geq 0.2$, independently from the initialization the zero level shrinks and finally vanishes. This undesired behavior happens if the weight μ for the term corresponding to motion in normal direction becomes too large.

4 CONCLUSIONS AND FURTHER WORK

In this paper, we report on our numerical experiments with the active contour algorithm for image segmentation. Extensions of the present work interesting for

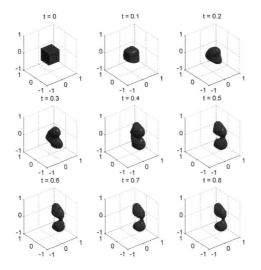

Figure 7. Example 3: Evolution of the zero level set in the iteration for $\mu = 0.05$.

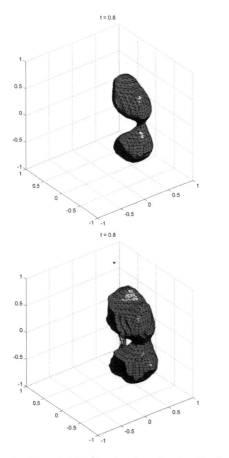

Figure 8. Example 3: Results of segmentation algorithm for $\mu = 0.05$ (top) and $\mu = 0.001$ (bottom).

doctors focus on the development of a low cost optical motion correction system for a SPECT/PET restraint free (without anesthesia) small animals. The problem of the implementation of this system can be divided in two major parts: motion detection and motion correction. One of the possibilities for using level sets in this context is to obtain the 4D contour (we must consider the time variable) of the animal.

REFERENCES

T. F. Chan, and L. A. Vese. Active contour without edges. *IEEE Trans. Image Processing*, 10(2): 266–277, 2001.

I. M. Mitchell, and J.A. Templeton. A toolbox of Hamilton-Jacobi solvers for analysis of nondeterministic continuous and hybrid systems. In *Lecture Notes in Computer Science (LNCS)*, volume 3414, pages 480–494. Springer Verlag, 2005.

S. Osher, and R. Fedkiw. *Level set methods and dynamic implicit surfaces*, volume 153 of *Applied Mathematical Sciences*. Springer-Verlag, New York, 2003.

S. Osher, and J. A. Sethian. Fronts propagating with curvature-dependent speed: algorithms based on Hamilton-Jacobi formulations. *J. Comput. Phys.*, 79(1):12–49, 1988.

J.A. Sethian. *Level set methods and fast marching methods*, volume 3 of *Cambridge Monographs on Applied and Computational Mathematics*. Cambridge University Press, Cambridge, second edition, 1999.

H.K. Zhao, T. Chan, B. Merriman, and S. Osher. A variational level set approach to multiphase motion. *J. Comput. Phys.*, 127:179–195, 1996.

Computational Vision and Medical Image Processing – João Tavares & Natal Jorge (eds)
© 2008 Taylor & Francis Group, London, ISBN 978-0-415-45777-4

Detection of synthetic singularities in digital mammographies using spherical filters

Céline Gouttière & Joël De Coninck

University of Mons-Hainaut, CRMM, Parc Initialis, Mons, Belgium

ABSTRACT: Since it has been demonstrated that spherical filters are efficient to detect synthetic singularities in both synthetic images and real textures, these filters are herewidth tested for the detection of the same singularities in digital mammographies. In fact, this kind of defects is rather similar to microcalcifications in mammography. We hereafter compare the efficiency of different spherical filters to detect synthetic microcalcifications in mammographies.

1 INTRODUCTION

Microcalcifications and masses are important indicators of breast cancer. Thus it is important to detect them as soon as possible. Many researches have been developed to improve the automated detection of anomalies in digital mammographies (Arodz et al. 2005; Mousa et al. 2005) particularly to detect microcalcifications (Arodz et al. 2006; Al-Qdah et al. 2005; Fu et al. 2005; Yu et al. 2006; Lemaur et al. 2003) and masses (Cheng et al. 2006; Kom et al. 2007).

Since it has been previously proved that spherical filters can efficiently detect synthetic singularities in synthetic images (Gouttière et al. 2007a; Gouttière et al. 2007b), which are somewhat similar to mammographies, and also in real reference textures (Gouttière et al.), these filters are now tested for digital mammographies. The same singularities and method of detection than the previous studies are chosen and are briefly explain respectively in section 2 and 3. Examples of tested images are presented in section 4. The results obtained for the synthetic images and the real reference textures are recalled in section 5. Latest investigations about digital mammographies are also described in this section.

2 CREATION OF THE SYNTHETIC SINGULARITIES

Two types of singularities have been put to the test in our previous studies: one represented by a cone, and another by a truncated cone. The advantage to work with synthetic defects is the possible tuning of the parameters like the radius or the height of the singularities. Another parameter is used for the detection of the truncated defects: the percentage of the height where

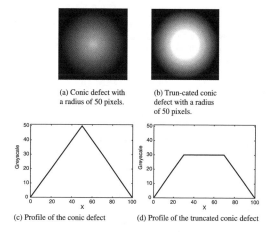

(a) Conic defect with a radius of 50 pixels.   (b) Trun-cated conic defect with a radius of 50 pixels.

(c) Profile of the conic defect   (d) Profile of the truncated conic defect

Figure 1. Examples of synthetic defects.

the truncation is performed. Examples of synthetic singularities are shown in Fig. 1.

The intensity of the defect can be adapted related to the image background with a parameter Ω, which multiplies the height of the defect. Ω varies from 0 to 1, where a closed value to 0 means that the defect becomes invisible.

In the following section, the detection method of these defects is briefly explain. More details about this can be found elsewhere (Gouttière et al. 2007a; Gouttiere et al.).

3 METHOD OF DETECTION

Spherical wavelets and filters have proved their efficiency about the detection of singularities in synthetic images and real reference textures (Lemaur and De

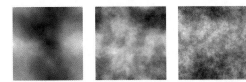

(a) Low roughness   (b) Mean roughness   (c) High roughness

Figure 2.   Examples of synthetic images.

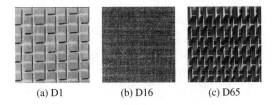

(a) D1   (b) D16   (c) D65

Figure 3.   Examples of Brodatz textures.

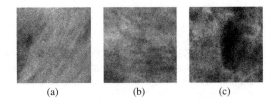

(a)   (b)   (c)

Figure 4.   Examples of digital mammography parts.

Coninck 2003; Gouttière et al. 2007a; Gouttiere et al.). Different families of filters and wavelets were tested for these experiments, and two wavelets and one filter, the most spherical, are kept for the detection in digital mammographies. The considered wavelets are the ϕ and ψ wavelets, developed in a previous study (Lemaur and De Coninck 2003), which are nearly spherical, and the considered filter is the Mexican hat filter, which is isotropic.

In order to evaluate the detection efficiency of the filters, the limiting intensity of the defects must be determined. This limiting value corresponds to the lowest intensity for which the defects are still detected by the filters, depending on a given error rate, fixed at 0.1% in our tests.

4   TESTED IMAGES

In a first step, the created defects are inserted in synthetic images, which have a tunable roughness. Only one parameter is needed to vary the roughness, it is denoted h. h can take values between 0 and 1, a closed value to 1 corresponds to a very rough image. Three levels of roughnesses were examined: 0.4, 0.6 and 0.8. An example of images for these values is shown in Fig. 2. The detection efficiency measure of the different filters was computed for this kind of images, and the most spherical wavelets and filters, *i.e.* the ϕ and ψ wavelets and the Mexican hat filter, were the most performant.

In a second step, the background chosen was the real reference textures from the Brodatz database[1]. Examples of these images are shown in Fig. 3. As for the synthetic images, the most spherical filters were the most efficient to detect the small singularities.

Latest tests were performed to evaluate the detection efficiency of the ϕ and ψ wavelets and the Mexican hat filter in the case of digital mammographies. A database with 30 parts of digital mammographies is constituted, three of these images are presented in Fig. 4.

Let us specify that the size of all the tested images is 256×256 pixels.

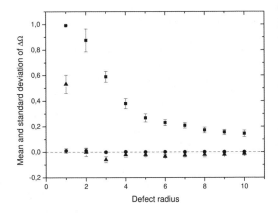

Figure 5.   Pointwise defect detection in synthetic images: comparison of filters with the ϕ wavelet: (▲) Mexican Hat; (●) ψ; (■) Gabor.

5   RESULTS AND INVESTIGATIONS

The results are summarized in this section. First the results for the detection in the synthetic images, and then for the Brodatz textures.

5.1   *Results for the synthetic images*

Since the results for each roughness come to the same conclusion, only those for the highest roughness are here presented. The efficiency of the filters is compared in the graph in Fig. 5. This graph presents the mean and the standard deviation of the differences between the limiting values of Ω obtained for two distinct filters and for each defect radius studied (from 1 to 10 pixels). The results obtained with the Gabor filters are also shown for information.

---

[1] http://www.ux.uis.no/~tranden/brodatz.html

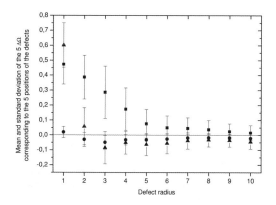

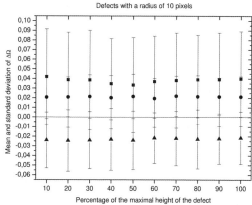

Figure 6. Pointwise defect detection in Brodatz textures: comparison of filters with the ϕ wavelet: (▲) Mexican Hat; (●) ψ; (■) Gabor.

Figure 7. Truncated singularities detection in Brodatz textures: comparison of filters with the ϕ wavelet: (▲) Mexican Hat; (●) ψ; (■) Gabor.

From this graph, we can conclude that the most spherical wavelets and filters are the most efficient to detect pointwise defects in synthetic images.

5.2 Results for the Brodatz textures

Since the textures are completely different from one to another, five positions were chosen for the conic defects in order to improve the strength of the results. The same comparison as the preceding section is performed and the graph is shown in Fig. 6, but the mean and the standard deviation of the differences between the limiting values of Ω is processed for the five positions. The same tendency than for the synthetic images can be observed in this graph. Thus, the conclusions are also the same: the ψ and ϕ wavelets and the Mexican hat filter are the most efficient to detect singularities because they are the most spherical.

Then the influence of the cone top on the efficiency detection of the filters was studied. To this end, the proposition was to truncate the cone at a variable height, and to evaluate the efficiency detection of the filters for this kind of defects. The truncation was performed from 10% to 100% of the height of the cone, which has a fixed radius of 10 or 20 pixels. Only the results for 10 pixels are presented since those for 20 pixels are similar. The comparison graph of the filters is shown in Fig. 7. From this graph, we can conclude that the efficiency detection of the filter is not influenced by the top of the cone.

5.3 Results for the digital mammographies

As said in section 4, the two types of defects (conic and truncated conic defects) are also inserted in the digital mammographies in order to evaluate the detection efficiency of the spherical filters in the case of medical images.

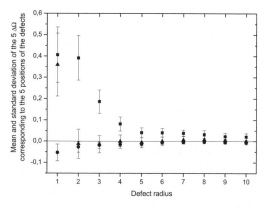

Figure 8. Pointwise defect detection in mammography parts: comparison of filters with the ϕ wavelet: (▲) Mexican Hat; (●) ψ; (■) Gabor.

The comparison of the filters about their detection efficiency is presented in Fig. 8 for the conic defects and in Fig. 9 for the truncated singularities.

Concerning the first graph, *i.e.* the conic defects, we can observe that the Mexican hat filters seem to be less efficient than for the other sets of images (the synthetic and the Brodatz images), but they are as much performant as the nearly isotropic wavelets ϕ and ψ. This is true for defects with a radius higher than 2 pixels. Indeed, when the defects are smaller than 2 pixels, the ϕ and ψ wavelets can adapt better to the background than the other filters through their form. Finally, the least efficient filters are the Gabor filters probably because these are oriented filters.

The tendency observed for the conic defects is confirmed for the truncated singularities. No matter what the percentage of the maximal height is truncated, the Mexican hat filters have a detection efficiency

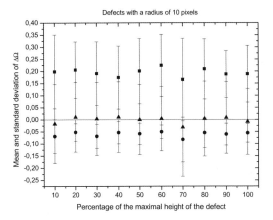

Figure 9. Truncated singularities detection in mammography parts: comparison of filters with the ϕ wavelet: (▲) Mexican Hat; (●) ψ; (■) Gabor.

similar to those of the nearly isotropic wavelets. The truncation of the defects has thus no effect on the detection efficiency of the examined filters. This probably means that the wavelets and filters respond in relation to the base rather than the top of the defects.

6 CONCLUSIONS

Spherical wavelets and filters have proved their efficiency about singularity detection in synthetic images and Brodatz textures. The results obtained for the detection in digital mammographies confirm the efficiency of this kind of filters for a new family of real images, namely medical images. It means that the spherical filters would be an interesting tool to detect microcalcifications in digital mammographies.

ACKNOWLEDGMENTS

The authors would like to acknowledge the support provided by the *Fonds pour la Formation à la Recherche dans l'Industrie et dans l'Agriculture* in Belgium.

REFERENCES

Al-Qdah, M., A.R. Ramli, and R. Mahmud (2005, December). A system of microcalcifications detection and evaluation of the radiologist: comparative study of the three main races in malaysia. *Computers in Biology and Medicine 35*(10), 905–914.

Arodz, T., M. Kurdziel, T.J. Popiela, E.O.D. Sevre, and D.A. Yuen (2006, January). Detection of clustered microcalcifications in small field digital mammograpy. *Computer Methods and Programs in Biomedicine 81*(1), 56–65.

Arodz, T., M. Kurdziel, E.O.D. Sevre, and D.A. Yuen (2005, August). Pattern recognition techniques for automatic detection of suspicious-looking anomalies in mammograms. *Computer Methods and Programs in Biomedicine 79*(2), 135–149.

Cheng, H.D., X.J. Shi, R. Min, L.M. Hu, X.R. Cai, and H. N. Du (2006, April). Approaches for automated detection and classification of masses in mammograms. *Pattern Recognition 39*(4), 646–668.

Fu, J.C., S.K. Lee, S. T. C. Wong, J.Y. Yeh, A.H. Wang, and H.K. Wu (2005, September). Image segmentation feature selection and pattern classification for mammographic microcalcifications. *Computerized Medical Imaging and Graphics 29*(6), 419–429.

Gouttiere, C., G. Lemaur, and J. De Coninck. Detection of singularities in natural textures using nearly spherical wavelets and filters. *Submitted in the International Journal for Computational Vision and Biomechanics*.

Gouttière, C., G. Lemaur, and J. De Coninck (2007a, March). Influence of filter sphericity on the detection of singularities in synthetic images. *Signal Processing 87*(3), 552–561.

Gouttière, C., G. Lemaur, and J. De Coninck (2007b). Influence of sphericity parameter on the detection of singularities in synthetic images. In *Computational Modelling of Objects Represented in Images*, pp. 211–214.

Kom, G., A. Tiedeu, and M. Kom (2007, January). Automated detection of masses in mammograms by local adaptive thresholding. *Computers in Biology and Medicine 37*(1), 37–48.

Lemaur, G. and J. De Coninck (2003, July). Sphericity of wavelets may improve the detection of singularities in images. In *Proceedings of Computing Engineering in Systems Applications*, Lille, France.

Lemaur, G., K. Drouiche, and J. De Coninck (2003, March). Highly regular wavelets for the detection of clustered microcalcifications in mammograms. *IEEE Transactions on Medical Imaging 22*, 393–401.

Mousa, R., Q. Munib, and A. Moussa (2005, May). Breast cancer diagnosis system based on wavelet analysis and fuzzy-neural. *Expert Systems with Applications 28*(4), 713–723.

Yu, S.-N., K.-Y. Li, and Y.-K. Huang (2006, April). Detection of microcalcifications in digital mammograms using wavelet filter and markov random field model. *Computerized Medical Imaging and Graphics 30*(3), 163–173.

Computational Vision and Medical Image Processing – João Tavares & Natal Jorge (eds)
© 2008 Taylor & Francis Group, London, ISBN 978-0-415-45777-4

The use of computational fluid dynamic in swimming research

D.A. Marinho, A.J. Silva, A.I. Rouboa, L.T. Leal, L.S. Sousa & V.M. Reis
University of Trás-os-Montes and Alto Douro, Vila Real, Portugal

F.B. Alves
Technical University of Lisbon, Lisbon, Portugal

J.P. Vilas-Boas
University of Porto, Porto, Portugal

ABSTRACT: The aim of the present study was to apply Computational Fluid Dynamics in the study of the hand/forearm forces in swimming using a 3-D model. Models used in the simulations were created in CAD, based on realistic dimensions of a right adult human hand/forearm. The governing system of equations considered was the incompressible Reynolds averaged Navier-Stokes equations implemented in the fluent® commercial code. The drag coefficient was the main responsible for propulsion, with a maximum value of force propulsion corresponding to a pitch angle of 90°. The lift coefficient seemed to play a less important role in the generation of propulsive force with pitch angles of 0° and 90° but it is important with a pitch angle of 45°.

1 INTRODUCTION

The creation of propulsive force in human swimming has been studied recently using numerical simulation techniques with computational fluid dynamics (CFD) models (Bixler & Schloder 1996, Bixler & Riewald 2002, Silva et al. 2005, Rouboa et al. 2006, Gardano & Dabnichki 2006).

Some limitations still persist, regarding the geometrical representation of the human limbs. In the pioneer study of Bixler & Schloder (1996), these authors used a disc with a similar area of a swimmer hand, while Gardano & Dabnichki (2006) used standard geometric solids to represent the superior limb. Rouboa et al. (2006) tried to correct and to complement the backward works using a 2-D model of a hand and a forearm of a swimmer, situation that seems to be an important step forward in the application of CFD to the human propulsion. However, it seems that it is possible to go forward, reason why we propose to apply CFD to the studied of the hand and forearm propulsion with three-dimensional (3-D) models, as it was already experimented by Lyttle & Keys (2006) to analyse the dolphin kicking propulsion.

In this sense, with this work we want to continue using CFD as a new technology in the swimming research, applying CFD to the 3-D study of the propulsion produced by the swimmer hand and forearm. Therefore, the aim of the present study is twofold. First, continuing to disseminate the use of CFD as a new tool in swimming research. Second, to apply the method in the determination of the relative contribution of drag and lift coefficients resulting from the numerical resolution equations of the flow around the swimmers hand and forearm using 3-D models under the steady flow conditions.

2 METHODS

2.1 *Mathematical model*

The dynamic fluid forces produced by the hand/forearm, lift (L) and drag (D), were measured in this study. These forces are function of the fluid velocity and they were measured by the application of the equations 1 and 2.

$$D = C_D \; \tfrac{1}{2} \; \rho \; A \; V^2 \tag{1}$$

$$L = C_L \; \tfrac{1}{2} \; \rho \; A \; V^2 \tag{2}$$

In equations 2 and 3, V is the fluid velocity, C_D and C_L are the drag and lift coefficients, respectively, ρ is the fluid density and A is the projection area of the model for different angles of pitch used in this study (0°, 45°, 90°).

CFD methodology consists in a mathematical model applied to the fluid flow in a given domain that replaces the Navier-Stokes equations with discretized algebraic expressions and solved by iterative calculations. This domain consists in a three-dimensional grid

Figure 1. Hand and forearm model inside the domain with 3-D mesh of cells.

or mesh of cells that simulate the fluid flow (Fig. 1). The fluid mechanical properties, the flow characteristics along the outside grid boundaries and the mathematical relationship to account the turbulence were considered.

The incompressible Reynolds averaged Navier-Stokes equations with the standard k-epsilon (k-ε) model was considered and implemented in CFD commercial code Fluent® (Moreira et al. 2006).

2.2 Resolution method

The whole domain was meshed with 400,000 trapezoidal elements of 4 nodes each. The numerical method used by Fluent is based on the finite volume approach. The steady solutions of the governing system equations are given in each square element of the discretized whole domain. In order to solve the linear system, Fluent code adopts an AMG (Algebraic Multi-Grid) solver. Velocity components, pressure, turbulent kinetic energy and turbulent kinetic energy dissipation rate are a degree of freedom for each element.

The convergence criteria of AMG are 10^{-3} for the velocity components, the pressure, the turbulent kinetic energy and the turbulent kinetic energy dissipation ratio.

The numerical simulation was carried out in three-dimensions (3-D) for the computational whole domain in steady regime.

2.3 Application

In order to make possible this study we analysed the numerical simulations of a 3-D model of a swimmer hand and forearm. Models used in the simulations were created in CAD, based on realistic dimensions of a right adult human hand/forearm.

Angles of pitch of hand/forearm model of 0°, 45° e 90°, with a sweep back angle of 0° (thumb as the leading edge) were used for the calculations (Schleihauf 1979).

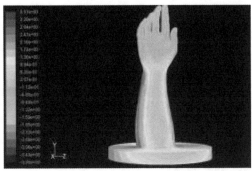

Figure 2. Computational vision of the relative pressure contours on the hand/forearm surfaces.

On the left side of the domain access, the x component of the velocity was chosen to be within or near the range of typical hand velocities during freestyle swimming underwater path: from 0.5 m/s to 4 m/s, with 0.5 m/s increments. The y and z components of the velocity were assumed to be equal to zero. On the right side, the pressure was equal to 1 atm, fundamental pre requisite for not allowing the reflection of the flow.

Around the model, the three components of the velocity were considered as equal to zero. This allows the adhesion of the fluid to the model.

It was also considered the action of the gravity force ($g = 9.81$ m/s^2), as well as the turbulence percentage of 1% with 0.1 m of length.

The considered fluid was water, incompressible with density ($\rho = 996.6 \times 10^{-9}$ kg/mm^3) and viscosity ($\mu = 8.571 \times 10^{-7}$ kg/mm/s).

The measured forces on the hand/forearm model were decomposed into drag and lift components. The combined hand and forearm drag (C_D) and lift (C_L) coefficients were calculated, using equations 2 and 3. The independent variables were the angle of pitch and fluid boundary velocity. The dependent variables were pressure and velocity of the fluid within the dome. Post-processing of the results with Fluent allowed the calculation of component forces through integration of pressures on the hand/forearm surfaces (Fig. 2).

3 RESULTS

In table 1 it is possible to observe the C_D and C_L values produced by the hand/forearm segment as a function of pitch angle. It is presented the values found for a flow velocity of 2.00 m/s with a sweep back angle of 0°.

The C_D and C_L values were almost constant for the whole range of velocities (for a given pitch angle).

According to the obtained results, hand/forearm drag was the coefficient that accounts more for propulsion, with a maximum value of 1.10 for the model with an angle of pitch of 90°. C_L seems to play a residual influence in the generation of propulsive

Table 1. Values of C_D and C_L of the hand/forearm segment as a function of pitch angle. Sweep back angle $= 0°$ and flow velocity $= 2.00$ m/s.

Pitch angle	C_D	C_L
0°	0.35	0.18
45°	0.63	0.32
90°	1.10	0.05

force by the hand/forearm segment at angles of pitch of 0° and 90°, but it is important with an angle of pitch of 45°.

4 DISCUSSION

C_D was the main responsible for propulsion, with the maximum value of force production corresponding to an angle of pitch of 90°, as expected. C_L has a residual influence in the generation of propulsive force by the hand/forearm segment for angles of attack of 0° and 90°, but it is important with an angle of pitch of 45°.

These data confirm recent studies reporting reduced contribution of lift component to the overall propulsive force generation by the hand/forearm in front crawl swimming, except for the insweep phase, when the angle of attack nears 45° (Berger et al. 1995, Sanders 1999, Bixler & Riewald 2002, Rouboa et al. 2006).

Although in this study we had tested only flow in steady regime and this situation does not represent truly what happens during swimming, the present study allowed as to apply CFD in the study of propulsive forces in swimming, using a more realistic model of a human hand/forearm. By itself, this situation seems to be an important step to the advancement of this technology in sports scope.

The results of the values of C_D and C_L are similar to the ones found in experimental studies (Wood 1977, Schleihauf 1979, Berger et al. 1995, Sanders 1999), important fact to the methodological validation of CFD, giving as well conditions to the primary acceptation to the analysis of hydrodynamic forces produced through unsteady flow conditions and through different orientations of the propelling segment.

For the three different orientation models and for the whole studied velocity range, the C_D and C_L remain constant. Similar results were as well observed in other studies using CFD (Bixler & Riewald 2002, Silva et al. 2005, Rouboa et al. 2006).

5 CONCLUSION

This study tried to apply CFD in the analysis of swimming propulsion. As conclusions we can state that the computational data found seem to demonstrate an important role of the drag force and a

minor contribution of the lift force to propulsive force production by the swimmer hand/forearm segment.

On the other hand, it was demonstrated the utility of using CFD in the propulsive force measurements, using a more realistic model (3-D) of a human segment. This situation is an additional step forward to the necessary continuation to keep developing this technology in sport studies, in general, and in swimming, as a particular case.

ACKNOWLEDGEMENT

This work was supported by the Portuguese Government by a grant of the Science and Technology Foundation (SFRH/BD/25241/2005; POCTI/10/58872 /2004).

REFERENCES

Berger, M.A., de Groot, G. & Hollander, AP. 1995. Hydrodynamic drag and lift forces on human hand arm models. *Journal of Biomechanics*, 28(2): 125–133.

Bixler, B.S. & Riewald, S. 2002. Analysis of swimmer's hand and arm in steady flow conditions using computational fluid dynamics. *Journal of Biomechanics*, 35: 713–717.

Bixler, B.S. & Schloder, M. 1996. Computational fluid dynamics: an analytical tool for the 21st century swimming scientist. *Journal of Swimming Research*, 11: 4–22.

Gardano, P. & Dabnichki, P. 2006. On hydrodynamics of drag and lift of the human arm. *Journal of Biomechanics*, 39: 2767–2773.

Lyttle, A. & Keys, M. 2006. The application of computational fluid dynamics for technique prescription in underwater kicking. Biomechanics and Medicine in Swimming X. *Portuguese Journal of Sport Sciences*, 6 (Suppl. 2): 233–235.

Moreira, A., Rouboa, A., Silva, A.J., Sousa, L., Marinho, D., Alves, F., Reis, V., Vilas-Boas, J.P., Carneiro, A. & Machado, L. 2006. Computational Analysis of the turbulent flow around a cylinder. Book of Abstracts of the Biomechanics and Medicine in Swimming X. *Portuguese Journal of Sport Sciences*, 6 (Suppl. 1): 105.

Rouboa, A., Silva, A., Leal, L., Rocha, J. & Alves, F. 2006. The effect of swimmer's hand/forearm acceleration on propulsive forces generation using Computational Fluid Dynamics. *Journal of Biomechanics*, 39(7): 1239–1248.

Sanders, R.H. 1999. Hydrodynamic characteristics of a swimmer's hand. *Journal of Applied Biomechanics*, 15: 3–26.

Schleihauf, R.E. 1979. A hydrodynamic analysis of swimming propulsion. In J. Terauds & E.W. Bedingfield (eds), *Swimming III*: 70–109. Baltimore: University Park Press.

Silva, A., Rouboa, A., Leal, L., Rocha, J., Alves, F., Moreira, A., Reis, V. & Vilas-Boas, J.P. 2005. Measurement of swimmer's hand/forearm propulsive forces generation using computational fluid dynamics. *Portuguese Journal of Sport Sciences*, 5(3): 288–297.

Wood, T.C. 1977. *A fluid dynamic analysis of the propulsive potential of the hand and forearm in swimming*. Master of Science Thesis. Halifax, NS: Dalhouise University Press.

Computational Vision and Medical Image Processing – João Tavares & Natal Jorge (eds)
© 2008 Taylor & Francis Group, London, ISBN 978-0-415-45777-4

A Drusen volume quantification method based on a segmentation algorithm

Fernando Moitinho
Intelligent Robotics Center, Uninova, Portugal

André Mora
Intelligent Robotics Center, Uninova, Portugal
Faculty of Sciences and Technologies, New University of Lisbon, Portugal

Pedro Vieira
Faculty of Sciences and Technologies, New University of Lisbon, Portugal

José Fonseca
Intelligent Robotics Center, Uninova, Portugal
Faculty of Sciences and Technologies, New University of Lisbon, Portugal

ABSTRACT: Age Related Macular Degeneration (ARMD) is a frequent disease appearing mainly in elderly people, commonly diagnosed by ophthalmologists during patient examination. ARMD is characterized by the accumulation of extra cellular materials under retina seen in retina images as yellowish spots and called Drusen spots. Until now, ophthalmologists analysis has done manually, based only on qualitative aspects. However this process is highly dependent on the ophthalmologist and is very difficult to reproduce.

In this paper we propose a methodology for an automatic Drusen analysis, which models each Drusen spot calculating important information like Drusen area, volume and location. The proposed methodology is a reproducible process independent of the clinician allowing the comparison of, images taken in different instants to evaluate treatment effectiveness and disease evolution.

To achieve our objectives four different algorithms were used: the first divides the original image in smaller images grouping Drusen neighbours, the second algorithm is based in a Gaussian Blur filter and corrects the non-uniform illumination. The third algorithm based on labelling gradient paths and is responsible for giving Drusen spots location and intensity. The last algorithm is based in Levenberg-Marquardt methodology and is responsible for this modelling of each Drusen spot.

Keywords: Drusen Detection, Medical Image Processing, Image Modelling

1 INTRODUCTION

A very frequent disease appearing mainly in elderly people is called Age-Related Macular Degeneration (ARMD). It is considered the leading cause of irreversible blindness in developed countries [1] and therefore is object of attention of ophthalmologists during patient examination.

A great improvement in computer processing capacity can be seen in modern computers, therefore it was obvious to use this technology for helping the ophthalmologist community analysing retina images.

Age Related Macular Degeneration is characterized by the accumulation of extra cellular materials under retina. This abnormality is characterized by yellowish spots. In Figure 1 an example of an ARMD affected retina image can be seen.

Currently ARMD analyses are done manually, based only on qualitative aspects. Also the analyses are very depended on the ophthalmologist and are not a reproducible process.

To improve manual analysis, a software application is proposed in this paper. This software application is able to automatically detect and model Drusen spots on retina images. The benefits of using this application are a more accurate analysis, a reduction of image non-uniform illumination problems, independence of the image model to noise and the achievement of a reproducible process independent of the clinician.

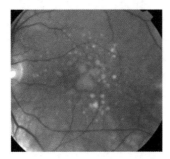

Figure 1. ARMD affected retina.

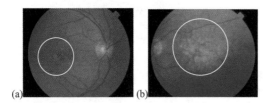

(a)   (b)

Figure 2. Retina images with Soft and Hard Drusen a) Hard Drusen spots; b) Soft Drusen spots.

Another important feature is that it gives important information about each Drusen spot like area and volume. This information can be useful to evaluate the disease progress in a sequence of images taken during a long-term treatment, helping ophthalmologists on the evaluation of therapy effectiveness.

There are two phenotypes of Drusen spots: hard and soft. These phenotypes are used in grading systems such as Wisconsin age related maculapathy grading system [2] or the Alabama Age-Related maculopathy grading system [3].

The developed application returns a modelled image containing the relevant Drusen spots and the Drusen spots information (area, volume, location and gaussian parameters).

This paper starts with a brief presentation of selected related works on Drusen detection. The following section is dedicated to the modelling process describing the following algorithms: non-uniform illumination correction, Drusen spots isolation, Drusen location, Drusen spots modelling and Drusen spots quantification. The remaining sections are dedicated to the developed application, experimental results and conclusions.

2 RELATED WORK

The first important work was published by Peli and Lahav [4] in 1986 and further developed be Sebag et al. [5] from New England Medical Center and Tufts University School of Medicine. It consisted in dividing the image in 8×8 pixel windows, computing all local windows threshold value, interpolate these values

using a two-dimensional linear interpolation and then with the computed values applying a local window threshold. The results can be considered acceptable, but it has detected several small image irregularities as Drusen (false positives).

An adaptive local threshold that considers the local histogram skewness for detecting the presence of pixels belonging to Drusen was presented in 1999 by Shin et al. [6] from University of Pennsylvania School of Medicine. The proposed methodology exhibits good results as previous works but it tends to over-detect Drusen spots.

With a study on the light reflectance of the macula, Smith et al. [7] from Colombia University proposed in 2003 an algorithm for levelling the background based on the elliptically concentric geometry of the reflectance on a normal macula. The Drusen segmentation and area measurements were performed by global threshold. The algorithm for background levelling produced good results and is well supported since it is based on the physical geometry of the macula. The results produced by the Drusen segmentation method are similar to the previous works that used image thresholds.

This small survey on some previous works on Drusen detection shows that the use of other techniques rather then adaptive or global thresholds has not been studied yet. It is also important to notice that all these threshold methods have the same tendency for producing false positives, especially when the image has small irregularities. Computing Drusen total area (number of pixels) is the only proposed method for evaluating Drusen evolution along time using threshold methods

3 MODELLING PROCESS

The Modelling process has several stages, beginning with non-uniform illumination correction. After illumination correction Drusen spots are isolated. The next procedure is to locate Drusen spots, ending in the Optimization algorithm. The following sections will describe each algorithm.

Before beginning the modelling process is important to do a pre-processing stage, in order to improve images quality. The use of these algorithms is justified by the need to reduce processing time and achieve a more reliable and accurate result.

The first step is to reduce the original image to the Region Of Interest (ROI). Because Drusen spots appear mainly in the macula region, the image to be modelled can be centred on the macula region cropping the original image and resulting on an image of 500×500 pixels instead of one with approximately 1200×800 pixels.

Other important goal of this work was to obtain a tridimensional modelling of the Drusen spots. Using the

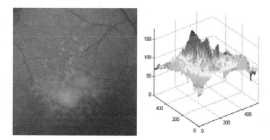

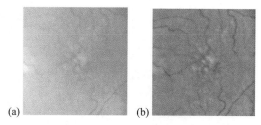

(a)   (b)

Figure 5.   Non-uniform illumination correction and normal-ization (a) Original image; (b) Corrected image.

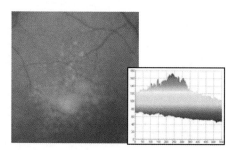

Figure 3.   Example of 2D view and 3D view of the same image.

Figure 4.   Example of illumination correction.

image two-dimension coordinates and pixel intensity values it is possible to estimate a tri-dimensional surface where spots elevation can be perceived (Figure 3). It should be noticed that it is an estimation and not a real tri-dimensional view since the depth coordinate is very dependent on illumination.

3.1   Non-uniform illumination correction

The objective of the first algorithm is to correct the non-uniform illumination characteristic of retina images. This phenomena appears in the images as brighter areas near optic disc and darker areas far from it.

Other issue that affects Drusen quantification is that images taken on the same patient, using similar conditions may have different brightness values. This problem difficults Drusen spots analyses and therapy evaluation. Therefore, a method to normalize images has to be taken under consideration to achieve reliable results.

Figure 4 shows an example of an image with non-uniform illumination, a 2D graph of XZ axis where the Z axis represents the pixels intensity value.

To overcome these problems we propose two algo-rithms: one to achieve a uniform illumination and other to normalize image intensity values based on image contrast.

To correct the non-uniform illumination we use a Gaussian Blur algorithm. The algorithm is similar to applying a low-pass filter. It has two starting

parameters: r that represents the mask dimension (r × r mask) and σ that is the standard deviation. To obtain a smoother image those values are increased. The weights of the mask are calculated by Equation (1).

$$GB(x, y) = \frac{1}{2\pi\sigma^2} \times e^{-\frac{x^2+y^2}{2\sigma^2}} \times 100 \qquad (1)$$

Each pixel value is calculated by the r by r neighbours.

Next the background image is removed from the original one, by dividing the both images (original, background).

$$flatten = \frac{original}{background} \qquad (2)$$

Once obtained this flatten image the normalization process based in the image standard deviation is applied. The image is divided into several windows and the standard deviation of each window is calculated. The median value of the calculated standard deviation ($M\sigma$) is returned. The final image is achieved using Equation (3).

$$Flatten\_out = \left(\left(\frac{nf}{M\sigma} \times flatten\right) - (offset)\right) \times ill \qquad (3)$$

where,

Nf	Normalization factor
Offset	Mean of flatten image
ill	Intensity centre value
Mσ	Median standard deviation

In Figure 5 the result of the normalization and illumination correction algorithm can be seen.

3.2   Isolation Drusen spots

In our previous work [8] the optimization algorithm was applied to the original image. That methodology has shown that processing time increments propor-tionally to the image size. In order to improve that, we chosen to model smaller images, so Drusen spots were grouped in small groups of islands and each one representing an image that is later optimized.

59	59	62	62	63	55	55	55
59	62	65	65	62	55	55	69
59	62	62	62	59	55	55	69
55	55	55	55	55	55	55	55
62	62	59	55	59	59	55	55
62	69	62	69	69	62	59	55
62	62	69	70	69	62	59	55
55	62	62	62	62	59	59	55

(a)

0	0	1	2	3	0	0	0
0	4	5	6	7	0	0	8
0	9	10	11	12	0	0	13
0	0	0	0	0	0	0	0
14	15	0	0	0	0	0	0
16	17	18	19	20	21	0	0
22	23	24	25	26	27	0	0
0	28	29	30	31	0	0	0

(b)

0	0	1	1	1	0	0	0
0	1	1	1	1	0	0	8
0	1	1	1	1	0	0	8
0	0	0	0	0	0	0	0
14	14	0	0	0	0	0	0
14	14	14	14	14	14	0	0
14	14	14	14	14	14	0	0
0	14	14	14	14	0	0	0

(c)

Figure 6.   Connected components labelling a) Original image; b) Labelling pixeis >60; c) Propagate labels.

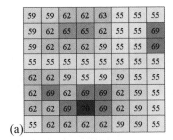

Figure 7.   Isolate Drusen spots.

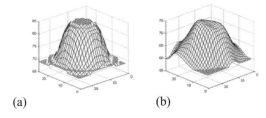

(a)   (b)

Figure 8.   Example of original Drusen spot and Gaussian function a) Original Drusen spot; b) Gaussian function.

Using this methodology the processing time has significantly decreased, improving the overall analysis.

This was accomplished using a labelling process grouping all objects bigger than a predefined value (image intensity centre value ill Equation 3). A modified version of the iterative connected components labelling algorithm [9] was used.

The first step is the labelling of all pixels where intensity value is bigger than the ill value. Then propagate the first label to all adjacent pixels. First scan top-down left-right, next scan down-top right-left. Repeat this procedure until there are no changes. An example is shown in Figure 6. In this example three objects were found. The next procedure of the algorithm is to divide the image into three images that are modelled later by the optimization algorithm.

An example of the application of this new algorithm can be seen in Figure 7. The original image was divided into four smaller images containing four possible Drusen spots or group of islands. Those images are later introduced in the optimization algorithm to be modelled.

3.3   Drusen location algorithm

To achieve reliable spots detection we propose an algorithm based on labelling the maximum gradient path. The proposed algorithm consists in a labelling procedure similar to connect components labelling or to watershed segmentation (described in [9]), but optimized to maximums search [10].

This information will be used by the optimization algorithm in order to decrease processing time.

3.4   Optimization function

To achieve a accurate and reliable result for the modelling process, it was important to choose an adequate mathematical function. Requisites for that choice were similarity with Drusen spots geometry and sufficient amount of adjusting parameters.

After searching and analyzing several mathematical functions, the one that better fulfilled the requisites was the Gaussian function.

In Figure 8 it is presented an example of an isolated Drusen spot and next to it is an example of a Gaussian function. It can be seen the resemblance between both images.

These observations motivated the use of a tri-dimensional model of the Gaussian function. The tri-dimensional Gaussian function is given by the equation below.

$$G(x,y) = a \times e^{\left(\frac{X^2}{S_x} + \frac{Y^2}{S_y}\right)^{\frac{2}{d}}} + z_0 \tag{4}$$

where,

$$X = (x - x_0)\cos(\theta) + (y - y_0)\sin(\theta) \tag{5}$$

$$Y = -(x - x_0)\sin(\theta) + (y - y_0)\cos(\theta) \tag{6}$$

with parameters:

a – amplitude	θ – rotation
(x_0, y_0) – centre coordinates	D – shape factor
S_y – scale factor in y	S_x – scale factor in x
Z_0 – background value	

It is a modified version of the Gaussian function since it uses a shape factor d that allows another degree of freedom, in order to achieve a more accurate model.

3.5 Optimization algorithm

To accomplish the proposed objective of Drusen spots modelling, an optimization algorithm was chosen. The main criteria for this algorithm choice were accuracy and processing time. The algorithm that better satisfied these requisites was the Levenberg-Marquardt optimization algorithm. This algorithm is consider a standard of nonlinear least-squares routines [11].

Levenberg–Marquardt is a combination of steepest descent and the Gauss–Newton Method. When the current solution is far from the correct one the algorithm behaves like a steepest descent method. When the current solution is close to the correct solution, it becomes a Gauss-Newton method.

For a given mathematical function depending on several parameters the algorithm adjusts those parameters in order to minimize the chi-square error function (χ^2).

$$\chi^2(a) = \sum_{i=1}^{N}\left[\frac{z_i - z(x_i; y_i; a)}{\sigma_i}\right]^2 \qquad (7)$$

This equation represents the distance between the current solution ($z(x_i, y_i, a)$) and the correct one (z_i), with σ representing the standard deviation.

The algorithm has presented some convergence problems, that small variations make difference between actual chi-square and its previous. So it was decided to impose two stop conditions: the first condition is if the chi-square value doesn't improve for four consecutive iterations and the second condition is if the chi-square value reaches a predefined minimum value.

3.6 Drusen spots quantification

The Drusen quantification algorithm takes two important measures: area and volume. This area value is calculated summing the number of pixels influenced by each Drusen spot. The volume is achieved by summing pixels intensity value from each Drusen spot.

In order to achieve a physical measure, a conversion to μm must to be made. To convert pixels to μm it is necessary to measure the diameter 'of the optic disc. Because there is a typical value of approximately 1850 μm for the optic disc diameter, after calculating the ratio pixel-μm (pμ), the area of each Drusen spot is multiplied by that ratio.

The volume value cannot be converted into a physical value (μm), because relation between pixel intensity and physical dimensions is unknown. Therefore, the volume value is just a reference that can be used only for comparison proposes.

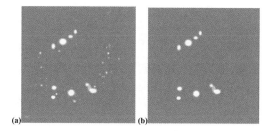

Figure 9. Example of removing non relevant Drusen spots a) Modelled image; b) modelled image after removing non releant Drusen spots.

The third step consists on discarding all Drusen spots where the volume doesn't reach a predefined value. This value was chosen analysing several images and concluding that if a spot volume is smaller than 1000 then that Drusen spot is not relevant. In Figure 9 is presented an example of a modelled image before (Figure 9.a) and after (Figure 9.b) removing non relevant Drusen spots. Images contrast was enhanced to facilitate visualization.

4 DEVELOPED APPLICATION

In the developed application all the algorithms described previously are included. The application has been developed in Builder C++. It has a main panel where images can be loaded and then are several tabs that implement these different functions such as isolate Drusen spots, calculate Drusen location, correct non-uniform illumination, optimize multiple images and Drusen information.

5 EXPERIMENTAL RESULTS

The first step in the optimization process is background correction, executed by the algorithm explained in section 3.1 (non-uniform illumination correction). Results can be seen in Figure 10. As it can be observed the new resulting image has a uniform illumination.

The algorithm has shown good results on tested images, making an accurate correction, preserving all important details and also increasing image contrast.

Processing time spen by the algorithm depends on the image size but in images with 500×500 pixeis the algorithm took approximately 2 seconds to make the illumination correction.

Figure 11 shows the final result of applying the proposed methodology. Modelling is carried out was made after non-uniform illumination correction. After this procedure the corrected image is divided into smaller images, grouping Drusen spots. This is

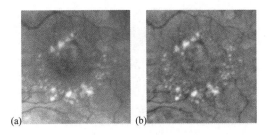

(a)   (b)

Figure 10.   Examples of non-uniform illumination correction a) Original image; b) Image after illumination correction.

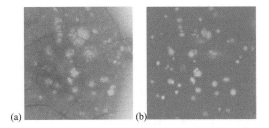

(a)   (b)

Figure 11.   Example of optimized image a) Original image; b) Modelled image.

done by the isolate Drusen spots algorithm. In this case the image has been divided in 202 images. The Drusen location algorithm detected 269 possible Drusen spots. This information is passed to the Levenberg-Marquardt algorithm that took approximately 400 seconds to reach its final results. Finally the volume segmentation algorithm is executed, for this step just Drusen spots that reached a predefined volume value are considered. In Figure 11.b the resulting image after AD3RI analysis is presented.

The achieved results can be considered a good approximation of the real image and the processing time has improved significantly since our previous work [8]. Tests are being done to evaluate the analysis efficiency made by the developed application (AD3RI). Analyses are compared with clinicians analyses, so that a standard on Drusen spots analysis can be reached.

6   CONCLUSIONS

Proposed methodology is divided in five steps. The first algorithm is responsible for the non-uniform illumination correction and normalization, which has presented good results even in images with accentuated background.

The second step divides the image in smaller images so that the optimization algorithm can reach a more accurate result. This algorithm has shown good results

with all processed images divided accurately and no Drusen spots missed.

The third algorithm (Drusen location algorithm) demonstrated that all Drusen spots were detected correctly. Information gathered by this algorithm is a great help for the next step, the Levenberg-Marquardt optimization algorithm. This algorithm has optimized tested images correctly and the processing time did not exceed 400 seconds, which is a good value considering that some images had more than 200 Drusen spots.

The last step is responsible for the quantification and Drusen spots characteristics like area and volume. The quantification algorithm has shown good results and quantification process has eliminated non relevant Drusen spots, quantifying just the relevant ones.

Future work will be concentrated in tuning the optimization algorithm in order to improve the modelling quality and speed. An automatic spot classifier for detecting non-Drusen spots and to classify Drusen spots in soft and hard classes will be the focus of our future work.

REFERENCES

1.   Pauleikhoff, D., et al., *Drusen as risk factors in age-related macular disease.* Am J Ophthalmol, 1990. **109**(1): pp. 38–43.
2.   Klein, R., et al., *The Wisconsin age-related maculopathy grading system.* Ophthalmology, 1991. **98**(7): p. 1128–34.
3.   Curcio, C.A., N.E. Medeiros, and C.L. Millican, *The Alabama Age-Related Macular Degeneration Grading System for donor eyes.* Invest Ophthalmol Vis Sci, 1998. **39**(7): pp. 1085–96.
4.   Peli, E. and M. Lahav, *Drusen Measurement from Fundus Photographs Using Computer Image Analysis*, in *Ophtalmology.* 1986. pp. 1575–1580.
5.   Sebag, M., E. Peli, and M. Lahav, *Image analysis of changes in drusen area*, in *Acta Ophtalmologica.* 1991. pp. 603–610.
6.   Shin, D., N. Javornik, and J. Berger, *Computer-assisted, interactive fundus image processing for macular drusen quantitation.* Ophthalmology, 1999. **106**(6): p. 1119–1125.
7.   Smith, R.T., et al., *A Method of Drusen Measurement Based on the Geometry of Fundus Reflectance.* BioMedical Engineering OnLine, 2003. **2**(10).
8.   Moitinho, F., et al. *AD3RI a Tool for Computer – Automatic Drusen Detection.* in *CompIMAGE – Computational Modelling of Objects Represented in Images: Fundamentals, Methods and Applications.* 2006. Coimbra, Portugal.
9.   Gonzalez, R. and R. Woods. *Digital Image Processing.* 1992: Addison–Wesley.
10. Mora, A., P. Vieira, and J. Fonseca. *Drusen Deposits on Retina Images: Detection and Modeling.* in *MEDSIP-2004.* 2004. Malta.
11. William H. Press, S.A.T., William T. Vetterling, Brian P. Flannery, *Numerical Recipes in C.* Second Edition ed. 1999: Cambridge University Press.

Automatic extraction and classification of DNA profiles in digital images

C.M.R. Caridade
Instituto Superior de Engenharia de Coimbra, Portugal

A.R.S. Marçal & T. Mendonça
Faculdade de Ciências, Universidade do Porto, Portugal

ABSTRACT: This paper presents an automatic system to extract and classify DNA profiles in a digital image. The system uses image segmentation and cumulative histogram analysis to identify both the number of layers in the image, and the number of profiles present in each layer. Once the individual profiles are located, their signatures are extracted and used to perform a hierarchical classification of the individual DNA profiles.

1 INTRODUCTION

There is a growing interest in the development of automatic systems for the analysis of DNA profiles obtained by electrophoresis and registered in digital images.

Electrophoresis is the procedure by which charged molecules are allowed to migrate in an electric field. The rate of migration is determined by the size of the molecules and their electric charge (Fig. 1). In gel electrophoresis small molecules or compact molecules migrate more rapidly than large or loose molecules. After a defined period of migration time (usually a few hours), the locations of the DNA molecules in the gel are assessed by making the DNA molecules fluorescent and observing the gel with ultraviolet radiation (Madigan et al. 1997).

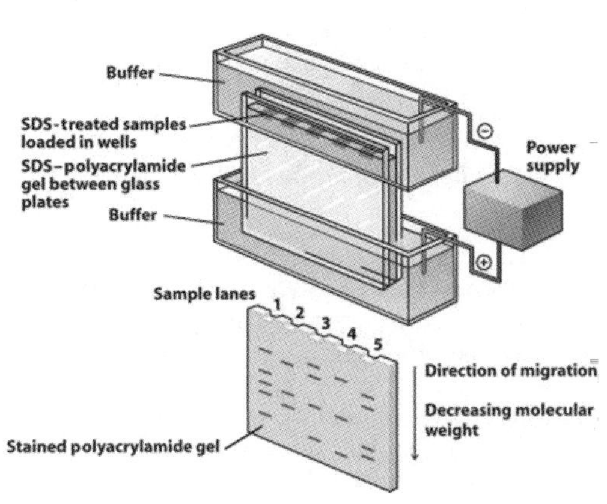

A computer scanner or a direct photograph of the ultraviolet radiation can be used to register the DNA profiles (lanes) for subsequent analysis. Figure 2 shows examples of DNA images, both in colour and in grey scale formats. As it can be seen in these images, the individual DNA profiles exhibit a characteristic signature or pattern.

Nearly all of the human DNA (99.9%) is identical amongst all individuals. Only the remaining 0.1% can differ, in several ways, between individuals. DNA profile (signature of one lane) analysis focuses on hyper variable regions of DNA and has proven to be extremely useful (Rademarker & Bruijn 1997).

The purpose of this work was to develop an automatic methodology, based on image processing techniques, to identify the number and location of

Figure 1. An electrophoresis apparatus for a single layer with 5 DNA profiles (lanes). (Adapted from Davis, 2003).

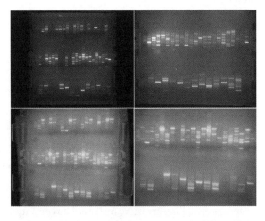

Figure 2. Examples of DNA images obtained from direct greyscale photograph (top) and as RGB colour images obtained using a scanner (bottom).

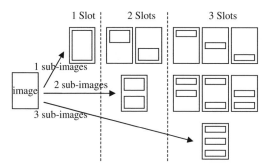

Figure 3. Schematic representation of the 11 cases possible with up to 3 layer slots.

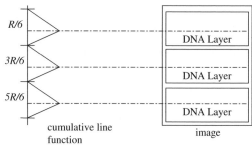

Figure 4. Schematic representation of the cumulative line function used to determine the number of layers in an image.

layers and lanes, to extract the DNA signatures and to perform hierarchical classification of the DNA profiles.

2 METHODOLOGY

The method developed has four separate stages:

(1) Layer identification and image division.
(2) Identification of the number of profiles in each layer.
(3) Extraction of the profile signatures.
(4) Hierarchic classification of the DNA profiles.

2.1 Layer identification and image division

The images used in this work are similar to those presented in Figure 2 (top row). The image on the top left corner of Figure 2 for example, has 3 sub-images, or layers, each one with a different number of lanes. However, there are cases where only 1 or 2 layers are present.

The purpose of the layer identification stage is to identify and extract the individual layers from the original image, resulting in 1 to 3 sub-images. In some images there is a vacant slot, which results in a different number of layers and slots. There are 11 possibilities altogether, as illustrated in Figure 3. There is therefore a need to identify not only the number of used layers but also the total number of slots present in the image.

The original greyscale image is converted to a binary image using segmentation by thresholding. Initially the morphological operator opening is applied to the original image, using as structuring element a 30 x 30 pixel square (Gonzalez & Woods 2002). This operation is performed in order to obtain the background, which is subsequently subtracted from the original image. The resulting greyscale image is then converted to a binary image using a global thresholding operation, with the threshold value obtained by the Otsu method (Otsu, 1979).

The fact that the central areas of each layer tend to have higher grey levels than the areas near the layer edges can be used to identify the location of the layers in the image. A cumulative line function is produced from the binary image, with the number of ON pixels per line. The central areas of the layers will result in higher cumulated values, as it is illustrated in Figure 4. In this example, the image has 3 layers, thus the peaks should appear roughly around $R/6$, $3R/6$ and $5R/6$, where R represent the number of rows.

The plot of the cumulative line function is used to identify which one of the 11 possible cases best matches the image being processed. Each of these cases will have a particular signature in the plot, as the location of the layers in the image is different for each case. These reference signatures are computed beforehand.

Initially the locations of the local maxima are determined in the cumulative plot. Three different possibilities are considered, where there are 1, 2 or 3 slots. For each case, the image is divided by the number of slots, and the average of the local maxima falling in each slot computed. Each of the three cases will have a signature (or vector), which is compared with the reference values, using the Euclidean distance as a metric. The reference signature that best matches the test image is used as the correct solution.

2.2 Identification of the number of profiles in a layer

Once a layer is extracted as a sub-image of the original image, the next stage is to identify how many DNA profiles, or lanes, are present in that layer, and to obtain the exact location of each lane. The initial step is to perform image segmentation by automatic thresholding. The thresholding strategy is the same as described in section 2.1., but this time applied only to the sub-image extracted. The resulting binary image will have the background set to 0 (OFF) and some of the pixels in the layers set to 1 (ON). This is illustrated in Figure 5 for a test image with one layer. The test image is

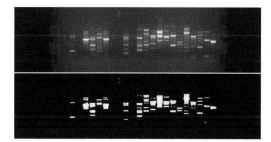

Figure 5. Test image of a layer in the original 8 bit greyscale format (top) and converted to a binary image (bottom).

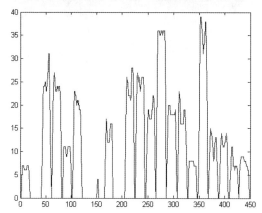

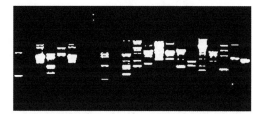

Figure 6. Binary interest area extract from the image (top) and the corresponding cumulative function f (bottom).

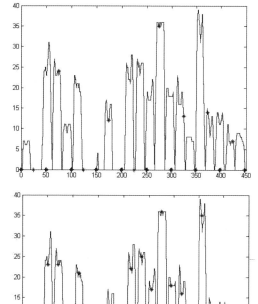

Figure 7. Lane location predicted for n=9 (top) and n=22 (bottom) for the test image. The red marks indicate the lane centres and the black marks the edges between lanes.

presented in its original form (8 bit greyscale) on the top of Figure 5, and as a binary image on the bottom of Figure 5.

The next step is to select the interest area (central area with profiles), using the binary image. This is achieved by removing all the columns of the image that have no profiles, starting both from left and right and moving towards the centre. The first columns that are not void define the left and right limits of the interest area. The result of the application of this procedure to the test image is presented in Figure 6 (top).

The binary image of the interest area is used to produce a function with the number of pixels ON for each column. The resulting function for the test image is presented in Figure 6 (bottom) in the form of a plot.

The x-axis of the plot is the column number and the y-axis the number of pixels ON per column. This cumulative function (f) is used to find the number of lanes, making use of the fact that all lanes have equal width. The number of lanes is searched for between N_{min} (e.g. 9) and N_{max} (e.g. 30). The interest image width is divided by n, with n being an integer between N_{min} and N_{max}. For each value of n, there is a predicted lane width (W_n), which will normally not be an integer number of pixels.

The estimated lane width (W_n) is used to predict the location of the n lanes in the image. The central column of a lane i (X^i) is given by Equation 1, while the edge between lane i and $i+1$ is given by Equation (2)

$$X^i = \left(i - \frac{1}{2}\right)W_n \tag{1}$$

$$X'^i = iW_n \tag{2}$$

The n lanes will have n central columns and $n-1$ columns corresponding to lane edges, considering that the extreme edges are not used. This process is illustrated in Figure 7 for $n = 9$ (top) and $n = 22$ (bottom),

113

where the red marks correspond to the predicted central columns of the lanes (X^i, for $i = 1, \ldots, n$) and the black marks to the lane edges. For this particular example, the correct value for n is 22, which is the case illustrated in the bottom plot of Figure 7.

The cumulative function (f) is used to evaluate the adequacy of the lane location predictions produced for n. The average value of the cumulative function is computed for the predicted centres ($F_c(n)$) and edges ($F_e(n)$), using equations (3) and (4). If the prediction is correct, there will be high values of f in the predicted central lanes (red marks on the plots of Figure 7) and low values of f for the predicted lane edges. A total of 3 columns are used for lane edges, centred on the predicted X^i for the edge, while for lane centres the number of columns used is the lane width minus 6 (Equations 3 and 4).

$$F_e(n) = \frac{\sum_{i=1}^{n} \sum_{j=4}^{W_n-3} f(W_n * (i-1) + j)}{(W_n - 6) * n} \qquad (3)$$

$$F_c(n) = \frac{\sum_{i=1}^{n-1} f(W_n * i - 1) + f(W_n * i) + f(W_n * i + 1)}{3 * (n-1)} \qquad (4)$$

A function (Φ) based on the difference between $F_c(n)$ and $F_e(n)$ is used to evaluate the likelihood of the solution n. Two versions of function ? are used, one that is simply the difference (Eq. 5) and another that uses an additional factor to penalize the solutions for low values of n (Eq. 6).

$$\Phi_1(n) = F_c(n) - F_e(n) \qquad (5)$$

$$\Phi_2(n) = (F_c(n) - F_e(n)) \times \sqrt{n} \qquad (6)$$

Both these functions will have high values if the solution is plausible – high values of f in the central areas of the lane and low values of f in the edges between lanes. The number of lanes predicted by function Φ is the n that has maximum value of Φ.

A plot of functions Φ_1 and Φ_2 for the test image of Figure 6 is presented in Figure 8. For this particular example, both Φ_1 and Φ_2 predict the correct number of layers ($n = 22$).

2.3 Extraction of DNA signatures profiles

The next processing stage consists of extracting the signature of each DNA profile in the layer. This is a straight forward task, once the location of the individual lanes is determined. The intensity profile along the 7 columns on the centre of each lane is averaged to produce a signature. This is illustrated in Figure 9 for the first 7 DNA profiles of the test image.

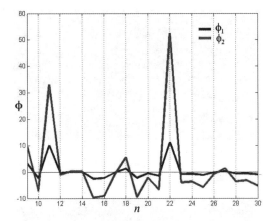

Figure 8. Plot of functions Φ_1 and Φ_2 for the test image. In this example both functions have a maximum for $n = 22$, the correct number of lanes.

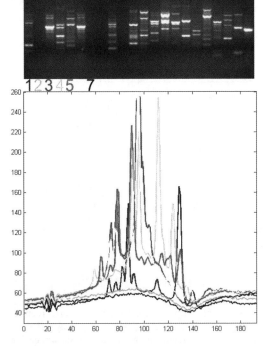

Figure 9. Representation of the first seven DNA profiles for the test image.

2.4 Hierarchic classification of the DNA profiles

The final processing stage consists of applying a hierarchical classification to the signatures of individual profiles. The basis of this classification is the construction of a similarity (or distance) matrix between the

114

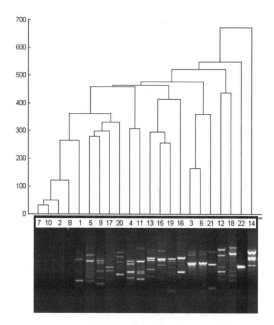

Figure 10. The hierarchical binary cluster tree for the test image.

profiles being evaluated. Once the distance between pairs (Euclidean or any other metric) of DNA profiles is computed, the similarity amongst the profiles is determined using the Dice coefficient, and the hierarchical cluster analysis is carried out using the Unweighted Pair-Group Mean Arithmetic (UPGMA) method. This algorithm is based on different ways of measuring the distance between two profiles. If n_r is the number of objects in profile r and n_s is the number of objects in profile s, and x_{ir} is the ith object in profile r, the definitions of the measurements is represented in equation (7).

$$d(x, y) = \frac{1}{n_r n_s} \sum_{i=1}^{n_r} \sum_{j=1}^{n_s} dist(x_{ri}, x_{sj}) \qquad (7)$$

Finally a dendrogram is produced by connecting lines between pairs of profiles. The height of each line represents the distance between the two profiles being connected. The result for the test image is presented in Figure 10. The order of the DNA lanes was changed in order to match the dendrogram.

3 RESULTS

The method proposed to automatically extract and classify DNA profiles were tested using a set of 21 images. The images are all greyscale images obtained with a digital camera. Figure 1 (top) shows 2 examples of the test images used.

Table 1. Performance of the method in the layer identification stage.

	No images	Correct detections (%)
3 layers	11	11 (100.0%)
2 layers	10	9 (90.0%)
TOTAL	21	20 (95.2 %)

Table 2. Performance of the proposed method in the identification of the number of lanes in a layer.

	Using Φ_1		Using Φ_2	
	#cases	(%)	#cases	(%)
Failure	15	28.8%	6	11.5%
Success	37	71.2%	46	88.5%

The method proved to be effective in identifying the number of layers and extracting the corresponding sub-images. The results are presented in Table 1. In only one of the 21 images there was a failure in the identification of the number of layers, which corresponds to a success rate of 95.2%.

The performance of the method for the identification of the number of lanes in a layer was also tested, using the 52 sub-images extracted from the 21 test images. The number of lanes in these images varied from 10 to 22. A summary of the results is presented in Table 2, using functions Φ_1 and Φ_2. The success rate for Φ_2 is 88.5%, while for Φ_1 is only 71.2%. There are 9 cases that were correctly identified using Φ_2 but not using Φ_1. In these images the correct number of lanes is a multiple (usually by a factor of 2 or 3) of the number predicted by Φ_1. The additional factor introduced in Φ_2 penalizes these solutions and permits therefore to have a better accuracy overall.

4 CONCLUSIONS

A method for the automatic extraction and classification of DNA profiles in digital images was proposed. The method was tested with 21 images, each with 2 or 3 layers. It managed to correctly identify and extract sub-images for each layer in all but one test image (95.2% of success). The correct identification of the number of lanes and its location was achieved in 88.5% of the 52 layers tested.

The DNA profiles are extracted using a simple method, although there are other more sophisticated approaches (Akbari & Albregtsen, 2004). Work is currently being carried out in order to follow a similar path to extract the DNA signatures. Also greater effort will be done in the classification algorithm selection and evaluation. Other plans for future work include the inclusion of colour images and an improvement of the segmentation process.

Overall it can be said that the results are encouraging but further work is still required in order to have an operational system for the automatic identification and classification of DNA profiles in digital images.

REFERENCES

Akbari, A. & Albregtsen, F., 2004. Automatic Segmentation of DNA bands in one dimensional gel images produced by hybridizing techniques. Proceedings of the 26th Annual international Conference of the IEEE EMBS. San Francisco. September 1–5.

Davis, R.E., 2003. Molecular Biology, Fall 2003 Biology 327/751, City University of New York, http://www.library.csi.cuny.edu/~davis/Bio_327/, visited May 2007.

Gonzalez, R.C. & Woods, R.E. (2ª ed.) 2002. Digital image processing, Prentice Hall International Inc.

Madigan, M.T., Martinko, J.M. & Parker, J. (8ª ed.) 1997. Brock Biology of Microorganisms. Prentice Hall International, Inc.

Otsu, N. 1979. A treshold selection method from Gray-Level histograms. IEEE Trans. on Systems Man Cybernetics, 9(1), 62–69.

Rademarker, J.L.W. & Bruijn, F.J. 1997. Characterization and classification of microbes by REP-PCR genomic fingerprinting and computer-assisted pattern analysis. *DNA Markers: Protocols, Applications and Overviews*: 151–171.

Computational Vision and Medical Image Processing – João Tavares & Natal Jorge (eds)
© 2008 Taylor & Francis Group, London, ISBN 978-0-415-45777-4

Measurement of the IMT using a semi-automated software (ThickSoft): A validation study

Anton Vernet

ECoMMFiT, Department of Mechanical Engineering, University Rovira i Virgili, Tarragona, Spain

Honorio Pallas, Lluis Masana & Blai Coll

Servei de Medicina Interna, Departament de Medicina Vascular. Hospital Universitari Sant Joan, Reus. Spain

ABSTRACT: This article presents a semi-automatic software (ThickSoft) for the intima-media thickness (IMT) measurement. A brief explanation of the working characteristics of the software is introduced and the validation results are shown. The validation have been performed over the analysis of 204 images coming from 34 participants. Results obtained with ThickSoft have been compared with the manual measures of the IMT obtained with AnaliSYS™ software.

1 INTRODUCTION

In the last decades cardiovascular diseases are becoming the leading cause of death in industrialized countries. Primary preventive strategies are based in the measurement of future risk focused on the measurement of risk factors. However, these strategies do not detect up to 60% of patients who will suffer a cardiovascular event in the future (Akosah et al. 2003), and then, new approaches should be warranted. One of these approaches is the measurement of the intima-media thickness (IMT) from ultrasound images of the carotid artery. This measure has revealed as a surrogate marker of cardiovascular events in several clinical trials (Lorenz et al. 2007). Further, the measurement of IMT is advised in males over 45 years of age, in order to better stratify the cardiovascular risk. However, it should be done in an specialized core-laboratory, using validated software (Greenland et al., 2000).

Figure 1 sketch the carotid artery as it could be obtained by ultrasound, where three different anatomic segments can be described: internal, bulb and common carotid artery (ICA, BCA and CCA respectively). Artery walls are made up of three different layers that are (from inside to outside) the intima, the media and the adventitia. Each of these layers have different internal structure that involve different properties. Some of these properties allows each layer to exhibit a different response to the ultrasound impulses. Thus, while adventitia and intima layers reflect the ultrasound waves and are viewed as light zones in the ultrasound image, lumen and media do not reflect them

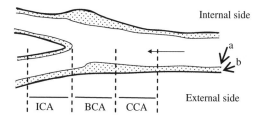

Figure 1. Drawing of the carotid artery and interfaces detected with ThickSoft, a) intima-lumen interface and b) adventitia-media interface.

and are displayed as dark zones in the image obtained. Therefore, the intima-lumen interface and adventitia-media interface could be detected allowing the IMT measurement (Tahmasebpour, 2005).

There are two separate phases on the process of the measure of the IMT, 1) the carotid image capturing and 2) the IMT measurement from the image obtained in phase 1. This article is focused on the second phase of this process. Specifically, the article describes a new semi-automatic software developed for the measurement of the IMT (ThickSoft) and its validation.

2 METHODS

Ultrasound examination have been performed on 34 participants under the same protocol, that consist in the acquisition of images of the right and left carotid artery centered in the ICA, BCA and CCA segments. Thus, each participant generates a set of 6 images, accordingly a total of 204 images have been analysed. All images have been obtained with an Acuson Sequioa ultrasound device, with a vascular probe ranging from 7 to 11 MHz. All the device characteristics have been previously set-up in order to standardize the acquisition of images, which have been stored in JPEG format for its later processing. All the 204 images have been analysed with the ThickSoft software and with the image processing manual software AnaliSYS™ (Soft Imaging System, Münster, Germany). All the statistical analysis were performed with SPSS 14.0. To evaluate the reliability and consistency between the two methods we calculate the intraclass correlation coefficient, in each anatomic segment and in a composite variable that was the IMT average of all segments.

3 AUTOMATIC MEASUREMENT SYSTEM

ThickSoft is a new software developed under the MATLAB® environment that allows the semi-automatic measurement of the IMT from ultrasound images. The main steps followed by the software are,

- Pixel calibration
- Selection of the region of interest (ROI)
- Detection of the intima-lumen interface
- Detection of adventitia-media interface
- IMT measurement and statistics.

Furthermore, and only in case that the quality of the ultrasound image is very poor, the software offers the possibility to obtain the IMT measure using a manual method. In that case the intima-lumen and adventitia-media interfaces are obtained selecting several points in the interfaces. Then, a cubic spline is used to adjust the points into a curve that represents the interfaces.

3.1 Pixel calibration

Previous to start with the analysis, the image is converted from its original RGB format to a grey-scale image. Thus, the image that the program will uses to compute the IMT is a matrix containing values ranging from 0 to 255. The next step is to calibrate the image (obtain the value millimetre/pixel), that will allow to express the IMT in millimetres instead of pixels. To do the calibration the software automatically detects the ticks in the axis of the original ultrasound image (figure 2), and compute the number of pixels between to consecutive ticks, that will correspond to a value in millimetres given by the operator that has obtained the images (usually this value is 10 mm between major ticks). To validate the result, this measurement is done in both horizontal and vertical axis.

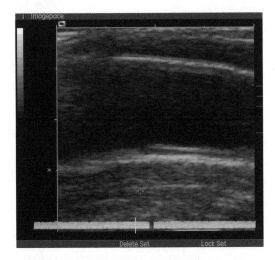

Figure 2.   Original CCA ultrasound image.

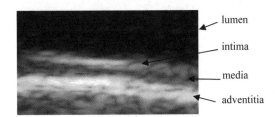

Figure 3.   ROI from ultrasound image in figure 2.

For all the images analysed in the validation procedure a value of 0.042 mm/pixel has been obtained.

3.2 ROI selection

As it could be seen in Figure 2, a considerable portion of the original ultrasound image is not used to compute the IMT. Therefore, the user is asked to draw a rectangle that define the region of interest used to obtain the IMT. Figure 3 show the ROI selected from the ultrasound image in figure 2.

The intima and the adventitia layers appear has zones with high intensity while the lumen and the media layers correspond to zones less reflective. Although, the image characteristics of the lumen and the adventitia are very different. Thus, the detection of the intima-lumen interface and the adventitia-media interface has to be done necessarily using different techniques.

3.3 Detection of intima-lumen interface

The detection of the intima-lumen interface is easier than the detection of the adventitia-media interface, due to the characteristics of the lumen. In general,

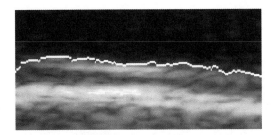

Figure 4. Intima-lumen interface.

Figure 6. Adventitia-media interface.

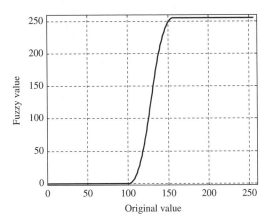

Figure 5. S-shaped membership function used to transform ROI pixels.

this layer corresponds to the darkness zone of the ROI and only some dark-grey spots generated by the blood turbulence could be expected to appear on it. Thus, the first step for detecting this interface is to binarize the ROI image, setting the pixels smaller than a given threshold equal to zero and the rest equal to one. This threshold is smaller enough to produce a big black zone in the lumen layer and to make the rest of the ROI appear as white with some smaller black spots. Taking the space derivatives for each column of the binarized matrix will set the intima-lumen interface with a value of 1 wile the rest will be set to zero. Figure 4 shows the intima-media interface detected with the software.

3.4 *Detection of adventitia-media interface*

As the media layer is not as well defined as the lumen layer, the procedure to detect le adventitia-media interface must be based in different strategies. In this case the ROI is not binarized using 0 or 1 to define the different zones of the image. Instead of this, the image is binarized using a fuzzy function that is shown in Figure 5. This function allows to transform the original values of the ROI pixels in a new set of values that enhance the adventitia-media interface.

After the transformation of the pixel values, the computation of the columwise derivatives allow the detection of a first approximation of the adventitia-media interface. The points obtained for the interface are analyzed one by one, comparing them with the corresponding points in the intima-lumen interface and with their neighbors. This procedure allows to eliminate the points with low probability to truly be in the adventitia-media interface and replace them with a new point obtained from the interpolation of the neighbor points (Figure 6).

Finally, once the intima-lumen and the adventitia-media interfaces have been detected the IMT could be computed for each pixel-column of the ROI image and the statistics could be computed. The software supply the number of measures of the IMT for each pixel column, the IMT mean, the standard deviation and the maximum and minimum values.

4 RESULTS

A total of 204 images were analysed with both methods, manual (with AnaliSYS™) and semi-automatic (with ThickSoft). The mean number of measurements performed by ThickSoft was 97.8 measurements/segment and with the manual method 3.7 measurements/segment (ANOVA, $p < 0.001$). Absolute differences in the average IMT was 0.08 (0.1) mm, corresponding the highest differences to the bulb [0.13 (0.1) mm], and the lowest to the common carotid [0.02 (0.07) mm]. When we analyse the reliability between both methods, we found a highly significant intraclass correlation coefficient ranging from 0.71 to 0.87. Table 1 shows the intraclass correlation coefficients (ICC) and the confidence interval (CI) between both methods in each anatomic segment and in the composite average of IMT.

5 CONCLUSIONS

The results obtained in the validation process shows that ThickSoft is a highly reliable software for the measurement of carotid IMT, specially in the common

Table 1. ICC and CI (95%) between both methods.

Variable	ICC	CI (95%)	p value
BCA	0.76	0.49–0.88	<0.001
CCA	0.87	0.73–0.93	<0.001
ICA	0.71	0.39–0.86	0.001
Average IMT	0.80	0.61–0.90	<0.01

carotid artery, but also in the bulb and internal. Further, it allows the performance of a significant higher number of measurements in each arterial segment, in a reduced amount of time. Besides these advantages, we have to take into account that ThickSoft has the ability to set a quality control application, that is of utmost importance when considering large scale or follow-up based studies.

In summary, we have developed an easy-to-use, highly reliable software for the measurement of carotid IMT, that eventually should lead us to the generalization of highly efficacious preventive strategies, based on carotid IMT, to fight against the incidence of cardiovascular events.

REFERENCES

Akosah. K., Schaper, A., Cogbill, C., & Schoenfeld, P. Preventing myocardial infarction in the young adult in the first place: how do the national cholesterol education panel III guidelines perform? J. Am. Coll. Cardiol., 2003; 41: 1475–1479.

Lorenz, S.M.W., Markus, H.S., Bots, M.L., Rosvall, M., & Sitzer, M. Prediction of Clinical Cardiovascular Events With Carotid Intima-Media Thickness. A Systematic Review and Meta-Analysis. Circulation. 2007; 115: 459–467.

Greenland, P., Abrams, J., Aurigemma, G.P., Bond, M.G., L.T. Clark, Criqui, M.H., Crouse, J.H., Friedman, L., Fuster, V., Herrington, D.H., Kuller, L.H., Ridker, P.M., Roberts, W.C., Stanford, W., Stone, N., Swan, H.J., Taubert, K.A., & Wexler, L. Prevention Conference V: Beyond Secondary Prevention: Identifying the High-Risk Patient for Primary Prevention: Noninvasive Tests of Atherosclerotic Burden : Writing Group III. Circulation 2000; 101; 16–22.

Tahmasebpour, H.R., Buckley, A.R., Cooperberg, P.L., & Fix, C.H. Sonographic Examination of the Carotid Arteries. RadioGraphics 2005; 25: 1561–1575.

Computational Vision and Medical Image Processing – João Tavares & Natal Jorge (eds)
© 2008 Taylor & Francis Group, London, ISBN 978-0-415-45777-4

Volume properties of the hippocampal region as predictors of DAT development

A. Chincarini & G. Gemme
INFN, Sezione di Genova, via Dodecaneso & Genova

P. Calvini, M.A. Penco & S. Squarcia
Laboratorio di fisica e statistica medica, Dipartimento di Fisica, Università di Genova,
via Dodecaneso, Genova

F.M. Nobili & G. Rodriguez
Neurofisiologia Clinica – DTC e DISEM, Università di Genova – Azienda Ospedaliera San Martino,
largo R. Benzi, Genova

ABSTRACT: In this work we look at the statistical distribution of geometrical observables computed in volumes centered on the right and left hippocampus in MR images. The observable distributions are then correlated with the clinical diagnosis in patients potentially affected by the Alzheimer disease. We then evaluate the discrimination ability of the geometrical observables combined with the MMSE test results.

1 INTRODUCTION

In the framework of the Alzheimer disease (AD) researches, a great deal of effort is devoted to finding "biomarkers" (Kantarci 2004). These biomarkers should be tested with minimally invasive techniques and should provide a sure and prompt diagnosis of the disease.

Nowadays, several biomarkers have been found in the analysis of Magnetic Resonance (MR) and Positron Emission Tomography (PET) images, while the protein fraction dosage in the Cerebro-Spinal Fluid (CSF), although a good indicator requires a non trivial intervention on the patient (Csernansky 2005).

On the other hand, the neuropsychological assessment is very sensitive but poorly specific in addressing different forms of dementia (Murphy 1993, Laakso 1995). The goal for a good test would be an early diagnosis on Mild Cognitive Impairment (MCI) patients, since only about 50% of them evolve towards a form of dementia, the Alzheimer type (DAT) being the most frequent (Bennett 2002).

The usual macro-classification in Control, MCI and DAT subjects is performed on the actual cognitive evaluation and cannot therefore provide an early diagnosis. Furthermore, the cognitive status must be evaluated again in time, both because it can change significantly, and because a longitudinal study may reveal other form of dementia or even psychological pathologies.

The last decade saw a wealth of initiatives both international and local, which sponsored the research for DAT early assessment. Among several other indications, these studies pointed out that the onset of the disease is correlated with atrophy in specific areas of the brain. The hippocampus seems to be one of the earliest affected areas (Kesslak 1991, Scheltens 1992, Killiany 1993, Ikeda 1994, Convit 1997, Fox 1996, Jack 1997).

In this work we use a fully automatic procedure to extract a small volume surrounding both hippocampi. On these volumes (hippocampal boxes, HBs) we calculate several observables whose distribution will be used as a supplementary tool for the cognitive evaluation.

Since automatic hippocampus segmentation is a non trivial task, we wanted to check whether the HBs alone could already bring useful information to the clinical assessment of DAT.

This analysis is nevertheless the forerunner for a fully automated hippocampus extraction and analysis tool which is currently under testing.

2 DATABASE

Our database consists of 122 subjects, whose age, sex, neuropsychological evaluation and clinical evaluation (Controls, MCI and DAT) is known (see Table 1).

Table 1. Ensemble properties of our subjects' database. The error on the parameters is the standard deviation.

	Controls	MCI	DAT
No of subjects	49	21	52
Age [y]	75 ± 5	75 ± 5	76 ± 8
MMSE score	29 ± 1	28 ± 2	24 ± 3
Sex	35% M, 65% F		

Table 2. Statistics for San Martino hospital patients.

	Controls	MCI	DAT
No of subjects	14	21	28
Age [y]	73 ± 6	75 ± 5	77 ± 7
MMSE score	29 ± 1	28 ± 2	25 ± 3
Sex	22% M, 78% F		

Table 3. Properties of the ADNI patients.

	Controls	MCI	DAT
No of subjects	35	//	24
Age [y]	76 ± 5	//	75 ± 9
MMSE score	29 ± 1	//	24 ± 2
Sex	49% M, 51% F		

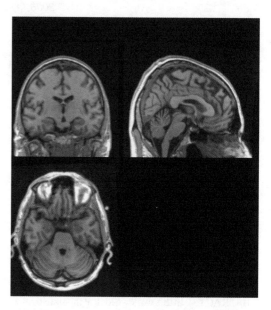

Figure 1. Oriented and ICBM152-aligned MRI example.

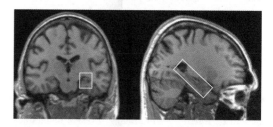

Figure 2. Hippocampal box position. The box dimensions are 30 × 70 × 30 mm. The box is positioned so that the hippocampus lies approximately in the geometrical center of the extracted volume.

Of the 122 subjects, 63 come from San Martino Hospital, neurophysiology department in Genoa, Italy. This subset consists of patients who have not undergone any particular selection as they are those who ask for a neurophysiologic check-up to the local hospital. They are therefore not necessarily screened to exclude other neuropsychiatric disorders that could have confounded the diagnosis of DAT.

Since those who approach the hospital for evaluation do show some symptoms (although not necessarily DAT), there are very few controls and their age is rather on the elderly side (see Table 2). In addition, the clinical evaluation coming from the hospital has not been reviewed by other independent clinicians.

For this reason, we decided to include other subjects to compensate for the disproportion. These extra subjects come from the ADNI consortium database (ADNI 2004). These have been selected among the mp-rage series and their ensemble properties are shown in Table 3.

Only Controls and DAT subjects have been selected from the ADNI database. The purpose of this choice was to improve the Controls/DAT statistics, using then those categories to classify the available MCI into either one of them.

3 HIPPOCAMPAL REGION EXTRACTION

3.1 Brain alignment and normalization

All raw MRIs underwent a processing which consisted in: scale normalization, re-orientation (if needed), spatial resampling and co-registration.

Scale normalization and image reorientation were used to counteract possible differences arising from different acquisition instruments and/or user settings. For the reorientation, when needed, we chose RPI (Right to Left-Posterior to Anterior-Inferior to Superior, see Figure 1).

Spatial resampling was performed to compare all sub-volumes extracted from the MRIs on a voxel by voxel base. We used a bicubic interpolator to match 1 mm cubic voxel side.

All MRIs have than been registered to the ICBM152 (Montreal Neurological Institute) template of the in order to normalize the brain volume and facilitate the box extraction. The registration was performed by an affine transformation on a Gaussian smoothed and under-sampled version of the original MRI, whereas the final transformation matrix was applied to the original MRI.

3.2 *Box coordinate determination*

Once the MRIs are properly aligned, we apply an algorithm that uses co-registration of a set of templates to the MRI as a means to extract a volume box around the hippocampus.

The template boxes have been carefully selected to represent hippocampal regions exhibiting various grade of atrophy. They have been extracted among a subset of patients (n = 83) with an iterative procedure, starting from a first positioning by hand and then looking for the best position match in the remaining n − 1 patients.

To find the box coordinate on a new image, the algorithm uses a 6 degree of freedom rigid transformation with a normalized correlation metric, matching the set of template boxes onto the proper MRI region (see Figure 2).

For each right and left side, 8 template boxes are separately registered onto the MRI and the transformation parameter together with the final metric is recorded. The best match is selected on the base of the metric and its coordinates (after the rigid transform) are used to extract a new box from the MRI. The 6 coordinate parameters of the best template match are checked against a statistics, which triggers an alarm if the newly found parameters are "too far" from the expected values. The purpose of this test is to find inconsistencies which may be due to mis-registration or poor MRI data quality.

The extracted hippocampal box (HB) has the same size as the template boxes, that is $30 \times 70 \times 30 \, mm^3$. This size was chosen to be large enough to accommodate all hippocampi without including too large fraction of adjacent tissues.

4 FEATURE EXTRACTION

4.1 *Gray levels segmentation and observables*

Hippocampal boxes are used to see whether one can extract relevant clinical information without the need of segmenting the hippocampus (Chincarini 2006).

While hippocampus segmentation on a limited number of subjects can be carried out manually (although with the help of a well trained operator and in a considerable amount of time), automatic segmentation of a considerable number of subjects with a varied

atrophy grade is all but a trivial task. One could ask whether a statistical analysis on a small region around the hippocampus conveys good enough information to be clinically relevant.

The problem can be formulated as follows: we want to look for one or more observables whose distribution effectively separates the DAT from the Controls.

The observables are single value functions of the position and gray level of the voxels in the HB. They can be any kind of property, from integral information such as the number of voxels in a given gray range, to geometrical properties such as the local curvature.

Since the hippocampus consists mainly of gray matter (GM), it is only natural to assume that GM based observables will carry the relevant information. The first task therefore consists in separating GM, White Matter (WM) and Cerebro-Spinal fluid (CSF) within the boxes.

The gray level segmentation is carried out with a K-means clustering algorithm working on the gray level values only (no spatial information coded), and imposing 3 clusters. A more sophisticated segmentation algorithm, which takes into account the spatial distribution and domain connectivity, is currently being tested.

Once the binary mask for GM, WM and CSF are found, one can calculate several quantities, for example: the GM percent, the WM thickness, the CSF average curvature, the GM asymmetry (between the right and left HB), etc.

4.2 *Masking*

The aforementioned quantities can also be calculated on a sub-volume of the HB. Extracting a box sub-volume means enhancing the observable values in that sub-volume. We extract a sub-region by masking the HB with a gray level 3D matrix (of the same size of the HB).

We tested 6 different masks, two of the most relevant being a Gaussian ellipsoid placed in the box geometrical center, and a half-Gaussian tail covering the back of the HB (see Figure 3 for an example).

The observables are first computed on the whole HB and subsequently masked with the 3D matrix. The observable normalization similarly follows the masking process.

4.3 *Choosing the observable*

We tested many different observables and masking (approx. 350 combinations) and for each of them we calculated the statistical distribution and correlation with the clinical evaluation. The area under the Receiver Operator Characteristic (ROC) curve separating Controls from DAT was used as parameter to discriminate "good" and "bad" observables.

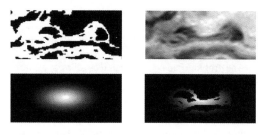

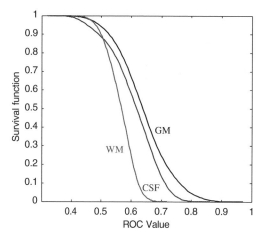

Figure 3. HB processing example: (top right) hippocampal volume after a spatial resampling of factor 2; (top left) GM voxels (white) after the segmentation in GM, WM e CSF; (bottom left) 3D mask (Gaussian shaped ellipsoid); (bottom right) GM after applying the mask.

Figure 4. Cumulative distribution (survival function) of the ROC tests for the observables related to GM (black curve), WM (red curve) e CSF (blue curve).

The ensemble statistics for the ROC area can be shown as cumulative distribution, that is a curve that plots the probability to find an observable whose ROC area is greater than a certain value (survival function, see Figure 4).

Plotting the survival function and splitting the observables in GM, WM and CSF reveals that the GM related observables are the most efficient in separating Controls from DAT. This behavior should be expected if we recall that DAT pathologies involve atrophy in the GM and, in second order, an enlargement of the CSF volume.

The two most effective observables were found to be the mean GM thickness masked with a Gaussian ellipsoid (Ω_1, see Figure 5) and the mean GM percent calculated on the whole HB volume.

These observables add useful information to the already good classification based on the MMSE score.

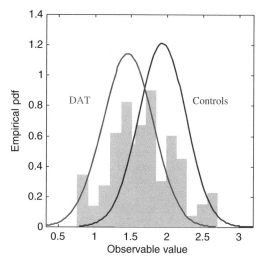

Figure 5. Histogram and Gaussian estimate of the probability density function (pdf) for Ω_1, that is the mean GM thickness in the Gaussian masked box volume (red curve = DAT, blue curve = Controls). The ROC area value is 84%.

5 CLINICAL CLASSIFICATION

5.1 Single and combined observables

The single observable statistics shows how well Controls and DAT can be considered separate populations. Each observable distribution (grouped by the clinical evaluation) expresses a statistic which may (or may not) be significant with respect to the Controls/DAT classification.

It is therefore quite intriguing whether observable combinations might yield a better classification. Obviously we cannot hope to simply get a better classification by combining two or more randomly chosen observables. Three main issues should be considered: (a) most observables are (at least) partially linearly dependent from the others, (b) some observables are not meaningful at all (in the sense of discriminating between Controls and DAT) and (c) there are a number of ways to mathematically combine them.

A simple way to combine observables is to add their normalized (z-scored) values. As it is shown in Figure 6, the combined performance is better than either single performance.

5.2 Improving over MMSE score

It is though more interesting to check whether the geometrical observables calculated on the HBs convey more information to the neurophysiologic evaluation.

The MMSE test is a widely used measure for DAT assessment and is available for all the subjects. Its ROC

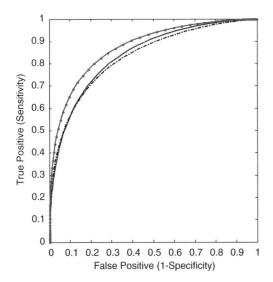

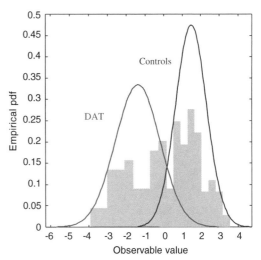

Figure 6. Area under the ROC curve for the single observable (blue curve and black dashed curve) and combining two observables (red dotted curve). In this example the first observable is Ω_1 and the second one is the mean CSF percentage in the whole HB. The combined ROC area is 87%.

Figure 7. Histogram and Gaussian estimate of the probability density function for the joint observables Ω_1 and MMSE score. The ROC area value is 98% (for DAT and Controls curves).

area value (calculated using the same parameters as the other observables) yields 90%. A simple combination (z-scored sum) with the Ω_1 observable readily improves the ROC area to 98% (Figure 7). To calculate the ROC area value for these observables, we assumed a Gaussian probability distribution whose parameters are estimated from the real data.

This result hints to using all possible features for the classification and a better screening for linear dependency.

The grand total of the available information on the subjects, that is their age, sex, MMSE score and a group of relevant geometrical observables (27 in total) are processed by Principal Component Analysis (PCA) to extract the 13 abstract components responsible for the 99.5% sample variation. These components are then fed into a feed-forward neural network classifier trained to separate the DAT from the Control patients.

Once the network performs properly on the DAT and Control subjects, it is presented with the PCA components of the 21 MCI patients for classification.

The network therefore classifies the MCI subjects as if they belonged either to the Control or to the DAT class.

A clinical feedback with the responsible neurologist is carried through to look for updated neurological evaluation on the MCI subjects and to establish whether the network correctly predicts the onset of the DAT. Results are presented in Figure 8.

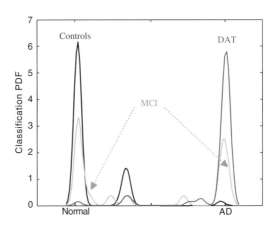

Figure 8. Neural network classification of the Controls (blue curve), DAT (red curve) and MCI (green curve) subjects. The MCI curve is approximately divided in 2 major areas, belonging to the DAT and Control classes respectively.

Of the 21 MCI subjects presented to the network, 7 were classified as possible DAT developers. The clinical follow-up showed that 4 of them really developed DAT, 2 are still classified MCI and 1 seems to suffer from another type of dementia.

6 CONCLUSIONS

The goal of this work was to ascertain whether volume properties of the hippocampal region convey additional information which, when added to the standard neurological tests, can help predicting the MCI evolution.

The advantage of analyzing HBs instead of hippocampus shapes lies mainly in avoiding the intricacies of an automatic segmentation.

Although the results seem encouraging, this work is a complementary step to the automatic segmentation and classification of the hippocampus. In addition, we are constantly adding new MRIs to our database in order to improve the statistical significance of the classification. We are also testing more sophisticated algorithms for the grey level segmentation and the neural network classification.

ACKNOWLEDGEMENTS

This work was developed in the framework of the Magic-5 INFN project.

We would like to acknowledge the Alzheimer's disease Neuroimaging Initiative (ADNI) for accessing to the data used in the preparation of this article. As such, the investigators within the ADNI contributed to the design and implementation of ADNI and/or provided data but did not participate in analysis or writing of this report.

ADNI investigators include: (complete listing available at www.loni.ucla.edu\ADNI\Collaboration\ADNI_Citation.shtml).

This work benefited from the use of the Insight Segmentation and Registration Toolkit (ITK), an open source software developed as an initiative of the U.S. National Library of Medicine and available at http://www.itk.org.

REFERENCES

ADNI consortium, (Alzheimer's Disease Neuroimaging Initiative), study starting date: 2004, anticipated end date: 2009, http://www.loni.ucla.edu/ADNI/.

Bennett D.A. et al. (2002), *Natural history of mild cognitive impairment in older persons*, Neurology 59,198–205.

Chincarini A. et al. (2006), *Unsupervised Extraction of the Hippocampal Formation and Adjacent Structures from the MR for Atrophy Quantification*, Human Brain Mapping Conference, Firenze 11–15 June.

Convit A, et al. (1997), *Specific hippocampal volume reductions in individuals at risk for Alzheimer's disease*. Neurobiol Aging; 18:131–138.

Csernansky J.G. et al. (2005), *Preclinical detection of Alzheimer's disease: hippocampal shape and volume predict dementia onset in the elderly*, NeuroImage 25, 783–792.

Fox N.C. et al. (1996), *MRI-based quantitative assessment of the hippocampal region in very mild and moderate Alzheimer's disease*. Lancet; 348:94–97.

Ikeda M, et al. (1994), *MRI-based quantitative assessment of the hippocampal region in very mild and moderate Alzheimer's disease*. Neuroradiology; 36:7–10.

Jack C.R. Jr et al. (1997), *Medial temporal atrophy on MRI in normal aging and very mild Alzheimer's disease*. Neurology; 49:786–794.

Kantarci K. et al. (2004), *Quantitative Magnetic Resonance Techniques as Surrogate Markers of Alzheimer's Disease*, NeuroRx 1, 196–205.

Kesslak et al. (1991), *Quantification of magnetic resonance scans for hippocampal and parahippocampal atrophy in Alzheimer's disease*. Neurology; 41:51–54.

Killiany R.J, et al. (1993), *Temporal lobe regions in magnetic resonance imaging identify patients with early Alzheimer's disease*. Arch Neurology; 50:949–954.

Laakso M.P et al. (1995), *Volumes of hippocampus, amygdala and frontal lobes in the MRI-based diagnosis of early Alzheimer's disease: correlation with memory functions*. J Neural Transm; 9:73–86.

Murphy D.G, et al. (1993), *Volumetric magnetic resonance imaging in men with dementia of the Alzheimer's type: correlations with disease severity*. Biol Psychiatry; 34:612–621.

Scheltens P, et al. (1992), *Atrophy of the medial temporal lobes on MRI in probable Alzheimer's disease and normal aging: diagnostic value and neuropsychological correlates*. J Neurosurg Psychiatry; 55:967–972.

Computational Vision and Medical Image Processing – João Tavares & Natal Jorge (eds)
© 2008 Taylor & Francis Group, London, ISBN 978-0-415-45777-4

Fuzzy morphology based on uninorms: Image edge-detection opening and closing

Manuel González-Hidalgo
Computer Graphics and Vision Group. Maths and Computer Science Dept. University of the Balearic Islands, Spain

Arnau Mir Torres, Daniel Ruiz-Aguilera & Joan Torrens Sastre
Fuzzy Logic and Information Fusion Group. Maths and Comp. Sci. Dept. University of the Balearic Islands, Spain

ABSTRACT: In this paper a fuzzy mathematical morphology based on fuzzy logical operators is proposed and the Generalized Idempotence (GI) property for fuzzy opening and fuzzy closing operators is studied. It is proved that GI holds in fuzzy mathematical morphology when the selected fuzzy logical operators are left-continuous uninorms (including left-continuous t-norms) and their corresponding residual implications, generalizing known results on continuous t-norms. Two classes of left-continuous uninorms are emphasized as the only ones for which duality between fuzzy opening and fuzzy closing hold. Implementation results for these two kinds of left-continuous uninorms are included, they are compared with the classical umbra approach and the fuzzy approach using t-norms, proving that they are specially adequate for edge detection.

1 INTRODUCTION

The fuzzy mathematical morphology is an alternative extension of binary morphology to gray-scale morphology (Serra 1988), using concepts and techniques from fuzzy set theory, see for example (Bloch and Maître 1995), (DeBaets 1997), (DeBaets, Kerre, and Gupta 1995a), and (DeBaets, Kerre, and Gupta 1995b). The basic tools of mathematical morphology are the morphological operations on an image A, which are defined relatively to a fuzzy structuring element B, the size and shape of which can be chosen by the morphologist in order to analyse the structure of A. Several researchers have introduced alternative morphological operations, a detailed account can be found in (Nachtegael and Kerre 2000), and references therein.

The most usual conjunctions used in order to define the fuzzy mathematical operators are t-norms and their residual implications. Recently, conjunctive uninorms (as another particular case of conjunctions) have also been used for the same purpose, see (DeBaets, Kwasnikowska, and Kerre 1997) and (González, Ruiz-Aguilera, and Torrens 2003). In particular, it is studied in (González, Ruiz-Aguilera, and Torrens 2003) which

conjunctive uninorms need to be chosen in order to preserve the algebraic and morphological properties needed to obtain a "good" mathematical morphology.

Our goal is to extend the results presented in (González, Ruiz-Aguilera, and Torrens 2003), in two directions. On one hand, we want to study some properties of fuzzy opening and fuzzy closing as well as open and closed fuzzy objects, when conjunctive uninorms are used. In particular, we mainly deal with the generalized idempotence property. On the other hand, we present some experimental results based on these morphological operators, using left-continuous conjunctive uninorms.

The paper is organized as follows. In Sections 2 and 3 we recall the basic definitions and properties of fuzzy logical operators and those of fuzzy morphological operators, respectively. In Section 4 we discuss the properties of open and closed fuzzy objects and its representations. The results presented in this section generalize those presented in (DeBaets 2000) where continuous t-norms are used. All the proofs are omitted to enlarge the last section, where we display some comparative experimental results using several conjunctive uninorms.

2 FUZZY LOGICAL OPERATORS

Definition 2.1 *A decreasing and involutive unary operator* $\mathcal{N}$ *on* [0, 1] *with* $\mathcal{N}(0)=1$ *and* $\mathcal{N}(1)=0$ *is called a* strong negation.

Definition 2.2 *An increasing binary operator* $\mathcal{C}$ *on* [0, 1] *is called a* conjunction *if it satisfies*

$$\mathcal{C}(0,1) = \mathcal{C}(1,0) = 0 \quad and \quad \mathcal{C}(1,1) = 1.$$

Definition 2.3 *A binary operator* $\mathcal{I}$ *on* [0,1] *is called an* implication *if it is decreasing with the first partial map, increasing with the second one, and it satisfies*

$$\mathcal{I}(0,0) = \mathcal{I}(1,1) = 1 \quad and \quad \mathcal{I}(1,0) = 0.$$

Given a conjunction $\mathcal{C}$, two implications can be constructed

$$\mathcal{I}_{\mathcal{C},\mathcal{N}}(x,y) = \mathcal{N}(\mathcal{C}(x,\mathcal{N}(y)))$$

called *strong implication* of $\mathcal{C}$ and $\mathcal{N}$, and

$$\mathcal{I}_{\mathcal{C}}(x,y) = \sup\{z \in [0,1] \mid \mathcal{C}(x,z) \le y\}$$

this last one is called the *residual implication* of $\mathcal{C}$.

A special kind of conjunctions are the well known t-norms. In fact, fuzzy morphological operators are usually constructed from t-norms and, a special kind of them, the nilpotent ones, has been proved to be the most useful in this framework, see for instance (Nachtegael and Kerre 2000). However, a generalization of t-norms appeared in (Fodor, Yager, and Rybalov 1997), the uninorms.

Definition 2.4 *A binary operator* U *on* [0,1] *is called a* uninorm *if it is associative, commutative, increasing in each place and such that there exists some* $e \in [0,1]$, *called the* neutral element, *such that* $U(e,x)=x$ *for all* $x \in [0,1]$.

It is clear that function U becomes a t-norm when $e = 1$ and a t-conorm when $e = 0$. For any uninorm we have $U(0,1) \in \{0,1\}$, if $U(1,0)=0$, U is called *conjunctive* and if $U(1,0)=1$ it is called *disjunctive*. Moreover, a uninorm U is called *idempotent* whenever $U(x,x)=x$ for all $x \in [0, 1]$.

This kind of operators is specially interesting because of their behaviour: like a t-norm in $[0,e]^2$, and like a t-conorm in $[e,1]^2$. Note that conjunctive uninorms are particular cases of conjunctions and consequently they can be used in fuzzy mathematical morphology.

In order to have "good" morphological properties, it is essential to use left-continuous conjunctions. In this paper we will use two classes of conjunctive uninorms: left-continuous representable and left-continuous idempotent ones. Their definitions and characterizations can be found in (Fodor, Yager, and Rybalov 1997) and (DeBaets 1999), respectively.

3 FUZZY MORPHOLOGICAL OPERATORS

In the following sections, $\mathcal{I}$ will denote an implication, $\mathcal{C}$ a conjunction, $\mathcal{N}$ a strong negation, U a conjunctive uninorm with neutral element e, $\mathcal{I}_U$ its residual implication, A a gray-scale image, and B a gray-scale structuring element.

From the definition of classical erosion and dilation (Serra 1988) it is clear that the intersection and inclusion of sets play a major role. The idea of De Baets (DeBaets 1997) was to fuzzify the underlying logical operations, i.e. the Boolean conjunction and the Boolean implication, to obtain a successful fuzzification. An n-dimensional gray-scale image is modeled as an $IR^n \to [0, 1]$ function. It is required that the gray values of the image belong to the real unit interval in order to consider an image as a fuzzy object. Thus, we have the following definitions.

Definition 3.1 *The* fuzzy dilation $D_{\mathcal{C}}(A,B)$ *and* fuzzy erosion $E_{\mathcal{I}}(A,B)$ *of A by B are the gray-scale images defined by*

$$D_{\mathcal{C}}(A,B)(y) = \sup_x \mathcal{C}(B(x-y),A(x))$$

$$E_{\mathcal{I}}(A,B)(y) = \inf_x \mathcal{I}(B(x-y),A(x)).$$

As in classical morphology, the difference between the fuzzy dilation and the fuzzy erosion of a gray-scale image, $D_U(A,B) \setminus E_{\mathcal{I}_U}(A,B)$, called the *fuzzy gradient operator*, can be used in edge detection.

Definition 3.2 *The fuzzy closing* $C_{\mathcal{C},\mathcal{I}}(A,B)$ *and fuzzy opening* $O_{\mathcal{C},\mathcal{I}}(A,B)$ *of A by B are the gray-scale images defined by*

$$C_{\mathcal{C},\mathcal{I}}(A,B)(y) = E_{\mathcal{I}}(D_{\mathcal{C}}(A,B),-B)(y)$$

$$O_{\mathcal{C},\mathcal{I}}(A,B)(y) = D_{\mathcal{C}}(E_{\mathcal{I}}(A,B),-B)(y).$$

Note that the reflection $-B$ of a n-dimensional fuzzy set B is defined by $-B(x)=B(-x)$, for all $x \in IR^n$. Given a strong negation $\mathcal{N}$, we define by $(co_{\mathcal{N}}A)(x)=\mathcal{N}(A(x))$ the $\mathcal{N}$-complement $co_{\mathcal{N}}A$ of a fuzzy set A.

It is known that the fuzzy dilation and the fuzzy erosion are $\mathcal{N}$-dual if and only if $I=I_{\mathcal{C},\mathcal{N}}$. And then, the fuzzy closing and fuzzy opening are also $\mathcal{N}$-dual (DeBaets 1997).

Obviously, we can use conjunctive uninorm and residual implications to define fuzzy morphological operators following the previous definitions. It is investigated in (González, Ruiz-Aguilera, and Torrens 2003) which conjunctive uninorms need to be chosen in order to preserve the algebraic and morphological properties satisfied by the classical morphological operators.

Among this uninorms, two special kinds satisfying $\mathcal{I}_U = \mathcal{I}_{U,\mathcal{N}}$, and therefore duality, are given in the following proposition, see (DeBaets and Fodor 1999) and (Ruiz-Aguilera and Torrens 2004).

Proposition 3.1 *Let h be a strictly increasing, continuous function $h : [0,1] \to [-\infty, \infty]$ with $h(0) = -\infty$, $h(0) = e$, $h(1) = \infty$, and $\mathcal{N}$ a strong negation. The identity $\mathcal{I}_U = \mathcal{I}_{U,\mathcal{N}}$ is satisfied in each one of the following situations*

(i) *If U is the conjunctive representable uninorm given by $U_h(x,y) =$*

$$\begin{cases} h^{-1}(h(x) + h(y)) & \text{if } (x,y) \notin \{(1,0),(0,1)\} \\ 0 & \text{otherwise.} \end{cases}$$

and $\mathcal{N}(x) = h^{-1}(-h(x))$. Its residual implicator I_{U_h} is given by $I_{U_h}(x,y) =$

$$\begin{cases} h^{-1}(h(y) - h(x)) & \text{if } (x,y) \notin \{(1,0),(0,1)\} \\ 1 & \text{otherwise.} \end{cases}$$

(ii) *If U is the conjunctive left-continuous, idempotent uninorm $U^{\mathcal{N}}$, given by*

$$U^{\mathcal{N}}(x,y) = \begin{cases} \min(x,y) & \text{if } y \leq \mathcal{N}(x) \\ \max(x,y) & \text{elsewhere.} \end{cases}$$

Its residual implicator $I_{U^{\mathcal{N}}}$ is given by:

$$I_{U^{\mathcal{N}}}(x,y) = \begin{cases} \min(\mathcal{N}(x),y) & \text{if } y < x \\ \max(\mathcal{N}(x),y) & \text{if } y \geq x. \end{cases}$$

Thus, these two kinds of conjunctive uninorms guarantee duality between fuzzy morphological operators. Consequently, they are the most suitable in our framework.

The algebraic properties of the fuzzy morphological operators defined from uninorms are studied in (Nachtegael and Kerre 2000), and in (González, Ruiz-Aguilera, and Torrens 2003). In particular if U is a left-continuous uninorm, the corresponding fuzzy dilation and the fuzzy closing are extensive, whereas the fuzzy erosion and the fuzzy opening are anti-extensive. Moreover, the idempotence of the fuzzy closing and the fuzzy opening are ensured by next proposition.

Proposition 3.2 *Let $\mathcal{U}$ be a left-continuous conjunctive uninorm and $I_{\mathcal{U}}$ its residual implicator, let A be a gray-scale image and let B be a gray-scale structuring element, then it holds:*

1. *The fuzzy closing $C_{\mathcal{U},\mathcal{I}_{\mathcal{U}}}$ is extensive and the fuzzy opening is anti-extensive:*

$$O_{\mathcal{U},\mathcal{I}_{\mathcal{U}}}(A,B) \subseteq A \subseteq C_{\mathcal{U},\mathcal{I}_{\mathcal{U}}}(A,B).$$

2. *The fuzzy closing and the fuzzy opening are idempotent, i.e.:*

$$C_{\mathcal{U},\mathcal{I}_{\mathcal{U}}}(C_{\mathcal{U},\mathcal{I}_{\mathcal{U}}}(A,B),B) = C_{\mathcal{U},\mathcal{I}_{\mathcal{U}}}(A,B),$$

$$O_{\mathcal{U},\mathcal{I}_{\mathcal{U}}}(O_{\mathcal{U},\mathcal{I}_{\mathcal{U}}}(A,B),B) = O_{\mathcal{U},\mathcal{I}_{\mathcal{U}}}(A,B).$$

4 CLOSED AND OPEN FUZZY OBJECTS

The previous fact motivates, as in the classical mathematical morphology, the following definitions.

Definition 4.1 *Let A and B be two gray-scale images. We say that A is B-closed (resp. B-open) if $C_{U,\mathcal{I}_U}(A,B) = A$ (resp. $O_{U,\mathcal{I}_U}(A,B) = A$).*

Observe that, using Prop. 3.2, it is clear that $C_{U,\mathcal{I}_U}(A,B)$ is B-closed and $O_{U,\mathcal{I}_U}(A,B)$ is B-open. Moreover, it can be proved that all B-open and B-closed objects are the opening and the closing of some image, respectively. We have the following proposition that was advanced in (Kerre and Nachtegael 2000), without proof (see (González, Ruiz-Aguilera, and Torrens 2004) for the proof).

Proposition 4.1 *If U is left-continuous, then it holds:*
(a) A is B-open if and only if there exists a fuzzy object F such that $A = D_U(F, -B)$.
(b) A is B-closed if and only if there exists a fuzzy object F such that $A = E_{\mathcal{I}_U}(F, -B)$.

As it was pointed out in (Bodenhofer 2003) opening and closing operators only make sense if the opening always gives an open result, and the closing operator gives a closed result. Moreover, it is desirable to have "extremal properties". We see now that this last requirement is also satisfied by our opening and closing fuzzy operators.

Proposition 4.2 *If U is left-continuous, then it holds:*
(a) $O_{U,\mathcal{I}_U}(A,B)$ is the largest B-open fuzzy subset of A.
(b) $C_{U,\mathcal{I}_U}(A,B)$ is the smallest B-closed fuzzy superset of A.

Now let us consider the preservation of B-openness and B-closedness by intersections and unions, respectively.

Proposition 4.3 *Consider U a left-continuous uninorm, and A_1 and A_2 two gray-levels images. Then, it holds:*
(a) If A_1 and A_2 are both B-open then, $A_1 \cup A_2$ is B-open.
(b) If A_1 and A_2 are both B-closed then, $A_1 \cap A_2$ is B-closed.

Previous propositions are valid for any left-continuous conjunctive uninorm. However, if we want to have duality between closed and open fuzzy objects we need again the two kinds of uninorms stated in Proposition 3.1.

129

Proposition 4.4 *Consider U satisfying the condition (i) or (ii) from proposition 3.1. Then, A is B-open if and only if $co_N A$ is B-closed.*

Let us now prove that, using left-continuous uninorms like in Prop. 3.1, we obtain the so-called generalized idempotence laws for fuzzy closing and fuzzy opening. First we need several previous results that we done without proofs (for details of the proofs see (Ruiz-Aguilera 2007)).

Proposition 4.5 *Let U be a uninorm with neutral element $e \in [0,1]$. The following statements are equivalent*

(i) *U is left-continuous;*

(ii) *for all $(a,b) \in [0,1]^2$, $U(a,\mathcal{I}_U(a,b)) \leq b$;*

(iii) *for all $(a,b,c) \in [0,1]^3$,*

$$U(\mathcal{I}_U(a,b),\mathcal{I}_U(b,c)) \leq \mathcal{I}_U(a,c).$$

Proposition 4.6 *Let U be a left-continuous uninorm. Then U and $\mathcal{I}_U$ satisfy*

$$\mathcal{I}_U(x,U(y,z)) \geq U(\mathcal{I}_U(x,y),z)$$

for all $x,y,z \in [0,1]$.

Proposition 4.7 *Let U be a left-continuous uninorm. For any $a,b,c,d,e,f \in [0,1]$, if $U(a,\mathcal{I}_U(b,c)) \geq d$, and $U(e,\mathcal{I}_U(f,b)) \geq a$ then*

$$U(e,\mathcal{I}_U(f,c)) \geq d.$$

Proposition 4.8 *Let U be a left-continuous uninorm, and $\mathcal{I}_U$ be its residual implication. If for any real numbers $a,b,c,d,e,f,g,h \in [0,1]$, it is satisfied $U(a,\mathcal{I}_U(b,c)) \geq d$, $U(c,\mathcal{I}_U(e,f)) \leq g$ and $U(d,\mathcal{I}_U(e,f)) \geq h$, then*

$$U(a,\mathcal{I}_U(b,g)) \geq h.$$

Now, using left-continuous uninorms like in Proposition 3.1, and taking into account the inequalities satisfied by uninorms and their residual implications, we obtain the so-called generalized idempotence laws for fuzzy closing and fuzzy opening. The proof of inclusions concerning fuzzy opening are quite similar to those given by De Bates in (DeBaets 2000) for continuous t-norms, using in our case Prop. 4.7 and Prop. 4.8. With respect to inclusions concerning fuzzy closing, their proofs follow from duality, guaranteed by Prop. 3.1.

Proposition 4.9 *Consider U satisfying the condition (i) or (ii) from proposition 3.1. If A is B-open and rang(A) and rang(B) are finite sets, then for any fuzzy object F it holds:*

$$O_{U,\mathcal{I}_U}(F,A) \subseteq O_{U,\mathcal{I}_U}(F,B) \subseteq F$$

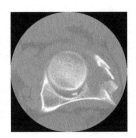
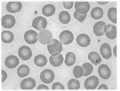
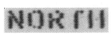

Figure 1.   Input images used in the experiments.

and dually, $F \subseteq C_{U,\mathcal{I}_U}(F,B) \subseteq C_{U,\mathcal{I}_U}(F,A)$.

Proposition 4.10 *(Generalized Idempotence) Consider U satisfying the condition (i) or (ii) from proposition 3.1 and $\mathcal{I}_U$ its residual implication. If A is B-open and rang(A) and rang(B) are finite sets, then for any fuzzy object F it holds:*

$$O_{U,\mathcal{I}_U}(O_{U,\mathcal{I}_U}(F,B),A)$$
$$= O_{U,\mathcal{I}_U}(O_{U,\mathcal{I}_U}(F,A),B) = O_{U,\mathcal{I}_U}(F,A)$$

and dually for the fuzzy closing.

Remark The previous propositions are valid for any left-continuous uninorm, and in particular, for any left-continuous t-norm, improving the results in (DeBaets 2000). Moreover, it is possible to omit the condition of finite rang in the case of continuous t-norms (see Ruiz-Aguilera 2007) for details).

5   EXPERIMENTAL RESULTS

In this section we present some experiments showing the differences between basic fuzzy morphological operators using different uninorms. The examples presented illustrate the influence of the choice of the pair $(U,\mathcal{I}_U)$ using both, idempotent and representable conjunctive uninorms. Some of our input images, A, are depicted in Figure 1. The structuring element chosen, B, used for the fuzzy operators is represented by the matrix

$$B = e \cdot \begin{pmatrix} 0.86 & 0.86 & 0.86 \\ 0.86 & 1.00 & 0.86 \\ 0.86 & 0.86 & 0.86 \end{pmatrix} \quad (1)$$

where e is the neutral element of the uninorm.

130

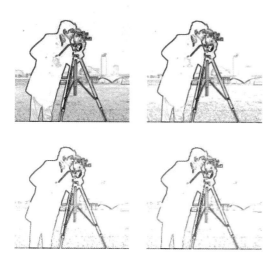

Figure 2. Top left: using $U^{\mathcal{N}}$ with $\mathcal{N}(x) = \sqrt{1-x^2}$. Top right: using U_h with $h(x) = \ln\left(\frac{x}{1-x}\right)$. Down left: using $(T_L, \mathcal{I}_{T_L})$. Down right: umbra approach.

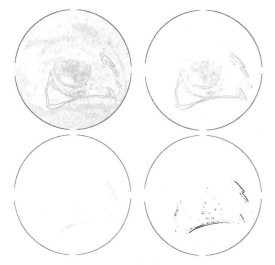

Figure 3. Top left: using $U^{\mathcal{N}}$ with $\mathcal{N}(x) = \sqrt{1-x^2}$. Top right: using U_h with $h(x) = \ln\left(\frac{x}{1-x}\right)$. Down left: using $(T_L, \mathcal{I}_{T_L})$. Down right: umbra approach.

In the next figures, Figure 2 to Figure 5, we show the fuzzy gradient operator corresponding to each one of the input images displayed in Fig. 1, using several left-continuous representable and idempotent uninorms, compared with the fuzzy gradient using the pair $(T_L, \mathcal{I}_{T_L})$, the Łukasiewicz t-norm and its residual implication, and the umbra approach, with the same structuring element but taking $e = 1$ in (1). Recall that the pair $(T_L, \mathcal{I}_{T_L})$ is the representative of the only class of t-norms (nilpotent ones) that guarantees

Figure 4. Top left: using $U^{\mathcal{N}}$ with $\mathcal{N}(x) = 1 - x$. Top right: using U_h with $h(x) = \ln\left(\frac{x}{1-x}\right)$. Down left: using $(T_L, \mathcal{I}_{T_L})$. Down right: umbra approach.

Figure 5. Top left: using $U^{\mathcal{N}}$ with $\mathcal{N}(x) = 1 - x$. Top right: using U_h with $h(x) = \ln\left(\frac{x}{1-x}\right)$. Down left: using $(T_L, \mathcal{I}_{T_L})$. Down right: umbra approach.

the fulfillment of all the properties in order to have a good fuzzy mathematical morphology, including duality (Nachtegael and Kerre 2000).

In Figure 2 we display some of the results obtained using the classical cameraman image. It can be observed that, while the hard edges (see the person in the foreground and some buildings in the background) are detected very well in all the cases, it can be observed that the gradient obtained using conjunctive uninorms (top) detect some soft edges better than the gradient obtained with $(T_L, \mathcal{I}_{T_L})$ and the umbra approach (bottom) (see some of the buildings in the background).

Figure 3 shows the edges obtained for the hip image displayed in Fig. 1 (top-right). In case of uninorms (top), we can observe how the bone structure and a little fracture is preserved. Otherwise, in the umbra approach and t-norms, these structures are lost. So, we cannot use umbra approach and t-norms if we want to make a subsequent analysis.

The edge-images displayed in Figure 4 corresponds to the red bloods cells image shown in Fig. 1, also know as erythrocytes image. The edge-images obtained using idempotent and representable uninorms (top) improve the results corresponding to t-norms and are quite similar to umbra approach results (bottom). Observed that the boundary and structure of some cells disappear when we used t-norms, but in the boundary

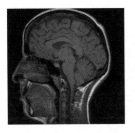
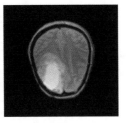

Figure 6. MR images used as input images in the experiments.

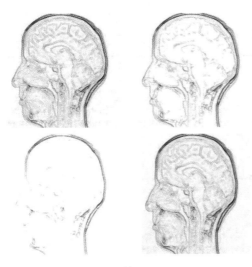

Figure 7. Top left: using $U^{\mathcal{N}}$ with $\mathcal{N}(x) = 1 - x$. Top right: using U_h with $h(x) = \ln\left(-\frac{1}{\ln 2}\ln(1-x)\right)$. Down left: using $(T_L, \mathcal{I}_{T_L})$. Down right: umbra approach.

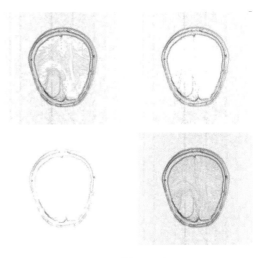

Figure 8. Top left: using $U^{\mathcal{N}}$ with $\mathcal{N}(x) = 1 - x$. Top right: using U_h with $h(x) = \ln\left(\frac{x}{1-x}\right)$. Down left: using $(T_L, \mathcal{I}_{T_L})$. Down right: umbra approach.

images obtained with uninorms these structures are preserved.

Figure 5 is an artificial image. We can observe, againe, that the results based on uninorms improve the results based on t-norms. In the last case, we can observe a very soft boundary, although it is darker in the umbra approach.

In Figure 6 we shown two slices of different MR images. On the left we display a saggital slice of an MRI and, on the right, we display an axial slice. In this last image the existence of a pathology can be observed. The gradients obtained from these two images are displayed in Figures 7 and 8 respectively. As we can see, using uninorms (top) we obtain better results than using the Łuckasiewicz t-norm (bottom left). Note also that, in the case of idempotent uninorm (top left) the shape of the pathology and the brain are better delimited than in the umbra approach case.

6 CONCLUSIONS AND FUTURE WORK

We have proposed a fuzzy mathematical morphology based on left-continuous conjunctive uninorms, that fulfills the same good morphological properties that the one based on nilpotent t-norms. Some of this properties, known for continuous t-norms are generalized to left-continuous uninorms.

Respect to the experimental results, it is important to note that, in all cases, the results obtained in edge-detection with the approach presented here, are equal to or better than the ones obtained with the umbra approach and the nilpotent t-norm fuzzy approach, because they detect some soft edges more accurately. Depending of the application, this feature can be either an advantage or a drawback. On the other hand, note that, in general the edge-image with uninorms contains the same (or more) information as the edge image derived from the umbra approach. Note also that, from a practical point of view, the choice of the uninorm and its residual implicator will also depend upon the specific problem or application. The same remark holds for the choice of the structuring element. We are adressing currently some experiments in this aspect. We have also implemented the opening and closing operators (not included here because of space reasons), and it is our intention to use them for defining top-hat transformations and fuzzy morphological filters, in order to follow with our comparative study with the umbra and t-norms approaches.

ACKNOWLEDGEMENTS

This work has been supported by the project MTM2006-05540, of the Spanish Government.

REFERENCES

Bloch, I. and H. Maître (1995). Fuzzy mathematical morphologies: a comparative study. *Pattern Recognition 28*, 1341–1387.

Bodenhofer, U. (2003). A unified framework of opening and closure operators with respect to arbitrary fuzzy relations. *Soft Computing 7*, 220–227.

DeBaets, B. (1997). *Uncertainty Analysis in Engineering and Sciences: Fuzzy Logic, Statistics, and Neural Network Approach*, Chapter Fuzzy morphology: a logical approach, pp. 53–68. Kluwer Academic Publishers.

DeBaets, B. (1999). Idempotent uninorms. *European J. Oper. Res. 118*, 631–642.

DeBaets, B. (2000). *Fuzzy techniques in image processing*, Chapter Generalized idempotence in fuzzy mathematical morphology, pp. 58–75. Heidelberg: Springer-Verlag.

DeBaets, B. and J. Fodor (1999). Residual operators of uninorms. *Soft Computing 3*, 89–100.

DeBaets, B., E. Kerre, and M. Gupta (1995a). The fundamentals of fuzzy mathematical morfologies part i: basics concepts. *International Journal of General Systems 23*, 155–171.

DeBaets, B., E. Kerre, and M. Gupta (1995b). The fundamentals of fuzzy mathematical morfologies part ii: idempotence, convexity and decomposition. *International Journal of General Systems 23*, 307–322.

DeBaets, B., N. Kwasnikowska, and E. Kerre (1997). Fuzzy morphology based on uninorms. In *Proceedings of the seventh IFSA World Congress*, Prague, pp. 215–220.

Fodor, J., R. Yager, and A. Rybalov (1997). Structure of uninorms. *Int. J. Uncertainty, Fuzziness, Knowledge-Based Systems 5*, 411–427.

González, M., D. Ruiz-Aguilera, and J. Torrens (2003). Algebraic properties of fuzzy morphological operators based on uninorms. In *Artificial Intelligence Research and Development*, Volume 100 of *Frontiers in Artificial Intelligence and Applications*, Amsterdam, pp. 27–38. IOS Press.

González, M., D. Ruiz-Aguilera, and J. Torrens (2004). Opening and closing operators in fuzzy morphology using conjunctive uninorms. In *IPMU'2004, Tenth International Conference on Information Processing and Management of Uncertainty in Knowledge-Based Systems*, Perugia, Italy, pp. 13–14. (Extended version: Technical Report DMI-UIB A-01-2004).

Kerre, E. and M. Nachtegael (2000). *Fuzzy techniques in image processing*, Volume 52 of *Studies in Fuzziness and Soft Computing*. New York: Springer-Verlag.

Nachtegael, M. and E. Kerre (Eds.) (2000). *Fuzzy techniques in image processing*, Volume 52 of *Studies in fuzziness and soft computing*, Chapter Classical and fuzzy approaches towards mathematical morphology, pp. 3–57. Heidelberg: Springer-Verlag.

Ruiz-Aguilera, D. (2007). *Contribució a l'estudi de les uninormes en el marc de les equacions funcionals. Aplicacions a la morfologia matemàtica*. Ph. D. thesis, University of Balearic Islands.

Ruiz-Aguilera, D. and J. Torrens (2004). Residual implications and co-implications from idempotent uninorms. *Kybernetika 40*, 21–38.

Serra, J. (1982,1988). *Image analysis and mathematical morphology, vols. 1, 2*. London: Academic Press.

Nonlinear 3D foot FEA modelling from CT scan medical images

P.J. Antunes & G.R. Dias
IPC – Institute for Polymers and Composites, University of Minho, Guimarães, Portugal

A.T. Coelho
AIS – Amorim Industrial Solutions, Corroios, Portugal

F. Rebelo
Ergonomics Laboratory, Faculty of Human Kinetics, Technical University of Lisbon, Lisbon, Portugal

T. Pereira
The Health Sciences School, University of Minho, Braga, Portugal

ABSTRACT: A 3D anatomically detailed non-linear finite element analysis human foot model is the final result of density segmentation 3D reconstruction techniques applied in Computed Tomography (CT) scan DICOM standard images in conjunction with 3D Computer Aided Design operations and finite element analysis (FEA) modelling. Density segmentation techniques were used to geometrically define the foot bone structure and the encapsulated soft tissues configuration. The monitoring of the contact pressure values at the foot plantar area assumes a vital role on the human comfort optimization. The contact pressure distribution at the plantar area and stresses at the bone structures are calculated for this article for a rigid and direct contact between the plantar foot area and the ground support. Linear and non-linear elastic constitutive material models were implemented to mechanically characterize the behaviour of the biological materials. Furthermore, an experimental validation of the FEA rigid-based contact pressure results is presented.

1 INTRODUCTION

Finite element analysis (FEA) can be a very powerful tool in the foot biomechanical study. The human foot comfort can be related with the contact pressure generated at the plantar/insole(soil) interface.

Large values of contact pressure can generate pain or pathologies due to the obstruction of blood circulation in areas with peak values of pressure. The comfort enhancement at the foot region can be achieved by the application of shoe insoles that must be mechanically optimized to simultaneously support the body weight without foot deviations and act as contact pressure reducers in the precarious plantar zones.

The geometrical complexity of the foot structure implies the use of reverse engineering tools in order to obtain a model that can accurately simulate the biomechanical behaviour of the human foot, namely soft tissues and bone structure.

This article describes the methodology applied in the development of an anatomically detailed three-dimensional foot model for non-linear finite element analysis from medical image data obtained from a CT scan.

2 MODELLING METHODOLOGY

The complex mechanical behaviour of the foot and the necessity of obtaining accurate results for posterior validation with experimental values implies an adequate modelling of the foot structure in terms of 3D anthropometrical characteristics and material constitutive modelling.

The initial step concerning the foot anthropometrical definition was a CT scan of the foot region of a 26 years old male. The DICOM images generated in the CT scan were processed with a medical imaging and editing software (MIMICS® 9.1) that was used to obtain the primary 3D models using density segmentation techniques. The generated primary 3D models were exported as geometrical files for a CAD system (CATIA®) that allowed the assembly and some 3D geometrical operations. Finally, the CAD model was exported to a non-linear FEM/FEA package (ABAQUS® 6.6.1). The model was then prepared for the non-linear structural analysis, namely, through the definition of loads, boundary conditions, material constitutive models, kinematic constraints and finite element mesh generation.

3 MEDICAL IMAGE DATA GENERATION

A CT scan was performed in a 26 years old male, height 175 cm and 75 kg weight in a Phillips® Brilliance 16 CT scan equipment. The scan was realized for both foot at the neutral posture and was defined by 482 cross-sectional cuts with a slice distance of 0.4 mm and a field of view (FOV) of 346 mm. The medical images were exported in the DICOM format with an image area of 1024 × 1024 pixels. The high image resolution associated with the reduced distance between slices assures a good geometrical definition of the primary 3D models in the future density segmentation operations.

4 3D MODELLING

4.1 3D reconstruction (density segmentation)

For the reconstruction of the primary 3D anthropometrical models (bone structure and encapsulated soft tissues) was used the MIMICS® 9.1 medical imaging density segmentation software. Thresholding based on Hounsfield units was used to separate each bone from the bone structure (Fig. 1a) and also for the definition of the encapsulated soft tissues volume (Fig. 1b). In order to include all the cortical and trabecular bone at the foot bone structure and exclude the cartilage regions, a lower limit of 250 HU and an upper limit of 2000 HU were defined. The soft tissues region was generated accounting a lower limit of −200 HU and an upper limit of 3071 HU.

4.2 CAD modelling

The cartilages that were not reconstructed in the segmentation process were then modelled in order to connect the bones and fill the cartilaginous space. After the cartilage modelling process, volume boolean operations were performed to achieve a volume of soft tissues that corresponds to the subtraction of the bone structure coupled with the cartilages. This approach, guarantees the perfect alignment of the models exterior surfaces, what is an important condition for the future finite element model generation.

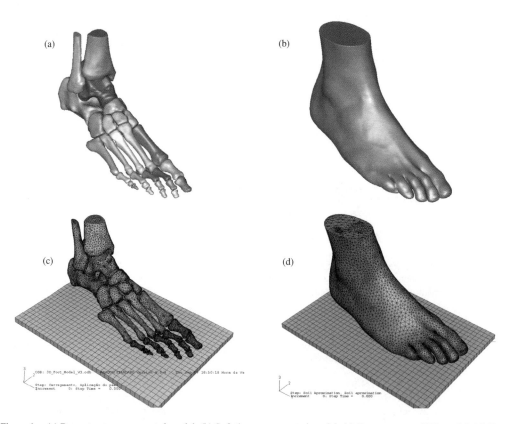

Figure 1. (a) Bone structure segmented model, (b) Soft tissues segmented model, (c) Bone structure FEA model, (d) Bone structure + soft tissues FEA model.

4.3 FEA modelling

The FEA software package ABAQUS® 6.6.1 was used to define the foot FEA bone structure as shown in Figure 1c, consisted of 29 bone parts and cartilaginous regions, that includes all the *distal, medial and proximal phallanges*, 3 *cuneiforms, talus, calcaneus, cuboid, navicular, tibia* and *fibula* bones. As shown in Figure 1d, the soft tissues region was also defined and involves the bone structure.

The bone and cartilage structure were bonded together forming a unique structure with different material regions as shown in Figure 1c. This structure was then bonded to the soft tissues volume, through the definition of mesh tie kinematic constrains as can be seen in Figure 1d.

The foot/ground interface was defined through contact surfaces, what allow the load transmission between support and foot model and consecutively the generation of a contact pressure field at the foot plantar area. A small-sliding tracking approach associated with a surface to surface contact formulation was defined to model the interaction tangential behaviour. An augmented lagrangian constraint enforcement method was implemented in the definition of the interaction normal behaviour. The friction coefficient between the foot and soil was set to 0.6, using the Coulomb friction model (Zhang et al. 1999).

For the present case, two different types of loading were considered. The first case, consider a pure vertical compression load of the foot defined only by a vertical force (375 N) applied in the ground reference point. The second loading case considers simultaneously the force applied in the calcaneus bone through the Achilles tendon and the ground reaction force, in order to simulate the balanced standing. The plantar fascia and Achilles tendon were included in the

FEA model through the definition of truss and axial connectors elements respectively.

The plantar fascia is one of the major stabilization structures of the longitudinal arch of the human foot and sustains high tensions during the weight application (Cheung et al. 2004). In the FEA model the plantar fascia was geometrically simplified and divided into 5 separated sections (rays) modelled with truss elements that only supports tensile stress.

The geometrical definition of the Achilles tendon through axial connector elements, allows the simulation of the load applied in the *calcaneus* zone for a foot during balanced standing. The load at the posterior aspect of the *calcaneus* bone is generated by the involuntary contraction of the triceps *surae* muscle group in order to stabilize the foot during standing (Gefen 2002). The study of Simkin (1982), who calculated that the Achilles tendon force should be approximately 50% of the body load during balanced standing, was considered for the present foot computational model.

The upper surfaces of the soft tissues, *tibia* and *fibula* were fixed through the analysis time via a kinematic constraint, while the boundary conditions applied at the soil reference point load, allowed uniquely the plate movement in the vertical (upper) direction.

A wide variety of continuum finite elements topology and formulations were used to descritize the foot model 3D structure. The foot geometrical complexity do not allows the use of hexahedral elements that usually provides higher accuracy with less computational cost.

Tetrahedral elements that are more versatile to capture the irregularly shapes of the bone structure and the encapsulated soft tissues, were used to mesh the model. Hybrid element formulation was used to assure the almost-incompressible constraint for the soft tissues non-linear elastic mechanical behaviour. Rigid elements were implemented to define the ground support.

All the materials (Table 1) were considered isotropic and linear-elastic except the soft tissues that are mechanically characterized by a non-linear elastic

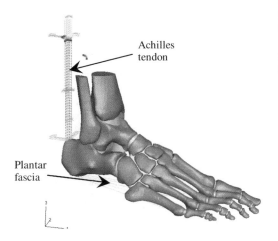

Achilles tendon

Plantar fascia

Figure 2. Plantar fascia and Achilles tendon FEA modelling.

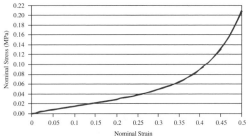

Figure 3. Soft tissues non-linear uniaxial mechanical behaviour.

Table 1. Material properties and finite element topology/formulation.

Components	Element Topology	Formulation	Young's modulus (MPa)	v	Cross-sectional area (mm²)
Bone structure	3D-Tetrahedra	Linear	7300	0.3	–
Cartilage	3D-Tetrahedra	Linear	10	0.4	–
Soft tissues	3D-Tetrahedra	Linear, Hybrid	Hyperelastic	≈0.5	–
Achilles tendon	1D	Axial Connector element	∞	–	–
Plantar fascia	1D	Truss element (No compression)	350	–	58.6
Soil	Quadrilateral	Rigid element	∞	–	–

Table 2. Soft tissues hyperelastic material parameters.

C_{10}	C_{01}	C_{20}	C_{11}	C_{02}	D_1	D_2
0.08556	−0.05841	0.03900	−0.02319	0.00851	3.65273	0

behaviour. The bone material behaviour was linearly defined with a Young's Modulus and Poisson's ratio (v), equal to 7300 MPa and 0.3, respectively. These values were obtained by weighing cortical and tra-becular bone elasticity according to Nakamura et al. (1981). The mechanical properties of the cartilage (Shanti et al. 1999 & Gefen 2003) and plantar fascia (Cheung et al. 2005), were selected from the literature. Specifically, the bulk soft tissues non-linear elastic mechanical behaviour definition was based on the uni-axial stress-strain data obtained from *in vivo* tests of the heel (Lemmon et al. 1997). The bulk soft tissues non-linear mechanical behaviour was defined through a hyperelastic model based on a second order poly-nomial strain energy function (Cheung et al. 2005 & Lemmon et al. 1997), given by the expansion of Equation 1.

$$\psi\left(J,\bar{I}_1,\bar{I}_2\right)=\sum_{i+j}^{N}C_{ij}\left(\bar{I}_1\text{-}3\right)^i\left(\bar{I}_2\text{-}3\right)^j+\psi_{vol}\left(J\right) \qquad (1)$$

Setting $N=2$ and considering that the pure vol-umetric response is given by the strictly convex function, given by Equation 2.

$$\psi_{vol}(J)=\frac{1}{D_i}\left(J\text{-}1\right)^{2i} \qquad (2)$$

where ψ is the overall strain energy per unit of ref-erence volume; J is the volume ratio; C_{ij} and D_i are material dependent parameters obtained from the experimental data; $\bar{I}_1$ and $\bar{I}_2$ are the modified strain invariants. The material parameters used for the def-inition of the hyperelastic model associated with the non-linear mechanical definition of the soft tissues, are presented in Table 2.

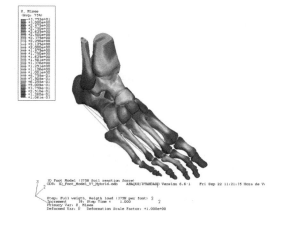

Figure 4. Von Mises stress at the bone structure.

5 NON-LINEAR FEA RESULTS

The anatomically detailed 3D FEA foot model was developed from CT scan images using density segmen-tation techniques and CAD manipulation. Kinematic constrains between bone structures, cartilages and soft tissues were defined. The load transmission between ground support and the foot structure was defined by the introduction of contact pairs, namely, at the foot plantar area/ground support interface.

Large deformations and non-linear geometrical analysis associated with material nonlinearities were considered.

The created FEA model allows the output of several results that can be used for comfort evaluation of shoe

Table 3. Contact pressure distribution.

Load case	1st metatarsal	2nd metatarsal	3rd metatarsal	4th metatarsal	5th metatarsal	Calcaneus
Pure compression	0.073 MPa	0.042 MPa	0.073 MPa	0.074 MPa	0.018 MPa	0.131 MPa
Balanced Standing	0.083 MPa	0.051 MPa	0.084 MPa	0.083 MPa	0.041 MPa	0.111 MPa

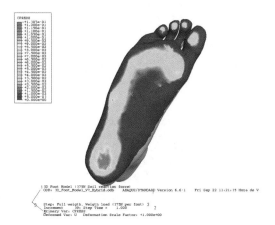

Figure 5. Contact pressure values at the foot plantar are (pure compression).

insoles or to study other biomechanical aspects of the foot. The monitoring of contact pressure values at the foot plantar area assumes a vital role on this study. The results were obtained considering two different load cases, namely, pure compression load (weight load) and balanced standing load (weight load + Achilles tendon load).

The FEA model predicts a maximum plantar contact pressure value of 0.131 MPa (13.1 N/cm^2) and 0.111 MPa (11.1 N/cm^2) for the pure compression (Fig. 5), and balanced standing case respectively, at the heel region.

The contact pressure under the *metatarsal* heads and the *distal phalanges* increases with the load application at the posterior *calcaneus* (through the Achilles tendon). The load at the *calcaneus* compared with the pure compression load, displaces the centre of pressure and increases the load-bearing at the forefoot reducing consequently the load-bearing at the rearfoot. At the bone structure, peak of stress are present at the *metatarsal* and *talus* bones (Fig. 4). The insertion points of the fascia plantar truss elements at the *phalanges/metatarsal* connection region and *calcaneus* bone, experienced large stress due to the generated fascia plantar tension. In Table 3 are presented indicative nodal contact pressure values at the same nodes beneath the *metatarsal* heads and heel regions. The FEA predicted centre of pressure for the balanced standing model was approximately located at the 2nd cuneiform.

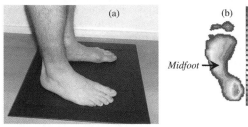

Figure 6. (a) Podologic pressure measuring test, (b) experimental foot plantar contact pressure distribution.

6 EXPERIMENTAL VALIDATION

6.1 Methodology

For a quantitative evaluation of the contact pressure generated at the plantar foot area, the same individual that volunteered for the CT scan was used to experimentally evaluate the pressure distribution during barefoot balanced standing. For this experimental study a podologic pressure measuring equipment from Eclipse 2000® was used (Fig. 6a).

A barefoot balanced standing posturologic test was conducted during 10 s, where the contact pressure values were evaluated 200 times in that time range, corresponding to a results output frequency of 20 Hz. The mean contact pressure value was automatically calculated by the equipment software. In Figure 6b is showed the experimental contact pressure values distribution for the right barefoot balanced standing situation.

6.2 Comparison of experimental and FEA results

The obtained experimental results were compared with the FEA results obtained for the balanced standing case. An experimental peak pressure for the foot plantar area of about 0.071 MPa (7.1 N/cm^2) is measured at the heel region. The FEA model predicts a maximum value for the same area but greater and equal to 0.111 MPa (11.1 N/cm^2). In a midfoot indicative point as displayed in Figure 6b, the experimental contact pressure is approximately equal to 0.038 MPa (3.8 N/cm^2), while the correspondent FEA predicted pressure for approximately the same point is equal to 0.049 MPa (4.9 N/cm^2). Beneath the *metatarsal* heads, indicative experimental contact pressure values of about 0.033, 0.039, 0.045, 0.387 and 0.045 MPa from the 1st to 5th *metatarsal* head were measured, while the correspondent FEA nodal contact pressure values were about the following ranges: [0.053, 0.103], [0.041, 0.051], [0.051, 0.088], [0.039, 0.098] and [0.041, 0.067]MPa, for approximately the same foot regions. Notice that the FEA values are taken in a min/max form due to the difficulty in obtaining a single contact pressure value for those zones. The FEA predicted contact area was approximately equal to 89 cm^2, compared to 138 cm^2 obtained from the experimental measurements.

7 DISCUSSION

The FEA capability in predicting the stress state in the foot, namely the contact pressure at the foot plantar area, makes suitable the application of the present model in the comfort optimization of shoe insoles or other foot support devices. This model describes the geometry of the ankle-foot complex and contains information about the bony, cartilaginous, plantar fascia and soft tissues materials mechanical behaviour. The experimental plantar contact pressure distribution is qualitatively comparable with the predicted FEA results, nominally, the peak pressure values zones at the centre of the heel region and beneath the *metatarsal* heads. However in the quantitative point of view the FEA results are higher than the experimental results. This difference may be caused by the resolution of the pressure sensors that report an average pressure over a sensor area while the FEA model reports the contact pressure as calculated from a nodal force per element's surface area. Therefore is expected that the FEA contact pressure results be higher than the experimental ones. The predicted FEA results showed that at the heel region and beneath the *metatarsal* heads the contact pressure values were maximum what indicate the regions that most probably ignite foot pain and ulcer development. These results are in accordance with the observation of foot ulcers appearance at the medial forefoot and heel regions of the diabetic patients (Mueller et al. 1994 & Raspovic et al. 2000). These regions must be protected with deformable insole materials that accommodates deformation and reduce the contact pressure trough the local increase of the contact area.

The FEA predicted contact area values for balanced standing (89 cm^2) are proportionally comparable with the results obtained by Cheung et al. (2005), who reported a contact area equal to 68 cm^2 for a 1.74 m and 70 kg male with normal soft tissues stiffness. The higher contact area values obtained in our experimental results can be due to the contact area acquisition mode and sensors resolution at the podologic pressure measuring equipment.

The centre of pressure predicted with the FEA model is approximately in accordance with the results obtained by Simkin (1982) who reported a centre of pressure located at the 3rd cuneiform or for some cases, between the 2nd and 3rd cuneiform.

The difficulty in obtaining a completely static posture that replicates the FEA model load condition was a major experimental difficulty encountered and an aspect that must be accounted during the results evaluation and discussion. In fact, the continuous individual displacement originated by the involuntary muscle contraction during balanced standing can affect the foot load bearing and consequently the experimental contact pressure distribution and its values.

8 CONCLUSIONS AND FUTURE WORK

The finite elements method can be a very powerful method to understand the foot mechanical behaviour and its implications to human foot comfort. The present non-linear FEA model intends to be a tool for the design optimization process of shoe insoles. For that purpose, an anatomically detailed foot model was generated from CT scan image data using segmentation reconstruction techniques and 3D CAD modelling operations.

In the present model, several material constitutive models were considered. Kinematic constraints and parts interactions, namely, at the foot plantar/soil interaction were implemented. Achilles tendon and plantar fascia were introduced considering some geometrical simplifications.

The FEA contact pressure values were experimentally verified by the use of a podologic pressure measuring equipment. The effect of the load application at the Achilles tendon was also studied to understand its effect on the contact pressure distribution at the foot plantar area.

A wide variety of insole geometries and materials can in the future be tested in order to study and improve the foot comfort through the modification of insole geometrical design and/or insole materials formulation.

The introduction of the foot ligament and muscular structure is predicted to be incorporated in future FEA works to dynamically model the human gait.

REFERENCES

Camacho, D.L.A., Ledoux, W.R., Rohr, Eric S., Sangeorzan, B.J. & Ching, Randal P. 2002. A Three-Dimensional, anatomically detailed Foot Model: A Foundation for a Finite Element Simulation and Means of Quantifying Foot-Bone Position; *Journal of Rehabilitation Research & Development* 39:401–410.

Cheung, J.T, Zhang, M., Leung, A.K.L. & Fan, Y.B. 2005. Three Dimensional Analysis of the Foot During

Standing – A Material Sensitivity Study; *Journal of Biomechanics* 38:1045–1054.

Cheung, J.T., Zhang, M. & An, K.N. 2006. Effect of Achilles Tendon Loading on Plantar Fascia Tension in the Standing Foot. *Clinical Biomechanics* 21:194–203.

Cheung, J.T., Zhang, M. & An, K.N. 2004. Effects of Plantar Fascia Stiffness on the Biomechanical Responses of the Ankle-Foot Complex. *Clinical Biomechanics* 19:839–846.

Gefen, A. 2003. Plantar Soft Tissue Loading Under the Medial Metatarsals in the Standing Diabetic Foot. *Medical Engineering & Physics* 25: 491–499.

Gefen, A. 2002. Stress Analysis of the Standing Foot Following Surgical Plantar Fascia Release. *Journal of Biomechanics* 35: 629–637.

Lemmon, D., Shiang, T.Y., Hashmi, A., Ulbrecht, J.S. & Cavanagh, P.R. 1997. The Effect of Shoe Insoles in Therapeutic Footwear – A Finite Element Approach. *Journal of Biomechanics* 30:615–620.

Mueller, M.J., Sinacore, D.R., Hoogstrate, S. & Daly, L. 1994. Hip and ankle walking strategies:effect on peak plantar pressures and implications for neuropathic ulceration. *Archives of Physical Medicine and Rehabilitation* 75: 1196–1200.

Nakamura, S. & Crowninshield R.D., & Cooper, R.R. 1981. An Analysis of Soft Tissue Loading in the Foot: a preliminary report. *Bulletin of Prosthetics Research* 18:27–34

Raspovic, A., Newcombe, L., Lloyd, J. & Dalton, E. 2000. Effect of customized insoles on vertical plantar pressures in sites of previous neuropathic ulcerations in the diabetic foot. *The Foot* 10: 133–138.

Shanti, J. & Mothiram K.P. 1999. Three Dimensional Foot Modelling and Analysis of Stresses in Normal and Early Stage Hansen's Disease with Muscle Paralysis *Journal of Rehabilitation Research & Development* 36.

Simkin, A. 1982. Structural Analysis of the Human Foot inStanding Posture. PhD Thesis, Tel Aviv University.

Zhang, M. & Mak, A.F.T. 1999. In vivo skin frictional properties. *Prosthetics and Orthotics International* 23: 451–456.

Computational Vision and Medical Image Processing – João Tavares & Natal Jorge (eds)
© 2008 Taylor & Francis Group, London, ISBN 978-0-415-45777-4

Alternative lossless compression algorithms in X-ray cardiac images

D.R. Santos, C.M.A. Costa, A.J. Silva, J.L. Oliveira & A.J.R. Neves
DETI/IEETA, Universidade de Aveiro, Portugal

ABSTRACT: Over the last decade, the use of digital medical imaging systems has increased greatly in health-care institutions. Today, Picture Archiving and Communication System (PACS) is one of the most valuable tools supporting medical profession in both decision making and treatment procedures. It reduced the costs associated with the storage and management of image data and also increased both the intra and inter-institutional portability of data. One of the most important benefits of the digital medical image is that it allows the widespread sharing and remote access to medical data by outside institutions. PACS presents an opportunity to improve cooperative workgroups taking place either within or with other healthcare institutions.

Storage and transmissions costs are continuously decreasing, but, as individual digital medical studies become significantly larger, further improvements on transmission performance and on storage efficiency are critical. Image compression algorithms offer the means to reduce storage cost and to increase transmission speed.

Following previous methodologies [1], this paper provides a comparison about the application of DICOM (Digital Imaging and Communications in Medicine) lossless compression standards on Angiography cine acquisition images. A new lossless compression approach that exploits time redundancy between successive frames is also presented. Finally, the standards codecs values are compared with results obtained with the proposed method and with video lossless codecs.

1 INTRODUCTION

The compression of digital medical images is of vital importance in Picture Archiving and Communications Systems (PACS) and teleradiology applications, due to the typical huge volume of patients' data. This is especially true for grayscale images used in radiology applications (e.g. a single 240 frames angiographic cine acquisition represents approximately 60 MB of storage and a Multislice cardiac Computerized Tomography study of 2500 images per patient, may reach approximately 1.3 GB).

Digital storage of medical images can be challenging specially because of the requirement to preserve the image quality and also because the images can be very large in size and number [2, 3]. Several reasons can be pointed to stress the importance of compression [3]:

- Digital medical images databases are normally very large repositories;
- Patients' data must be stored for long periods resulting in a continuous growth of medical imaging databases;
- High resolution and many bits per pixel results in large volumes;

- The image transmission time is volume driven. The usefulness of a PACS depends greatly on appropriate transfer waiting times.

Common images compression algorithms, such as JPEG, are data-aware, i.e. they explore redundancy that is specific of this kind of two dimensional data source. In dynamic images (videos), such as angiographic cine acquisition, we should be able to take advantage of redundancy in the third dimension (time) and achieve greater compression ratios [2].

Various methods, both for lossy and lossless image compression, are proposed in the literature. Lossy compression techniques can achieve high compression ratios but they do not allow one to reconstruct exactly the original version of the input data. The use of lossy compression in medical imaging is still controversy for particular applications since it can result on decreased diagnostic accuracy and confidence [4]. The only possibility to obtain a completely reversible compression scheme is to use lossless compression methods but, in this case, the achievable compression ratios are only in the order 2:1 up to 4:1 [3].

The lossless image compressors allowed nowadays by the DICOM standard are lossless JPEG, JPEG2000

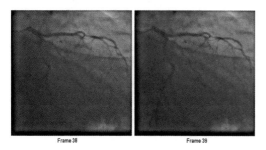

Frame 38   Frame 39

Figure 1.   Two successive XA frames.

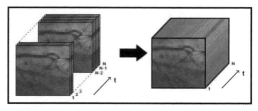

Figure 2.   Volume creation.

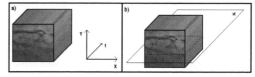

Figure 3.   Xt images extraction from volume.

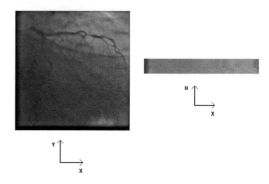

Figure 4.   Original frame (left) and xt frame (right).

(lossless mode) and JPEG-LS (lossless mode). The last two are state of the art image compressors and achieve better performance than lossless JPEG sv 1 which is, however, the most used image compressor in DICOM images. This is due to the fact that only the most recent equipments have the capability to decode JPEG-LS and JPEG2000 image codestream.

Within the Cardiac Imaging domain, Coronary Angiography (XA) is currently a prominent X-ray image modality for coronary heart disease diagnostic and treatment [4].

Our angiographic experiment used a set of 278 DICOM digital cine angiograms, acquired at 15 fps at a matrix size of 512×512 pixels with an 8-bit grayscale (256 gray levels) resolution. The lengths of sequences varied from approximately 16 to 277 images, corresponding to about 1 to 18 seconds.

Observing successive XA frames (Figure 1) it is possible to recognize that two consecutive images are very similar to each other and the exploitation of this redundancy would possibly lead to a better compression approach. On the other hand, the majority of data volume produced in an angiographic study is relative to dynamic captures (i.e. videos). So, the alternative compression method described in this paper is based on the hypothesis that the compression of the differences between successive images would obtain better results than the compression of the individual images. This solution is very similar to Differential Pulse Code Modulation (DPCM) used, for example, in digital coding and also on some video encoders/decoders (codec). The improvement is obtained since the differences between pixels of consecutive images are smaller than the actual pixel amplitudes.

The normal approach would be to obtain the *differential images* from the natural image acquisition order ($\text{dif}_{i,j} = \text{im}_i - \text{im}_j$ in which i and j are the images acquisition number and $i = j + 1$). Several compression experiments were carried out using different compression algorithms but results were not satisfactory mainly due to noise presence. Instead using the difference between image that are separated in time, we have aggregated all frames together (maintaining the normal frame order) to form a tridimensional volume (x, y, t) of $512 \times 512 \times N$ where N is the number of frames of the XA cine acquisition (Figure 2). The acquisition number (N) of each frame is equivalent to the instant (t) it was acquired so a higher acquisition number imply a later acquisition. This volume process was performed using Matlab® software. Successive images are obtained from the volume but in the *xt* plane – Figure 3b (in opposition to the normal temporal image sequence – *xy* plane – Figure 3a).

It can be observed, in Figure 4, the two types of images, the original frame (left) and the *xt* plane (right). The later is the one used on *differential method*. *Differential images* are obtained by subtracting an *xt* plane image from the previous one. The first *xt* plane is the top of the volume and the last *xt* plane is the bottom of it.

The pixel values of the difference between *xt* plane images use a direct method $\text{pixel\_dif}_{i,j}(m,n) = \text{pixel}_i(m,n) - \text{pixel}_j(m,n)$ but since differential images have, as original ones, 8 bits resolution the difference value can not be directly assigned because of negative

Figure 5. Xt plane division result for 64 frames cine angiogram (152 × 64).

differences. A similar coding scheme to JPEG-LS was used – Equation 1 [5]:

$$pixel(\varepsilon) = 2|\varepsilon| - \mu(\varepsilon) \qquad (1)$$

where $\mu(\varepsilon) = 1$ if $\varepsilon < 0$ or 0 otherwise

This scheme may, in some cases, seem to have the disadvantage that the *differential image* needs 2 bytes per pixel but this has never happened in all the procedures tested because their pixel values are relatively small. The reconstruction operation is straightforward since an even $pixel(\varepsilon)$ value indicates a positive difference and an odd indicates a negative difference. It is important to focus that the *differential images* consider both spatial and temporal redundancy since the original *xt* images have intrinsic temporal information. Two consecutive difference images of a 64 frames cine angiogram are presented on Figure 5. The *differential images* were compressed with several compressors, including state of the art image compression algorithms in order to obtain the best compression ratio possible and to compare their performance when *differential images* are provided to them. Because the *differential images* are not typical images we also decide to include non-image non commercial lossless compression algorithms in our analysis.

Finally, the analysis was extended to video lossless codecs to compare the compression ratios with the ones obtained with still image lossless compressors in both methods mentioned above.

2 RESULTS

Several compression algorithms were tested in this work in order to compare the best image and non-image based lossless compressors. Implementations relied on publicly available software libraries. The compression effectiveness presented in Table 1 and Table 2 was obtained using the ratio between the raw pixel data size (usually 262144 Bytes in XA) and the compressed file size since it is more indicative of compression performance in the real world.

The tables present the compression results achieved by applying state of art lossless compressors in the original XA frames (Table 1) and general lossless compressors in *differential images* (Table 2). The general

Table 1. Angiographic compression results (Original images).

Number of procedures	278
Number of frames	23152
Total of space volume	6.077.106.590 bytes (6 GB)
Compressor	Average compression ratio
Lossless JPEG sv1	2,67
PPMd	2,96
JPEG 2000	3,37
JPEG-LS	3,52
PAQ8i	3,78
BMF2.0	3,80

Table 2. Angiographic compression results *(Differential images)*.

Compressor	Average compression ratio
BMF2.0	2,74
BZ2	2,99
PPMd	3,38
PAQ8f	3,43

PPMd (Prediction by Partial Matching) compressor has been chosen as the reference non-image based lossless compression algorithm.

From Table 1 and Table 2 observation we can conclude several aspects:

– The JPEG-LS is the best lossless compressor between the three image compressors adopted by DICOM standard on Angiographic studies.
– However, there are two image lossless codecs (BMF2.0 and PAQ8i) that compress more than actual codecs supported by the DICOM standard. Yet, their performance decreased on *differential images* (see BMF2.0 in Table 2)
– The *differential method* with PPMd compressor achieves, in average, 27% higher compression than JPEG lossless sv1. It also has a slightly better compression ratio than JPEG2000 and loses only 4% on JPEG-LS.
– The general predictive compressor PPMd (Bold), i.e. a non-image based compressor, performs excellent results with *differential images*, approximately equal to the best DICOM standard codec. We must emphasize to the fact that JPEG-LS was specially developed for image compression.
– The PAQ8 family general compressors were among the best on both strategies. However, they were excluded as practical solution because they are heavy time and resource consuming compressors so they are used in this test only as benchmark compressors.

After several tests we have concluded that expected higher compression ratios were not possible with the

Table 3. Video lossless codec results.

Encoder/Decoder	Average compression ratio
Alparysoft lossless [6]	3,02
Lagarith lossless [7]	3,37
MSU lossless [8]	3,43

differential image method because of noise presence in the images. Some previous experiences with other image modality, namely Multislice cardiac Computerized Tomography, strength that idea.

Finally, to realize the video experiment three non commercial and freely available lossless video codec were applied on original XA DICOM procedures. The video codec results are presented in Table 3.

MSU lossless codec has a slightly better average compression ratio but the compression time is very high, disabling the usage of this codec (with maximum compression configuration) in real environment. This codec can be tuned so that the time will be similar (but still slower) than the others codecs but the compression ratio falls between Lagarith and Alparysoft codec. Lagarith codec was noticeably faster than the other two.

3 CONCLUSION

This study shows that better compression ratios can be obtained using other lossless image compressors (PAQ8 family and BMF2.0) than the ones allowed by the DICOM standard (JPEG family). The purpose of this work was to test several compressors including the ones used nowadays by Picture Archiving and Communications Systems.

This paper also presents an alternative compression strategy based on preprocessing the pixel raw data of XA DICOM images and using publicly available non-image based compressor libraries. Relevant compression ratios were obtained with this method, taking in consideration that we are working with lossless products. The results obtained suggest that lossless compression of this kind of images is reaching a limit with little improvement from previous compression schemes (compression ratios between 3 and 4) and that general purpose compression schemes are acceptable for image compression depending on the way that data is provided.

This study also suggests that image noise removal is crucial to obtain higher compression ratio but this must be accomplished without compromising the full diagnostic information contained in the image. Therefore, if noise bits removal doesn't affect image diagnostic accuracy and confidence, storage space can be saved. The method, however, will become visually lossless. Bit discarding techniques such as the one described in [9, 10] could result in noise removal without affecting the visual content and permitting a considerable increment in compression ratio.

Combined strategies of lossy and lossless compression can also play a significant role in the compression of medical images and achieve much greater impact than pure lossless methods [11].

Available lossless video codecs do not improve compression ratio of XA since they do not take into account the fact that XA files are 8 bits grayscale and it is not necessary to save three colorspace components. The development of a grayscale video codec is of great importance and could lead to better compression ratios. Since X-ray imaging is necessarily grayscale, the applications of this codec could be extended to almost every X-ray medical imaging modality.

REFERENCES

1. Costa, C.M.A., et al., *Himage PACS: A new approach to storage, integration and distribution of cardiologic images.* Proceedings of the SPIE, 2004. **5371**: pp. 277–287.
2. Clunie, D.A., *Lossless compression of grayscale medical images – effectiveness of traditional and state of the art approaches.* Proceedings of the SPIE, 2000. **3980**.
3. Kivijarvi, J., et al., *A comparison of lossless compression methods for medical images.* Computerized Medical Imaging and Graphics, 1998. **22**: pp. 323–339.
4. Kerensky, R.A., et al., *American College of Cardiology/European Society of Cardiology international study of angiographic data compression phase I: The effects of lossy data compression on recognition of diagnostic features in digital coronary angiography.* Journal on the American College of Cardiology, 2000. **35**(5): pp. 1370–1379.
5. Weinberger, M.J., G. Seroussi, and G. Sapiro, *The LOCO-I Lossless Image Compression Algorithm: Principles and Standardization into JPEG-LS.* IEEE Transactions on Medical Imaging, 2000. **9**(8): pp. 1309–1324.
6. http://www.alparysoft.com/.
7. http://lags.leetcode.net/codec.html.
8. http://www.compression.ru/video/.
9. Keh-Shih, C., et al., *Content Analysis in Clinical Digital Images.* Radiographics 1994. **14**(2): pp. 397–403.
10. Chen, T.-J., et al., *A Blurring Index for Medical Images.* Journal of Digital Imaging, 2006. **19**(2): pp. 118–125.
11. Ashraf, R. and M. Akbar, *Absolutely lossless compression of medical images.* Proceedings of the 2005 IEEE Engineering in Medicine and Biology 27th Annual Conference, 2005: pp. 4006–4009.

Computational Vision and Medical Image Processing – João Tavares & Natal Jorge (eds)
© 2008 Taylor & Francis Group, London, ISBN 978-0-415-45777-4

Direct approach for constructing volumetric meshes from 3D image data

V. Bui Xuan, P.G. Young, G.R. Tabor & T. Collins
University of Exeter, Exeter, United Kingdom

R. Said
Simpleware Ltd., Exeter, United Kingdom

M. Gomes & V.M.B. Martins-Augusto
Norcam, Porto, Portugal

ABSTRACT: This paper presents an advanced approach for directly generating volumetric meshes from 3D images, i.e. constructing CAD representation and/or surface mesh of the computational domain. The approach combines the geometric detection and mesh creation stages in one process; it also allows the construction of hybrid tetrahedral-hexahedral or tetrahedral dominant meshes.

1 INTRODUCTION

Although a wide range of mesh generation techniques are currently available these, on the whole, have not been developed with meshing from segmented 3D imaging data in mind. Meshing from 3D imaging data presents a number of challenges but also unique opportunities for presenting more realistic and accurate geometrical description of the computational domain. The majority of techniques developed up-to-date involve generating a surface model (either in a discretized format, e.g. triangular mesh, or continuous format, e.g. bi-cubic patches) from the scan data first and then using a third-party mesh generation package to create the volume mesh. Such an approach, referred to as 'CAD-based approach', can be very time consuming and not very robust and virtually intractable for the complex topologies. A more 'direct approach' is to combine the geometric detection and mesh creation stages in one process. This approach involves identifying volumes of interest (segmentation of 3D image) and then directly generating the volumetric mesh based on an orthotropic grid intersected by interfaces defining the boundaries.

2 MESH GENERATION FROM 3D IMAGING DATA

Automatic generation of meshes from image data is a very significant area in bioengineering research, and there are a number of different methods being investigated. In general, two types of approaches can be distinguished: 'CAD-based approach' and 'direct- approach'.

2.1 CAD-based approach

One approach is to use the scan data to define the surface of the domain and then create elements within this defined boundary (Cebral et al, 2001). The element creation process was initially quite simplistic, but automatic meshing has been an area of great interest in the FEA/CFD community, with applications well beyond the biomedical field, and so reasonably robust algorithms are now available (Antiga et al, 2002).

These techniques do not easily allow for more than one domain to be meshed as multiple surfaces generated are often non-conforming with gaps or overlaps at interfaces where one or more structures meet.

2.2 Direct approach

A more direct approach would be to combine the geometric detection and mesh creation stages in one process. In this paper we use a methodology that generates 3D hexahedral or tetrahedral elements throughout the volume of the domain, thus creating the mesh directly. This technique was originally developed for FE analysis of bones, for both stress and vibration analysis (Young, 2003), and has been implemented as a set of computer codes (ScanIP, +ScanFE). Since FE and FV meshes are conjugate structures, the same techniques can be used to output a FV mesh (cell/face representation, rather than point/edge representation).

3 GENERATING A MESH

The steps involved in the generation and processing of finite element models based on medical imaging data are:

3.1 *Image processing*

The process of generating a mesh involves first segmenting the different volumes of interest (VOI) from the 3D data. Both semi-automated and manual techniques are available within ScanIP, as well as a range of alternative image processing packages, to provide segmented masks. Techniques include noise filters, three dimensional thresholding tools through to bitmap painting.

An extensive range of image processing and meshing tools can be used to generate highly accurate models based on data from any 3D medical imaging modality such as MRI, Ultrasound and CT. Features of particular interest include: Segmentation tools including Level Set Methods; Metal Artifact Reduction (MAR) Algorithms; Volume and topology preserving smoothing; Robust multi-part surface mesh/STL generation; Accuracy of meshed topology/morphology is only contingent on image quality. The geometry of the structure is reproduced in the finite element mesh at sub-voxel accuracy.

3.2 *Volumetric mesh generation*

The segmented volumes of interest are then simultaneously meshed based on an orthotropic grid intersected by interfaces defining the boundaries. In effect a base Cartesian mesh of the whole volume defined by the sampling rate is tetrahedralised at boundary interfaces based on cutting planes defined by interpolation points. Smooth boundaries are obtained by adjusting the interpolation points in one, or a combination, of two ways: by setting points to reflect partial volumes or by applying a multiple material anti-aliasing scheme. The process results in either a mixed tetrahedral/hexahedral mesh or a pure tetrahedral mesh and incorporates an adaptive meshing scheme. The adaptive meshing scheme preserves the topology but reduces the mesh density where possible towards the interior of the mesh by agglomerating hexahedra into larger hexahedra and generating transitional tetrahedra. The approach is fully automated and robust, creating smooth meshes with low element distortions regardless of the complexity of the segmented data. The approach adopted allows for an arbitrary number of different volumes to be meshed. As neighboring submeshes share a common cutting surface, this ensures a node to node correspondence at the boundaries between different meshed volumes, thus trivially satisfying the geometrical constraints at the boundary (in other word no gaps or overlaps).

In addition the numbering scheme will be the same for the mesh interface as seen from either side – a topological requirement on the meshes. This is also easily fulfilled using this approach. Surface models suitable for rapid prototyping can also be generated which are exact representations of the meshed domains. Amongst other things this would allow experimental tests, possibly at different scales using dimensional analysis, to be carried out to provide experimental corroboration of numerical results.

The approach implemented in the mesher has several advantages over traditional finite element model development techniques, such as: Mesh generation from data sets of arbitrary geometric complexity; Topology and volume preserving smoothing algorithms; Meshing of multiple structures/regions of interest; Conforming contact surfaces/interfaces; User definable adaptive meshing; Material properties assigned to mesh based on signal strength. For volume data obtained from 3D imaging techniques the signal strength within an inhomogeneous medium (such as variable density foams, bone...) can, in some cases, be related to the material properties. Well-established and corroborated relationships have been obtained and used in the case of CT scan data of bone where the Hounsfield number can be correlated to the apparent density, which in turn can be mapped to the Young's Modulus (Schmitt et al, 2001; Zannoni et al, 1998; Ploeg et al, 2001; Taylor et al, 2002; Zannoni et al, 1998).

4 EXAMPLES

4.1 *Computer model of the human eye*

4.1.1 *Methodology*

High resolution in vivo MRI scans of a 29 year Caucasian female was obtained using both a surface and a head coil on a Philips Gyroscan 1.5 Tesla imager. The following structures were segmented from the 3D data set by a Physician: the globe and optic nerve, the bony orbit, the eyelids and facial soft tissues, the extra-ocular muscles (Figure 1).

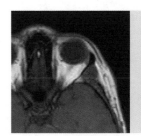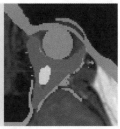

Figure 1. MRI scan (left) and segmentation (right) of the eye.

A number of finite element models were generated based on the segmented image data (Figure 2). Each structure was meshed with mixed hexahedral and tetrahedral elements. Interfaces between the different biological sub-structures can be either defined as merged (shared nodes across common boundaries/interfaces) or nodes can be de-coupled and contact surfaces spawned automatically at these interfaces. The contact surfaces are particularly robust as the master and slave contact faces are paired. Structures were either exported as volumetric meshes or as surface meshes as required (e.g. the bony orbit can most likely be modeled as a rigid structure defined by surface shell rather than as a volumetric mesh thereby providing some computational saving). Material properties within each and every structure could be straightforwardly assigned based on signal strength – in other words globe properties could be based on parent signal strength in image. However this facility is contingent on the user providing an appropriate relationship or mapping function between signal strength and material properties of interest – e.g. for the globe a relationship between signal strength and Young's modulus. Such relationships can be obtained empirically through combined experimental and imaging tests. A number of analyses were carried out to demonstrate the robustness of the models for simulation purposes and these demonstrate the remarkable sophistication of biological models which can now be generated based on in vivo data.

4.2 Compression in the human spine

4.2.1 Overview

Work related injuries to the spine are becoming more commonplace in the UK with over one million people suffering from work induced musculoskeletal disorders of the lower back, each year. New image processing and mesh generation techniques can now aid in the analysis of such biological structures, particularly where the geometry is complex. Novel proprietary techniques have been developed for the automatic generation of volumetric meshes from 3D image data including image datasets of complex structures composed of two or more distinct materials.

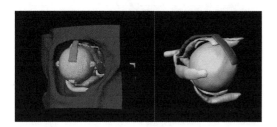

Figure 2. FE mesh of the eye with bone and only the eye ball.

4.2.2 Methodology

A 3D volumetric FE mesh of the human lumbar spine was generated from in vivo high resolution MRI scan data of 1 mm in-plane and slice-to-slice separation. A mesh of 535 610 elements was generated in just twenty minutes and the anatomical details segmented in the model included five vertebrae, the annulus fibrosus, nucleus pulposus and the cartilaginous end plates (Figure 3).

A contact surface was created on the surface of the vertebrae between the superior and inferior articular processes. With a fixed boundary condition applied to the lower end of the model, a compressive strain was applied to the top of the spine in order to simulate a healthy young adult carrying a heavy load. The FE analysis took just under two hours on a pc and the results presented the individual pressure response of the components of the spine (Figure 4).

In this case, the tools have enabled the accurate analysis of compression to the human spine and will contribute significantly to the continued understanding of lower back pain.

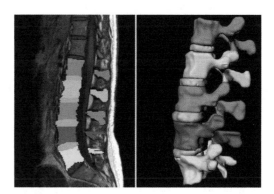

Figure 3. Segmentation and 3D view of lumbar spine.

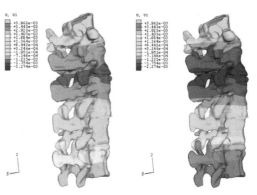

Figure 4. Compression of lumbar spine.

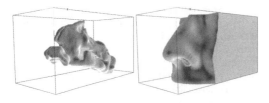

Figure 5.   Mesh of nasal cavities and the whole face.

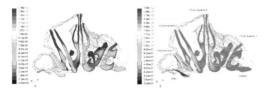

Figure 6.   Velocity magnitude and relative humidity.

4.3   *Air flows through the human nasal cavities*

4.3.1   *Overview*
This study reports on air flow, relative humidity and absolute temperature using a geometrically accurate model of the human nasal cavity. The aim of this analysis was to investigate the air conditioning capacities of the nasal passages using computational fluid dynamics (CFD) and image-based meshing using actual patient-specific geometric data.

4.3.2   *Methodology*
Image-based meshing technology was used to construct a physiologically realistic geometry of the human nasal passages from MRI scans. The data was segmented into three masked regions: the nasal passages, the face, and the outside (Figure 5).

   A computational mesh of the nasal cavities was then created using and exported to Fluent. During the export operation the contact surfaces between different masks and between masks and the exterior of the data field were different types of boundary. The mask of the nasal cavities had three contact surfaces. The contact between the nasal cavities and the face was defined as a wall, the contact between the nasal cavities and the outside was defined as an inlet, and the contact between the nasal cavities and the exterior was defined as an outlet. In Fluent, a steady state aspiratory flow calculation was performed with an inlet velocity of $u = 1.45$ m/s, appropriate for quiet tidal breathing. The inlet air boundary conditions were set equal to those of ambient air with a $\varnothing = 20\%$ and $T = 298$ K, giving $Y_\upsilon = 0.004$.

4.3.3   *Results*
The computed results (Figure 6) were found to be insensitive to the number of cells in the mesh, demonstrating mesh independence of the solution. Both steady state and transient calculations have been performed and presented, with the results being in good agreement with existing experimental data and physically realistic. During this analysis it was assumed that the temperature and relative humidity of the nasal walls remained constant no matter how much heat or moisture was transfers to or from them and the respiratory air. The present work has demonstrated that the human nasal cavities are more than capable of conditioning aspiratory air to alveolar conditions. The flow calculations also demonstrated that significant amounts of heat and moisture are recovered by the posterior section of the nasal cavities during expiration.

5   CONCLUSIONS

The ability to automatically convert any 3D image dataset into high quality meshes, is becoming the new modus operandi for anatomical analysis. The techniques guarantee the generation of robust, low distortion meshes from 3D data sets for use in finite element analysis (FEA), computational fluid dynamics (CFD), computer aided design (CAD), and rapid prototyping (RP).

REFERENCES

Antiga, L., Ene-Iordache, B., Caverni, L., Cornalba, G.P., & Remuzzi A. Geometric reconstruction for computational mesh generation of arterial bifurcations from ct angiography. Computerized Medical Imaging and Graphics, 26:227–235, 2002.

Cebral, J.R., & Loehner, R. From medical images to anatomically accurate finite element grids. Int.J.Num.Methods Eng., 51:985–1008, 2001.

Ploeg, H., Taylor, W., Warner, M., Hertig, D., & Clift, S. FEA and bone remodeling after total hip replacement. BENCHmark, 2001.

Schmitt, J., Meiforth, J., & Lengsfeld, M., Development of a hybrid finite element model for individual simulation of intertrochanteric osteomies. Medical Engineering and Physics, 2001, V. 23, pp.529–539.

Taylor, W.R., Roland, E., Ploeg, H., Hertig, D., Klabunde, R., Warner, M.D., Hobatho, M.C., Rakotomanan, L., & Clift, S.E. Determination of orthotropic bone elastic constants using FEA and model analysis. Journal of Biomechanics, 2002, V. 35, pp. 767–773.

Young, P. Automating the generation of 3D finite element models based on medical imaging data: application to head impact, Conference Paper 3D Modelling, Paris, France, 2003.

Zannoni, C., Mantovani, R., & Viceconti, M. Material properties assigned to finite element models of bones structures: a new method. Medical Engineering and Physics, 1998, V. 20, pp. 735–740.

Zannoni, C., Viceconti, M., Pierotti, L., & Cappello, A. Analysis of titanium induced CT artifacts in the development of biomechanical finite element models. Medical Engineering and Physics, 1998, V. 20, pp. 653–659.

Computational Vision and Medical Image Processing – João Tavares & Natal Jorge (eds)
© 2008 Taylor & Francis Group, London, ISBN 978-0-415-45777-4

Analysis of manual segmentation in medical image processing

K. Tingelhoff, K.W. Eichhorn, I. Wagner & F. Bootz
Department of Otolaryngology, Head and Neck Surgery, University of Bonn, Germany

M.E. Kunkel, A.I. Moral, R. Westphal, M. Rilk & F. Wahl
Institute for Robotics and Process Control, Technical University of Braunschweig, Germany

ABSTRACT: Manual segmentation is often used as gold standard for evaluation of automatic segmentation. The purpose of this paper is to describe the problems of manual segmentation and the dubiety of manual segmentation as a gold standard. We realized two experiments in which we determined inter- and intraindividual variability. In the first one ten ENT surgeons and ten medical students segmented the right maxillary sinus and ethmoid sinuses manually on a standard CT dataset of a human head. In the second experiment two participants outlined maxillary sinus and ethmoid sinuses five times consecutively. The manual segmentation was accomplished with custom software. The first experiment shows the variability of manual segmentation which can be caused by components like experience, different interpretations of CT data or different levels of accuracy. The second experiment shows varying results even if one ENT specialist or one manual segmenta-tion expert segments the dataset five times consecutively.

Keywords: Manual segmentation; evaluation; paranasal sinuses; computed tomography.

1 INTRODUCTION

Segmentation of medical image data achieves increasing importance in different medical fields. It is used for diagnosis of diseases, surgical planning, surgical simulation, radiotherapy or for guiding minimal invasive surgery.

1.1 Segmentation of medical image data

Segmentation algorithms often analyze medical image data like computed tomography (CT), magnetic resonance tomography (MRT) or x-ray image data. Problems of medical image processing are poor contrast, noise or artifacts. Besides these difficulties which are caused by the image data itself the pathological deformation, complexity and variability of anatomical structures are basic problems for segmentation algorithms.

Current research and industry developments for ENT surgery explore robot-assisted systems (Tingelhoff et al. 2007) or develop navigated control (Strauss et al. 2005). These assisting systems need a workspace definition which is realized by segmentation.

1.2 Manual Segmentation

Manual segmentation is often believed to be accurate if the specialist segmenting the object is very experienced. Beyond this, manual segmentation delivers individual, not reproducible results. Manual segmentation is very time-consuming and it is not suitable for the surgical workflow. Therefore automatic segmentation algorithms are developed for medical image data with increasing success.

The evaluation of segmentation results is getting more and more important over the last years. Especially in medical image processing the segmentation results must be evaluated before they can be used for e.g. radiotherapy.

The evaluation of segmentation results can be realized by analytical or empirical analysis (Haralick 2000, Hong et al. 1998). Another evaluation method is the quality review by human specialists (Salah et al. 2005). In medical image processing segmentation results are usually compared with a reference image, the gold standard (Pohle et al. 2001) which is often generated by one manual segmentation.

1.3 Goals of the study

The following aims should be reached by our experiments:

1. Interindividual variability
2. Intraindividual variability
3. Possible reasons for the variability
4. Consequences for generating gold standards in medical image processing.

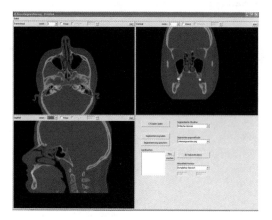

Figure 1.   Manual segmentation software.

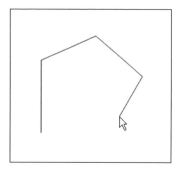

Figure 2.   Segmentation line with sparse points.

2   MATERIALS AND METHODS

2.1   *Software*

We used custom software for manual segmentation (Fig. 1). The software can load and display DICOM datasets in coronal, transversal and sagittal view. The user can click through the dataset slice by slice for each view. For manual segmentation the software provides line segmentation. The user marks several points and the software draws straight lines between two points each (Fig. 2).

2.2   *CT dataset*

For the experiments we used a CT dataset which was acquired by a 16-slice spiral CT from Philips. The CT dataset of the paranasal sinuses consists of 167 transversal slices, each 2.0 mm thick, with a resolution of 512×512 pixels. The pixel spacing is 0.457 mm $\times\ 0.457$ mm. The dataset was acquired from a female patient age 27. The dataset shows swollen mucosa of the posterior ethmoidal sinuses. The roots of teeth protrude into the maxillary sinus and the patient has aplastic frontal sinuses.

2.3   *Participants*

20 participants took part in the first study, ten ENT surgeons (ENT Surg) and ten medical students (Stud). Five of ten ENT surgeons are specialist whereas the other five were junior doctors. Four students were in their last year at university and six were in the advanced study period.

2.4   *Experimental procedure*

Two different experiments were realized. On the one hand ten ENT surgeons and ten students segmented the borders of the right maxillary and ethmoidal sinuses. During the second experiment two participants outlined the border of the right maxillary and ethmoidal sinuses five times consecutively.

After a short software demonstration all participants got the same written task with explicit segmentation criteria to ensure the comparability of the results. The segmentation should be realized as bone segmentation, so the interior bony border which limits the sinuses should be segmented with closed outlines. All participants outlined the borders in the coronal view, which makes it easier to compare the segmentation results. The cells of the ethmoidal sinuses should be outlined en block. All participants segmented on their own without communicating to each other.

2.5   *Analysis of manual segmentation*

After manual segmentation the data needed to be postprocessed. The contour is closed and afterwards the maxillary and ethmoidal sinuses were separated and filled.

To evaluate manual segmentation results we propose four indices: a) volume, b) extension in x-, y- and z-direction c) visual analysis and d) time.

Volume is computed separately for both sinuses. For each sinus the number of voxels which belong to the sinus is counted and the number of voxels is multiplied by the volume of one voxel in cm^3. The extension in x-, y- and z-direction is measured parallel to the object coordinate system of the CT dataset. For the extension in z-direction we searched the first and last voxel in z-direction and computed the distance of the z-coordinate. For the visual analysis we used 3D-reconstruction and exemplary 2D-slices.

2.6   *Visualization*

For 3D-visualization we used the postprocessed manual segmentation results. Gaussian filter was applied five times consecutively in order to smooth the binary segmentation results. Afterwards marching cubes (Lorensen et al. 1987) is used for generating the 3D-mesh (Fig. 8). The Marching cubes algorithm computes a polygon mesh from a 3D-voxel-based graphic.

3 RESULTS

20 participants segmented right maxillary and ethmoidal sinuses. The segmentation took between 30 and 120 minutes in average 73.7 minutes. Two participants segmented the CT dataset five times. It took between 40 and 70 minutes in average 52.5 minutes with a standard deviation of 9.3 minutes (Fig. 7).

3.1 Experiment 1

Figures 3 and 4 present the results of experiment one. The total volume of maxillary and ethmoidal sinuses varies between 15.4 cm^3 and 23.6 cm^3. The average size of maxillary sinus is 13.7 cm^3 with a standard deviation of 0.7 cm^3. The size of ethmoidal sinuses varies between 2.6 cm^3 and 9.5 cm^3. The mean size of ethmoidal sinuses is 4.3 cm^3 with a standard deviation of 1.5 cm^3.

Figure 4 shows the extension of ethmoidal sinuses in x-, y- and z-direction parallel to the object coordinate system of the CT dataset. The major variance appears in y- and z-direction. The mean extension in y-direction is 37.2 mm with a standard deviation of 3.6 mm. In z-direction the values for mean extension and standard deviation are 28.9 mm and 3.7 mm.

Visual analysis shows different segmentations of hiatus semilunaris. One participant did not segment hiatus semilunaris as part of maxillary or ethmoidal sinuses. Another participant outlined hiatus semilunaris as part of maxillary sinus and another one segmented it separately. Furthermore there are different ways to segment ethmoidal sinuses as a small bloc, as a big bloc or separately.

Our experiments show the change of accuracy of one subject during the segmentation procedure. The first slices were segmented accurately and the last slices are characterized by few corners and long straight lines. The segmentation results during the last slices are not very accurate.

3.2 Experiment 2

Figures 5, 6 and 7 present the result of experiment two. Figure 5 shows the volume of maxillary and ethmoidal sinuses of both participants who segmented the CT dataset five times. The maxillary sinus volume was segmented with an average size of 13.4 cm^3 and a standard deviation of 0.4 cm^3. The mean size of ethmoidal sinuses is 4.0 cm^3 with a standard deviation of 0.5 cm^3. The variability of the volumes in experiment two is smaller than the one in experiment one.

Figure 6 shows the extension of ethmoidal sinuses. The mean extension in z-direction is 24.0 mm with a standard deviation of 1.5 mm.

The visual analysis of experiment two shows three different segmentation results of the ENT surgeon during first, second and fifth segmentation. During the

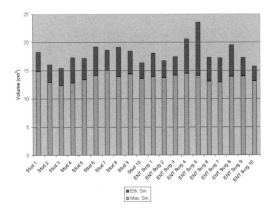

Figure 3. Experiment 1: volume of maxillary sinus and ethmoidal sinuses.

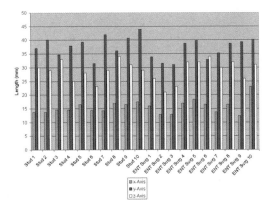

Figure 4. Experiment 1: extension of ethmoidal sinuses parallel to each of the three axes of the CT coordinate system.

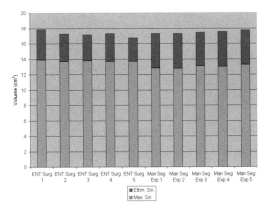

Figure 5. Experiment 2: volume of maxillary and ethmoidal sinuses of an ENT surgeon and a manual segmentation specialist who segmented the dataset five times each.

153

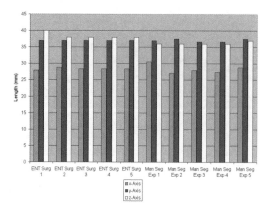

Figure 6.   Experiment 2: extension of ethmoidal sinuses in x-, y- and z-direction.

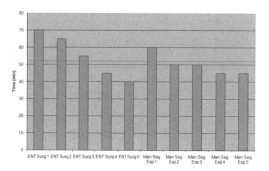

Figure 7.   Experiment 2: time which was needed by the ENT surgeon and the manual segmentation expert who segmented the dataset five times each.

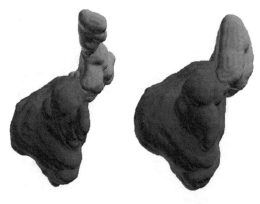

Figure 8.   3D-reconstruction of two different segmentation results: (a) student 7; (b) ENT surgeon 4. Maxillary sinus (light gray); ethmoidal sinuses (dark gray).

first segmentation he outlined hiatus semilunaris as part of maxillary sinus. In the second run he segmented it separately and during the last segmentation he did not outline it.

4   DISCUSSION

In this paper we present two types of experiments which show the variability of manual segmentation.

4.1   Aim 1: Interindividual variability

For the manual segmentation we decided to concentrate on the right maxillary and ethmoidal sinuses because it would have taken too much time to segment all paranasal sinuses (8–10 hours).

The results (Figs. 3, 4) show great variability of manual segmentation between all participants. The maxillary sinus volume varies marginally for all participants (Fig. 3) but they are in the range of $20.5 \pm 9.2\,\mathrm{cm}^3$ which is described as maxillary sinus volume in literature (Ikeda et al. 1998). The standard deviation of the ethmoidal ($1.5\,\mathrm{cm}^3$) and maxillary volume ($0.7\,\mathrm{cm}^3$) shows that the variance of the ethmoidal volume is higher. The higher variance can be caused by different data interpretation due to complex anatomy. Another problem are thin bony structures between the ethmoidal cells. Either they can not be seen in CT data because the bony structure is smaller than the resolution in z-direction or these structures can be seen with little contrast which makes segmentation difficult. However the contour of the maxillary sinus is characterized by high contrast of the bony border, so it is easier to segment.

The segmentation variability of ENT surgeons and students is similar whereas the maximal deviations were caused by two ENT surgeons (ENT Surg 4 and 5).

4.2   Aim 2: Intraindividual variability

Analyzing the segmentation times of experiment two we noticed decreasing time for both participants. The time of the inexperienced participant, the ENT surgeon, decreased about 42.9% whereas the time of the manual segmentation expert decreased about 25.0%.

In experiment two we measured smaller volume variances of ethmoidal and maxillary sinuses (Fig. 5) than in experiment one (Fig. 3). Ashton et al. also describes that the interindividual variability is considerably higher than intraindividual variability, which is equal to our results (Ashton et al. 2003). The manual segmentation expert segmented the CT-dataset without great variability of the volume (fig. 5) or the x-, y-, z-extension (fig. 6). The variability is smaller than the one segmented by the ENT surgeon, because the manual segmentation expert has much experience with the line segmentation tool which makes it easier to draw accurate segmentation borders.

The ENT surgeon produced a volume variability because he segmented hiatus semilunaris in different ways. These differences were not generated by the manual segmentation specialist. She segmented paranasal sinuses several times before, so the learning effect is lower than for the ENT surgeon who segmented right maxillary and ethmoidal sinuses for the first time.

The segmentation expert segmented the data with closed contours, so there was no contour closing necessary. The standard deviation (fig. 3, 5) of ethmoidal sinuses ($0.5\,cm^3$) and maxillary sinus ($0.4\,cm^3$) are lower than the one of experiment one ($1.5\,cm^3$ for ethmoidal sinuses and $0.7\,cm^3$ for maxillary sinus). Furthermore the extension variances are low too (fig. 4). The results of the manual segmentation expert are characterized by very accurate and curved contours and they visually seem to be perfect segmentation results.

Nevertheless the segmentation expert produces little variances during five segmentations, so the results are not reproducible. We think that they could not be used as gold standard.

4.3 Aim 3: Possible reasons for the variability

There are multiple reasons for the variability of the segmentation results. Basic problems arise from software handling, from line segmentation and problems based on the mouse interaction. These problems can have different influences on the segmentation results depending on the person who segments the image data.

Furthermore the results depend on the accuracy of each participant. Some candidates outlined the borders accurately and took more time than others. Manual segmentation is monotone and exhausting which also affects the segmentation accuracy.

Besides these general difficulties of manual segmentation there are special problems in medical image data. Experiment two shows that even an ENT surgeon analyzes CT data in different ways. These different interpretations are not wrong, but the participants segmented the image data with different intentions. Some candidates segmented the ethmoidal sinuses from a surgical and some from the ontogenetic point of view.

Most problems evolved by segmenting hiatus semilunaris and ethmoidal sinuses. These problems cause different segmentation results and great variability in the volume of ethmoidal sinuses. The variability is caused by complex anatomy of the ethmoidal sinuses and the junction between maxillary sinus and ethmoidal sinuses.

Certainly the complex anatomy is a great problem for automatic segmentation algorithms.

Inter- and intraindividual variability is caused by multiple reasons. Manual segmentation problems are influenced by accuracy, interaction problems with the mouse, anatomy approximation by shape primitives, exhaustion and image resolution. Furthermore the variability is caused by the complex anatomy. Different interpretation of the anatomy and different point of views (e.g. surgical or ontogenetic) affect different segmentations.

At the moment we can not evaluate which factor basically causes the manual segmentation variability.

4.4 Aim 4: Consequences for generating gold standards in medical image processing

For the evaluation of automatic segmentation in medicine reference images are often used.

Considering these different interpretations of medical image data, it is very difficult to generate a gold standard. Therefore manual segmentation by one radiologist or physician is often used for evaluation purpose in literature. Inter- and intraindividual variances of our experiments show that manual segmentation is not adequate as gold standard.

Possible solutions are described in (Niessen et al. 2000, Chalana et al. 1997). They compute gold standards based on multiple expert segmentations.

5 CONCLUSION

In our study we approved that manual segmentation is time-consuming and not reproducible, therefore automatic and standardized segmentation algorithms are needed. Standardized segmentation results are absolute necessary for medical applications like radiotherapy or robot-assisted surgery.

The visually best results of manual segmentation were produced by the manual segmentation expert. She segmented the dataset five times consecutively and delivers least variances. Nevertheless variances appear during five segmentations, so the results are not reproducible and they could not be used as gold standard. Therefore standardized and reproducible segmentation results are absolutely necessary for medical applications.

For the evaluation of medical image segmentation reference images based on multiple manual segmentation results could be used, even though they are no real gold standard.

REFERENCES

Ashton, E.A., Takahashi, C., Berg, M.J., Goodman, A., Totterman, S. & Ekholm, S. 2003. Accuracy and reproducibility of manual and semiautomated quantification of MS lesions by MRI. J. Magn Reson Imaging, vol. 17, no. 3, pp. 300–8.

Chalana, V. & Kim, Y.M. 1997. A methodology for evaluation of boundary detection algorithms on medical images, IEEE Transactions on Medical Imaging, vol. 16, no. 5, pp. 642–652.

Eichhorn, K.W.G., Tingelhoff, K., Wagner, I., Westphal, R., Rilk, M., Wahl, F.M. & Bootz, F. 2007. Evaluation of Force Data with a Force/Torque Sensor during FESS. A Step towards Robot Assisted Surgery, ESBS Prag.

Haralick R.M. 2000. Validating image processing algorithms, Proc. of SPIE, Medical Imaging, vol. 3979, part 1, pp. 2–16.

Hong, L., Wan, Y. & Jain, A. 1998. Fingerprint Image Enhancement: Algorithm and Performance Evaluation, in K.W. Bowyer, P. J. Phillips: Empirical Evaluation Techniques in Computer Vision, pp. 117–134.

Ikeda, A., Ikeda, M. & Komatsuzaki, A. 1998. A CT Study of the Course of Growth of the Maxillary Sinus: Normal Subjects and Subjects with Chronic Sinusitis. ORL, vol 60, no. 3, pp. 147–152.

Kunkel, M.E., Moral, A.I., Westphal, R., Rode, D., Rilk, M. & Wahl F.M. 2007. Using robotic systems in order to determine biomechanical properties of soft tissues, Proc. 2nd Symposium on Applied Biomechanics, Medicine Meets Engineering, Regensburg, Germany.

Lorensen, W.E. & Cline, H.E. 1987. Marching Cubes: A high resolution 3D surface construction algorithm. In: Computer Graphics, vol. 21, no. 4, pp. 163–169.

Niessen, W.J., Bouma, C.J., Vincken, K.L. & Viergever, M.A. 2000. Viergever: Performance Characterization in Computer Vision, Kapitel Error metrics for quantitative evaluation of medical image segmentation, pp. 275–284. Kluwer Academic Publishers.

Pohle, R. & Toennies, K.D. 2001. A New Approach for Model-Based Adaptive Region Growing in Medical Image Analysis, Lecture Notes in Computer Science, vol. 2124, pp. 238–245.

Salah, Z., Bartz, D., Dammann, F., Schwaderer, E., Maassen M. & Strasser W. 2005. A Fast and Accurate Approach for the Segmentation of the Paranasal Sinus, in: Proc. of Workshop Bildverarbeitung in der Medizin, pp. 93–97.

Strauss, G., Koulechov, K., Richter, R., Dietz, A. & Lueth, T.C. 2005. Navigated Control in functional endoscopic sinus surgery. Int J Medical Robotics and Computer Assisted Surgery, vol. 1 nr. 3, pp. 31–41.

Tingelhoff, K., Wagner, I., Eichhorn, K., Rilk, M., Westphal, R., Wahl, F.M. & Bootz, F. 2007. Sensor-based force measurement during FESS for robot assisted surgery. *GMS CURAC*, vol 2, nr. 1.

Computational Vision and Medical Image Processing – João Tavares & Natal Jorge (eds)
© 2008 Taylor & Francis Group, London, ISBN 978-0-415-45777-4

Customized implant optimization for maxillo-mandibular osteotomy

C. Pereira
Dep.º Eng.ª Mecânica, I.S.E.C./I.P.C., Coimbra, Portugal

F. Ventura
Dep.º Eng.ª Mecânica, F.C.T.U.C., Coimbra, Portugal

M.C. Gaspar
Dep.º Eng.ª Industrial, E.S.T./I.P.C.B., Castelo Branco, Portugal

A. Mateus
Dep.º Eng.ª Mecânica, E.S.T.G./I.P.L., Leiria, Portugal

ABSTRACT: Anatomical three-dimensional modeling invariably improves both the surgical planning and the surgery itself, which is particularly important in complex reconstructions, such as maxillo-mandibular osteotomies. The goal of this kind of surgery is to restore the proper anatomical form and functional relation in patients with congenital or acquired anomalies of the face. However, a customized implant is required if the bone defect is to fit exactly, with external contours adjusted to compensate for overlying soft-tissue disparities. This paper describes the procedure used in the design and optimization of customized implant for a specific patient who is to undergo maxillo-mandibular reconstruction after tumor ablation. Several key aspects, such as the generation of a leaner, less heavy geometry, the accurate overall positioning of implant components and the selection of the most suitable materials were considered.

1 INTRODUCTION

The combination of medical imaging, finite element analysis (FEA), computer-aided design/computer-aided manufacture (CAD/CAM) and, more recently, rapid prototyping/rapid manufacturing (RP/RM) technologies in medical applications has led to an improvement in such areas as the three-dimensional (3D) visualization of a specific anatomy, surgical planning, design and verification of implants/prostheses and their production (Jonov & Kaltman 1997, D'Urso & Redmond 2000, Chelule et al. 2000, Trikeriotis & Diamantopoulos 2006). The accurate construction of complex virtual and physical models of a patient's anatomy directly from images provided by a hospital scanner is a technique that has been used extensively in various medical applications. It has been especially useful in orthopedics to investigate certain issues with implants and the design of custom-made implants (Lohfeld et al. 2006, Xia et al. 2006) These precise models allow surgeons to study the bony structures separately from the body and to manipulate their shapes as required to achieve the desired result. Implant designs can thus be combined, viewed, manipulated, modeled, analyzed and compared with real patient data within a single environment (Jonov & Kaltman 1997).

The outcome of complex surgery depends not only on the surgical procedure but also on a multitude of factors that begins long before the actual operation, not to mention the control of the variables long after surgery. Therefore, a careful preparation of a pre-operative planning and simulation facilitates and increases the success rate of complex surgeries such as maxillo-mandibular osteotomy (Vos et al. 2006). In this kind of surgical intervention the extent of the resection, the type of missing tissue, and the precise function and contour, are all-important considerations. The generation of virtual and physical models of the patient's maxillofacial skeleton is therefore crucial, since it allow the simulation of surgical conditions in an environment that closely reproduces the actual condition.

The reconstruction of the maxillofacial skeleton implies replicating the normal volume and contour of both soft and hard tissue to produce normal form (to preserve the facial aesthetics) and functional relationship of the face, mouth, and jaws (speech and masticatory function) (Papageorge & Karabetou 1997,

Al-Sukhun et al. 2006a). Each customized implant is generated to fit the bone defect exactly, with external contours adjusted to compensate for overlying soft-tissue disparities, but there may be an exaggerated level of mechanical stability and it could use excessive amounts of material. Consequently, the patient's natural tissues will have to bear higher loads than is physically necessary (Lohfeld et al. 2006). In addition, anatomical limitations and bony morphology may compromise implant number, length, and inclination (Sadowsky 2001). Functional mandibular deformation is a multifaceted phenomenon involving an irregular structure with a complex external and internal anatomy. The deformation is of considerable significance in implant treatment where the essentially rigid and elastic osseointegrated implant/bone interface, often combined with a rigid superstructure, can be associated with high stress gradients due to jaw deformation (Al-Sukhun et al. 2006b). Furthermore, as the maxilla consists of a looser arrangement of trabecular bone, it is less capable of stabilizing and supporting implants (Sadowsky 2001).

By linking FEA to the design process, relative variations of both, in cross-section and in depth, can be analyzed and it is then feasible to generate a leaner and more lightweight design of the implant. In addition, having a finite element model makes it possible to determine the loading of the bone through the implant and, the best placement for its fixation, and to perform analyses to evaluate alternative designs or the selection of a more suitable surgical technique (Anglin et al. 1999, Stolk et al. 2002, Al-Sukhun et al. 2006a, Lohfeld et al. 2006). Moreover, the material properties of the digital model can easily be changed and the mechanical performance of the implant evaluated. This information can be incorporated into the design of the entire implant. The purpose of this work is to generate a lightweight customized implant for a specific patient who is to undergo maxillo-mandibular reconstruction after tumor ablation. Relevant aspects such as the accurate overall positioning of implant components and the selection of the most suitable material are considered in the design and optimization of the customised implant.

2 MAXILLO RECONSTRUCTIONS STATE-OF-ART

If no obturators are used neurosurgeons traditionally use bone grafts to restore the proper anatomical form and functional relationship in patients with congenital or acquired anomalies of the face. The bone graft can be harvested from several donors sites of patient body, such as the iliac crest. This medical practice has a number of drawbacks, however: at the first level there is the risk of tissue rejection and appreciable donor site morbidity. In addition, longer surgery time

and bone graft closure have been reported (Ali et al. 1995, Marx 1996, Malchiodi et al. 2006). In fact, an insufficient volume of available bone tissue coupled with the difficulty of ensuring secure fixation of the osteosynthesis plates and screws to guarantee the graft retention makes this a complex procedure. But the advent of biomodelling has made it possible to create a biomodel as a template on which a graft may be directly shaped intra-operatively. With this approach, the operating time can be dramatically reduced because the surgeon can shape the graft on the biomodel while the assistant prepares the exposure of donor site (D'Urso et al. 1999). When the graft requires a large and fairly complex shape, molding it to the desired shape is not an easy task, however. Beside this, there may not be enough autogenous bone to repair a large resected area available. In such cases, bone replacement materials are needed as an alternative to autogenous bone grafts.

Reverse engineering techniques are used to create a biomodel – customized implant – to fit the bone defect exactly, with external contours adjusted to compensate for overlying soft-tissue disparities. The previous biomodel of the defect could be made into a mould from which the implant could be cast, i.e. the model could act as a template on which some biocompatible materials are molded to generate an implant (D'Urso 2005). Rapid Prototyping and Rapid Manufacturing techniques can efficiently assist the production of customized implants using different materials. Titanium, cold-cured acrylic and hydroxyapatite are the materials traditionally recommended to replace bone (D'Urso et al. 2000, Wright et al. 2006). Naturally, each of these materials has advantages and disadvantages: hydroxyapatite exhibits excellent biocompatibility and bioactivity, since it allows bone ingrowth, but it is extremely expensive; acrylic is cheaper and easily moldable, but it presents some level of toxicity; titanium has the advantage of being biologically inert, though this is offset by its cost. However, like hydroxyapatite, is difficult to mold or shape (D'Urso 2005, Wright et al. 2006). The use of these materials in the particular case under study is analyzed and discussed below.

3 3D BIOMODELING

An important task in reverse engineering and CAD for biomedical applications, particularly for the design of customized orthopaedic implants, is the accurate mathematical modelling of 3D complex surfaces based on measured data. Here, the bone geometry can be captured as a sequence of two-dimensional (2D) cross-sections using computed tomography (CT), magnetic resonance imaging (MRI), or ultrasound imaging (UI). Besides their diagnostic application for clinical practice, these techniques provide detailed information about the geometry and physical properties of

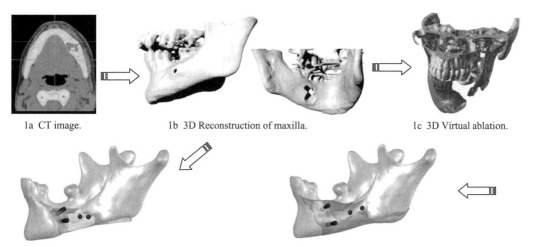

1a CT image.   1b 3D Reconstruction of maxilla.   1c 3D Virtual ablation.

1d First approach of customized implant geometry.   1e Optimized customized implant geometry.

Figure 1.   Reverse Engineering applied to generate a customized implant.

skeletal structures (Lattanzi et al. 2004). In our work, the maxilla geometry was captured using Siemens SOMATOM Sensation CT equipment. A sequence of 153 parallel images of 2D cross-sections was obtained (see Figure 1a). The slice spacing was 0.5 mm and the pixel size was 0.334 mm/dot. Each acquired image was sequentially numbered and converted into Non-Uniform Rational B-Splines (NURBS) representations of the surfaces to achieve the subsequent 3D reconstruction, which is done using specific software. The 3D model obtained can be exported as a Standard Triangulated Language (STL) file, to be used directly for manufacture on RP machines. This means that at this stage, and through Rapid Prototyping techniques, a physical model can be generated in order to improve the subsequent surgical intervention of tumor ablation, since this would make more accurate diagnosis and better surgical planning possible. In this work the NURBS on each cross-sectional slice were imported into commercially available 3D CAD software, aligned using a fixed reference frame and joined together by means of several surface/solid reconstruction strategies in order to generate the corresponding 3D CAD model (Figure 1b), where the accurate location and extent of the tumor can be observed. This procedure was adopted because this kind of file is better suited to performing the tumor ablation simulation and to designing the customized implant. In addition, the generated 3D CAD model can be easily exported as a standard neutral file, (e.g. IGES, PARABOLIC, SAT, SETP) which in turn can be imported by several FE packages for biomechanical analysis.

Once the 3D CAD model has been built up, the tumor was removed virtually with a resection margin of 1.5 cm of normal bone beyond the lesion margin

(Zemann et al. 2006). Figure 1c illustrates the result of virtual resection, where the large size of the resected area and its anatomical complexity can be seen. With the purpose of maintaining optimal aesthetics and function of the maxilla, and to fit the bone defect exactly, with external contours adjusted to replace the resected bone, the customized implant geometry was generated in a virtual environment, using a conventional CAD software package. Figure 1d illustrates the 3D CAD model of the assemblage between the maxilla and the first approach for the geometry of customized implant -GEO1-. This figure shows that the resected area was completely filled, but from both the anatomical and aesthetics aspects the desired external contours were not yet achieved. In order to cover the defect with similar contours to those of the other side of maxilla, to keep the face symmetry and to generate a leaner and more lightweight design, a new geometry was therefore designed -GEO2- as shown in Figure 1e. This figure shows that minor anatomical differences were found between the model and the resected area and both maxilla sides present a similar aesthetic appearance. The geometry of each created geometry and the geometric differences between them are shown in Figure 2.

At this stage the best implant fixation strategy and means to provide the necessary retention of maxilla have to be determined. Two strategies were developed: as a first approach, the maxilla and implant were attached by means of four titanium miniplates, each fixed with two screws; in the second, maxilla retention is ensured by means of four screws, strategically inserted as shown in Figures 3a and 3b, respectively. To determine the loading of the bone through the implant and the most suitable fixation strategy for this

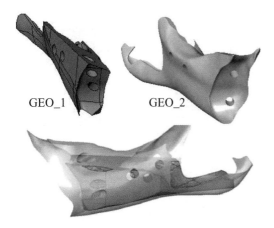

GEO_1 GEO_2

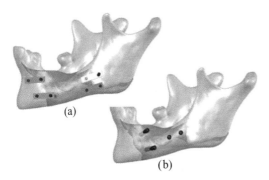

(a)

(b)

Figure 2. Geometric differences between the 3D CAD models of the two created customized implant geometries.

Figure 3. Possible implant fixation strategies considered. a) Titanium miniplates. b) Four screws.

particular case, a biomechanical analysis was performed, the results of which are presented in Section 4. Others positions were considered for fixing the prosthesis, but the results are not presented here.

4 BIOMECHANICAL EVALUATION

Three-dimensional finite element models can provide meaningful predictions of the stress field in a bone structure, since they appropriately reflect the mechanical properties as well as the exact geometry of the different structures (Zannoni et al. 1998, Anglin et al. 1999). This information is considered by a number of investigators to be a key factor for understanding bone's functional behaviour in many research and clinical applications. Fracture risk assessment, design and validation of prosthetic implants, are examples of the possible applications of 3D FEA in clinical studies (Zannoni et al. 1998, Couteau et al. 2000, Stolk et al. 2002). This work sets out to create and optimize the geometry of a customized implant to repair the

Table 1. Material properties considered in FEA analysis.

	Elastic modulus GPa	Poisson's ratio
Cancellous bone	0.74	0.2
Titanium	110	0.3
Acrylic	2.4	0.35
Hydroxyapatite	138	0.3

maxillary defect herein presented; the relative variations in both the cross-section and the depth geometry, as well as the best placement for fixation, were evaluated. The biomechanical behavior of titanium, cold-cured acrylic and hydroxyapatite as materials to replace bone grafts was evaluated and the results are presented in subsection 4.3.

4.1 Materials properties

Human bone is a natural hybrid nanocomposite of plate-shaped hydroxyapatite mineral particles in oriented collagen polymer, with a nice balance of stiffness, toughness, and vibrational damping properties (Porter 2004). However, its mechanical properties' variations are difficult to describe and quantify, particularly for trabecular bone, considering its anisotropy and the inhomogeneity (Couteau et al. 2000, Adam et al. 2003). The apparent mechanical properties of cancellous bone are a function of both the hard tissue behavior and its microstructural organization, which determine its significant anisotropy (Jacobs et al. 1999, Rietbergen et al. 2002). Any attempt at modelling bones with isotropic constitutive properties is only an approximation to the complexities of the real material (Adam et al. 2003). Nevertheless, because of the lack of a comprehensive data bank that includes the material properties of bone as a function of the orthotropic load directions, bones have been modeled as a homogeneous, isotropic and linear elastic material (Wirtz et al. 2000). This assumption is also applied here. A large number of studies have established ranges for the different mechanical properties of bone. But since the maxilla consists of a looser arrangement of trabecular bone (Sadowsky 2001), the elastic constants considered here were those reported by Adam et al. (2003), for cancellous bone. For the hydroxyapatite the values taken are those suggested by (Chen et al. 2006), while for both cold-cured acrylic and titanium the values were taken from the library of finite element software. Table 1 summarizes the elastic properties considered in this study. The difference between the Young's moduli must be noted.

4.2 Boundary conditions

The biomechanics of the human masticatory system is anatomically and functionally complex. More than twenty muscles are involved in this process, and they interact to produce a large number of forces on a mandible morphologically constrained by irregularly shaped joints – temporomandibular joints. Furthermore, there are multiple contacts between the maxillary, mandibular teeth and soft tissues (Daumas et al. 2005, Peck & Hannam 2006). Despite the large amount of data in the literature on maxilla movement and muscle activation patterns, no unequivocal relationship can be found between morphology, muscle action and maxilla motion (Koolstra et al. 1997). This means that a number of relevant parameters, such as joint and muscles forces, their orientations and locations are not explicitly available. This is not surprising since these data cannot be measured directly and varies considerably between individuals. Furthermore, the masticatory system is mechanically indeterminate, i.e. a unique static solution is not possible because several different combinations of joint and muscle forces can produce static equilibrium (Koolstra et al. 1997, Nickel et al. 2002). Therefore, in this work the maxilla was constrained at the places that correspond to left and right temporomandibular joints and a resultant force of 900 N was applied at the middle section of the maxilla.

4.3 Numerical results

The 3D CAD geometries were imported into commercially available finite element software and meshed automatically, considering tetrahedral elements and a Voronoy-Delauney meshing scheme. This meshing strategy and the type of element are well-suited to approximating highly non-regular geometries, like bones, and are widely reported in the literature. Figures 4 and 5 present Von Mises equivalent stress for the two implant fixation solutions illustrated in Figures 3(a) and 3(b), respectively. In the two simulations titanium was the material considered for both the customized implant and fixation components. These Figures show that for the fixation with screws the maximum stress concentration is observed at the bone/implant interface (Fig. 5). This stress concentration is due to the geometry factors, which means that the implant geometry must be improved, especially at its boundaries, in order to avoid this stress concentration. But despite this the implant fixation without plates seems to be a more robust and thus more secure procedure, since the use of titanium miniplates (Fig. 4) leads to additional points with significant stress concentration, corresponding to plate's position. Regarding the different materials considered for bone substitute, it was observed that acrylic (Fig. 6) presents the lowest level of stresses compared with the hydroxyapatite (Fig. 7) and titanium (Fig. 5). In addition, contrary to what is observed for the titanium, for the customized implant made in acrylic the maximum stress values are reached at the joints and not

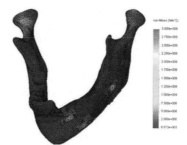

Figure 4. Equivalent Von Mises stress for the implant fixation with plates.

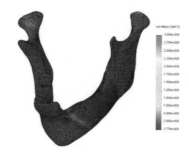

Figure 5. Results for the implant fixation with screws.

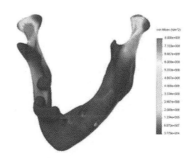

Figure 6. Results for the acrylic customized implant.

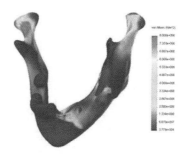

Figure 7. Results for the hydroxyapatite customized implant.

161

in the implant. This means that the implant is less loaded and so there is less probability of failure. This is not surprising, since the elastic properties of acrylic are the closer to those taken for bone (Table 1). As a result, the deformations are accommodated without significant stress concentrations. It can thus be concluded that the material used in the fabrication of customized implants should present an elastic modulus similar to that of natural bones. One way to achieve these desired physical and biological properties is through the use of scaffolds, which is the most promising method in the field of tissue engineering (Chu 2002, Lalan et al. 2001). The scaffolds serve as three dimensional templates for initial cell attachment and subsequent tissue formation – cell seeding and proliferation. Several factors are very important to the successful cellulization, vascularization and degradation of the scaffold, such as the material and the interior of scaffold design itself. The design of scaffold is chosen according to the needs of the particular medical application in terms of the desired mechanical properties and mechanical stability. The RP fabrication method offers flexibility and the ability to couple the design and development of a bioactive scaffold with the advances of cell-seeding technologies, in order to ensure the success of scaffold-based tissue engineering (Yang et al. 2002).

5 CONCLUSIONS

The major contribution of this work is the establishment of data and formulation of guidelines for the design, optimization and fabrication of customized implants to replace and/or repair bone defects caused by trauma, damage or bone loss. By using computer tomography data from each individual patient, implants with reasonable geometric and dimensional accuracy can be created using CAD software. Tailor-made 3D implants can be fabricated using biocompatible or biodegradable materials, which can shorten surgery time and meet the aesthetic needs of patients. Furthermore, based on the greater potential shown by Rapid Prototyping technologies in the field of tissue engineering, especially in the context of scaffold fabrication, we firmly believe that the use of scaffolds will be the future in this kind of medical application.

REFERENCES

Adam, C. et al. 2003. Stress analysis of interbody fusion-finite element modelling of intervertebral implant and vertebral body. *Clinical Biomechanics* 18: 265–272.
Ali, A. et al. 1995. Maxillectomy to reconstruct or obturate? Results of a UK survey of oral and maxillofacial surgeons. *British Journal of Oral and Maxiliofacial Surgery* 33: 207–210.

Al-Sukhun, J. et al. 2006a. Stereolithography and the use of pre-adapted or fabricated plates for accurate repair of maxillofacial defects. *Br. J. Oral Maxillofac. Sur.*: doi:10.1016/j.bjoms.2006.10.016.
Al-Sukhun, J. et al. 2006b. Biomechanics of the Mandible Part I: Measurement of Mandibular Functional Deformation Using Custom-Fabricated Displacement Transducers. *J. Oral Maxillofac. Surg.* 64: 1015–1022.
Anglin, C. et al. 1999. Glenoid cancellous bone strength and modulus. *Journal of Biomechanics* 32: 1091–1097.
Chelule, K.L. et al. 2000. Fabrication of medical models from scan data via rapid prototyping techniques. *Proceedings of the TCT (Time-Compression Technologies) Conference & Exhibition 2000*: 45–50. Cardiff: UK.
Chen, B. et al. 2006. Microstructure and mechanical properties of hydroxyapatite obtained by gel-casting process. *Ceram. Int.* doi:10.1016/j.ceramint.2006.10.021.
Chu, G. 2002. Mechanical and in vivo performance of hydroxyapatite implants with controlled architectures. *Biomaterials* 23(5): 1283–1293.
Couteau, B. et al. 2000. Morphological and mechanical analysis of the glenoid by 3D geometric reconstruction using computed tomography. *Clinical Biomechanics* 15: S8–S12.
Daumas, B. et al. 2005. Jaw mechanism modeling and simulation. *Mechanism and Machine Theory* 40: 821–833.
D'Urso, P.S. et al. 1999. Stereolithographic (SL) biomodelling in cranio-maxillofacial surgery: a prospective trial. *The Journal of Cranio-maxillofacial Surgery* 27(1): 30–37.
D'Urso, P.S. & Redmond, M.J. 2000. A method for the resection of cranial tumours and skull reconstruction. *Br. J. Neurosur* 14: 555–559.
D'Urso, P.S. et al. 2000. Custom cranioplasty using stereolithographic and acrylic. *British Journal of Plastic Surgery* 53(3): 200–204.
D'Urso, P.S. 2005. Real virtuality: Beyond the image. *Virtual modelng and rapid manufacturing. Paulo Bártolo et al. (eds). Taylor & Francis*: 29–37.
Jacobs, C.R. et al. 1999. The impact of boundary conditions and mesh size on the accuracy of cancellous bone tissue modulus determination using large-scale finite-element modelling. *Journal of Biomechanics* 32: 1159–1164.
Jonov, C. & Kaltman, S.I. 1997. The use of stereolithographic laser models in OMFS. *British Journal of Oral and Maxillofacial Surgery* 35(6): 444.
Koolstra, J.H. & Van Eijden, T.M.G.J. 1997. The jaw open – close movements predicted by biomechanical modeling. *Journal of Biomechanics* 30(9): 943–950.
Lalan, S. et al. 2001. Tissue engineering and its potential impact on surgery. *World J Surg.* 25: 1458–1466.
Lattanzi, R. et al. 2004. Specialized CT Scan Protocols for 3-D Pre-Operative Planning of Total Hip Replacement. *Medical Engineering & Physics* 26: 237–245.
Lohfeld, S. et al. 2006. Engineering Assisted Surgery™: A route for digital design and manufacturing of customized maxillofacial implants. *J. Mater. Process. Tech.*: doi:10.1016/j.jmatprotec.2006.10.028.
Malchiodi, L. et al. 2006. Jaw Reconstruction With Grafted Autologous Bone: Early Insertion of Osseointegrated Implants and Early Prosthetic Loading. *J Oral Maxillofac Surg.* 64: 1190–1198.

Marx, R.E. 1996. Mandibular and facial reconstruction rehabilitation of the head and neck cancer patient. *Bone* 19(1): 593–823.

Nickel, J.C. et al. 2002. Validated numerical modeling of the effects of combined orthodontic and orthognathic surgical treatment on TMJ loads and muscle forces. *American Journal of Orthod. and Dentofacial Orthopedics* 121(1): 73–83.

Papageorge, M.B. & Karabetou, S. 1997. Rehabilitation of patients with reconstructed jaws using osseointegrated implants. *British Journal of Oral and Maxillofacial Surgery* 35(6): 444.

Peck, C.C. & Hannam, A.G. 2006. Human jaw and muscle modeling. *Archives of Oral Biology* 52(4): 300–304.

Porter, D. 2004. Pragmatic multiscale modelling of bone as a natural hybrid nanocomposite. *Materials Science and Engineering* 365: 38–45.

Rietbergen, B.V. et al. 2002. High-resolution MRI and micro-FE for the evaluation of changes in bone mechanical properties during longitudinal clinical trials: application to calcaneal bone in postmenopausal women after one year of idoxifene treatment. *Clinical Biomechanics* 17: 81–88.

Sadowsky, S.J. 2001. Mandibular implant-retained overdentures: A literature review. *The Journal of Prosthetic Dentistry* 86(5): 468–473.

Stolk, R. et al. 2002. Finite Element and Experimental Models of Cemented Hip Joint Reconstructions Can Produce Similar Bone and Cement Strains in Pre-Clinical Tests. *Journal of Biomechanics* 35: 499–510.

Trikeriotis, D. & Diamantopoulos, P. 2006. Image-based biomechanical modelling in oral and maxillofacial practice. *Journal of Biomechanics* 39(1): S427.

Vos, W. et al. 2006. Functional imaging and CFD analysis to perform maxillo-mandibular osteotomy. *Journal of Biomechanics* 39: S427–S428.

Wirtz, D.C. et al. 2000. Critical evaluation of known bone material properties to realize anisotropic FE-simulation of the proximal femur. *Journal of Biomechanics* 33: 1325–1330.

Wright, S. et al. 2006. Use of Palacos® R-40 with gentamicin to reconstruct temporal defects after maxillofacial reconstructions with temporalis flaps. *British Journal of Oral and Maxillofacial Surgery* 44: 531–533.

Xia, J.J. et al. 2006. Cost-Effectiveness Analysis for Computer-Aided Surgical Simulation in Complex Cranio-Maxillofacial Surgery. *J. O. Max. Surg* 64: 1780–1784.

Yang, A. et al. 2002. The Design of scaffolds for Use in Tissue Engineering. Part II. Rapid Prototyping Techniques. *Tissue Engineering* 8: 1–11.

Zannoni, C. et al. 1998. Material properties assignment to finite element models of bone structures: a new method. *Medical Engineering & Physics* 20: 735–740.

Zemann, W. et al. 2006. Extensive ameloblastoma of the jaws: Surgical management and immediate reconstruction using microvascular flaps. *Oral Surg Oral Med Oral Pathol Oral Radiol Endod*: (in press).

Computational Vision and Medical Image Processing – João Tavares & Natal Jorge (eds)
© 2008 Taylor & Francis Group, London, ISBN 978-0-415-45777-4

Acrosome integrity classification of boar spermatozoon images using DWT and texture descriptors

Maribel González, Enrique Alegre & Rocío Alaiz
Department of Electrical, Systems and Automatic Engineering

Lidia Sánchez
Department of Mechanical, Computing and Aerospace Engineerings University of León,
Campus de Vegazana s/n, Spain

ABSTRACT: Automatic assessment of boar sperm head images according to their acrosome status is a challenge task in the veterinary field. In this paper we explore how much information texture features can provide for this task. A neural network-based machine that employs texture descriptors derived from the discrete wavelet transform is proposed as a classifier. Subbands are computed and for them two kind of statistical descriptors are calculated: first order statistics (mean and standard deviation) and second order ones (derived from the co-occurrence matrices). A Multilayer perceptron network with different number of neurons in the hidden layer as well as different activation functions are evaluated to categorize a set of sperm head images according to their acrosome status (damaged or intact). Experimental results point out that texture descriptors allow to further improve previous proposals (reaching an accuracy of 92.09%) what makes this approach attractive for the veterinary community.

Keywords: Discrete wavelet transform, neural networks, backpropagation, boar semen.

1 INTRODUCTION

The evaluation of the semen quality is an important problem in medical and veterinarian research. In the last years have been developed several computer assisted approaches to assess the quality of semen samples (1; 2). Firstly developed for human semen analysis, most of this approach have been currently adapted to other species. One crucial aspect in this assessment is the determination of the percentage of acrosome intact and damaged in a semen sample. It is usually carry out using stains because there are not computer tools for do it. The stains has several drawbacks as is its high cost in terms of time, specialized veterinarian staff and equipments required. Hence, it is very interesting to have at one's disposal a method for the automatic classification of the acrosomes as intact or damaged.

There are few works that address this problem. Some of them (3; 4) intends to automatically classify microscopic boar sperm head images according to its intracellular intensity distribution. The authors extract a model distribution from a training set of heads assumed as normal by veterinary experts. Then, they considerer two training sets, one with heads similar to the normal pattern and another formed by heads that

substantially deviate from that pattern, computing, for each spermatozoon head, a deviation from the model distribution. This produces a conditional probability distribution of that deviation for each set. Using a set of test images, the authors determine the fraction of normal heads in each image and compare it with result of expert classification.

Other one (5) performs a boar sperm acrosome classification by means of its integrity by using LVQ. In this case, the authors consider images of boar spermatozoa obtained with an optical phase-contrast microscope and they try to automatically classify single sperm cells as acrosome-intact (class 1) or acrosome-reacted (class 2). As a feature vector they use the gradient magnitude along the contour of the sperm head and they classify applying learning vector quantization (LVQ) to the feature vectors obtained. The training and test errors obtained in this work is of 0.165.

Texture analysis and classification implies to extract certain set of features from an image and is used widely in medicine to classify and recognize cells. Several authors have used texture descriptors to study pathological states of mouse liver cells (6), to recognize leukocytes (7), and to classify microorganisms by computing the Fourier transform (8).

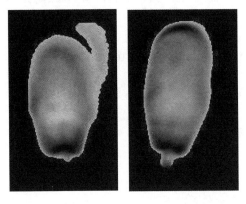

Figure 1. Image of a sperm head with the acrosome damaged (left) and intact (right).

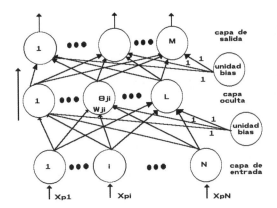

Figure 2. Image split in one and two subbands.

In this work, we use Discrete Wavelet Transform (DWT) to extract texture descriptors from boar sperm images. Once we apply the DWT, we compute first order statistics (mean and standard deviation), considering the Wavelet-Statistical Feature (WSF) (9), and we also calculate several features derived from the co-occurrence matrix using the Wavelet-Co-occurrence Features (WCF) (9). Later, we perform the classification step by backpropagation network (10).

2 METHODS

In this work, we have employed two methods to classify boar sperm images by means of their acrosome state (intact or damaged). We apply the discrete wavelet transform and we compute statistical features (WSF) and co-occurrence features (WCF) for the wavelet coefficients (9). So we classify sperm head images according to statistical texture properties that they present.

2.1 Preprocessing and segmentation

Using a digital camera connected to a phase-contrast microscope, boar semen images are captured with a resolution of 2560×1920 pixels. Information about the sample preparation can be found in (11). Then, sperm head images are cropped manually, obtaining images with an only sperm head. Finally, the head is segmented and the binary image is used to obtain a masked image with black background and the original grey level inside the head (5). (see Fig. 1).

2.2 Feature extraction

Information represented by spatial frequencies is frequently used for texture pattern recognition. We have employed a kind of descriptors of the frequency domain: discrete wavelet transform (DWT). Although in some cases, like in Fig. 1 it looks a morphological

approach could be successful, our experiments following this line did not show any result. Applying DWT to an image, a matrix of coefficients is obtained (see Fig. 2). There are four kinds of coefficients: approximations and horizontal, vertical and diagonal details. The first type holds almost all image energy, so it is the most important whereas the details take values close to zero. We compute the DWT coefficients to characterize the image texture and then we extract two kind of statistical features: first order and second order statistics.

In the first case, we consider the subbands obtained after three splits (LL1, LH1, HL1, HH1, LL2, LH2, HL2, HH2, LL3, LH3, HL3, HH3). Then, we compute first order statistics to provide information related to the image grey level distribution. Let be x a random variable representing the grey levels of an image $f(x)$, the fraction of pixels whose grey level is x is defined by $P(x)$ as follows:

$$P(x) = \frac{\text{number of pixels with grey level x}}{\text{number of image pixels}} \quad (1)$$

Considering this image histogram, the first order statistics are defined by:

$$\text{Mean: } u = \sum_{i=0}^{N-1} i . P_x(i) \quad (2)$$

$$\text{Standard deviation: } std = \sqrt{\frac{1}{N^2} \sum (p(i,j) - m)^2} \quad (3)$$

Computing the mean and the standard deviation for each obtained subband, we form a vector of 24 features (12 of each considered statistic).

For the second method, we consider the original image and the first subband of the wavelet transform of the image (LL1, LH1, HL1, HH1). For them, we calculate a set of descriptors proposed by Haralick derived

166

from the co-occurrence matrix to represent the image (12). Co-occurrence features provide texture information for a given image. In this work we have used four orientations (0°, 45°, 90° and 135°) and a distance of 1 since distances of 2, 3 and 5 did not improve the results. The considered features are the following:

$$\text{Energy} = \sum_i \sum_j c_{ij}^2 \tag{4}$$

$$\text{Homogeneity} = \sum_i \sum_j \frac{c_{ij}}{1 + (i-j)^2} \tag{5}$$

$$\text{Constrast} = \sum_i \sum_j |i-j|^2 c_{ij} \tag{6}$$

$$\text{Correlation} = \sigma_j = \sum_i \sum_j (j - u_j) c_{ij} \tag{7}$$

where u_x and u_y are the mean values for x and y and σ_x and σ_y the standard deviation:

$$u_i = \sum_i \sum_j i c_{ij} \tag{8}$$

$$u_j = \sum_i \sum_j j c_{ij} \tag{9}$$

$$\sigma_i = \sum_i \sum_j (i - u_i) c_{ij} \tag{10}$$

$$\sigma_j = \sum_i \sum_j (j - u_j) c_{ij} \tag{11}$$

Computing these four co-occurrence features for the original image and the four subbands, we obtain a vector of 20 features for each sperm head image.

3 EXPERIMENTAL RESULTS

Assessing to what extend texture descriptors extracted by means of the DWT allow to identify acrosome status is the goal of this section. The two target classes for this problem are: sperm head with damaged or intact acrosome. Samples were provided from a veterinary research group interested in this problem. The boar sperm head image dataset has 363 instances: 207 with intact acrosome and 156 with damaged acrosome.

We use a Multilayer Perceptron as classifier where previously described feature vectors are the inputs of the network. We evaluate a three layer network (one hidden layer) and two kinds of sigmoid activation functions for the hidden layer: logistic sigmoid (logsig, in the following) and hyperbolic tangent sigmoidal (tansig).

For both feature extraction methods, we compare the results of different neural network topologies: different number of neurons in the hidden layer (2, 3 and 20), different number of training cycles and different transfer functions for the output and hidden

Table 1. Hit rates for classification of damaged and intact acrosomes. [tansig logsig] with 3 neurons.

Cycles							
50	50	100	100	300	300	500	500
WSF	WCF	WSF	WCF	WSF	WCF	WSF	WCF
Classes							
Damaged 54,78	56,96	74,78	73,91	67,39	76,96	74,78	80,87
Intact 96,54	99,23	90,00	98,08	93,46	98,08	89,62	98,85
Hit rate (%) 76,93	79,38	82,85	86,73	81,22	88,16	82,65	90,40

Table 2. Hit rates for classification of damaged and intact acrosomes. [logsig logsig] with 3 neurons.

Cycles							
50	50	100	100	300	300	500	500
WSF	WCF	WSF	WCF	WSF	WCF	WSF	WCF
Classes							
Damaged 83,08	91,28	84,62	92,31	33,33	89,23	83,08	92,82
Intact 86,15	86,92	81,92	91,92	79,62	89,23	87,31	87,69
Hit rate (%) 84,70	88,96	83,18	92,09	57,89	89,22	85,31	90,09

layers ([logsig logsig], [tansig logsig]). Training was carried out with 75% of the images and the remaining 25% were used for test. Five runs were carried out and results averaged to get more representative estimations.

Tables 1 and 2 show empirical results (accuracy or hit rate) for the best neural network topology of those evaluated (three neurons in the hidden layer), different number of training cycles and different activation function for the hidden layer.

Co-occurrence (WSF) features provide the best accuracy of 92.09%, using a hidden layer of 3 neurons, 100 training cycles and a logistic transfer function in the hidden layer. This means a considerable reduction in error rate with previous approaches (from 20% to 8% in this work). On the other hand, classification with WCF features achieve a best accuracy of 88.96%

It is noticeable, however that misclassifications are lower in the case of boar sperm images with intact acrosome (see Table.1)-there are hit rates around 90% against 70% for the damaged class. Our understanding is that this may be due to the high variability presented in the samples of damaged acrosomes whereas the intact ones present a similar texture distribution. This leads us to explore in further work the use of unsupervised classifiers to find subclasses in the case of reacted acrosome images what could improve the recognition task.

4 CONCLUSIONS

We propose a method to classify boar sperm head images into two classes: those with acrosome damaged

167

or with acrosome intact. To deal with this, we apply the discrete wavelet transform to the boar sperm head images. For the obtained subbands, two kind of descriptors are computed: first order statistics (WSF) (mean and standard deviation) and features derived from the co-occurrence (WCF) matrices (energy, homogeneity, contrast and correlation). By means of a Backpropagation Neural Network, classification error rates of 8% are achieved with WCF features. This becomes a great improvement when compared with previous approaches and makes this automatic approach now more attractive for the veterinary field that demands an automatic and accurate discrimination system to replace the tedious manual process. This preliminary result with a simple individual classifier leads to further explore and improve the classification method by: (a) developing classifiers that employ both WSF and WCF features, but with feature selection techniques that protect against possible redundant information and (b) combining classifiers trained independently with WSF and WCF features.

REFERENCES

[1] J. Verstegen, M. Iguer-Ouada, and K. Onclin. Computer assisted semen analyzers in andrology research and veterinary practice. *Theriogenology*, 57:149–179, 2002.

[2] C. Linneberg, P. Salamon, C. Svarer, and L.K. Hansen. Towards semen quality assessment using neural networks. In *Proc. IEEE Neural Networks for Signal Processing IV*, pages 509–517, 1994.

[3] L. Sanchez, N. Petkov, and E. Alegre. Statistical approach to boar semen head classification based on intracellular intensity distribution. In A. Gagalowicz and W. Philips, editors, *Proc. Int. Conf. on Computer Analysis of Images and Patterns, CAIP 2005, Lecture Notes in Computer Science*, volume 3691, pages 88–95. Springer-Verlag Berlin Heidelberg, 2005.

[4] L. Sanchez, N. Petkov, and E. Alegre. Classification of boar spermatozoid head images using a model intracellular density distribution. In M. Lazo and A. Sanfeliu, editors, *Progress in Pattern Recognition, Image Analysis and Applications: Proc. 10th Iberoamerican Congress on Pattern Recognition, CIARP 2005, Lecture Notes in Computer Science*, volume 3773, pages 154–160. Springer-Verlag Berlin Heidelberg, 2005.

[5] N. Petkov, E. Alegre, M. Biehl, and L. Sanchez. Lvq acrosome integrity assessment of boar sperm cells. *CompIMAGES, Computational modelling of objects represented in images*, 2006.

[6] F. Albregtsen, H. Schulerud, and L. Yang. Texture classification of mouse liver cell nuclei using invariant moments of consistent regions. In *Computer Analysis of Images and Patterns*, pages 496–502, 1995.

[7] D.M.U. Sabino, L. da Fontoura, E.G. Rizzatti, and M.A. Zago. A texture approach to leukocyte recognition. 10(4):205–216, August 2004.

[8] T. Alvarez, Y. Martin, S. Perez, F. Santos, F. Tadeo, S. Gonzalez, J. Arribas, and P. Vega. Classification of microorganisms using image processing techniques. pages I: 329–332, 2001.

[9] S. Arivazhagan and L. Ganesan. Texture classification using wavelet transform. *Pattern Recognition Letters*, 24(9–10):1513–1521, June 2003.

[10] R. Masuoka. Neural networks learning differential data. Master's thesis, Department of Mathematical Sciences, The University of Tokyo, March 2000.

[11] L. Sanchez, N. Petkov, and E. Alegre. Statistical approach to boar semen evaluation using intracellular intensity distribution of head images. *Cellular and Molecular Biology*, 52(6):38–43, 2006.

[12] R.M. Haralick. Statistical and structural approaches to texture. In *Proceedings of the IEEE*, pages 45–69, 1978.

Computational Vision and Medical Image Processing – João Tavares & Natal Jorge (eds)
© 2008 Taylor & Francis Group, London, ISBN 978-0-415-45777-4

Computer-Aided Detection system for nodule detection in lung CTs

Gianfranco Gargano
On Behalf of MAGIC – 5 Collaboration
Dipartimento Interateneo di Fisica M.Merlin, Universit degli Studi di Bari, Italy
Istituto Nazionale di Fisica Nucleare (INFN) sez. Bari, Italy

ABSTRACT: A completely automated computer-aided detection (CAD) system for the selection of lung nodules in Computer Tomography (CT) images is presented. The system, based on Region Growing (RG) and Active Contour Model (ACM) algorithms, consists in three steps: 1) the lung parenchymal volume is segmented by means of a RG algorithm; nodules and vascular tree inside the lung are included through an ACM technique; 2) an RG algorithm is iteratively applied to the previously segmented volume in order to detect the candidate nodules; 3) a simple double-threshold cut and a neural network are applied to reduce the false positives (FPs). The system works on whole CTs, without the need for any manual selection. The CT database was recorded at the Pisa center of the ITALUNG-CT trial, the first Italian Randomized Controlled Trial for the screening of the lung cancer, following the Italung/I-ELCAP (International Early Lung Cancer Action Project) protocol. The detection rate of the system is 88.5% with 6.6 FPs/CT on 15 CT scans (about 4700 sectional images) with 26 nodules (15 internal and 11 pleural). A reduction to 2.47 FPs/CT is achieved at 80% efficiency.

1 INTRODUCTION

Lung cancer is the highest mortality kind of cancer all over the world. In U.S., a number of 172.570 new cases and 163.510 deaths have been reported in 2005 (1). The survival rate is estimated to be 14% after five years, with an increase up to 50% if the lung cancer is detected at an early stage (2). The use of the chest Computer Tomography (CT) strongly improves the radiologists' detection rate as well as the definition of the cancer type (3). A CT exam consists of a series of 2D images (of about 150 MB when reconstructed with thin slice thickness) to be visually examined. This task is particularly difficult and time-consuming, due to the fact that some nodules are hardly distinguished from non pathological structures (e.g. bifurcations of the vascular trees) by means of a slice-by-slice examination. These facts have encouraged in the last years the development of CADs for the automated detection of lung nodules in CT scans. A CAD system could provide valuable assistance to the radiologists and in many cases its use leads to a remarkable im provement of the detection rate at the cost of a low increase of the FPs (4; 5).

2 THE CT DATABASE

Chest Computer Tomography (CT) is actually considered as the best imaging modality for the detection of lung nodules. In the last years, low dose CT scans were shown to be effective for the analysis of the lung parenchyma (6), thus making possible the perspective of screening programs. The database used in this study consists of 15 low-dose CT (LDCT) scans recorded with a 4 slices spiral scanner Somatom Plus 4 VZ machine, with the following settings: 140 kVp, 20 mA, 1.25 mm collimation, and 1 mm reconstruction interval. The images were acquired at the Pisa center of the ITALUNG-CT trial, which is the first Italian Randomized Controlled Trial for the screening

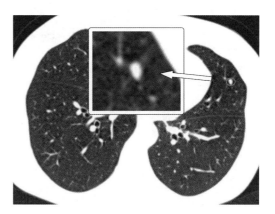

Figure 1. A typical example of a lung CT scan: the magnified window displays a nodule that is pathological structure.

of the lung cancer. Each image consists of a matrix of $512 \times 512 \times (314 \pm 23)$ voxels: it should be stressed that both the number of the voxels and their longitudinal size depend on the patient size.

The scans contain 26 nodules, so divided: 15 internal nodules (i.e. far from the pleura) and 11 pleural nodules (near the pleura). The nodule were diagnosed by experienced radiologists according to the ITALUNG/I-ELCAP (International Early Lung Cancer Action Project) protocol (7) which considers as pathological, structures of non calcified nodules with a diameter greater than 5 mm. The CT examinations were carried out independently by two radiologists by means of slice-by-slice visual inspection. The presence of a nodule is marked by a circle which completely encloses the nodule in its median slice. The mean diameter of the nodules in the whole database is $d = (6.7 \pm 1.5)$ mm.

3   LUNG VOLUME SEGMENTATION

The first step of the CAD system is the parenchymal volume segmentation. This step consists of two substages:

1. Internal lung volume segmentation by means of a RG algorithm;
2. Anatomic lung contour detection, to include both the pleural nodules and the vascular tree inside the lung, implemented slice by slice by means of an original ACM algorithm, named *Glued Elastic Band* (GEB).

3.1   *Internal lung volume segmentation*

The internal lung volume consists of air and bronchial tree that typically appears, in a CT slice, as low intensity voxels surrounded by high intensity voxels corresponding to the pleura. This suggests to segment the internal lung volume by means of a 3D RG algorithm. The choice of the inclusion rule with the optimal threshold and the selection of a proper seed point are of great relevance for the best working of the algorithm. We define the following rules:

1. *Simple Bottom/Top Threshold* (SBT or STT): if the intensity I is greater/lower that a certain threshold θ, the voxel is included into the growing region:

$$\begin{cases} I \geq \theta & \text{(SBT)}, \\ I \leq \theta & \text{(STT)}. \end{cases}$$

2. *Mean Bottom/Top Threshold* (MBT or MTT): the intensities of the voxel and its 26 neighbours are averaged; if the average $\langle I \rangle$ is greater/lower that the threshold θ, the voxel is included into the growing region:

$$\begin{cases} \langle I \rangle \geq \theta & \text{(MBT)}, \\ \langle I \rangle \leq \theta & \text{(MTT)}. \end{cases}$$

For this stage we use the MTT which allows to reduce the "noise" of the low dose CT, thus obtaining a volume with quite regular contours.

The threshold value $\bar{\theta}$ is automatically selected with the method adopted in Ref. (8).

The seed point of the RG is automatically selected as the voxel that satisfies the inclusion rule in a cubic region located as follows: the CT is divided lengthwise, thus obtaining two parts of equal sizes; the center of the cube is positioned at the crosspoint of the diagonals of, say, the left part (the choice of the right part would be equivalent). In this way, we are quite sure that the cubic region where the seed point is searched is inside the lung, for less than abnormal anatomical deformities. Once this voxel is found the growth of the internal lung volume is started, otherwise the search is repeated in a greater cubic region until a voxel satisfying the inclusion rule is found.

It should be stressed that the volume thus obtained includes the lung parenchyma, the bronchial tree and the trachea, while structures outside the lung, as bones, fat and vascular tree are ruled out. Also internal and pleural nodules are not included at this stage because they do not satisfy the MTT inclusion rule. To this purpose, the contour of the lung must be outlined and all voxels inside this contour must be considered.

3.2   *2D anatomic lung contour detection*

The anatomic lung contour selection is implemented slice-by-slice by applying the ACM to the external contours of the lung sections. In order to include concave parts with little bending radii (pleural nodules and sections of the vascular tree), we have developed a new local convex hull implemented by means of an original ACM algorithm, named *Glued Elastic Band* (GEB). The algorithm simulates the dynamics of a spline *glued* along the contour; in this way, the nodes of the spline feel the effects of the following forces:

1. Constant internal forces that the nodes exchange each one with the nearest neighbours (the previous and the following along the spline);
2. Constant external adhesive forces acting when the nodes are in contact with the section contour, as if there was some glue on the spline;
3. The constraint reactions acting when the nodes are pushed inside the contour.

As a result of this dynamics, in the concave parts with little bending radius where the sum of the elastic forces is strong enough to exceed the adhesive forces, the spline is pulled out and the concave parts are included inside the spline; on the other side, concave parts with great bending radius are not included by the spline which remains glued to the contour. The final effect is that the spline reaches an equilibrium position that includes concave parts with little bending radius

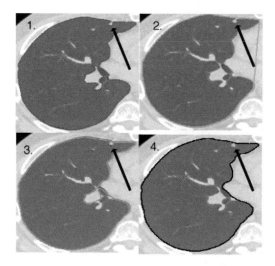

Figure 2. (1) Initial position of the spline corresponding to the sectional lung contour (case $k = +\infty$); (2) Result of a simple Convex Hull (case $k = 0$); (3) A configuration of the GEB spline obtained with a suboptimal value of the parameter k; (4) the position of the spline obtained with the best value of k.

as pleural nodules and sections of the vascular tree near the lung hilum, while concave regions with great bending radius are ruled out.

The only parameter of the GEB is the ratio between internal and adhesive forces:

$$k = \frac{F_{adhesive}}{F_{internal}}. \tag{1}$$

In the limit $k = +\infty$, the internal forces are irrelevant with respect to the adhesive forces and the spline is perfectly glued to the section contour: in this case, the action of the GEB causes no effect on the original contour (Figure 1). This case corresponds also to the initial position of the spline. On the other side, for $k = 0$, the adhesive force is zero and the result is that of a simple CH (Figure 2). Between these two cases, one can have intermediate values with different results: the highest k, the greater the bending radius of the concave regions that are ruled out. Figure 3 shows a possible configuration of the spline obtained with a suboptimal value of the parameter k. In Figure 4, the section contours, obtained with the best value of k, include only concave regions with small bending radii, as pleural nodules and sections of the vascular tree near the lung hilum, while concave regions with great bending radius are ruled out. As shown in the flowchart (Figure 3), the dynamics of the elastic band has been implemented in a simplified fashion: when the sum of the elastic forces exceeds the adhesive force

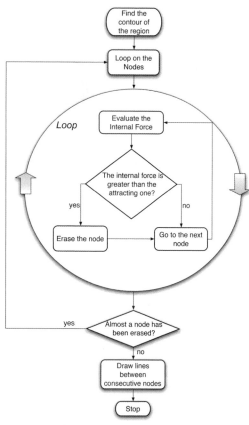

Figure 3. Flowchart of the GEB algorithm.

on a node, it is detached from the contour and its final position must be necessarily on the straight line joining its neighbors. Due to this fact, we automatically delete a node once detached from the contour, and join its neighbors with a straight line. This procedure reproduces only the final effects of the dynamics of the elastic band, thus avoiding the complexities of an oscillatory process. In fact, the complete simulation of the oscillations of a elastic band around its equilibrium position could converge only if we include dissipative processes, like friction, that generate dampened oscillations, but this kind of simulation is too complex and not necessary for our purpose.

The above described algorithm is applied to every slice of the CT scan. All voxels inside the final position of the spline are considered and combining together the regions inside the 2D contours of each slice we obtain the 3D segmented volume which contains: bronchial and vascular trees inside the lung, trachea, pleura, internal and pleural nodules. Nodules will be searched inside this *working* volume.

171

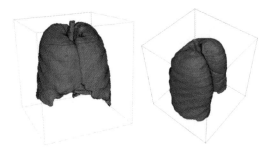

Figure 4. Two view of the 3D reconstructed lung volume. The pathological structures will be searched inside this volume.

4 NODULE CANDIDATE DETECTION

The second step of the CAD system consists in detecting the candidate nodules inside the working volume. This is implemented by a RG algorithm with an inclusion rule given by the *AND* combination of MBT and SBT rules. The thresholds are chosen in order to maximize the detection rate (or sensitivity, or efficiency) defined as the fraction of selected nodules with respect to the total number of nodules diagnosed by the radiologist. The seed points are searched automatically as follows: the segmented volume is scanned until a voxel satisfying the inclusion rule is found; this voxel is used as seed point and the growth is started. Once the region is completely grown, it is removed from the CT and stored for further analysis. Then the search for new seed points is restarted. The routine is iterated until no more seed points satisfying the inclusion rule are found.

To asses the efficiency of this step, we define as *true positives* (TPs) the candidate nodules that meet the radiologists' diagnosis according to the following condition:

$$\begin{cases} |X_{rad} - X| < R_{rad} \\ |Y_{rad} - Y| < R_{rad} \\ |Z_{rad} - Z| < R_{rad} \end{cases}$$

where $\{(X_{rad}, Y_{rad}, Z_{rad}), R_{rad}\}$ are the center coordinates and the radius of the radiologists' drawn circle, and $\{(X, Y, Z), R\}$ are the same quantities of CAD candidate nodule. All other candidates are considered as false positives (FPs).

5 FALSE POSITIVE REDUCTION

With the above mentioned definitions, the efficiency of the segmentation step on 15 CTs containing 26 nodules (15 internal and 11 pleural), is 88.5% (23/26), with about 2775 FPs/CT. Almost all FPs findings refer to candidates with too many or too few voxels and can be

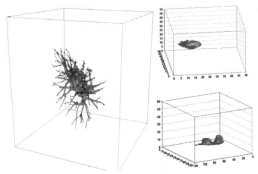

Figure 5. Examples of the structures found by the CAD system: part of the vascular tree (left) and nodules (right).

easily ruled out by a simple double-threshold cut on the volume V (expressed in number of voxels).

By considering nodule candidates with $V \in [30,800]$, the efficiency of the segmentation step (88.5%) is preserved, with a reduction to 48.1 FPs/CT. The values of the thresholds arise from general considerations on the dimensions of the nodules and do not depend on the analyzed database. Yet the stability of the CAD performance, measured as the area under the ROC curve (AUC), has been checked for different values of the thresholds in proximity of the above mentioned values ($V_{min} = 30$ and $V_{max} = 800$). Consequently, the results we will report in the following are the same for different thresholds close to the reported values.

A further reduction of the FPs can be obtained by means of a classification step carried out by a supervised two-layered (3 inputs, 7 hidden neurons and 1 output) feed-forward neural network, trained with gradient descent learning rule (9) (learning rate: $\eta = 0.01$) with momentum (10) (momentum term: $\alpha = 0.9$), and sigmoid transfer function (gain factor $\beta = 1$). These parameters have been set in order to obtain the best classification performance in term of AUC. The input features are:

1. Volume, **V**;
2. Roundness, $\mathbf{R} = V/V_S$, where V_S is the volume of the smallest sphere containing the segmented region;
3. Radius r, defined as the mean distance between the nodule center and the contour points.

To train the network, all 23 TPs and a subset of 69 FPs are used in *cross validation* modality (11). In particular, due to the low number of TP patterns at our disposal, we adopt the *leave-one-out* scheme to exploit the highest possible number of TPs during the training phase.

The outputs of the neural networks give rise to a 2 class distribution in the range [0,1], where t = 0,1 are, respectively, the FP/TP targets in the network train ing

172

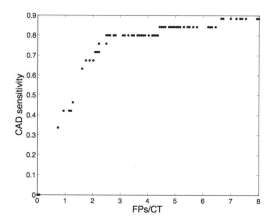

Figure 6. Overall CAD FROC curve reporting the sensitivity, evaluated with respect to the radiologists' diagnosis, against the FPs/CT.

phase. By shifting a decision threshold through out this interval, the sensitivity, the FP rate and the number of FPs can be evaluated, and, consequently, the ROC and FROC curves can be drawn.

6 RESULTS AND CONCLUSIONS

The AUC (Area Under the Curve) of the ROC is $A_z = 0.969 \pm 0.027$, where the error is computed as reported in Ref. (12). The use of the neural network allows to reduce the number of FPs/CT from 48.1 to 6.6 with the same efficiency (88.5%) of the double-threshold cut. A further reduction to 2.47 FPs/CT can be obtained at a lower detection rate (80%).

The nodule detection in lung CT scans is a hard task due to the fact that a sequence of slices must be analyzed. From this point of view, the CAD systems can be a useful tool to help the radiologists for lung cancer diagnosis. We have developed a CAD for nodule detection and its complete automatization

suggests that it can be used as a valuable support to the radiologists for the diagnosis of lung nodules in low dose CT scans.

REFERENCES

[1] Cancer Facts & Figures 2005, American Cancer Society
[2] Cancer Facts & Figures 1999, American Cancer Society
[3] D.F. Yankelevitz, A.P. Reeves, W.J. Kostis, B. Zhao, and C.I. Henschke, "Small pulmonary nodules: volumetrically determined growth rates based on CT evaluation", Radiology 217, 251–256 (2000).
[4] M.S. Brown, J.G. Goldin, S. Rogers, H.J. Kim, R.D. Suh, M.F. McNittGray, S.K. Shah, D. Truong, K. Brown, J.W. Sayre, D.W. Gjertson, P. Batra, and D.R. Aberle, "Computeraided Lung Nodule Detection in CT: Results of Large-Scale Observer Test", Academic Radiology 12 (6), 681–686 (2005).
[5] K. Peldchus, P. Herzog, S.A. Wood, J.I. Cheema, P. Costello, J. Schoepf, "Computer-Aided Diagnosis as a Second Reader: Spectrum of Findings in CT Studies of the Chest Interpreted as Normal", Chest 128, 1517–1523 (2005).
[6] J.G. Ravenel, E.M. Scalzetti, W. Huda, and W. Garrisi, "Radiation exposure and image quality in chest CT examinations", AJR 177, 279–284 (2001).
[7] http://www.ielcap.org/ielcap.pdf
[8] T.W. Ridler, and S. Calvard, "Picture thresholding using an iterative selection method", IEEE Transaction on System, Man and Cybernetics 8, 630–632 (1978).
[9] J. Hertz, A. Krogh, and R.G. Palmer, "Introduction to the theory of neural computation", Addison Wesley (1991).
[10] D.E. Rumelhart, and J.L. McClelland, "Parallel Distributed Processing" Vol.I MIT Press, Cambridge, MA (1986).
[11] M. Stone, "Cross-validatory choice and assessment of statistical predictions", Journal of the Royal Statistical Society B 36 (1), 111–147 (1974).
[12] J.A. Hanley, and B.J. McNeil, "The Meaning and Use of the Area under a Receiver Operating Characteristic (ROC) Curve", Radiology 143, 29–36 (1982).

Computational Vision and Medical Image Processing – João Tavares & Natal Jorge (eds)
© 2008 Taylor & Francis Group, London, ISBN 978-0-415-45777-4

Paranasal sinuses segmentation/reconstruction for robot assisted endonasal surgery

A.I. Moral, M.E. Kunkel, M. Rilk & F.M. Wahl
Institute for Robotics and Process Control, Technical University of Braunschweig, Germany

K. Tingelhoff & F. Bootz
Clinic and Policlinic of Ear, Nose and Throat/ Surgery, University of Bonn, Germany

ABSTRACT: The purpose of this study was to evaluate an approach for use of segmented computed tomography images in volumetric estimation of the paranasal sinuses cavities. For this purpose, 452 images were processed with the software Amira™ 4.1. These images were obtained from a dummy human head, which is used to rehearse the movements of the surgeon during endoscope nasal surgery, and will be used as test-bed in future works to simulate robot assisted endonasal surgery. The volumes of the frontal, maxilar, sphenoidal and ethmoidal sinuses were examined both by material injection and by 3D CT images. The volumes were all in the respective ranges compared with previous reports. Three-dimensional reconstruction on the paranasal sinuses will be used to create Finite Element Models for Endonasal surgery simulations. Therefore, the precise knowledge of the geometric configuration of these regions is an important step in our research.

Keywords: Paranasal sinuses; computed tomography; 3D-reconstruction; robot assisted endonasal surgery

1 INTRODUCTION

1.1 *Paranasal sinuses anatomy*

The nasal cavity is a large air-filled space above and behind the nose in the middle of the face. It is divided into right and left halves by the nasal septum. The paranasal sinuses are four pairs of hollow structures within the bones surrounding the nasal cavity (Fig.1). The sinuses are 0divided into subgroups named according to the bones they lie under. The frontal sinus is located over the eyes, in the forehead bone; the maxillary sinuses, under the eyes, in the upper jawbone; the ethmoidal sinuses are comprised of a variable number of air cells, ranging from 3 to 18 on each side, and they are between the nose and the eyes, backwards into the skull; and finally, the sphenoidal sinuses, are located in the centre of the skull base (Lang, 1989). The shape, size and position of each sinus differ between individuals and variation may even occur on either side of the head (Pérez-Pinas et al. 2000).

The physiologic roles of the nasal passages are humidification, warming, and removal of particulate mater from inspired air (Zinreich et al. 1987). All paranasal sinuses have their origin in the ethmoidal portion of the nasal cavity and communicate with

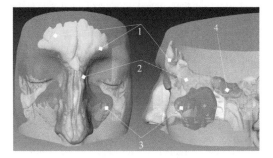

Figure 1. Frontal and lateral view of the human paranasal sinuses: frontal (1), ethmoidal (2) maxillary (3), and sphenoidal sinuses (4).

this above the inferior nasal concha via small orifices called ostia (Navarro, 2001).

1.2 *Functional endoscopic sinus surgery*

Endonasal surgery has become standard for the treatment of pathologies that can be reached via the nasal cavity. Functional endoscopic sinus surgery (FESS) is a minimal invasive approach adopted in case of chronic sinusitis (inflammation of the paranasal sinuses) and is aimed to restore normal physiology by reestablishing normal mucociliary drainage and ventilation of the sinuses (Kennedy et al. 1985; Stammberger et al. 1990). During FESS the surgeon needs to remove the localized mucosal disease obstructing the nasal pathways, for that it is necessary to move the endoscope and other surgical instruments within the nasal cavity and through the ostia to reach the paranasal sinuses (Zinreich et al. 1987; Moral et al. 2007).

Although most of the otolaryngologists accept FESS as the best treatment for chronic sinusitis (Becker, 2003; King et al. 1994), the technique has its clear limitations as well as its specific problems. The major drawback in this surgery is regarding the surgeons to be subject to fatigue due to handling of surgical instruments with one hand for long time periods while the other hand is guiding the endoscope. Surgical workflows demonstrate the need and the feasibility of automatic assistance in guiding the endoscope in FESS (Strauss et al. 2005).

Figure 2. 3D-Model of the nose and paranasal sinuses (Axel LANG, Zurich, Switzerland) to be used for the homogeneous model of the nasal cavity and paranasal sinuses.

1.3 Purpose

A project for robotic endoscope guidance in endonasal surgery is being developed for the Robotic Surgery Group at the Institute for Robotics and Process Control (Technical University of Braunschweig, Germany) in cooperation with the Clinic and Policlinic for Ear, Nose and Throat/Surgery (University of Bonn, Germany) (Rilk, 2006; Tingelhoff et al. 2006, Wagner et al. 2005). In that context, models of the nasal cavity and paranasal sinus, based on Finite Element Method (FEM), are being created to perform simulation of the mechanical behaviour of the inner nasal structures under endoscope loading during endonasal surgery. Finite Element Analysis (FEA) will provide important information for the safe control strategy of the robotic endoscope guidance system.

Some characteristics of the paranasal structures as inhomogeneity, anisotropy, viscoelasticity and non-linearity make difficult the task of producing its biomechanical model. Therefore, to produce an initial model, it is necessary to use simplified concepts, to gain more insight into the mechanical behaviour of this structure during endoscope interaction. Despite of the structures that form the human nasal structures present inhomogeneities due to the diversity of the biological tissues (bone, cartilage, mucosa and others), it is possible consider the hypothesis that a homogeneous model represents the deformation of the nasal structures under endoscopic loading. This is acceptable just if it is assumed that the inner nasal structures function as a linear system submitted to small deformation due to endoscopic mechanical contacts during FESS.

In this paper, we present results concerning to obtaining of the geometric configuration of the

paranasal sinuses of a replica of human head for the homogeneous FEM model. This homogeneous model should be simple but sufficiently accurate once it will be used as a background for the further more complex models. Our purpose was to perform segmentation of the sinuses regions for three-dimensional (3D-) reconstruction, visualization and volumetry. To calibrate the approach, the volume of the paranasal sinus was examined both by material injection and by 3D CT images.

2 MATERIALS AND METHODS

2.1 Dummy human head

In this study we have used CT images from a 3D-model of the nose and paranasal sinuses. It is a dummy human head from silicone (Axel LANG, Zurich, Switzerland), based on a cadaver specimen with normal anatomy and that was designed to reproduces the human nasal structures and to provide a realistic training environment for endoscopy of the nasal cavity and paranasal sinuses (Manestar et al. 2006). The dummy human head is for didactic purposes divided into 5 parts showing different level of the nasal structures. The anatomical model was commercially designed to provide a realistic training environment for endonasal sinus surgery, allowing the implementation of the robotic endoscope guidance, reproducing de movements of the endoscope as a real sinus surgery (Fig. 2). A total of 452 CT images from the dummy human head were acquired by a spiral CT from Philips. The images have high resolution ($0.0390625 \times 0.0390625 \times 0.04$) and a slice thickness of 0.625 mm.

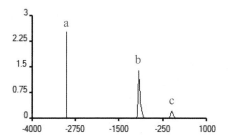

Figure 3. Representative histogram of CT images from the dummy human head showing the grey level of the black background (1), of the cavities (2), and the structures of the head.

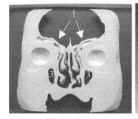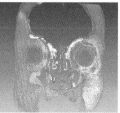

Figure 5. Frontal view of one layer of the dummy human head (left) and the partial 3D surface reconstructions of the nasal cavity and paranasal sinuses (right). The numbered structures are: Frontal (1), ethmoidal (2), maxillary (3) and sphenoidal sinuses (4) and nasal cavity (5).

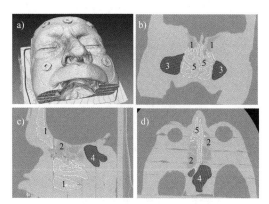

Figure 4. a) Dummy human head, b) Segmentation in a) Frontal, b) Lateral and c) Transversal view. The numbered structures are: Frontal (1), ethmoidal (2), maxillary (3) and sphenoidal sinuses (4) and nasal cavity (5).

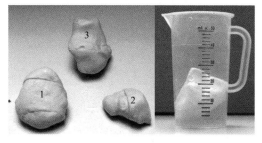

Figure 6. Left side: Solid casts of right (1) and left (2) maxillary and right (3) sphenoid sinuses of a human dummy head. Right side: Graduated cup filled with water used to perform volume measurement of the casts of the sinuses through immersion in water.

To verify the correspondence between the model and the segmented head, a 3D-reconstruction of surface of the nasal cavity and paranasal sinuses was performed (Fig. 5). The 3D volume is generated after the segmentation results are post processed.

2.2 CT Imaging

A semi-automatic segmentation of the CT images of the nasal cavity and paranasal sinuses was performed using Amira™ 4.1 software for medical images (Mercury Computer System Inc., USA). The same software was also used to perform the 3D-reconstruction and volumetry.

Since the material of the anatomical head is homogeneous only one type of material was used. The semi-automatic segmentation was performed, using different tools that segment the image following an image gradient or a growing region based on the grey level of the object (head), the background or the nasal cavities. These regions could be clearly differentiated in all CT images (Fig. 3).

The regions identified in the segmentation were: model of the dummy human head, nasal cavity and paranasal sinuses that were separately segmented (Fig. 4).

2.3 Paranasal sinuses volume measurements

The measurement of paranasal sinuses volume using the 3D reconstructions of CT images was performed using the tool tissue statistics from Amira™ 4.1. For each sinus the number of voxels which belongs to this sinus is counted and the numbers of voxels are multiplied by the volume of one voxel in 3 mm.

To perform the direct measurement of paranasal sinuses volume we have used a method considered to be reliable and accurate described for Uchida et al. (1998). The dummy human head, that is divided in 5 layers, was maintained closed in a box, than the nasal cavities and the sinuses were filled with molding plaster (Model Gips Krone, Osterode, Germany) using a syringe positioned in the left nostril. After one hour, the impression material had hardened, the dummy human head was opened, and the solid cast of the sinuses were easily removed (Fig. 6a). To measure the volume, the

Table 1. Estimated volumes of the paranasal sinuses by both 3D-reconstruction from CT images and direct measurements.

| Paranasal sinuses | | Estimated volume (ml) | |
		3D-reconstruction	Direct measurement
Maxillary	Right	12.85	12.2
	Left	5.46	5.8
	Total	18.32	18
Sphenoidal	Right	10.96	11.6
	Left	0.80	N/A
	Total	11.76	11.6
Ethmoidal	Right	3.02	N/A
	Left	3.00	N/A
	Total	6.02	N/A
Frontal	Total	12.04	11.8

N/A: not available

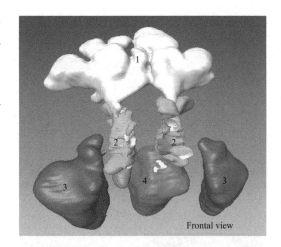

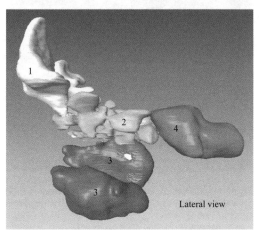

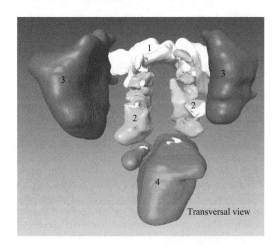

Figure 7. 3D-reconstructions of paranasal sinus from CT images using Amira software. Images in frontal, lateral and transversal views: The numbered structures are: Frontal (1), ethmoidal (2), maxillary (3) and sphenoidal sinuses (4).

casts were immersed in a graduated cup filled with water (Fig. 6b). The reproducibility of this method was assessed by repeating measurements. The cast of the ethmoidal sinuses as well as, the left sphenoid were despised.

3 RESULTS

3.1 3D paranasal sinuses reconstruction

The 3D-reconstruction of paranasal sinus CT images of the dummy human head is shown in Figure 7 in frontal (a), lateral (b) and transversal (c) view.

Ostia are small orifices that connect the sinuses to the nasal cavity. If the sinus ostium is blocked, the entire sinus thus becomes the pathologic cavity (sinusitis). Figure 8 shows the maxillary and sphenoidal ostium.

3.2 Estimated paranasal sinuses volume

The volume of paranasal sinuses measured by both 3D-reconstruction from CT images and direct measurements are given in Table 1.

3.3 Paranasal sinuses cavities mesh generation

Figure 9 shows the 3D mesh generated from the dummy human head without the paranasal sinuses volume.

4 DISCUSSION

Through this work we have evaluated an approach to obtain the geometric anatomy of the paranasal sinuses from CT images. 3D-representation of the head without these structures was transformed into tetrahedral meshes and will be used for the FE model

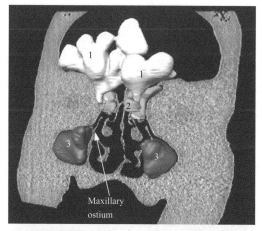

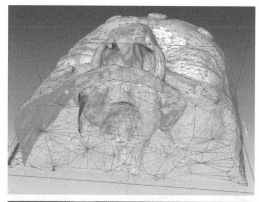

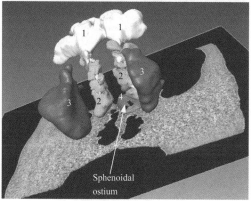

Maxillary
ostium

Sphenoidal
ostium

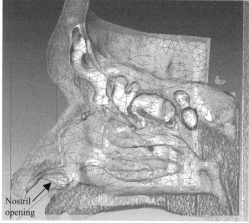

Nostril
opening

Figure 8.   The numbered sinuses are: Frontal (1), ethmoidal (2), maxillary (3) and sphenoidal sinuses (4).

Figure 9.   Transversal (top) and lateral (bottom) view of the 3D mesh from the dummy human head without the paranasal sinuses volume.

that will allow robotic assisted endonasal surgery simulations. Our modelling hypothesis was to consider the paranasal sinuses as a homogeneous, linear, isotropic and elastostatic material. For FE modeling it is very important the accurate estimation of the volume sinuses regions, because of the anatomical variation among individuals. Our results show the great asymmetry between left and right sides. For example, for the sphenoid and maxillary sinuses, a huge variation was seen between the right and left cavities (Table 1).

The results obtained from segmentation and 3D-reconstruction from paranasal sinuses CT images allow us to show the 3D morphologies of the sinuses from any viewpoint (Fig. 7). Moreover, we could also identify the ostiums that communicate nasal cavity and paranasal sinuses, as the maxillary and sphenoidal ostiums (Fig. 8).

The cavity volume of paranasal sinuses is an important index for the paranasal sinuses evaluation (Kawarai et al. 1999). Paranasal sinuses volumes

have conventionally been measured using cadavers by directly injecting a variety of material (Uchida et al. 1998 & Anagnostopoulous et al. 1991), while others employed CT images (Ariji et al. 1994; Kawarai et al. 1999 & Shi et al. 2006). In the current study paranasal sinuses volumes were estimated using commercial software (Amira). The accuracy of the 3D-recosntruction is an important prerequisite for a precise FE model of the nasal structures and it depends of the segmentation approach when the inner boundary of the paranasal cavity is semi-automatically traced. To validate the accuracy of the 3D reconstructions it was performed a comparison of these measurements with the real size of dummy human head sinuses (Table 1).

It was very difficult to guarantee full injection of the impression material in the ethmoidal sinuses, due to their anatomical complexity (Fig. 7). Moreover, the frontal sinus was damaged when attempting to remove it. For this reason, these casts were not used for the comparison. The morphology and the values

Table 2. Previous reports about the human paranasal sinuses volume. *Right + left sinuses.

Paranasal sinuses	Reference	Average* (ml)
Maxillary	Kikkawa, 1942	10.1
	Uchida et al. 1998	11.3 ± 4.60
	Oosugi, 1923	12.5
	Ariji et al. 1994	14.7 ± 6.33
	Yoshinaga, 1916	15.1
	Toida, 1937	15.6
	Anagnostopoulous et al. 1991	23.5
	Shi et al. 2006	24.76
	Kawarai et al., 1999	44.2 ± 6.6
Sphenoidal	Yoshinaga, 1916	4.1
	Toida, 1937	4.1
	Kikkawa, 1942	4.4
	Oosugi, 1923	4.5
	Kawarai et al. 1999	15.4 ± 6.9
Ethmoidal	Murai, 1937	4.9
	Kikkawa, 1942	5.3
	Kawarai et al. 1999	12.5 ± 1.6
Frontal	Kikkawa, 1942	2.8
	Oosugi, 1923	3.5
	Kawarai et al. 1999	8.1 ± 5.1

of the calculated sinuses volumes did not significantly differ from those obtained by other studies. The volumes were all in the respective ranges compared with previous reports (Table 2).

The generation of a FE mesh is critical since it affects both the accuracy and cost of subsequent numerical simulations. As shown in Figure 9, the paranasal sinuses geometry created with Amira software is simple enough to allow conversion into a mesh of finite elements that could be then later loaded by external forces. However, a refinement of the FE mesh to more accurately capture original anatomical detail is desirable. Currently, we have applied with success this approach for segmentation and 3D-reconstruction of sinuses cavities of *ex vivo* (Tingelhoff et al. 2007) and *in vivo* (Moral et al. 2007) CT images.

5 CONCLUSION

In this study, the approach proposed to obtain a precise geometric reconstruction of the human paranasal sinuses from CT images was evaluated. The results show accurate comparisons between volume estimation of the paranasal sinuses from an anatomical model by material injection and by 3D CT images. From the 3D-reconstruction of the sinuses regions, was obtained a simplified homogeneous model that resembled the general complex geometry of the paranasal sinuses cavities for FE models. It will to be used as

a background for the creation of further FEM models that should incorporate more structural features, such as inhomogenenity, different anatomical variations or pathological conditions.

REFERENCES

Anagnostopoulous S., Verieratos D., Spyropoulos N. 1991. Classification of human maxillar sinuses according to their geometric features. *Anat. Anz.* 173(3): 121–30.

Ariji Y., Kuroki T., Moriguchi S., et al. 1994. Age changes in the volume of the human maxillary sinus: A study using computerized tomography. *Dentomaxillofac. Radiol.* 23:163.

Becker, DG. 2003. The minimally invasive, endoscopic approach to sinus surgery. *Journal of Long-Term Effects of Medical Implants.* 13(3): 207–221.

Kawarai Y., Kkunihiro F., Teruhiro O., Kazunori N., Mehmet G., Masaaki F., Yu M. 1999. Volume quantification of healthy paranasal cavity by three-dimensional CT imaging. *Acta Otolaryngol.* 119(1), Supplement 540, 45–49.

Kennedy D.W., Zinreich S, Rosenbaum A.E., Johns M. 1985. Functional endoscopic sinus surgery: theory and diagnostic evaluation. *Arch. Otolaryngol.* 11: 576–82.

Kikkawa Y. Imaging for ethmoid sinusitis.1942. Practica Otologica 37: 701–62.

King J.M., Caladarelli D.D., Pigato J.B. 1994. A review of revision functional sinus surgery. *Laryngoscope.* 104: 4004–8.

Lang J. 1989. *Clinical anatomy of nose, nasal cavity and paranasal sinuses.* New York: Thieme Medical Publishers.

Manestar D., Manestar M., Groscurth P. 2006. *3D-model of the nose and paranasal sinuses. anatomy in coronal sections and corresponding CT-Images.* Tuttlinge: Endo-Press.

Moral A.I., Kunkel M.E., Rilk M., Wagner I., Eichhorn KWG, Bootz F, Wahl F.M. 3D Endoscopic Approach for Endonasal Sinus Surgery. Accepted at *29th Annual International Conference of the IEEE Engineering in Medicine and Biology Society*, Lyon, August, 2007.

Murai, Y. 1937. Anatomical structure of sphenoid sinus in Japanese. 56: 2267–388.

Navarro. J.A.C. 2001. *The nasal cavity and paranasal sinuses.* Berlin:Springer-Verlag.

Oosugi K. 1923. Anatomical study of nose in Japanese. II. Nose. *Practica Otologica,* 15: 1–300.

Pérez-Pinas I., Sabaté J., Carmona A., et al. 2000. Anatomical variations in the human paranasal sinus region studied by CT. *J. Anat.* 197: 221–227.

Rilk M., Winkelbach S. & Wahl F. 2006. Partikelfilter-basiertes Tracking chirurgischer Instrumente in Endoskopbildern. *Bildverarbeitung für die Medizin.* Springer, ISBN: 3-540-32136-5, 414–18.

Shi H., Scarfe W.C., Farman A.G. 2006. Maxilary sinus 3D segmentation and reconstruction from cone beam CT data sets. *Int J CARS.* 1:83–89.

Stammberger H., Posawetz W. 1990. Functional endoscopic sinus surgery. Concept, indications and results of the Messerklinger technique. *European Arch. of Otorhinolaryngol.* 247, 63–76.

Strauß G., Fischer M., Meixensberger J., Falk V., Trantakis C., Winkler D., Bootz F., Burgert O., Dietz A., Lemke HU. 2005. Bestimmung der Effizienz von intraoperativer technologie: Workflow-Analyse am Beispiel der endoskopischen Nasennebenhöhlenchirurgie. *HNO Springer Medizin Verlag*. 54:528–35.

Tingelhoff K., Wagner I., Eichhorn K., et al. 2006. Sensor-Based Force Measurement during FESS for Robot Assisted Surgery. *CURAC*. 2006.

Tingelhoff K., Moral A.I., Kunkel M.E., Rilk M., Wagner I., Eichhorn K.W.G., Wahl F.M., Bootz F. Comparison between Manual and Semi-automatic Segmentation of Nasal Cavity and Paranasal Sinuses from CT images. Accepted at *29th Annual International Conference of the IEEE Engineering in Medicine and Biology Society*, Lyon, August 2007.

Toida, N. Study of nasal cavity in Chinese. 1937. *J Med Mantura*, 27: 743–52.

Uchida Y., Goto M., Katsuki T., Akiyoshi T. 1998. A cadaveric study of maxillary sinus size as an aid in bone grafting of the maxillary sinus floor. *J Oral Maxillofac Surg* 56:1158–116, 1998.

Wagner I., Westphal R., Kunkel, E., et al. 2005. Measurement of Soft Tissue Properties in Ex Vivo Preparations for the Development of Robotic Assisted Functional Endoscopic Sinus Surgery. *CURAC* 2005.

Yoshinaga T. Anatomical study of sphenoid sinus in Japanese. *J Med Tokyo* 1915; 29: 397–437.

Zinreich S.J., Kennedy D.W., Rosenbaum A.E., et al. 1987. Paranasal Sinuses: CT Imaging Requirements or Endoscopic Surgery. *Radiolog.* 163:769–75.

Computational Vision and Medical Image Processing – João Tavares & Natal Jorge (eds)
© 2008 Taylor & Francis Group, London, ISBN 978-0-415-45777-4

Fluid-Structure interaction applied to blood flow simulations

J.S. Pérez, E. Soudah, J. García, E. Escolano & E. Oñate
International Center for Numerical Methods in Engineering, UPC, Barcelona, Spain

A. Mena, E. Heidenreich, J.F. Rodríguez & M. Doblaré
Group of Structures and Material Modeling. Aragon Institute of Engineering Research,
University of Zaragoza, Zaragoza, Spain

ABSTRACT: A coupled fluid-structure interaction model has being developed in order to study the vessel deformation and blood flow. This paper presents a methodology from which smooth surface are obtained directly form segmented data obtained from DICOM images. An integrated solution for segmentation-meshing-analysis is also implemented based on the GiD platform.

1 INTRODUCTION

Integration of different disciplines is an important aspects in the current developments of computer applications in biomedical engineering in order to go from imagining to computer simulations of tissue, organs, or biological system. Computed Tomography (CT), sometimes called CAT scan, or Magnetic Resonance Imaging (MRI for short) uses special equipment to obtain image data from different angles around the body, and then uses computer processing of the information to show a cross-section of body tissues and organs. Using this information, radiologists can more easily diagnose problems such as cancers, cardiovascular disease, infectious disease, trauma and musculoskeletal disorders (Fasel et al. 1998; Goldin et al. 2000). Due to their detailed information, these tools have become a useful tool in preventive medicine.

These images, on the other hand, can also be used to extract the geometry of the organs and tissues for computer analysis via segmentation of the DICOM image (Digital Imaging and Communications in Medicine). After performing the segmentation, a discretization of the domain is required for computer simulation. Generating a mesh for Finite Element simulations from a segmented image can be cumbersome due to the complicated geometry. To overcome this problem, methodologies which make direct use of the segmented data (voxel geometry) has been proposed (Heidenreich et al. 2006). Even though they result useful for electrophysiologic simulations, the non-smooth nature of the surface pose serious problems in solid mechanics and fluid-structure interaction simulations. Therefore, methodologies which provide smooth surface of the organs and tissues from biomedical images are desirable for computer simulations.

This paper presents a methodology for performing patient-specific computer simulations of cardiovascular systems, in particular fluid-structure interaction in an arterial bifurcation. In the example presented in the paper, the image of a femoral bifurcation is initially segmented and voxelized to defined the geometry. The voxel data is then used to produce computational meshes for the fluid and solid domains. Biomechanical data from the patient is performed for the simulations, given more realistic information regarding the performance of the particular cardiovascular system.

The remaining of the paper is organized as follows. Section 2 describes the image processing, segmentation and image voxelization. Sections 2.1 and 2.2 describe the meshing algorithm, while section 3.1, 3.2 and 3.3 explain the solid and fluid solvers and their interaction in FSI simulations. Finally section 4 presents an application in a femoral artery of a patient and section 5 gives some conclusions.

2 IMAGE PREPROCESSING

The objective of the image segmentation is to find and to identify objects with certain characteristics. The data segmentation will allow to visualize and to extract the part of the volume of interest. One of most widely used techniques is the grey thresholding segmentation. It is possibly the simplest and most direct method. The selection of the grey thresholds defining the object of interest is usually interactive, even though some alternative techniques have been proposed to determine it

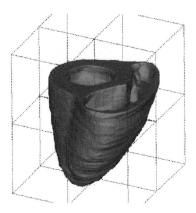

Figure 1. Segmented image of a human heart.

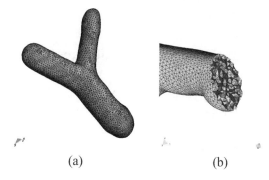

(a)   (b)

Figure 2. Mesh generated by marching cubes and advancing front.

in an automatic way. These threshold can be defined in either a local or global data set, and sometimes, over a three-dimensional data set.

In most applications, threshold segmentation is accompanied by manual segmentation which requires the physician expertise. Figure 1 shows a segmentation of a human heart.

The development of finite element simulations in medicine, molecular biology and engineering has increased the need for quality finite element meshes. After segmenting the medical image we end with a file with the image data and the value of the isosurface value defining the boundary of the volumen of interest.

The imaging data V is given in the form of sampled function values on rectilinear grids, $V = \{F(x_i, y_j, z_k) | 0 \leq i \leq n_x, 0 \leq j \leq n_y, 0 \leq k \leq n_z\}$. We assume a continuous function F is constructed through the trilinear interpolation of sampled values for each cubic cell in the volume. The data exchange between segmentation and meshing is done in VTK format (Kitware Inc 2006).

Given an isosurface value defining the boundary of the volume of interest we can extract a geometric model of it. We are interested in creating a discretization of the volumen suitable for finite element computation. In this work we have implemented the following methods to generate the finite element mesh to be used in the analysis stage: i) *Dual contouring*, ii) *Marching cubes*, iii) *Advancing front*, iv) *Volume preserving Laplacian smooth*. All this methods has being integrated into the general Pre/Post-processor GiD (CIMNE 2006).

2.1 *Tetrahedral mesh generation*

To build a tetrahedral mesh from voxels we combine the Marching Cubes method to generate first the boundary mesh first and then, after a smoothing, an Advancing Front (Lohner and Parikh 1988) method to fill the interior with tetrahedras.

The Marching Cubes (Lorensen and Cline. 1987) algorithm visits each cell in the volume and performs local triangulation based on the sign configuration of the eight vertices. If one or more vertex of a cube have values less than the user-specified isovalue, and one or more have values greater than this value, we know the voxel must contribute some component of the isosurface. By determining which edges of the cube are intersected by the isosurface, we can create triangular patches which divide the cube between regions within the isosurface and regions outside. By connecting the patches from all cubes on the isosurface boundary, we get a surface representation.

Some of the triangles generated by the Marching Cubes method does not exhibit good quality to be used in finite element computation. In order to improve the quality of those elements we apply a laplacian smooth which volume preserving. The smoothing algorithm implemented is simple: it try to preserve the volume after each application of the laplace operator by doing an offset of the vertices along the normals. Figure 2(a) shows the boundary mesh generated by Marching Cubes and smoothed.

The Advancing Front (Lohner and Parikh 1988) is an unstructured grid generation method. Grids are generated by marching from boundaries (front) towards the interior. Tetrahedral elements are generated based on the initial front. As tetrahedral elements are generated, the "initial front" is updated until the entire domain is covered with tetrahedral elements, and the front is emptied. Figure 2(b) shows a cut of the tetrahedral mesh generated by the Advancing Front method.

2.2 *Hexahedral mesh generation*

The dual countour method (Ju et al. 2002) generates a quadrilateral mesh aproximating the boundary of the body.

Dual contouring analyzes those edges that have endpoints lying on different sides of the isosurface, called sign change edge. Each sign change edge is shared by four cells, and one minimizer is calculated for

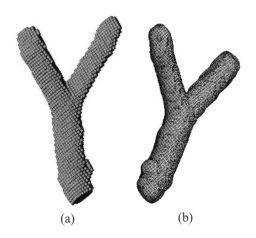

(a)   (b)

Figure 3.   (a) Difference finite mesh, (b) Dual contour in 3D.

each of them by minimizing a predefined Quadratic Error Function: $\mathbf{QEF}[x] = \sum (n_i(x - p_i))^2$, where p_i, n_i represent the position and unit normal vectors of the intersection point respectively. For each sign change edge, a quad is constructed by connecting the minimizers. These quads provide an approximation of the isosurface.

Here we generate hexahedral meshes following (Zhang et al. 2003) which is a variation of the original method adapted to generate hexahedral elements. The uniform hexahedral mesh extraction algorithm is simple: each interior vertex (a grid point inside the volume), which is shared by eight cells, is analyzed. One minimizer is calculated for each of the shared cells, and those eight minimizers construct a hexahedron. An example of a mesh generated by dual contour is shown in Figure 3(b).

Also a finite difference mesh, which is made of axis aligned orthogonal hexahedral, can be obtained. In this mesh each pair of neighbor voxels on the boundary of the body share a common face. Figure 3(a) shows and example of such a mesh.

3 ANALYSIS

3.1 Solid mathematical formulation

Biological soft tissues sustain large deformations, rotations and displacements, have a highly nonlinear behaviour and anisotropic mechanical properties and show a clear time and strain-rate dependency (Humphrey 2002). In order to capture the nonlinear anisotropic hyperelastic behavior, it is necessary to consider the formulation of finite strain hyperelasticity in terms of invariants with uncoupled volumetric-deviatoric responses (Simo and Taylor 1991), and employed for anisotropic soft biological tissues in (Weiss et al. 1996; Holzapfel et al. 2000).

Based on the kinematic decomposition of the deformation gradient, and following (Holzapfel et al. 2000) the free energy for an anisotropic hyperelastic soft tissue can be written in a decoupled form as:

$$\Phi(\mathbf{C}, \mathbf{a}_0, \mathbf{b}_0) = U(J) + \bar{\Phi}(\bar{I}_1, \bar{I}_2, \bar{I}_4, \bar{I}_6), \tag{1}$$

with

$$\begin{aligned} \bar{I}_1 &= \mathrm{tr}\bar{\mathbf{C}}, &\qquad \bar{I}_2 &= 1/2(\mathrm{tr}(\bar{\mathbf{C}})^2 - \mathrm{tr}\bar{\mathbf{C}}^2), \\ \bar{I}_4 &= \mathbf{a}_0 \cdot \bar{\mathbf{C}} \cdot \mathbf{a}_0, &\qquad \bar{I}_6 &= \mathbf{b}_0 \cdot \bar{\mathbf{C}} \cdot \mathbf{b}_0, \end{aligned} \tag{2}$$

$\bar{I}_1$ and $\bar{I}_2$ are the invariants of $\bar{\mathbf{C}}$, right Cauchy-Green tensor of the deformation gradient, and $\bar{I}_4$ and $\bar{I}_6$ are the square of the stretches along $\mathbf{a}_0$ and $\mathbf{a}_0$ respectively. Let $\mathbf{a}_0$ and $\mathbf{b}_0$ be the directions of collagen fibers within the tissue defining the transverse anisotropic behavior of the tissue (Holzapfel et al. 2000). In terms of the free energy, the total energy of the system is then given by the functional:

$$\Pi(\mathbf{u}) = \int_\Omega \Phi(\mathbf{X}, \mathbf{C}\mathbf{a}_0, \mathbf{b}_0)dV + \Pi_{\mathrm{ext}}(\mathbf{u}), \tag{3}$$

where the explicit dependence on $\mathbf{X}$ accounts for the heterogeneity, and Π_{ext} is the potential energy of the external loading. The finite element formulation is based in the minimization of (3), which first variation with respect to $\mathbf{u}$ along the direction η is given by

$$D_u\Pi(\mathbf{u}) \cdot \eta = \int_\Omega [\sigma : \nabla\eta - g_{\mathrm{ext}}(\eta)]dV, \tag{4}$$

where σ is de Cauchy stress and g_{ext} is the virtual work of the external loading. Even though (4) is valid for compressible solids and refer to this to keep the exposition simple, for a quasi-incompressible formulation the reader is referred to (Simo and Taylor 1991; Weiss et al. 1996).

For the solid simulations, the femoral artery was considered as an isotropic material with an strain energy function given by

$$\Phi(\mathbf{C}) = c_{10}(\bar{I}_1 - 3.0) + c_{20}(\bar{I}_1 - 3.0)^2 + U(J) \tag{5}$$

with $c_{10} = 174.0\,\mathrm{kPa}$, and $c_{20} = 1880\,\mathrm{kPa}$ (Ballyk et al. 98). The material has been treated as incompressible with the incompressibility constrained treated by means of the three variational principle introduced in (Simo and Taylor 1991 Weiss et al. 1996). Finite element calculations were carried with the finite element software MYDAS implemented within the DSS-DISHEART program.

All displacements were restricted at both ends of the artery in the solid domain.

3.2 Fluid mathematical formulation

Blood is a suspension of red and white cells, platelets, proteins and other elements in plasma and exhibits an anomalous non-Newtonian viscous behavior when exposed to low shear rates or flows in tubes of less than 1mm in diameter. However, in large arteries, vases of medium calibre as well as capillaries, the blood may be considered a homogeneous fluid, with "standard" behaviour (Newtonian fluid) (Perktold et al. 1991).

The governing equations for blood flow used are the Navier-Stokes equations with the assumptions of laminar, homogenous, incompressible (90% of the blood is water), and Newtonian flow. For the representation of the Navier-Stokes equations on the deforming fluid domain Ω of the model based on the arbitrary Lagrangian Eulerian (ALE) method (Formaggia et al. 1999), we adopt the following notation: Ω is a three-dimensional region denoting the portion of the district on which we focus our attention, and $x = (x_1, x_2, x_3)$ is an arbitrary point of Ω; $v = v(x, t)$ denotes the blood velocity. For $x \in \Omega$ and $t > 0$ the conservation of momentum and continuity in the compact form are described by the following equations (6) in $\Omega(0, t)$:

$$\rho \left(\frac{\partial u}{\partial t} + (u \cdot \nabla u) \right) + \nabla p - \nabla (\mu \triangle u) = \rho \cdot f \\ \nabla u = 0 \qquad (6)$$

where $u = u(x, t)$ denotes the velocity vector, $p = p(x, t)$ the pressure field, ρ density, μ the dynamic viscosity of the fluid and f the volumetric acceleration. Blood flow is simulated for average blood properties: molecular viscosity $\mu = 0.0035$ Pa.s and density $\rho = 1050$ kg/m3. The volumetric forces ($\rho \cdot f$) are not taken into account within this analysis.

The boundary conditions for the fluid domain, velocity boundary conditions were imposed at the main branch of the artery, the inflow mean velocity is time-dependent and the volumetric flow rate is oscillatory, as shown in Figure 4(a). The pulsatile velocity waveform is represented by a polynomial equation based on the in-vivo measurement by magnetic resonance imaging. This pulse is appropriate for normal hemodynamic conditions at the end of abdominal segment of the human aorta.

Where pressure boundary conditions were applied at both bifurcations ends. The pressure is also time-dependent and oscillatory, as show in figure 4(b). The pulsatile pressure waveform was calculated using a 1D model (E. Soudah et al. 2007) and validated with different analysis cases.

Tdyn code was used for solving the fluid problem (Compass 2006). Tdyn is a fluid dynamic (CFD) simulation environment based on the stabilized Finite Element Method, which uses an implicit Fractional Step Method based on finite element method (FEM) (Garcća 1999). A new stabilization method, known as

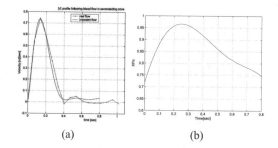

(a)   (b)

Figure 4. Boundary conditions used in the simulations. (a) Velocity waveform at the inlet (main branch of the bifurcation), (b) Pressure waveform at the outlet of the bifurcation.

"Finite Increment Calculus", has recently been implemented (Oñate 1999). By considering the balance of flux over a finite sized domain, higher order terms naturally appear in the governing equations, which supply the necessary stability for a classical Galerkin finite element discretisation to be used with equal order velocity and pressure interpolations.

3.3 Fluid-solid interaction

A distinctive feature of the fluid-structure problem is the coupling of two different sub-problems, the first referring to the fluid (whose solution is characterized by the velocity and pressure fields of the blood) and the second to the structure (whose unknown variable is the displacement field of the vascular wall). In this section, we illustrate an explicit algorithm for the coupling of fluid and structure.

To consider the problem arising when coupling fluid and structure models, let us restrict our analysis to a domain Ω. The boundary Γ is composed of a portion Γ_C, which is assumed to be compliant, and a part Γ_F, which is assumed to be fixed.

Let us consider the interface conditions between the fluid and the structure. The first condition ensures the continuity of the velocity field:

$$v = \dot{\eta} \qquad x \in \Gamma_C \qquad (7)$$

where $\dot{\eta}$ is the velocity field of the vessel.

The fluid exerts a surface force field over the vessel. These forces must be treated as a (Neumann) boundary data for the structure problem:

$$\Phi = -Pn + nS \qquad x \in \Gamma_C \qquad (8)$$

where Φ is the forces field vector applied in the vessel due to the blood flow.

The fluid-structure interaction problem we deal with is therefore specified by (6),(7),(8) and (5). In view of its numerical solution, the coupled problem

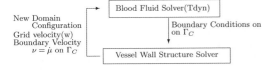

Figure 5.   Representation of the splitting in two sub-problem for our approach (coupled solver).

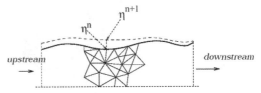

Figure 6.   Updating of the mesh.

ought to be split at each time step into two sub-problems, one in Ω , the other on Ω_s (the vessel domain), communicating to one another through the matching conditions (7) and (8). In particular, the structure problem provides the boundary data for the fluid problem; vice-versa, the fluid problem provides the forcing term for the structure.

V and **P** respectively as usual will denote the unknowns referring to the velocity and pressure, while the ones relative to the structure are denoted by **H**.

The algorithm iteration process for each time step can be resumed as follows:

a) Solving the structure problem (vessel wall) with the boundary terms due to the blood flow. At the first time level, the scheme is suitable modified, taking into account the initial data on the position and the velocity at time $t = 0$.
b) Updating domain configuration and boundary conditions for the fluid solver: Once $\mathbf{H}^{n+1}$ is known, we can compute the domain deformations and the movement of the nodes of the grid for the fluid. The new position of the boundary Γ_C is computed through the relation:

$$x_i^{n+1} = x_i^0 + \eta_i^{n+1} \qquad (9)$$

The displacement of the nodes of the grid for the fluid is obtained as a diffused into the fluid domain of the boundary displacement. Diffusion process is based on an arrangement by levels of the mesh nodes, where level 0 corresponds to the mesh nodes on the ship surface, level 1 to the nodes connected to level 0 nodes, and so on.

We compute the mesh velocity **w** by the equation:

$$w^{n+1} = \frac{1}{\Delta t} \cdot (x^{n+1} - x^n) \qquad (10)$$

The idea underlying this approach is to take advantage of the regularization due to the inversion of the Laplace operator in order to have an acceptable mesh. From time to time, however, it could be necessary to remesh the whole domain, if the grid is too distorted after a certain number of steps.

Finally, the mesh update is obtained by:

$$x^{n+1} = x^n + \Delta t \cdot w^{n+1} \qquad (11)$$

Solving the blood flow problem

c) The ALE formulation of the Navier-Stokes equations (6) is solved by implicit 2nd order accurate projection schemes. The choice of the time-advancing method satisfies the Geometric Conservation Laws.
d) Computing the force field applied as a boundary condition in the structure problem due to the fluid.

When the boundary nodes of the structure and the fluid are not coincident is necessary to make an interpolation of the nodal quantities during the interaction algorithm. The methodology used for this interpolation is based on an octtree search algorithm of elements and standard finite element techniques.

This algorithm performs a staggered coupling between the fluid and the structure problems; therefore, it should generally undergo stability limitations on the time step. These limitations could turn out to be restrictive in practical computations.

4   EXAMPLES: FSI in the Femoral Artery

An example of a human femoral bifurcation is considered in this section to demonstrate de methodology. Images of the femoral bifurcation were captured in a 16 Detector/16 Slice Toshiba Multidetector CT Scanner using a slice thickness of 3.2 mm with slices reconstructed every 1.6 mm to maximize longitudinal resolution. Images were reconstructed using Maximum-Intensity-Projection (MIP) algorithm in the frontal and sagittal views.

Arterial segmentation has been automatically performed by means of threshold segmentation using the Visual DICOM software within DSS-DISHEART.

From the voxel image given in figure 3(a), surface and solid meshes have been created for the fluid and solid domains using the software GiD within the DSS-DISHEART. The resulting meshes were 3003 hexahedral finite elements for the solid wall and 74125 tetrahedral elements for the fluid. Figure 7 show the meshes for the solid (arterial wall) and the fluid respectively. The solid mesh for the arterial wall has been generated by extruding the surface mesh a uniforme thickness of 1.5 mm.

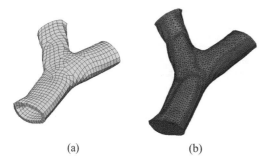

(a)   (b)

Figure 7. Finite element meshes obtained from the VTK file for the femoral artery. (a) Volume mesh for the arterial wall, (b) Volume mesh for the fluid.

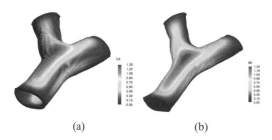

(a)   (b)

Figure 8. Displacement field for the couple problem. (a) Magnitude of the displacement of the solid wall in (mm), (b) Displacement of the ALE-mesh in (mm). The figure demonstrate the effective coupling achieved between the solid and fluid solvers.

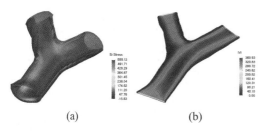

(a)   (b)

Figure 9. Stress and velocity fields in the artery at the time of maximum pressure. (a) Stress field in (kPa), (b) Velocity field in (mm/sec).

Figure 8 shows the displacement filed of the solid wall and of the ALE-mesh at the instant of maximum pressure. The figure demonstrate the effective coupling achieved between the solid and the fluid solvers.

Figure 9 shows the stress field in the arterial wall and the velocity field in the artery mid-plane. As expected with incompressible fluids, the maximum stress in the arterial wall occurs at the section were fluid velocity reduces since it implies an increment in the local pressure.

5 CONCLUSIONS

The results show the viability of applying the presented methodology to generate computational finite element meshes from segmentation files obtained from medical images. Also we have solved successfully the coupled fluid-structured interaction between the blood flow and the artery wall. We think this methodology/tool opens new possibilities for patient specific biomechanical applications.

REFERENCES

Ballyk, P., C. Walsh, J. Butany, and M. Ojha (98). Compliance mismatch my promote graft-artery intimal hyperplasia by altering suture-line stresses. *Journal of Biomechanics 31*, 229–237.

CIMNE (2006). Gid – the personal pre and postprocessor. Web address: http://www.gidhome.com/.

Compass (2001–2006). Tdyn: fluid dynamic (cfd) simulation environment. http://www.compassis.com. Compass Ingeniería y Sistemas, S.A.

E. Soudah, F. Mussi, and E. Oñate (2007). Validation of the one-dimensional numerical model in the ascending-descending aorta with real flow profile. In *III International Congress on Computational Bioengineering (ICCB 2007) Venezuela.*

Fasel, J., D. Selle, C. Evertsz, F. Terrier, H. Peitgen, and P. Gailloud (1998). Segmental anatomy of the liver: poor correlation with CT. *Radiology 206*, 151–156.

Formaggia, L., F. Nobile, A. Quarteroni, and A. Veneziani (1999). Multiscale modelling of the circulatory system: a preliminary analysis. *Computing and Visualisation in Science 2*, 75–83.

García, J. (1999). *A Finite Element Method for the Hydrodynamic Analysis of Naval Structures (in Spanish), Ph.D. Thesis, , Barcelona*. Ph. D. thesis, Univ. Politécnica de Cataluña.

Goldin, J. M., O. Ratib, and D. R. Aberle (2000). Contemporary cardiac imaging: An overview. *Journal of thoracic imaging 15(4)*, 218–22.

Heidenreich, E., A. M. A. J. Rodríguez, S. Olmos, and M. Doblaré (Noviembre 2006). Simulación de electrofisiología cardiaca de imagenes médicas. Modelos numéricos especificos a pacientes. In *Congreso Annual de la Sociedad Española de Ingenier'ia Biomédica.*, Pamplona: Spain.

Holzapfel, G., C. Gasser, and R. Ogden (2000). A new constitutive framework for arterial wall mechanics and a comparative study of material models. *Journal of Elasticity 61*, 1–48.

Humphrey, J. D. (2002). *Cardiovascular Solid Mechanics: Cells, Tissues, and Organs.* Springer.

Ju, T., F. Losasso, S. Schaefer, and J. Warren (2002). Dual contouring of hermite data. In *SIGGRAPH*, pp. 339–346.

Kitware Inc (2006). Vtk file formats. Online document:: http://www.vtk.org/pdf/file-formats.pdf.

Lohner, R. and P. Parikh (1988). Three dimensional grid generation by the advancing-front method. *International Journal for Numerical Methods in Fluids 8*, 1135–1149.

Lorensen, W. E. and H. E. Cline (1987). Marching cubes: A high reso-lution 3d surface construction algorithm. In *SIGGRAPH*, pp. 163–169.

Oñate, E. (1999). A stabilised finite element method for incompressible viscous flows using a finite increment calculus formulation. Technical report, CIMNE.

Perktold, K., M. Resch, and H. Florian (1991). Pulsatile non-newtonian flow characteristics in a three-dimensional human carotid bifurcation model. *Journal of biomechanical engineering 13*, 507–515.

Simo, J. C. and R. L. Taylor (1991, February). Quasi-incompressible finite elasticity in principal stretches.

continuum basis and numerical algorithms. *Computer Methods in Applied Mechanics and Engineering 85*(3), 273–310.

Weiss, J. A., B. N. Maker, and S. Govindjee (1996, August). Finite element implementation of incompressible, transversely isotropic hyperelasticity. *Computer Methods in Applied Mechanics and Engineering 135*, 107–128.

Zhang, Y., C. Bajaj, and B.-S. Sohn (2003). Adaptive and quality 3d meshing from imaging data. In *SM '03: Proceedings of the eighth ACM symposium on Solid modeling and applications*, New York, NY, USA, pp. 286–291. ACM Press.

Computational Vision and Medical Image Processing – João Tavares & Natal Jorge (eds)
© 2008 Taylor & Francis Group, London, ISBN 978-0-415-45777-4

Automatic accuracy evaluation for distance measurement by medical ultrasound scanners

F.P. Branca & A. D'Orazio*

Department of Mechanics and Aeronautics University of Rome "La Sapienza"

S.A. Sciuto & A. Scorza

Department of Mechanical and Industrial Engineering University of Rome

ABSTRACT: In recent years several ultrasound (US) scanners for medical diagnosis was presented and specific methods for ultrasounds images quality evaluation was developed to verify the apparatuses performances. The proposed methods still depend on subjective evaluation, performed by expert operators, of images produced by dedicated test objects (US phantoms). Distance accuracy is usually evaluated on screen by means of device settings adjustment and use of electronic caliper implemented in the scanner, both set by the operator, and with methods not standardized; therefore results are dependent on adopted procedure. In present work, an automatic accuracy evaluation method for distance measurement by medical US scanners is developed. It adopts Point Spread Functions phantoms as reference and allows accuracy evaluations in vertical and horizontal distance measurements. Analysis is performed on uncompressed two-dimensional images produced by the scanner, with a least squares estimation on results. An index for global distance measurement accuracy is proposed.

1 INTRODUCTION

In recent years the number of medical ultrasound (US) scanners in Diagnostic Imaging world market is grown, so specific methods for performance evaluation of these instrumentations have been studied. The methods for US scanners evaluation are not standardized and often are based on observations directly performed by human operator on monitor. The evaluation is performed by expert operators on US images produced by means of test objects and in particular the accuracy in distance measurement is usually evaluated on the US scanner screen by means of an adjustment of device settings and use of electronic caliper implemented in the scanner, both directly set by the. Also the method for distance accuracy evaluation is not standardized and, therefore, the result is definitely dependent on the adopted procedure. Some studies shown US distance measurements depend on acoustic velocity of the medium (Goldstein 2000), Digital Scan Converter (DSC) performances (Winter et al. 1985), spatial resolution and aberration correction (Flax & O'Donnel 1988). In scientific literature some works appear in which computer-based analysis for image quality evaluation in medical US is

proposed and developed (Gibson et al. 2001, Browne et al. 2004, Gupta et al. 2002, Goldstein 2004). Most of them do not describe an objective analysis of the distance accuracy, whereas subjective comparison between electronic caliper measurements conducted on the US scanner monitor and real distances (as declared by phantom manufacturer), performed by a human operator, are reported (Fornage et al. 2000, Goodsit et al. 1998). Moreover, it does not seem clear how the results of these methodologies must be handled to provide an accuracy measure, neither an exhaustive justification of the instrumentation acceptance criterions is expressed. By the way in this paper, measures on a large number of medical US scanners of recent production (from the 2003) and of different technological classes are performed: they have pointed out that, for medical ultrasound instrumentation acceptance, a limit of 1.5–2% on distance measurements accuracy, as reported in Goodsit et al. (1998), is often too restrictive, even for scanners of good performances and high quality images. Thus in this paper an accuracy index related to distances measure along a direction has been investigated and calculated by computer-based method for US image analysis. Procedure is based on analysis of B-mode images, produced by the scanner when a US phantom is used, in order to determine the accuracy in known distances measure for both vertical and horizontal directions. A second index is proposed summarizing in a single vector results in both directions.

2 METHOD AND EXPERIMENTAL SET UP

Error affecting vertical and horizontal distance measurements in US images is also due to the difference between effective acoustic velocities c_m of the medium and acoustic velocity c_t used to calibrate the scanner (design velocity of imaging equipment is usually the mean velocity in soft tissues $c_t = 1540$ m/s). Using also results by Goldstein (2000), Flax & O'Donnell (1988), an acceptance limit in distance measurement accuracy could be proposed for US scanners. Actually, this limit is a function of acoustic velocities difference and also depends on the image scan conversion in DSC. With regard to the mentioned causes of error, theoretical vertical and horizontal discards Δy and Δx between measured distance in US image and real distance can be computed, by considering as valid the following expressions reported in Goldstein (2000), where the distance x is measured from image center and distance y is measured from image top.

For linear, phased, vector and convex array respectively it results

$$\Delta y = y \cdot \left(\frac{c_t}{c_m} - 1 \right) \tag{1}$$

$$\Delta x = 0 \tag{2}$$

$$\Delta y = \sqrt{x^2 + y^2} \left(\frac{c_t}{c_m} \sqrt{1 - \left(\frac{c_t}{c_m} \right)^2 \frac{x^2}{x^2 + y^2}} - \frac{y}{\sqrt{x^2 + y^2}} \right) \tag{3}$$

$$\Delta x = x \left[\left(\frac{c_t}{c_m} \right)^2 - 1 \right] \tag{4}$$

$$\Delta y = \frac{y}{\cos \vartheta_m} \left(\frac{c_t}{c_m} \sqrt{1 - \left(\frac{c_t}{c_m} \right)^2 \sin^2 \vartheta_m} - \sqrt{1 - \sin^2 \vartheta_m} \right) \tag{5}$$

$$\Delta x = y tg \vartheta_m \left[\left(\frac{c_t}{c_m} \right)^2 - 1 \right] \tag{6}$$

$$\Delta y = (y + R) \left(\frac{c_t}{c_m} - 1 \right) \left(1 - \frac{R}{\sqrt{(y + R)^2 + x^2}} \right) \tag{7}$$

$$\Delta x = x \left(\frac{c_t}{c_m} - 1 \right) \left(1 - \frac{R}{\sqrt{(y + R)^2 + x^2}} \right) \tag{8}$$

Table 1. Theoretical discards Δy and Δx in distance measurements on ultrasound images due to difference between sound propagation speed c_m in the medium and calibration US scanner value c_t for different probe.

Probe	Range of values for c = 1480 ÷ 1650 m/s	
	Δy (cm)	Δx (cm)
Linear array	(y = 1 cm) −0.07 ÷ 0.040	(x = 1 cm) 0
Phased array	(x = 0 y = 1cm) −0.07 ÷ 0.040	(x = 1 cm) −0.13 ÷ 0.08
Vector array	($\theta_t = 0$ y = 1 cm) −0.07 ÷ 0.040	($\theta_t = 45°$ y = 1 cm) −0.15 ÷ 0.08
Convex array	(x = 0 y = 1 cm) −0.07 ÷ 0.04	(y = 10 cm R = 6 cm x = 1 cm) −0.04 ÷ 0.02

With regard to vector array, ultrasound beam is subjected to refraction phenomenon due to different acoustic velocities into probe and medium; refraction angle θ_m is related to image angle θ_t by

$$\sin \vartheta_m = \frac{c_m}{c_t} \sin \vartheta_t \tag{9}$$

In Table 1, provided c_m ranges from 1480 m/s to 1650 m/s and for each probe model is $c_t = 1540$ m/s, theoretical discard Δy of distances between objects in the image center (x = 0) along a vertical direction is between −7% and +4%, while the range of values in horizontal direction Δx could be grater and depends strongly on the probe model and settings. The range of values in table 1 are an example of relative accuracy in the distances measure which can theoretically affect any US scanner calibrated with $c_t = 1540$ m/s during diagnostic exams. Moreover, errors due to DSC interpolation and approximation must be also added. However an evaluation criterion of accuracy of US scanner in distances measure must have in some cases a discrimination threshold higher than 2%, to avoid excessive severity even with apparatuses of good performances as a whole. Besides such threshold must be function of probe type, measurement direction, US scanner settings and evaluation method. It must be pointed out that methods based on measurements by human operators of known distance directly on the monitor by means of electronic caliper provide uncertainty up to 30% of measured distance value due to eye sensitivity, monitor brightness, operator experience, etc. The development of computer-based methods appear useful, because they reduce the measure uncertainty due to higher objectivity, limited dependence on instrumentation settings and independence from room illumination. In particular the computer-based procedure here proposed can provide data for technical evaluation of US scanners, even to not experienced customers. Main steps of this procedure are described in the following.

1 Acquisition of ultrasound images by US phantoms. Generally an ultrasound phantom is an object containing test objects of known characteristics embedded in a material simulating reflection, absorption and diffusion properties of biological tissues to ultrasounds. Test objects of phantoms used in this work (Point Spread Function PSF phantoms) are extremely small in the transducer azimuthal plane (nylon wire 0.1 mm diameter) at a rigorously known distances one each other. Images are stored in loseless format (Tag Image File Format TIFF) in a directory of a pc.

2 Finding and loading the filed image from directory. On operator demand, software shows files in the directory, in order to select the previously stored image file.

3 Input of Field of View (FOV) size. Software asks the operator for the depth of displayed FOV (this information is always available on US image) and sets in the image pixel and mm correspondence (scale factor) Some scanners display images where each pixel represents same size both in vertical and horizontal directions; the implemented method can operate with systems where pixel does not have same size in vertical and horizontal direction: in this case software asks the operator for the actual width of displayed FOV.

4 Diagnostic image extraction. Software recognizes and extracts image diagnostic part (B-scan part).

5 Choice of vertical and horizontal distances number to process. Number of vertical and horizontal target pair for the test are chosen by operator..

6 Choice of Region of Interest (ROI). Operator chooses ROI size and software searches for each target. Operator also inputs location uncertainty of test objects, generally declared in US phantom specifications.

7 Start of measuring test. For all target couple, operator must roughly choose each test object location in diagnostic image and enclose the nominal distance between them. In particular, operator locates a ROI in proximity of each target without center the target since it is enough that it lies into ROI) and software individualize its barycentric coordinates using it for measuring distances for each couple of test object.

8 Evaluation of relative accuracy distances measure e_r and data presentation. For each test objects pair software calculates the nominal and measured distances difference in the image and shows it on a graph, where measured discards versus nominal distances are plotted. Then, data are interpolated with a least squares method to fit a linear model where relative accuracy e_r is the angular coefficient of least squares straight line.

Table 2. Ultrasound phantoms characteristics. For spacing (vertical group) location uncertainty is ±0.2 mm; for spacing (horizontal group) in CIRS50, it refers to axial resolution and lateral resolution test objects.

Material	CIRS 54 Zerdine™	CIRS 50 Zerdine™
Speed of sound	1540 ± 3 m/s	1540 ± 3 m/s
Attenuation Coefficient	0.5 dB/cm/MHz	0.5 dB/cm/MHz
Line targets	Monofilament nylon wires 0.1 mm in diameter	Monofilament nylon wires 0.1 mm in diameter
Depth Range (vertical group)	18 cm	8 cm
Spacing (vertical group)	2 cm	1 cm
Width Range) (horizontal group)	12 cm (at 9 cm depth)	15 mm and 30 mm
Spacing (horizontal group)	2 cm	1 to 6 mm

The developed method provides also information on the computed relative accuracy uncertainty. It is basically due to three causes: uncertainty of the reference (location of test object), pixelation (Goldstein 2000) and curve fitting method. Results are quite independent from spatial resolution (i.e. how closely two reflectors or scattering regions can be one to another while they can be identified as different reflectors), since measurements are done for distances greater than US scanner resolution (about 10 times or higher) and target spreading due to spatial resolution degradation do not appreciably change its barycentric position.

Measurement set up is composed by US scanner, probe on US phantom and notebook pc. Ultrasound phantoms used in present work contain nylon wires at known distances as test objects (PSF phantoms) and they are a CIRS54 model for low and middle frequencies (since US system performances depend on transducers bandwidth, US phantoms are often classified according to frequency range of US probes) and a CIRS50 model for high frequencies. Some of their characteristics are shown in table 2.

The probe is applied on the phantom by means of pliers assembled on a support (figure 1a) and its position is adjusted to clearly display the test objects (pin targets) in the US image (figure 1b). After that, image is acquired and directly stored in US scanner hard disk and then exported on a peripheral (MO disk, CD-ROM) in a loseless TIFF or DICOM format (Dicom images are converted to a TIFF format with a viewer software given from the manufacturers). Experimental data are collected on 14 different US scanner models with 3 probes each one (42 probes); results

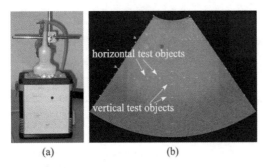

(a)   (b)

Figure 1. In (a) a convex probe on the PSF phantom, in (b) corresponding ultrasound image.

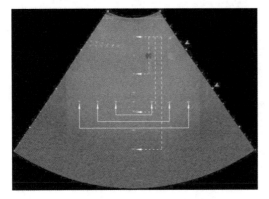

Figure 2. Distance measurement on the ultrasound image.

from two different probes on the same US scanner are considered as produced by two different US system with similar technology level.

The software was developed in Matlab to evaluate relative accuracy in vertical and horizontal distance measurements with high versatility. It also allows to choose any number of test objects pairs and evaluate corresponding distances (Fig. 2).

As mentioned, the method is computer-based. The operator initially chooses areas of image where test object lies within (he must take care that in each area only one object is present) and software calculates distance for every pair, after finding the barycentre for each target; then it computes least squares straight line from known distances and corresponding absolute discards, directly measured on the image, with an uncertainty evaluation, plots data (discards vs real distances) and their curve fitting and shows relative accuracy e_r values both for vertical and horizontal distances. Figure 3 shows an example of plotted results, where discard e is given as absolute difference $e = |d_r - d_i|$ between nominal d_r and measured d_i distances. Results uncertainty mainly depends on test objects location uncertainty in US phantom (datum provided by manufacturer), on the test object number

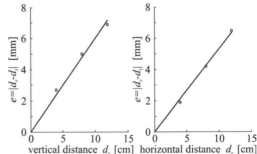

vertical distance d_r [cm]   horizontal distance d_r [cm]

Figure 3. Output graphs show discard e vs nominal distance. Relative accuracy is given by the least squares straight line angular coefficient.

used in the measure and on the instrumentation linearity (an estimation of linearity in distance measurement can be performed by means of correlation coefficient (R-square). Thus e may range from some percentage up to 30% or more.

3 RESULTS AND DISCUSSION

A first set of experimental data are collected from new US equipments. For vertical and horizontal distances respectively, percentage relative accuracy $e_\%$ is $1.6 \pm 0.8\%$ and $1.8 \pm 0.9\%$ for 14 linear array probes, $1.7 \pm 0.7\%$ and $1.3 \pm 0.6\%$ for 19 phased array probes, $4.7 \pm 3.1\%$ and $3.3 \pm 2.2\%$ for 9 convex array probes. Within each group (linear, phased and convex array probes) settings are similar in frequency, field of view, dynamic range, transmit power, gain and focal zones. First results are in agreement with both literature values and theoretical prediction. Moreover, they seem to be mainly dependent on image Field of View, test object number and distances (vertical measurement are collected from the top of the image, horizontal distance are collected from the centre of the image to its extremity). On the other hand, further tests seem to confirm a substantial independence from dynamic range, frequency, vertical test object centering, focal zone position and spatial resolution. Data do not present significant difference with regard to vertical and horizontal direction. Moreover measured data suggest that, for quality assurance test of medical US scanners, an upper limit of 2% found in literature for the admissible relative accuracy in distance measurements, could result excessively severe for the proposed method and it must be related to probe model and settings. Results for each evaluation (accuracy in vertical and horizontal direction) can be summarized in a complex number vector $\mathbf{m} = m_V + jm_H$ where m_V is the vertical accuracy while m_H is the horizontal accuracy. It can be plotted in the cartesian graph first quadrant whit m_H component on x-axis and m_V one

194

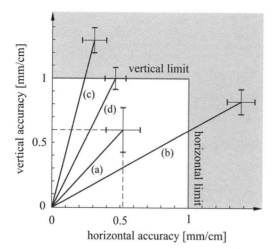

Figure 4. Example of representation of proposed accuracy index in distance measurement for ultrasound scanners: on x-axis there is the relative error in the horizontal direction while on y-axis is given the relative error on vertical direction. (a) the vector corresponds to a positive result (scanner accuracy is acceptable). (b) and (c) correspond to a negative result (scanner accuracy is not acceptable) while (d) represents a doubtful result.

on the y-axis as shown in Figure 4. The US scanner distance measurement accuracy is judged positively if vector **m** is within a rectangular area limited by maximum admissible accuracy in horizontal and vertical directions.

4 CONCLUSIONS

A computer-based method for accuracy evaluation in measuring vertical and horizontal distances for modern clinical US scanners. The procedure provides relative accuracy and its uncertainty, evaluated from a linear least square curve fitting on the measure of known distances versus absolute discards (calculated as nominal distance and measured one difference) in the US image. A single complex index keeping in account obtained results is proposed, as a vector with components corresponding to vertical and horizontal relative accuracy. The proposed method can be an objective and low cost instrument for performance evaluation. From a first set of experimental data, collected from new instrumentation of recent production, it is observed that percentage relative accuracy in distance measurement is a function of the probe model and ranges from 1.3% to 4.7%. Test will be performed on a higher number of US scanners to improve procedure and its results.

REFERENCES

Browne, J.E., Watson, A.J., Gibson, N.M., Dudley, N.J. & Elliott, A. T., 2004, Objective measurements of image quality, *Ultrasound in Med. & Biol.*, 30 (2): 229–237.

Carson, P. L. & Zagzebski, J. A., 1980 , Pulse Echo ultrasound imaging systems: performance test and criteria, *AAPM report*, 8 November.

Flax, S.W. & O'Donnel, M., 1988, Phase-Aberration Correction Using Signals From Point Reflectors and Diffuse Scatterers: Basic Principles, *760 IEEE Transactions on Ultrasonics, Ferroelectrics and Frequency Control*, 35 (6).

Fornage, B.D., Atkinson, E.N., Nock, L.F. & Jones, P.H., 2000, US with extended field of view: phantom-tested accuracy of distance measurements, *Radiology*, February,

Gibson, N.M., Dudley, N.J. & Griffith, K., 2001, A computerised quality control testing system for B-mode ultrasound, *Ultrasound in Med. & Biol.*, 27.(12): 1697–1711.

Goldstein, A., 2000, The effect of acoustic velocity on phantom measurements, *Ultrasound in Med. & Biol.*, 26 (7): 1133–1143.

Goldstein, A., Sharma, R.K. & Arterbery, E., 2002, Design of Quality Assurance for Sonographic Prostate Brachytherapy Needle Guides, *J Ultrasound Med* 21: 947–954.

Goldstein, A., 2004, Beam width measurements in low acoustic velocity phantoms, *Ultrasound in Med. & Biol.*, 30 (3): 413–416.

Goodsitt, M.M., et al., 1998, Real-time B-mode ultrasound quality control test procedures, Report of AAPM Ultrasound Task Group No. 1, *Med. Phys.* 25 (8), August.

Gupta Rajiv, et al., 2002, Quantitative analysis system and method for certifying ultrasound medical imaging equipment, *United States Patent* 6: 370–480.

Winter, J., Kimme-Smith, C. & King, W., 1985, Measurement Accuracy of Sonographic Sector Scanners, *AJR* 144: 645–648, March.

Computational Vision and Medical Image Processing – João Tavares & Natal Jorge (eds)
© 2008 Taylor & Francis Group, London, ISBN 978-0-415-45777-4

Use of statistic texture descriptors to classify boar sperm images applying discriminant analysis

Sir Suarez & Enrique Alegre
Department of Electrical and Electronics Engineering

Manuel Castejón & Lidia Sánchez
Department of Mechanical, Computing and Aerospace Engineerings, University of León,
Campus de Vegazana s/n, León, Spain

ABSTRACT: This work presents a boar semen image classification by analyzing the sperm head texture. We have considered four techniques: Laws, coocurrence matrices, Legendre and Zernike moments. We have applied them to classify sperm head images as damage or not damage, comparing their efficiency. From the coocurrence matrices, 4 features are computed (contrast, correlation, energy and homogeneity) in four orientations (0°, 45°, 90° and 135°) and distances 1, 3 and 5. In the case of Laws, Legendre and Zernike, 9 features are computed for neighborhoods of 5×5. To carry out the classification, we use Fisher lineal discriminant analysis (LDA) and quadratic discriminant analysis (QDA). In the classifier training stage, we use 75% of the images and 25% for the tests. Features derived from coocurrence matrices classified by QDA yield the best results with a error rate of 13.81%. It is followed by Laws, Legendre and Zernike.

Keywords: Classification, Laws, Legendre, Zernike, coocurrence matrix, texture, discriminant analysis, boar semen.

1 INTRODUCTION

Image classification considering its texture has been used in medicine to prevent and cure human and animal diseases (1; 2). Among all, researchers have been focused on breed improvement (like for instance, boars) by employing artificial insemination techniques. To deal with this, the main goal is to determine the semen quality and for this reason in this work we propose a system to classify images of boar spermatozoa as damage or not damage.

There are several works that support the quality estimation carried out by a veterinary expert. Petkov et al. classify, by using Learning Vector Quantization, the sperm head images according to the contour status of membrane of the sperm cells (3). In (4), a method to classify sperm head images by their intracellular intensity distribution is proposed.

According to Tuceryan, texture techniques can be grouped as structural, statistical, geometrical, based on models and based on signal processing (5). For statistical texture methods, there are several procedures to compute features like first and second order statistics, moments and so on. Classification can be also carried out by supervised and unsupervised methods, using neural networks or statistical classifiers.

In this work, we propose to represent images of damage and not damage boar spermatozoa by computing statistical texture descriptors like coocurrence matrices (6; 7), Laws, Legendre and Zernike moments (8; 9). We have used lineal and quadratic discriminant analysis to perform the classification. So, we develop an application using Matlab and the R Project for Statistical Computing.

Section 2 explains the proposed methodology. Section 3 shows the experiments and the obtained results. Finally, conclusions are presented in Section 4.

2 METHODS

To carry out this work, we have employed texture descriptors and statistical classifiers, since one has demonstrated that they offer excellent results. In this Section, these texture techniques are explained. For the classification stage, we use supervised methods-LDA and QDA- to use a parametric approach by means of the multivariate normal distribution of the classes.

2.1 Preprocessing and segmentation

Using a digital camera connected to a phase-contrast microscope, boar semen images are captured with a resolution of 2560×1920 pixels. Information about the sample preparation can be found in (10). After that, sperm head images are cropped manually, obtaining images with an only sperm head. Finally, the heads were automatically segmented (10) and the binary image obtained was multiply by the original grey level resulting the masked image.

2.2 Feature extraction methods

We have used four techniques to obtain texture features of boar sperm head images.

2.3 Coocurrence matrices

This method obtains quantifiable information of an image using the following normalized coocurrence matrix:

$$N_d(i,j) = \frac{C_d(i,j)}{\sum_i \sum_j C_d(i,j)} \qquad (1)$$

Cd is a coocurrence matrix.

This matrix depends on the pixel distribution across the image, the neighborhood relations and the pixel intensity. We compute 4 Haralick statistics derived from such matrix (11):

1. Contrast:

$$C = \sum_i \sum_j |i-j|^2 c_{ij} \qquad (2)$$

2. Correlation:

$$C_{rr} = \frac{1}{\sigma_i \sigma_j} \left| \sum_i \sum_j (i-u_i)(j-u_j)c_{ij} \right| \qquad (3)$$

where:

$$u_i = \sum_i \sum_j i c_{ij} \qquad (4)$$

$$u_j = \sum_i \sum_j j c_{ij} \qquad (5)$$

$$\sigma_i = \sum_i \sum_j (i-u_i)c_{ij} \qquad (6)$$

$$\sigma_j = \sum_i \sum_j (j-u_j)c_{ij} \qquad (7)$$

3. Energy:

$$u = \sum_i \sum_j c_{ij}^2 \qquad (8)$$

4. Homogeneity:

$$HL = \sum_i \sum_j \frac{c_{ij}}{1+(i-j)^2} \qquad (9)$$

2.4 Laws method

Laws method (12) consists in applying convolutions with several filters to images, yielding such many images as convolutions are carried out. Let I be the initial image and $g_1, g_2, \ldots, g_n$ a set of filters, a generic image resulted after the convolution is defined by $J_n = I g_n$. Kernels used are defined to neighborhoods of 5×5, so a 16 feature vector is obtained for each image pixel. Subsequently, vectors are reduced from 16 to 9 features. Let be $F_k[i,j]$ the result of applying the k-th mask on the pixel (i,j), then the map E_k of the texture energy for the filter k is define as:

$$E_k(r,c) = \sum_{j=c-7}^{c+7} \sum_{i=r-7}^{r+7} |F_k(i,j)| \qquad (10)$$

2.5 Legendre moments

These moments allow new applications as the reconstruction of an image from the mathematical features provided by these moments. Shu and Yu (13) present an efficient method for computation of Legendre moments. Legendre moments are defined by:

$$\lambda_{p,q} = \frac{(2p+1)(2q+1)}{4} \int_{-1}^{+1} \int_{-1}^{+1} P_p(x)P_q(y)f(x,y)dxdy \qquad (11)$$

where P_p y P_q are the Legendre polynomials. Employed images have to be normalized to define them over a unit squared.

2.6 Zernike moments

Zernike moments (9) are used to carry out an independent recognition of the order, number and orientation. They yield rotation, scale and translation invariance. As Zernike moments are orthogonal, the reconstruction of the original function from the obtained one is simplified. Complex Zernike polynomials are constructed using complex polynomials forming a complete orthogonal base defined on the unit disc:

$$(x^2 + y^2) \le 1 \qquad (12)$$

From these polynomials

$$A_{m,n} = \frac{m+1}{\pi} \int_x \int_y f(x,y)[Vm,n(x,y)]dxdy \qquad (13)$$

complex Zernike moments are computed.

2.7 Classification techniques

There are two kinds of classification methods. The unsupervised ones are based on classifying a set of

198

samples into an unknown number of classes. The supervised techniques know previously the number of different classes that a set of samples belongs to. The goal is to distinguish the samples from one class to another by using a set of features. In this work we have used lineal discriminant analysis (LDA) and quadratic discriminant analysis (QDA).

Discriminant analysis (14) consists in obtaining lineal functions over a set of independent variables in order to classify items of a population into subgroups determined by the dependent variable values.

2.8 *Lineal discriminant analysis*

Sample space is divided by straight lines in the two-dimensional space, planes in the three-dimensional space and, in general, hyperplanes in n-dimensional spaces. A line dividing two classes is defined to go across the straight line linking the centers of both classes. Line orientation depends on the point group shape. Straight lines, planes or hyperplanes are lineal combinations of the variables which characterize the examples.

Let be $\Sigma_1 = \Sigma_2 = \Sigma$, estimators S_1 and S_2 will be substituted for an estimator $S_w = [(n_1 - 1)S_1 + (n_2 - 1)S_2]/(n_1 + n_2 - 2)$. Replacing S_1 and S_2 with S_w this rule is based on assigning the observation x_o to P_1 only if $w'x_0 < w'(\frac{x_1+x_2}{2})$ being $w = s_w^{-1}(x_2 - x_1)$. From the geometric point of view, this method is the same that dividing the R^d space into two semispaces E_1 and E_2, separating by the hyperplane $w'x = w'(\frac{x_1+x_2}{2})$ so that $x_1 \in E_1$ and $x_2 \in E_2$. The observation $x_0 \in \hat{E_1}$ so the value $w'x_0$ is closer to $w'x_1$ than $w'x_2$.

2.8.1 *Quadratic discriminant analysis*

A quadratic discriminant is similar to a lineal one, but the border between two discriminated regions is a quadratic surface. If we consider that covariance matrices are not the same, then a quadratic surface is obtained (ellipsoid, hyperballoid, etc) in maximum likelihood terms under a normal distribution. This discrimination can tackle those situations in which members of one class are surrounded to a large extend by another class members. Quadratic discriminant function is defined in the simplest way as the logarithm of the proper probability density function, so that for each considered class, a discriminant is computed. For the normal distribution, the A_i class discriminant is defined as follows:

$$log\pi_i f_i(x) = log(\pi_i) - \frac{1}{2}log(|\sum_i|) - \frac{1}{2}(x - u_i)^T \sum_i^{-1}(x - u_i) \quad (14)$$

Posterior probabilities $P(A_i|x)$ are the following:

$$P(A_i|x) = exp[log\,\pi_i f_i(x)] \quad (15)$$

3 EXPERIMENTS AND RESULTS

3.1 *Experiments*

To carry out the experiments, we consider a set of samples formed by 453 images of two classes of boar spermatozoa: 216 damage sperm cells and 237 not damage. From this set of images, the 75% of them have been used for the classifier training and 25% of them for the tests. We first extract the set of descriptors using Matlab. To deal with this, we apply a texture filter to achieve a better identification rate by means of the standard deviation.

Then, we compute the normalized coocurrence matrices considering orientations of 0°, 45°, 90° and 135° for distances 1, 3 and 5. For each computed matrix, we calculate four features: contrast, correlation, energy and homogeneity, so a vector of 48 features is obtained. These features are averaged and the feature vector is reduced to 4 components.

For the Laws method, we apply the convolution filters proposed by Laws (12). The filter windows that use it is defined for the neighborhoods 5 × 5 obtaining for each pixel of the image a vector of 16 characteristics which become finally in 9. Consider the following convolution masks.

$$
\begin{array}{llllll}
L = [& 1 & 4 & 6 & 4 & 1 &] \\
E = [& -1 & -2 & 0 & 2 & 1 &] \\
S = [& -1 & 0 & 2 & 0 & -1 &] \\
R = [& 1 & -4 & 6 & -4 & 1 &]
\end{array}
$$

Each one of which has like object to stand out different aspects of the image: L5(level), E5(Edge), S5(spot), R5(Ripple). The 2D convolution masks are obtained by computing outer products of pairs of vectors For example, the mask L5S5 is computed as:

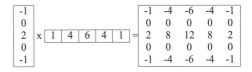

The first step in Laws procedure is to remove effects of illumination by moving a small window (say 15 × 15) around the image, and subtracting the local average from each pixel. This produces an image whose pixel values are around zero. Next, process the image with the 16 2D convolution filters. Assume Fk (i,j) is the result of filtering with the kth mask at pixel (i,j). The texture energy map Ek for filter k is then equation (10). Each texture energy map is a full image, representing the application of the kth filter to the input image. Then, combine energy maps of certain symmetric pairs of filters to produce the final nine maps. For instance, E5L5 and L5E5 measure horizontal and vertical edges of an image, the average of

Table 1. LDA – hit rate for damage and not damage sperm head images.

Classes	Laws	Haralick	Legendre	Zernike
Damage	64.81	77.78	81.48	75.33
Not damage	80.65	93.55	75.81	80.02
Hit rate	73.26	86.19	78.44	77.82

Table 2. QDA – hit rate for damage and not damage sperm head images.

Classes	Laws	Haralick	Legendre	Zernike
Damage	66.67	74.07	79.63	71.33
Not damage	75.81	88.71	74.19	82.06
Hit rate	71.54	81.88	76.71	77.05

the two maps indicates total edge content. The final nine maps are: L5E5/E5L5, L5S5/S5L5, L5R5/R5L5, E5E5, E5S5/S5E5, E5R5/R5E5, S5S5, S5R5/R5S5 and R5R5.

Besides these features, we have used 9 Legendre moments, the ones from ML00 to ML22, and the features derived from Zernike moments. Specifically, we have considered the absolute, imaginary and real value of 9 Zernike moments (until 4th order).

Once we have obtained the samples vectors or training patterns of each class for each considered set of descriptors, we compute the vector (direction of the projection) to differentiate between classes for both LDA and QDA using the statistic program R. Next, we compute the discriminant scores for each image, comparing them with the cross point, determining which class the image belongs to. We calculate the error rate by comparing the classification obtained by our method with the expected one. Considering these results, new samples can be classified.

3.2 Results

As we have used a reduced set of data, the validation process was carried out by applying cross-validation. Therefore, we train the system leaving out a sample, which is used afterwards to validate the classification.

The hit rate for damage and not damage is damage*46,55 + not damage*53,44. Where 46,55% and 53,44% are the percentage of the number of images of each category.

The experimental results are showed in Table 1.

According to the obtained results for the Lineal Discriminant Analysis, the best descriptors are the ones proposed by Haralick (hit rate of 76.38%). For the damage sperm head images, the hit rate is 98.46% with all descriptors but Zernike. Obtained hit rates using Quadratic Discriminant Analysis are showed in the Table 2.

Table 3. Hit rate of damage and not damage spermatozoon classification using Haralick descriptors and LDA.

Class	Damage	Not damage
Damage	42	4
Not damage	12	58

Best classification is carried out by using Haralick descriptors with 86.19% of successful identifications, reducing the hit rate difference between damage and not damage spermatozoom classification obtained by LDA (see Table 3).

4 CONCLUSIONS

In this work, we have used texture features to classify images of boar spermatozoa damage and not damage. From the fours kind of considered descriptors, the less number of misclassified images is yielded by using Haralick descriptors for both classifiers LDA and QDA. Best results are provided by LDA and Haralick with an error rate of 13.81%, which is quite acceptable for veterinary experts. We have also compared the computation times, being features derived from Laws and coocurrence matrices, the fastest methods. This work allows reduce time spent by veterinary experts to determine if a sperm head is damage or not damage. Apart from Haralick features, Legendre and Laws using LDA are the second and third best method, respectively. It is noticeable that Haralick, Legendre and Laws identify really good not damage spermatozoa using LDA but they fail in Damage sperm heads. We think this is due to the high dispersion and variability in the damage class.

REFERENCES

1. L. Jim. An automated malignant tumour localization algorithm for prostate cancer detection in trans-rectal ultrasound images. Master of applied science. Master's thesis, MASc in Electrical and Computer Engineering, Faculty of Engineering, University of Waterloo.
2. T. Olmez, and Z. Dokur. Classification of heart sounds using an artificial neural network. *Pattern Recogn. Lett.*, 24(1–3):617–629, 2003.
3. N. Petkov, E. Alegre, M. Biehl, and L. Sanchez. Lvq acrosome integrity assessment of boar sperm cells. *CompIMAGES, Computational modelling of objects represented in images*, 2006.
4. L. Sanchez, N. Petkov, and E. Alegre. Classification of boar spermatozoid head images using a model intracellular density distribution. In M. Lazo and A. Sanfeliu, editors, *Progress in Pattern Recognition, Image Analysis and Applications: Proc. 10th Iberoamerican Congress on Pattern Recognition, CIARP 2005, Lecture Notes*

in Computer Science, volume 3773, pages 154–160. Springer-Verlag Berlin Heidelberg, 2005.

5. M. Tuceryan and A. K. Jain. *Texture Analysis*, pages 207–248. In The Handbook of Pattern Recognition and Computer Vision (2nd Edition), by C. H. Chen, L. F. Pau, P. S. P. Wang (eds.), 1998.

6. V. Kovalev and M. Petrou. Multidimensional co-occurrence matrices for object recognition and matching. *Graphical Models and Image Processing*, Volume 58(3):187–197, May 1996.

7. M. Partio, B Cramariuc, and M. Gabbouj. Texture similarity evaluation using ordinal co-occurrence. In *Image Processing, 2004. ICIP '04. 2004 International Conference on*, pages III: 1537–1540, 2004.

8. S. Liao. *Image Analysis by Moments*. PhD thesis, Electrical and Computer Engineering, The University of Manitoba, Canada, 1993.

9. S. Liao and M. Pawlak. Image analysis with zernike moment descriptors. *Electrical and Computer Engineering. IEEE 1997 Canadian Conference on.*, 2:700–703, 1997.

10. L. Sanchez, N. Petkov, and E. Alegre. Statistical approach to boar semen evaluation using intracellular intensity distribution of head images. *Cellular and Molecular Biology*, 52(6):38–43, 2006.

11. R.M. Haralick. Statistical and structural approaches to texture. In *Proceedings of the IEEE*, pages 45–69, 1978.

12. K. Laws. Texture energy measures. In *In Image Understanding Workshop, DARPA*, 1979.

13. H. Shu, L. Luo, X. Bao, and W. Yu. An efficient method for computation of legendre moments. *Graphical Models*, 62:237–262, 2000.

14. S. Balakrishnama, A. Ganapathiraju, and J. Picone. *Linear Discrimant Analysis - A Brief Tutorial*. Institute for Signal and Information Processing, Mississippi State University, MS State, MS, USA, March 2 1998.

Computational Vision and Medical Image Processing – João Tavares & Natal Jorge (eds)
© 2008 Taylor & Francis Group, London, ISBN 978-0-415-45777-4

3D object reconstruction from uncalibrated images using a single off-the-shelf camera

T.C.S. Azevedo
INEGI – Instituto de Engenharia Mecânica e Gestão Industrial
LOME – Laboratório de Óptica e Mecânica Experimental
FEUP – Faculdade de Engenharia da Universidade do Porto, Portugal

J.M.R.S. Tavares & M.A.P. Vaz
INEGI, LOME
DEMEGI – Departamento de Engenharia Mecânica e Gestão Industrial
FEUP – Faculdade de Engenharia da Universidade do Porto, Portugal

ABSTRACT: Three-dimensional (3D) objects' reconstruction using just bi-dimensional (2D) images has been a major research topic in Computer Vision. However, it is still a hard problem to solve, when automation, speed and precision are required and/or the objects present complex shapes and visual properties. In this paper, we compare two Active Computer Vision methods commonly used for the 3D reconstruction of objects from image sequences, acquired with a single off-the-shelf CCD camera: *Structure From Motion* (*SFM*) and *Generalized Voxel Coloring (GVC)*. *SFM* recovers the 3D shape of an object using the camera(s)'s or object's movement, while *VC* is a volumetric method that uses photoconsistency measures to build a 3D model for the object. Both methods considered do not impose any kind of restrictions to the relative motion involved.

1 3D RECONSTRUCTION

1.1 Introduction

Computer Vision is continuously developing theories and methodologies to automatic extract useful information from images, in a way as similar as possible as humans do with their visual system.

Contactless methods usually used to recover the 3D geometry of objects are commonly divided in two categories: active techniques, that require some kind of energy projection or relative moment between the camera's and object's, and passive techniques, that only use ambient illumination and so is not considered any kind of projected energy.

The main goal of this work was the comparison of two active methods commonly used in Computer Vision for 3D objects reconstruction: *Structure From Motion* (*SFM*) and *Generalized Voxel Coloring* (*GVC*).

1.2 State of the art

3D reconstruction of objects has become an intensive research topic in Computer Vision. Digital 3D models are required in many applications, such as industrial inspection, biomedical, navigation, objects identification, etc.

Usually, high quality 3D models of static objects are obtained using *scanners* systems, generally expensive, but easy to handle. The explosive growth in computer's processing, in terms of computational power and memory storage, and its continuous reducing price, together with the development of more and more sophisticated and affordable digital cameras, allowed the practical use of photogrammetric methods for 3D reconstruction in the last decades. In fact, they appear now has a low-cost and portable alternative to the common range-based 3D reconstruction methods. However, image-based reconstruction is still a difficult task, in particular for large or complex objects and if uncalibrated images are used, or when a wide baseline between the acquired images is presented, [Remondino, 2006].

The next two subsections will focus on two commonly used image-based reconstruction methods: *Structure From Motion* (*SFM*), that belongs to the standard stereo-based methods and *Generalized Voxel Coloring* (*GVC*), that belongs to the recent volumetric reconstruction methods.

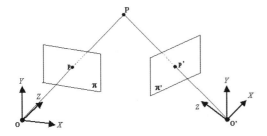

Figure 1. Stereo vision's principle: **P**'s 3D coordinates are determined through the intersection of the two lines defined by the optical centers **O** and **O'** and the matched 2D image points **p** and **p'**.

1.3 Structure from motion

Proposed in [Ullman, 1979], *SFM* is a stereo-based method, Figure 1. It uses the relative movement between the camera(s) used and the object to be reconstructed, to make assumptions about the 3D object's shape. Thus, by knowing the trajectories of object's feature points in the image's plane, this method determines the 3D shape and motion that better describes most of the point's trajectories.

This method has received several contributions and diverse approaches, e.g. [Chaumette, 1991], [Aans, 2002], [Chiuso, 2002] and [Hui, 2006]. In the present case, we do not pretend impose any kind of restrictions to the movement involved.

1.4 Generalized voxel coloring

Stereo-based methods, like *SFM*, fail to capture shapes with complicated topology, due to occlusion or smooth surfaces.

For smooth object's, 3D reconstruction using volumetric or voxel-based methods have been quite popular for some time, [Seitz, 1997]. These methods assume that there is a bounded volume in which lays the object of interest. The 3D space model is then represented or sampled by *voxels* (regular volumetric structures also known as *3D pixels*).

First volumetric methods combine silhouette images of the object to be reconstructed with camera's calibration information to set the visual rays in 3D space for all silhouette points, which define a generalized cone within lays the same object. The intersection of these cones defines the *visual hull*, [Laurentini, 1994], a volumetric space in which the object to be reconstructed is guaranteed to be, Figure 2. The accuracy of the reconstruction obtained depends on the number of images used, on the positions of each viewpoint considered, on the camera's calibration quality and on the complexity of the object's shape.

Generalized Voxel Coloring (*GVC*) is a volumetric method that does not require a matching process between the object's feature points along the image

Figure 2. Left to right: from the original object to its visual hull ([Yang, 2003]).

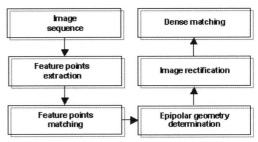

Figure 3. *SFM* methodology followed to obtain the 3D reconstruction of objects from uncalibrated images.

sequence used as the *SFM* needs, [Slabaugh, 1999]. Instead, starting with a sequence of calibrated images from the object to be reconstructed, *GVC* uses photo-consistency measures to determine if a certain voxel belongs or not to the object being reconstructed. This technique simultaneously builds and colors a 3D model for the object to be reconstructed.

2 METHODOLOGY FOLLOWED

In this work, both methods were tested on two objects with different topological properties: a simple parallelepiped and a human's hand model.

The parallelepiped has a straightforward topology, with flat orthogonal surfaces, whose vertices are easily detected in each image and simply matched along the image sequence. On the contrary, the hand model has a smooth surface and complicated topology.

To test the *SFM* method, we follow the methodology proposed in [Pollefeys, 2004], Figure 3. Thus, the first step is to acquire two uncalibrated images, from the object to be reconstructed, using a single off-the-shelf digital camera. Then, image feature points are extracted and matched, followed by the determination of the epipolar geometry, image rectification and, finally, dense matching.

In other hand, to test the *GVC* method we follow the methodology proposed in [Azevedo, 2007], Figure 4. Thus, a single off-the-shelf CCD camera is used, to acquired object's image sequences, and the *Zhang's* calibration method was used to calibrate the same, [Zhang, 2000]. Then, to obtain the object's silhouettes from the input images, image segmentation is performed. Combining the original image sequence

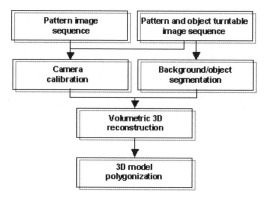

Figure 4. *GVC* methodology followed to obtain the 3D reconstruction of objects from uncalibrated images.

Figure 5. Stereo image pairs of the objects used to test the *SFM* reconstruction method.

and associated silhouette images, and considering the camera's calibration parameters, both objects' models are built using the *GVC* volumetric method, and polygonized and smoothed using the *Marching Cubes* algorithm, [Lorensen, 1987].

3 EXPERIMENTAL RESULTS

3.1 *SFM method*

Figure 5, shows the acquired stereo image pairs of both objects used in this work.

For both objects, 200 image features were extracted using the *Harris* corner detector, [Harris, 1988], imposing a minimum distance of 10 pixels between each detected feature. Robust matching of features between the stereo images was made using the *RANSAC* algorithm, [Fischler, 1981]. The results obtained can be observed in Figure 6 and Figure 7. Since the hand model presents a smooth surface, obviously many wrong matches were detected and, consequently, the determined epipolar geometry was incorrect.

After, both stereo pairs were rectified – Rectification transforms a stereo image pair in such a way that epipolar lines became horizontal, using the algorithm presented in [Isgrò, 1999]. This step allows an easier dense matching process. The results obtained with this step can be observed in Figure 8 and Figure 9.

Figure 6. Results of the feature points (robust) matching with the stereo image pair of the parallelepiped object (matched feature points of first image are marked with green crosses and the red crosses represent the matched feature points of second image).

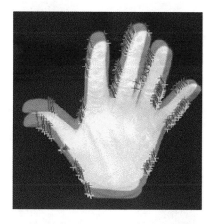

Figure 7. Results of the feature points (robust) with the stereo image pair of the parallelepiped object (matched feature points of first image are marked with green crosses and the red crosses represent the matched feature points of the second image).

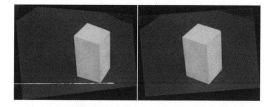

Figure 8. Rectification results for the stereo images of the parallelepiped object.

205

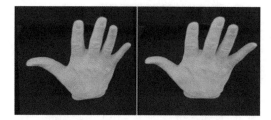

Figure 9. Rectification results for the stereo images of the hand model object.

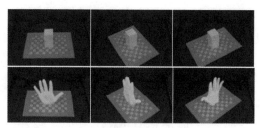

Figure 12. Three images used for the 3D reconstruction of the parallelepiped (above) and the hand model (below).

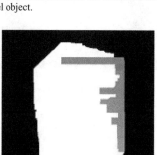

Figure 10. Disparity map obtained for the parallelepiped object.

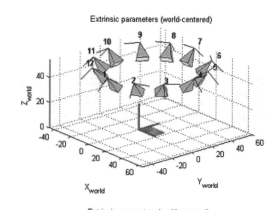

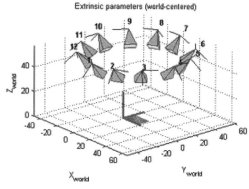

Figure 13. 3D graphical representation of the extrinsic parameters obtained from the camera's calibration process: above, parallelepiped object case; below, hand model case.

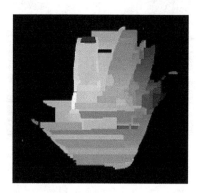

Figure 11. Disparity map obtained for the hand model object.

Dense matching was performed using the *Stan Birchfield* algorithm, [Birchfield, 1999]. This algorithm returns a disparity map, that gives some depth information about the objects present in a rectified pair of images: far objects will have zero disparity (black regions in the disparity map) and the closest objects will have maximum disparity instead (white regions in the disparity map). The results obtained for both objects considered in this work can be observed in Figure 10 and Figure 11. Given the incorrect results

obtained in the previous steps, when compared with the parallelepiped object case, the dense matching for the hand model was, consequently, of poor quality.

3.2 GVC method

Figure 12, shows some examples of the images acquired to reconstruct the objects considered in this work using the *GVC* method. For an idea of the viewpoints considered in the image acquisition process,

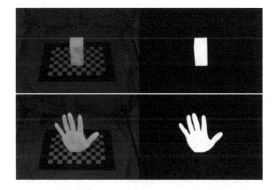

Figure 14. One example of image segmentation for the parallelepiped (above) and the hand model (below): on the left, the original image; on the right, the binary image obtained.

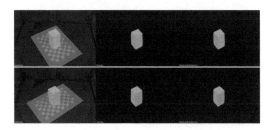

Figure 15. Two different viewpoints of the 3D model obtained for the parallelepiped case: left, original image; middle, voxelized 3D model; right, polygonized and smoothed 3D model.

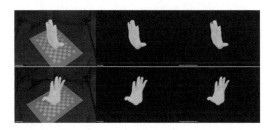

Figure 16. Two different viewpoints of the 3D model obtained for the hand model case: left, original image; middle, voxelized 3D model; right, polygonized and smoothed 3D model.

Figure 13 has a 3D graphical representation of the obtained extrinsic parameters for both objects.

The results obtained from the camera calibration were very accurate for both cases.

Some segmentation results can be observed in Figure 14. Those results were achieved by first removing the red and green channels from the original RGB images and, finally, using image binarization by threshold value.

Figure 15 and Figure 16 shows the results of the 3D reconstruction obtained for both objects using the *GVC* implementation in [Loper, 2002]. Both reconstructed models are very similar to the real 3D object, even in the case of the hand model. Comparing these results with the previous obtained by the *SFM* methodology, *GVC* has no problem to reconstruct objects with smooth surfaces or with complicated morphology. On the other hand, the accuracy of the 3D models built by this last methodology is highly dependent on the previous calibration and segmentation steps. Thus, *GVC* puts some restrictions, such as a background with low color variation and suitable calibration apparatus, making it unfit for real-world object reconstruction.

4 CONCLUSIONS

The main goal of this paper was to compare experimentally two commonly used image-based methods for 3D object reconstruction: *Structure From Motion* (*SFM*) and *Generalized Voxel Coloring* (*GVC*).

To test and compare the both methods, two objects with different topological properties were used: a parallelepiped and a human's hand model.

Our adopted *SFM* methodology gave fine results when the objects presents strong feature points, easy to detect and match along the input images. However, we can conclude that even small errors in the matching and epipolar geometry estimation can seriously compromise the remaining steps. On the other hand, this is a flexible method for real-world scenes, because it does not require an independent camera calibration process, as the features matched along an image sequence can be used in an auto-calibration process.

The models built using the *GVC* method were quite similar and closer to the real objects, as in terms of shape as in color. Even though, the reconstruction accuracy was highly dependent on the quality of the results' of the camera calibration procedure and of image segmentation. These can be two major drawbacks in real-world scenes because they limit the application of the *GVC* method.

Thus, when comparing the two methods, we can conclude that, on one hand, *GVC* performs better the 3D reconstruction of objects with complex topology and, on the other hand, *SFM* is better for unconstrained real-world objects reconstruction.

ACKNOWLEDGEMENTS

This work was partially done in the scope of project "Segmentation, Tracking and Motion Analysis of Deformable (2D/3D) Objects using Physical Principles", with reference POSC/EEA-SRI/55386/2004, financially supported by *FCT – Fundação para a Ciência e a Tecnologia* from Portugal.

REFERENCES

Aans, H. & Kahl, F., *Estimation of Deformable Structure and Motion*, Vision and Modelling of Dynamic Scenes Workshop, Copenhagen, Denmark, 2002.

Azevedo, T. C. S., Tavares, J. M. R. S., et al., *3D Volumetric Reconstruction and Characterization of Objects from Uncalibrated Images*, 7th IASTED International Conference on Visualization, Imaging, and Image Processing, Palma de Maiorca, Spain, 2007.

Birchfield, S., *Depth Discontinuities by Pixel-to-Pixel Stereo*, International Journal of Computer Vision, vol. 35, no. 3, pp. 269–293, http://vision.stanford.edu/~birch/p2p/, 1999.

Chaumette, F. & Boukir, S., *Structure from motion using an active vision paradigm*, Int. Conference on Pattern Recognition, The Hague, Netherlands, vol. 1, pp. 41–44, 1991.

Chiuso, A., Favaro, P., et al., *Structure from motion causally integrated over time*, IEEE Transactions on Pattern Analysis and Machine Intelligence, vol. 24, no. 4, pp. 523–535, 2002.

Fischler, M. A. & Bolles, R., *RANdom SAmpling Consensus: a paradigm for model fitting with application to image analysis and automated cartography*, Communications of the Association for Computing Machinery, vol. 24, no. 6, pp. 381–395, 1981.

Harris, C. G. & Stephens, M. J., *A combined corner and edge detector*, Forth Alvey Vision Conference, University of Manchester, England, vol. 15, pp. 147–151, 1988.

Hui, J., *A holistic approach to structure from motion*, Computer Science Dissertation, University of Maryland, USA, 2006.

Isgrò, F. & Trucco, E., *Projective rectification without epipolar geometry*, IEEE International Conference on Computer Vision and Pattern Recognition, Fort Collins, Colorado, USA, vol. 1, pp. 94–99, 1999.

Laurentini, A., *The visual hull concept for silhouette-based image understanding*, IEEE Transactions on Pattern Analysis and Machine Intelligence, vol. 16, no. 2, pp. 150–162, 1994.

Loper, M., *Archimedes: Shape Reconstruction from Pictures – A Generalized Voxel Coloring Implementation*, http://matt.loper.org/Archimedes/Archimedes_docs/html/index.html, 2002.

Lorensen, W. E. & Cline, H. E., *Marching cubes: A high resolution 3D surface construction algorithm*, International Conference on Computer Graphics and Interactive Techniques, ACM Press, New York, USA, vol. 21, no. 4, pp. 163–169, 1987.

Pollefeys, M., Gool, L. V., et al., *Visual Modeling with a Hand-Held Camera*, International Journal of Computer Vision, vol. 59, no. 3, pp. 207–232, 2004.

Remondino, F. & El-Hakim, S., *Image-based 3D Modelling: a review*, The Photogrammetric Record, vol. 21, no. 115, pp. 269–291, 2006.

Seitz, S. N. & Dyer, C. R., *Photorealistic Scene Reconstruction by Voxel Coloring*, IEEE Conference on Computer Vision and Pattern Recognition Conference, San Juan, Puerto Rico, pp. 1067–1073, 1997.

Slabaugh, G. G. & Culbertson, W. B., et al., *Generalized Voxel Coloring*, Workshop on Vision Algorithms, Corfu, Greece, pp. 100–115, 1999.

Ullman, S., *The Interpretation of Visual Motion*, Massachusets MIT Press, Cambridge, USA, 1979.

Yang, L., *3D Surface Reconstruction from 2D Images*, Center for Visual Computing, Instructional Computing, Stony Brook University, EUA, 2003.

Zhang, Z., *A Flexible New Technique for Camera Calibration*, IEEE Transactions on Pattern Analysis and Machine Intelligence, vol. 22, no. 11, pp. 1330–1334, 2000.

Computational Vision and Medical Image Processing – João Tavares & Natal Jorge (eds)
© 2008 Taylor & Francis Group, London, ISBN 978-0-415-45777-4

Eye detection using a deformable template in static images

Fernando Jorge Soares Carvalho
Departamento de Matemática, Instituto Superior de Engenharia do Porto, R. Dr. Bernardino de Almeida, Porto, Portugal

J.M.R.S. Tavares
Lab. de Óptica e Mecânica Experimental, Instituto de Eng. Mecânica e Gestão Industrial Dep. de Engenharia Mecânica e Gestão Industrial, Faculdade de Engenharia da Universidade do Porto Rua Dr. Roberto Frias, s/n – Porto, Portugal

ABSTRACT: In this paper, we present a methodology that uses a deformable template to detect the human eye in static images. Geometrically, the used template is represented by two distinct entities: a circumference, that defines the iris contour; and two parabolas, one concave and other convex, that define respectively, the above and below contours of the eye. The geometrical shape of the used template is controlled by a set of eleven control parameters that allows its change in scale, position and orientation. To process the matching between the eye, represented in the input image, and the template used, we consider information obtained from four energy fields. Then, to process the dynamic update of the template parameters, we use an energy function, based on those energy fields, that characterizes the cost of the template deformation. During each of the seven processing phases, we search iteratively for the best combination parameters values that minimize the energy cost of the template deformation. In this paper, we illustrate the functionality of this method by presenting some experimental results, and present some conclusions and perspectives of future work as well.

1 INTRODUCTION

In Computational Vision domain, one of the main development areas is related with the detection of faces, either in static images or along image sequences. In this area, the face can be detected as a whole, using, for example, a geometric contour that defines it in a global manner, like a balloon, or the same detection can be accomplished by identifying certain facial features, as the eye, the mouth, the eyelids or the chin.

The methodologies developed for face detection « are usually defined in function of the manner the face is expected to appear in the input images. Thus, for example, some existent methodologies consider that the face should appear in the input images without facial hair and glasses, in a frontal position, without partial occlusion, etc.

In (Yang et al. 2002), is presented one of the most known surveys about the existent works related with face detection in images. In that work, the existent methodologies are divided into four categories: 1) based on knowledge; 2) based on invariant features; 3) based on appearance; and 4) based on template matching.

In the present work, we had a special interest in exploring the methodologies based on template matching. Commonly, these particular methodologies use templates that try to represent the face to be detected completely or by using just some features of it.

Usually, the detection process is lead by a dynamic process for matching the used template in the input image; some examples of related works that can be considered in this field are: cross correlation, see for example (Carvalho & Tavares 2005b); deformable templates, as in (Yuille et al. 1992), (Yuwen Wu et al. 2003) and (Carvalho & Tavares 2005a), for instances; and circular patterns detection using Hough transform, as is done, for example, in (Toennies et al. 2002).

In the deformable templates area, (Yuille et al. 1992) describes a method that allows the detection and extraction of facial features, like the eye and the mouth. Thus, considering the similarity between the contours of those features and some predefined mathematical models, it is possible to represent those facial features using geometrical shape contours that are usually known as deformable templates. Once the used deformable templates are parameterized, the update of their parameters allows their global deformation, and so the flexibility to change their scale, position and orientation in the image space.

Frequently, to lead the dynamic matching process of the deformable template in the input image, four different types of the energy fields are used, that are obtained from the original image. The four energy fields considered are: intensity edges, intensity valleys, intensity peaks, and intensity grey levels. The fields considered are the result of the application of specific morphologic operators on the input image, in order to enhance some important aspects of the features to be detected. For example, the iris of the human eye has a strong intensity valley, being very easy to identify its presence in that field. Therefore, this field has the capacity to attract the template used to its correct position. Thus, the eye and the mouth detection in the input image consist on the dynamic and iterative update of the parameters of the used template by matching the same one in the energy fields governed by an energy function. This energy function is defined by a sum of several primitives that uses information obtained from the energy fields, and allows also the characterization of the matching cost during the template deformation. Thus, a low matching cost assures that the correspondence between the template used and the facial feature to be detected is adequate, and consequently that a correct detection was achieved. Otherwise, if the matching cost is high, then the template geometry is very different from the facial feature to be detected and so the matching is not appropriate.

Usually, the template matching process in the input image is strategically defined by a set of processing phases, and given an initial set of parameters values, is it very probably that the set of final values obtained is considerably different of the initial one.

In this work, we consider the method proposed in (Yuille et al. 1992), for the detection of the human eye in static images, and present the description of the most relevant steps of this method as well as some illustrative experimental examples.

This paper is organized as follows: in the next section, we presented the template and energy fields used in our work; then, in section 3, we described the used energy function, the matching process, and the update of the template parameters; in section 4, some experimental results are presented; and finally, in the last section, we addressed some conclusions and perspectives of future work.

2 EYE TEMPLATE AND ENERGY FIELDS

2.1 Eye template

The template used for the eye detection in an input image is geometrically represented by: a circumference, that defines the iris contour and is controlled by three parameters; and two parabolas, one concave and other convex, that define respectively, the above and

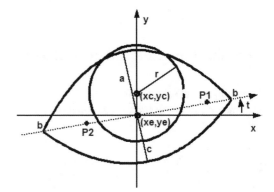

Figure 1. The deformable template used in this work is defined by a set of eleven control parameters: a, b, c, xe, ye, xc, yc, P, P_2, t.

below contours of the eye, and are controlled by eight parameters, Figure 1.

Considering a total of eleven parameters, and the image space (x,y), the template parameterization consists of the following elements: a circle, of radius r and centre (xc,yc), that defines the boundary between the iris and the white part of the eye (sclera); two parabolic sections, one above, Equation 7, and other below, Equation 8, centered in (xe,ye), being their maximum width $2b$, their maximum height a, measured between the above contour and the centre, their maximum height c, calculated between the below contour and the centre, and angle t their orientation; finally, two peak points, P_1 and P_2, that are used in Equations 5 and 6, and correspond to the centers of the sclera. The position of these points, P_1 and P_2, allow a correct determination of the template orientation. Moreover, for the determination of the template orientation, is considered two units vectors:

$$\vec{e_1} = (\cos(t), sen(t)),\qquad(1)$$

$$\vec{e_2} = (-sen(t), \cos(t)).\qquad(2)$$

Thus, any point $\vec{u}$ in the image space (x,y), can be represented as:

$$\vec{u} = x\,\vec{e_1} + y\,\vec{e_2},\qquad(3)$$

then, around the template centre (xe,ye), the peak points P_1 and P_2, can be used in Equations 5 and 6:

$$\vec{u_1} = (x_e + P_1\cos(t), y_e + P_1 sen(t)),\qquad(5)$$

$$\vec{u_2} = (x_e - P_2\cos(t), y_e - P_2 sen(t)).\qquad(6)$$

210

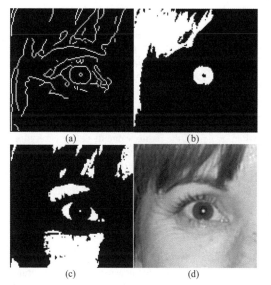

Figure 2. Representative images of the energy fields considered: intensity edges (a), intensity valleys (b), intensity peaks (c) and intensity grey levels (d).

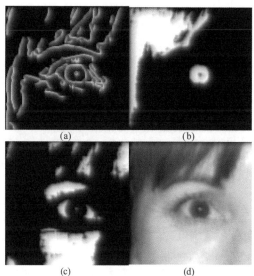

Figure 3. Representative image of the energy fields obtained after using a Gaussian filter: intensity edges (a), intensity valleys (b), intensity peaks (c) and intensity grey levels (d).

Finally, to represent the two parabolic sections, we use the equations:

$$y = y_e + a - \frac{a}{b^2}(x - xe)^2, \tag{7}$$

$$y = y_e - c + \frac{c}{b^2}(x - xe)^2. \tag{8}$$

2.2 Energy fields

In this work, are used four types of energy fields, all obtained from the input image: intensity edges, intensity valleys, intensity peaks, and intensity grey levels, Figure 2.

In this work, we use the Canny edge detector for enhancing the areas of the input image that present highest intensity contrasts; that is, the detection of the contours presented in the same images.

The template used is then attracted by the edges of the energy fields considered; which, usually, allows fine adjustments of the template used. Thus, the iris template is attracted by the valleys energy field; being characterized to contain a higher gradient inside the iris area, which was used to find the correct iris location. Therefore, the peaks of the energy fields contain strong gradients inside the sclera, which allow the attraction of the peak points P_1 and P_2 to the centre of the eye, giving consequently the correct orientation of the template. Finally, the intensity grey levels energy

field contains generic information about the brightness intensities distribution, and is generically used to update all template parameters.

The four energy fields considered, Figure 2, are then filtering using a Gaussian filter, in order to spread the energy values, enabling therefore the attraction of the template from larger distances, Figure 3.

3 DEFORMABLE TEMPLATES METHOD

In this section, is presented a description of the used method and the definition of the energy function, as well as the followed strategy to detect the eye iteratively in the input image by using the principle of energy minimization.

3.1 The energy function for eye template

The total energy E, is the sum of several primitives involving the energy fields ϕ, described in the previous section, and given by:

$$E = E_v + E_e + E_p + E_i + E_{prior}, \tag{9}$$

where E_v is the intensity valleys energy, E_e the intensity edges energy, E_p the intensity peaks energy, E_i the intensity grey levels energy, and E_{prior} the internal energy.

During the matching process, the energy E_v should be maximized inside a circumference, and its value is calculated by the sum of the energy valleys inside the same circumference divided by its area:

$$E_v = -\frac{c_1}{A_1} \iint_{Rc} \phi_v(\vec{u}) dA,$$ (10)

where C_1 represents a weight constant, A_1 the circumference area, and Rc the interior domain of the circumference centered in (xc, yc).

The energy E_e, should also be maximized and its value is the sum of the energy edges over the boundaries of the parabolas and the circumference used in the template divide by their lengths:

$$E_v = -\frac{c_2}{L_1} \int_{Cb} \phi_e(\vec{u}) ds - \frac{c_3}{L_2} \int_{Pb} \phi_e(\vec{u}) ds,$$ (11)

where C_2 represents a weight constant, L_1 the circumference perimeter, L_2 the lengths of the parabola arcs, and Cb and Pb, represent the boundaries of the circumference and of the parabolas, respectively.

The energy E_i, has two distinct terms: a first one, related with the minimization of the total brightness inside the circumference, divide by its area, and a second one, associated with the minimization of the total brightness inside the sclera zone, divide by its area as well. That is:

$$E_i = \frac{c_4}{A_1} \iint_{Rc} \phi_i(\vec{u}) dA - \frac{c_5}{A_2} \iint_{Rs} \phi_i(\vec{u}) dA,$$ (12)

where C_4 and C_5 represent weight constants, A_1 and A_2 the areas of the circumference and of the sclera respectively, and Rc and Rs the correspondent areas domain.

Considering the position of the two peaks points P_1 and P_2, the energy E_p, should be minimized using:

$$E_p = c_6 \left(\phi_p(\vec{u} + P_1\vec{e}_1) + \phi_p(\vec{u} - P_2\vec{e}_1) \right),$$ (13)

where C_6 represents a weight constant.

Finally, the energy E_{priori} is used to avoid that the template used closes on itself:

$$E_p = \frac{k_1}{2} \left((x_e - x_c)^2 + (y_e - y_c)^2 \right)$$
$$+ \frac{k_2}{2} \left(P_1 - \frac{1}{2}(r+b) \right)^2$$
$$+ \frac{k_2}{2} \left(P_2 + \frac{1}{2}(r+b) \right)^2 + \frac{k_3}{2}(b - 2r)^2$$ (14)

where k_1, k_2, k_3 represent weight constants.

3.2 Matching process

The matching process used is defined by the dynamic update of the template parameters, which is processed by an optimal method: the steepest descendent gradient. For example, during the update of the iris radius we use:

$$r\_old = r\_new + dt \times r'_t,$$ (15)

where r_new is the radius value of the new iteration, r_old the radius of the previous iteration, dt the time step, and $r'(t)$ the variation rate of the radius in time, translated by the symmetrical variation rate of the total energy considering a small radius variation:

$$r'_t = \frac{dr}{dt} = -\frac{\partial E}{\partial r}.$$ (16)

Finally, for each descent step is calculate de total energy E, considering the new set of obtained parameters, and the best matching is found when the variation between two consecutives energy steps is less than a predefined tolerance. That means that the variation parameters are very small, which could indicate the correct eye detection in the input image. In Computational Vision, this procedure is usually known as the "minimization energy principle".

3.3 Parameters update

To update all parameters of the template used, the computational algorithm developed is divided into seven processing phases.

First, it is necessary to define a set of initial parameters values to be used as input in our computational algorithm. Thus, we consider that: $xe = xc$, $ye = yc$, $P1 = r$, $P2 = -r$, $t = 0$, $a = r$, $b = 2r$, $c = r$. We should note that the initial values chosen for xc, yc and r, must be carefully defined reflecting properly the size of the faces in the input image.

Then, in the first phase of our computational matching algorithm, we use the intensity valleys field to allow that the template takes a close position to the iris centre, being the energy weighted by constant C_1.

In the next processing phase, the radius and the centre of the iris are update, using the valleys, the grey levels, and edges intensity fields, keeping constant C_1 with the pervious value and increasing the weight of constants C_2 and C_4. This procedure permits the scaling of the circumference of the template used to the correct size of the iris.

In the third phase, previous constant values are increased again allowing the fitting of the template to the iris by a fine tuning approach.

To determine the correct angle rotation of the template used, we consider the fourth phase of our algorithm, using only the intensity peaks field, weighted

Table 1. Results of the parameters update during the phases 1, 2 and 3.

Parameters		Update processing phases			
	Units	0	1	2	3
t	Radians	0.00	–	–	–
xe		60.00	77.46	–	–
ye		21.00	51.47	–	–
xc		60.00	77.46	77.13	–
yc		21.00	51.47	50.13	–
P_1	Pixels	20.00	–	–	–
$-P_2$		20.00	–	–	–
a		20.00	–	–	–
b		40.00	–	–	–
c		20.00	–	–	–
r		20.00	–	18.00	17.96

by constant C_6. In this phase, the parameters to update are: the template centre (xe, ye), the peak points P_1 and P_2, and the rotation angle t.

The fifth phase, of our computational procedure, uses the edges, grey levels, and peaks intensity fields. This phase begins the process of scaling the parabolas to their correct position. The parameters that are update during this phase are: peak points P_1 and P_2, rotation angle t, eye centre (xe, ye), and eye width b.

In the sixth phase, using the same energy fields of the last phase, three parameters are update: eye width b, and heights a and b.

Finally, in the seven and last phase of our computational algorithm, all considering energy fields are used in order to update the eleven parameters of the template used simultaneously.

4 EXPERIMENTAL RESULTS

The Table 1, presents the obtained results considering the first three processing phases of our computational algorithm, using as input image the one presented in Figure (4a).

In Figure (4b), is represented the path followed by the template used during the first update processing phase, which allows the determination of the iris centre.

The path followed by the template used during the next two processing phases is represented in Figure 5. Thus, in the second phase, Figure (5a), the parameters related to the iris centre and to its radius are determined. In the third phase, Figure (5b), it is just done the fine adjustment of the iris radius.

During the fourth processing phase, as is indicated in Table 2, we update iteratively five parameters: orientation angle t, peak points P_1 and P_2, and centre of the template. The results obtained in this phase are

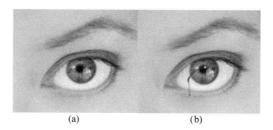

Figure 4. An input image (a) and the path followed by the template used in the iris centre detection during the first processing phase (b). (The start point is represented in red and the end point in green.)

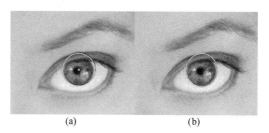

Figure 5. Update of the iris centre and its radius during the second processing phase (a), and the radius fine adjustment during the third processing phase (b).

Table 2. Results of the parameters update during the phases 4, 5, 6 and 7.

Parameters		Update processing phases			
	Units	4	5	6	7
t	Radians	0.17	0.30	–	0.28
xe		80.11	78.60	–	78.78
ye		42.38	42.10	–	42.06
xc		–	–	–	77.50
yc		–	–	–	52.05
P_1	Pixels	21.92	24.80	–	24.81
$-P_2$		21.94	19.14	–	20.72
a		–	–	19.75	19.45
b		–	30.95	31.49	32.06
c		–	–	16.39	17.58
r		–	–	–	17.96

showed in Figure (6a). In addition to the five parameters already considered in the fourth phase, is also update the template width, through the parameter b, during the fifth phase processing. This update accomplishes the first approximation of the template to the eye corners.

In the sixth phase of our computational processing, are only update the parameters that allow the scaling of the above and the below contours of the parabola included in the template. This is just an operation of

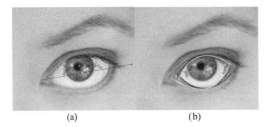

(a)   (b)

Figure 6.   Updating of the parameters of the template used during the fourth processing phase (a – location and orientation) and during the fifth processing phase (b – location, orientation and width). (The red colour represents the start positions and the green colour the final position.)

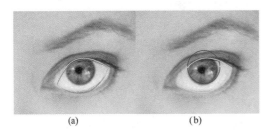

(a)   (b)

Figure 7.   Fine update of the template used during the sixth processing phase (a) and the final matching obtained after the last phase (b).

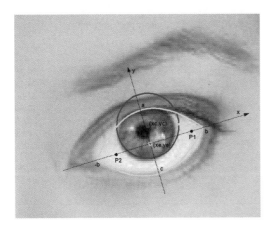

Figure 8.   Template used matched in the input image using our computational algorithm with its all control parameters identified.

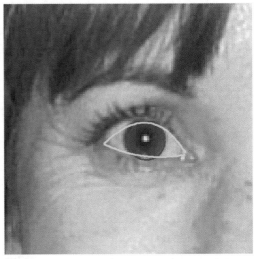

Figure 9.   Eye template match found using our detection algorithm on image of Figure 3.

We tested also our matching algorithm on the image used to illustrate the energy fields in the second section of this paper, Figure 3, considering the same constants values and time steps used in the previous experimental example. Although, the difference in scale that the two input images present, the detection obtained can still be consider as satisfactory, as can be verify in Figure 9. Of course, that a better result can be obtained on this input image if we selected a more suitable set of constants values and time steps.

5   CONCLUSIONS AND PERSPECTIVES OF FUTURE WORK

Having in mind the obtained results, we can consider that the methodology presented in this paper allows the adequate detection of the human eye in input images.

The algorithm developed converges very fast and gives good results when are used adequate energy weight constants. But, it still works fine even when the input images have faces with different scales and the initial parameters are not specially tuned for them. The disadvantages that the methodology presents consist on the necessity to define an adequate set of initial parameters for the template, and that template is initially placed near the eye to be detected in the input image.

As future works, we intend to develop a mechanism to automatically adjust the energy weight constants, applying the methodology considered in this work to images with faces presenting different scales, and

fine adjustment of the template to the eye represented in the input image, as is illustrated in Figure 7a). In Figure (7b), we show the best fit founded during the seven an last phase of our matching algorithm, and in the Figure 8, we present the template matched in the input image with all its control parameters identified.

apply this methodology in the tracking of the human eye in image sequences.

ACKNOWLEDGMENTS

The presented work was partially done in the scope of the project "Segmentation, Tracking and Motion Analysis of Deformable (2D/3D) Objects Using Physical Principles", with reference POSC/EEA-SRI/ 55386/2004, financially supported by *FCT – Fundação para a Ciência e a Tecnologia in Portugal.*

REFERENCES

Carvalho, Fernando Jorge Soares & Tavares, João Manuel R. S. 2005a. Detecção e Extracção de Características do Olho Humano a partir de um Modelo Protótipo Deformável. At Encontro Nacional de Visualização Científica, Espinho, Portugal.

Carvalho, Fernando J. S. & Tavares, João M. R. S. 2005b. Metodologias para identificação de faces em imagens: introdução e exemplos de resultados. At Congresso de Métodos Numéricos en Ingeniería 2005, Granada, España.

Toennies, K. & Behrens, F. & Aurnhammer, M. 2002. Feasibility of hough-transform based iris localisation for real-time-application. In *Proceedings of the 16th International Conference on Pattern Recognition (ICPR'02),* Quebec, vol. 2, pp. 299–305.

Yuille, A. & Hallinan, P. & Cohen, D. 1992. Feature Extraction from Faces Using Deformable Templates. *International Journal of Computer Vision* 8, pp. 99–111.

Yang, Ming & Kriegman, David J. & Ahuja, Narendra 2002. Detecting Faces in Images: A Survey. IEEE Transactions on Pattern Analysis and Machine Intelligence 248(1), pp. 34–58.

Wu, Y. & Liu, H. & Zha, H. 2003. A New Method of Human Eyelids Detection Based on Deformable Templates. In Proc. 2003 Sino-Korea Symp. On Intelligent Systems, pp.49–54, Guangzhou, China, Nov. 18.

Computational Vision and Medical Image Processing – João Tavares & Natal Jorge (eds)
© 2008 Taylor & Francis Group, London, ISBN 978-0-415-45777-4

A geometric modeling pipeline for bone structures based on computed tomography data: A veterinary study

D.S. Lopes, J.A.C. Martins & E.B. Pires
Instituto Superior Técnico and ICIST, Technical University of Lisbon, Lisbon, Portugal

L.B. Rodrigues
State University of Bahia Southwest – Campus of Itapetinga, Itapetinga, Brazil Mechanical
Engineering Graduate Program, Federal University of Minas Gerais, Belo Horizonte, Brazil

E.B. de Las Casas & R.R. Faleiros
Federal University of Minas Gerais, Belo Horizonte, Brazil

ABSTRACT: Computed Tomography (CT) is an imaging modality that reveals the inner parts of a body in a non-invasive fashion, providing the geometrical data suitable for the development of three-dimensional (3D) models. Due to the high signal contrast between hard and soft tissues, CT images are appropriate for bone structure modeling, visualization and manufacturing. In this paper, a mesh-based geometric modeling pipeline, capable of generating accurate surface meshes of bone structures for visualization and prototyping, is presented. A CAD-based modeling pipeline is also presented to provide computational finite element meshes for bone structures. The pipeline is composed by several software tools, mainly freeware, each with specific functionalities in the overall modeling scheme: image restoration and enhancement, image segmentation, mesh generation and adjustment. A veterinary application is considered aiming at the development of an intramedullary interlocking nail to be used for treatment of fractures in long bones of large animals.

1 INTRODUCTION

For a great number of pathologies, medical images are the starting point for clinical evaluation and the establishment of diagnosis. Imaging techniques such as computed tomography (CT) allow an accurate visualization of the inner body in a non-invasive fashion, providing relevant structural information.

The proposed geometrical modeling pipeline extracts 3D data from several contiguous tomographic images necessary to generate surface and volumic finite element meshes, and also boundary-representations (B-REP) of anatomical structures. The resulting meshes can be further used for visualization, measurement, biomechanical simulation, rapid prototyping and prosthesis design. Figure 1 presents the three fundamental blocks, connected in series, which compose a generic pipeline: image acquisition, digital image processing, mesh processing and CAD processing.

This paper describes techniques and enumerates the corresponding software tools that are used in mesh-based and CAD-based approaches that were designed to obtain 3D models of bones. The software tools

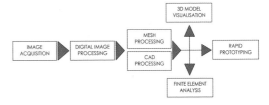

Figure 1.   Schema of a simplified pipeline for geometric modeling of tissue structures and some of its applications.

incorporate several algorithm blocks, each with specific functionalities in the overall modeling scheme: image restoration and enhancement, image segmentation, mesh generation and adjustment and boundary generation techniques (Cebral & Löhner 1999, Du et al. 2005, Sullivan et al. 2000, Viceconti et al. 1998, Sun et al. 2005).

Based on the results obtained for human bones (Lopes 2006), a veterinary application is considered for a calf femur, aiming at the development of an intramedullary interlocking nail for the treatment of fractures in large animals. For this purpose, a left femur

Figure 2. Intramedullary interlocking nail.

from an abated male Holstein breed calf with less than 15 days of life was obtained from a slaughterhouse. To facilitate the process of image acquisition, it was previously dissected from the attached soft tissues (although the periosteum and flesh vestiges remained).

In orthopedic veterinary, the treatment of long bone fractures in large animals such as horses and bovines is still a challenge to clinicians and surgeons. The available products that are used in surgeries to fix bone fragments are too expensive and are adapted from human devices. The most widely used implants are plates and intramedullary nails, which are inserted and/or fixed to bone tissue with screws. The intramedullary interlocking nail (Fig. 2) is an improvement of the formerly cited nail (Stiffler 2004) and since the 50's has been used in human orthopedics.

These implants are easier to use, they perform better and yield better mechanical and biological results than other implants on the treatment of fractures of the femur, tibia and humerus (Beale 2004). In the early 90's, this device was introduced in the treatment of long bone fractures in small animals (e.g. cats and dogs) (Dueland et al. 1999, Durall & Diaz-Bertrana 2005). Approximately in the same period, some studies on the use of intramedullary interlocking nails in fractures of long bones in large animals were performed (Watkins 1990, McDuffee et al. 2000). Some promising results were obtained (De Marval 2006, De Marval et al. 2006) with *ex vivo* and *in vivo* studies using a polypropylene interlocking nail.

The final goal of an ongoing veterinary research is the study of the process of bone healing in contact with an implant. For this purpose, computational simulation is an important tool for developing devices for medical purposes, since it provides good and reliable results and reduces the experimental cost and the amount of animals used in *in vivo* studies. Basically, it consists on two steps: first the development of a computational 3D model of a long bone, then the development and implementation of an algorithm to study the process of bone healing.

2 METHODOLOGY

In order to obtain surface and volumic meshes of a bone with the mesh-based approach, the steps described in sections 2.3 to 2.7 were followed.

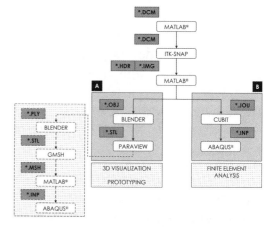

Figure 3. Diagram of the software pipeline. The white boxes present the software name and the dark grey boxes contain the data file extension. The pipeline outputs (A) a surface mesh for 3D visualization and prototyping or (B) a boundary representation and its associated volumic mesh used for finite element analysis. The dashed boxes represent the modeling approach described in sections 2.7 and 2.9.

A CAD-based modeling pipeline, developed to surpass mesh-based modeling limitations, is described in section 2.8.

2.1 *Computational system requirements*

The workstation used for the geometrical modeling was a PC computer running Windows® XP (Service Pack 2), 2 GB RAM and 2.8 GHz CPU dual-core, with a 256 MB NVIDIA GeForce 7800 GTX graphical card.

2.2 *Software tools*

The software pipeline used to model the anatomical structures is presented in Figure 3. Most of the software is freeware and open source, entirely licensed for academic research and education. This commercial software is commonly used in the academic community.

Each software tool has a role in the pipeline: MATLAB® and the Image Processing Toolbox® perform spatial image filtering, surface generation and file conversion; ITK-SNAP is responsible for image segmentation; Blender, although being an advanced modeling software, is only used for file conversion; surface adjustments and surface model visualization are assured by ParaView; GMSH automatically generates a volumic mesh; CUBIT Mesh Generation Toolkit was used to create boundary representations and volumic meshes; and ABAQUS® is a powerful finite element solver.

The file formats in Figure 3 correspond to medical images (*.dcm, *.hdr, *.img), 3D models (*.obj, *.stl, *.ply) and native files (*.jou, *.inp, *.msh).

2.3 Image acquisition

The process begins with a CT scan of the subject providing the input data for the pipeline. The CT images must have a high resolution, present high intensity contrast between hard and soft tissues and be promptly acquired to avoid motion blur artifacts. The main parameters that should be taken under consideration for CT acquisition are the voltage and current intensity of the X-ray lamp, voxel dimensions, slice thickness, image dimensions and radiation exposure.

Image resolution is directly proportional to radiation dose. On the other hand, radiation exposures must be limited in order to prevent major cellular damage. Hence, a compromise between radiation dose and image resolution must be established when data is collected from living subjects.

Contrary to clinical practice, in case of *ex vivo* structures, it is advisable to use higher radiation doses granting better image resolution. This is the case of the calf bone in the application considered in the present work.

2.4 Image restoration and enhancement

Anisotropic diffusion filtering (Perona & Malik 1990) was performed along the sagittal planes of the data volumic. Although the tomographic acquisition planes were axial, the option for a sagittal filtering proved to be more effective. What motivated this filtering orientation was the simpler noise structure presented in the sagittal plane: a point-like structure rather than the streak-like noise, which is characteristic of the axial plane.

The anisotropic diffusion filter smoothes constant regions and preserves the edges, therefore simultaneously improving tissue contrast and reducing noise.

To perform image restoration and enhancement, a script was developed in MATLAB®, requiring functions from the Image Processing Toolbox®.

2.5 Image segmentation

Segmentation plays the main role in the designed pipeline, as it consists in extracting the anatomical information, establishing the transition between image and mesh data.

Most of the employed segmentation algorithms for medical images are based on deformable models (Pham et al. 2000). The region competition snakes proposed by Zhu & Yuille (1996) are appropriate deformable models for image segmentation when the structure of interest has a high contrast relatively to the image background. This is the case with hard and soft tissue in CT images. Only bone tissue was segmented.

The segmentation process was performed with the ITK-SNAP software following the indications presented by Yushkevich et al. (2006). Manual segmentation is required to correct eventual over- and sub-segmented regions.

2.6 Surface mesh generation and adjustments

The segmentation step was followed by triangle surface mesh generation via the marching cubes algorithm (Lorensen & Cline 1987). Laplacian smoothing (Vollmer et al. 1999) and mesh simplification techniques, involving decimation (Schroeder et al. 1992) and quadric clustering (Lindstrom 2000), were applied in order to improve mesh quality and subsequent computational efficiency. The initial mesh was smoothed, simplified and then smoothed again (Sullivan et al. 2000) providing a surface mesh suitable for clinical visualization and rapid prototyping purposes.

The surface mesh was generated in MATLAB® and adjustments were performed with ParaView.

2.7 Volume mesh generation

From the triangle surface mesh, a volumic mesh is then generated by boundary constrained Delaunay tetrahedralization (Watson 1981). Both surface and volumic meshes are adaptive and unstructured.

The GMSH software was used to generate the volumic mesh.

2.8 CAD modeling

Due to the excessive number of nodes and a large number of bad quality elements, which are common features in mesh-based models, a CAD-based approach (Viceconti et al. 1998, Sun et al. 2005) was developed.

The endostium and periostium curves were extracted from a set of equally spaced segmented images. Each curve was interpolated with a spline. A skinning operation was performed to link consecutive curves forming the surfaces. A solid object was defined by the surfaces which bound it. Finally, a tetrahedral mesh was generated.

The MATLAB® and CUBIT software were used to create the boundary representation and volumic mesh.

2.9 Finite element analysis

The volumic meshes obtained from the pipeline were inserted into ABAQUS® code in order to perform finite element analyses.

With the model obtained by the mesh-based approach, ABAQUS® returned warnings regarding the presence of several distorted elements. On the other hand, with the CAD-based approach no such warnings appeared.

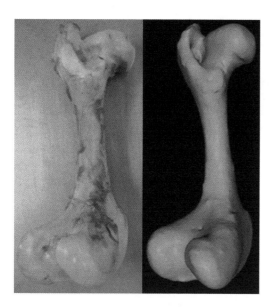

Figure 4. Comparison between a photograph of the left calf femur and the corresponding surface mesh (posterior view).

3 MODELING RESULTS

For the veterinary study considered in this work, the first result of the pipeline was a preliminary adaptive surface mesh of the left femur shown in Figure 4. The visual surface features of the femur are similar to the ones obtained by Lopes (2006) with human bones.

Due to computational limitations in generating the femur volumic mesh with the mesh-based approach, the diaphysis's cortical tissue was considered. This structure was also modeled via CAD approach to obtain the volumic mesh.

Quantitative analyses were performed to assure quality inspection, using the edge ratio quality measure (maximum edge length/minimum edge length). A quality visual analysis of the meshes obtained with the mesh-based and CAD-based approaches is presented (Fig. 5), in which each triangle element is colored with its corresponding quality value: 1.0 for an equilateral triangle and higher values for degenerate cases. Mesh quality histograms were plotted (Figs 6–7) showing the number of elements and its corresponding quality value.

The volumic meshes of the proximal region of cortical bone tissue obtained with both modeling approaches are presented (Figs 8–9). The differences are clearly apparent.

4 DISCUSSION

Image filtering can improve image segmentation by providing signal noise reduction and artifact

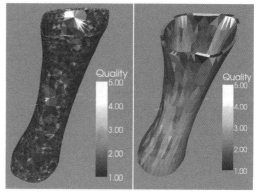

Figure 5. A quality visual analysis of the meshes obtained with the mesh-based (left) and CAD-based (right) approaches.

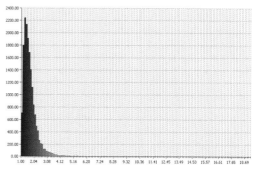

Figure 6. Mesh quality histogram (edge ratio) of the cortical bone of the diaphysis obtained with the mesh-based approach.

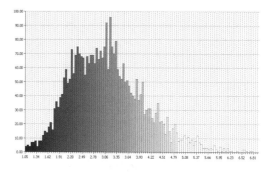

Figure 7. Mesh quality histogram (edge ratio) of the cortical bone of the diaphysis obtained with the CAD-based approach.

suppression, hence improving tissue segmentation. Other filtering techniques, such as the ones described by Kachelriess et al. (2001) and Westin et al. (2000) may also reduce the amount of noise and the presence of artifacts.

In medical image analysis, two-dimensional active contours are commonly used to segment each image

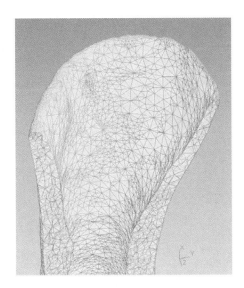

Figure 8. Volumic mesh obtained with the boundary constrained Delaunay tetrahedralization.

Figure 9. Volumic mesh obtained with the modeling CAD approach.

slice in the data volumic. Three-dimensional segmentation with spherical active contours is more efficient and effective as segmentation takes place at several image slices. However, segmentation is one of the most difficult tasks to be performed within the entire pipeline. One should bear in mind that a CT image comprises the information related to the tissue distribution in the human body, but also undesired signals. Artifacts, noise, blurred edges are some of the features common to any medical image and impose severe segmentation difficulties. Segmentation errors are normally corrected manually. This is the most time consuming task in geometrical modeling.

In the mesh-based modeling approach, the stage of surface mesh adjustment induces the most important geometric and topologic errors of the pipeline. Mesh simplification, which is required for further computational efficiency, may even have a harsher effect than smoothing, if done inappropriately.

Visual inspection of the surface mesh (Fig. 4) showed very positive results. Note that this type of check is qualitative and prone to subjective interpretation, so that it must be validated by medical or veterinary specialists. That was the case with the calf femur. Thus, mesh-based modeling approach outputs 3D models suitable for visualization and rapid prototyping (Sun et al. 2005).

The overall quality of the obtained surface meshes are shown in Figure 5. Ideally, the triangles should present quality values between 1.0 and 2.0. The histogram of the mesh-based surface (Fig. 6) exhibits a tail ranging from 2.0 to near 30.05, indicating the presence of several distorted elements. So, it is clear that the overall mesh quality is not ideal. For CAD-based modeling, the histogram ranges from 1.0 to 9.69, indicating better element quality relatively to the mesh-based surface.

The excessive number of nodes and, consequently, of elements, is another feature that limits the application of the mesh-based modeling for computational simulations.

Both of these mesh-based surfaces characteristics compromise the possibility of creating proper volumic meshes for finite element analysis: (i) since the volumic mesh generation algorithm used is boundary constrained, surface elements with poor quality (Fig. 8) will deeply affect the quality of the tetrahedra, hence, compromising the finite element analysis; (ii) the excess of nodes introduces a computational burden which limits processing and memory capabilities.

Solving these difficulties is not a trivial task. Therefore the CAD-based modeling pipeline was developed. This pipeline relies on the previous one (Lopes 2006) and incorporates the CUBIT Mesh Generation Toolkit for boundary representation, and surface and volumic mesh generation. The first result obtained (Fig. 9) is quite promising, presenting a regular volumic mesh with a much smaller number of nodes and an overall mesh quality greatly improved, contrary to the adaptive unstructured mesh-based volume (Fig. 8).

5 CONCLUSIONS

The surface meshes obtained with the mesh-based pipeline are suitable for visualization and prototyping purposes. The use of these surface meshes to generate volumic meshes for finite element analysis yielded poor elements with excessive geometric information. In this manner, a CAD-based geometrical modeling pipeline was also considered. This software pipeline incorporates the CUBIT Mesh Generation Toolkit® and has shown quite promising results on generating a volumic mesh of the cortical tissue of the calf femur.

The next step is to improve the pipeline in order to obtain the full bone volumic mesh with both cortical and spongy tissue. In the future, the CAD-modeling will be applied to obtain B-REP and meshes for 3D visualization, prototyping and finite element analysis.

ACKNOWLEDGMENTS

The first three authors would like to acknowledge the support of the Fundação para a Ciência e Tecnologia, the POCI 2010 and the Pluriannual Base Program POCTI-SFA-9-76.

The three remaining authors acknowledge the support of the Coordenação de Aperfeiçoamento de Pessoal de Nível Superior, that provides a doctoral scholarship for the fourth author during his studies at Instituto Superior Técnico, in Lisbon, and Fundação de Amparo à Pesquisa do Estado de Minas Gerais.

REFERENCES

ABAQUS®. http://www.hks.com/

Beale, B. 2004. Orthopedic clinical techniques femur fracture repair. *Clinical Techniques in Small Animal Practice*, 19 (3): 134–150.

Blender. http://www.blender.org/cms/Home.2.0.html

Cebral, J.R. & Löhner, R. 1999. From Medical Images to CFD Meshes, *Proceedings of the 8th International Meshing Roundtable*.

CUBIT Mesh Generation Toolkit. http://cubit.sandia.gov/

De Marval, C.A. 2006. *Estudo ex vivo e in vivo de polímero biocompatível como material alternativo na confecção de haste bloqueada para redução de fraturas em úmeros de bezerros*. (Dissertation, in Portuguese). Pós-Graduação em Medicina Veterinária. Escola de Veterinária da Universidade Federal de Minas Gerais. Brasil.

De Marval, C.A., Rodrigues, L.B., Jordão, L.R., Las Casas, E.B. & Faleiros, R.R. 2006. Haste bloqueada de polipropileno em bezerros: avaliação ex vivo (in Portuguese). *Arquivo Brasileiro de Medicina Veterinária e Zootecnia*, Vol 58, n. supl. 2, 108.

Du, J., Yang, X. & Du, Y. 2005. From Medical Images to Finite Grids System, *Engineering in Medicine and Biology 27th Annual Conference*.

Dueland, R.T., Johnson, K.A., Roe, S.C., Engen, M.H. & Lesser, A.S. 1999. Interlocking nail treatment of diaphyseal long-bone fractures in dogs. *J Am Vet Med Assoc.*, 214 (1): 59–66.

Durall, I. & Diaz-Bertrana, M.C. 2005. Fracture Fixation Using Interlocking Nails. *Proc. of 30th World Small Animal Veterinary Association World Congress.* City of México, Mexico.

GMSH. http://www.geuz.org/gmsh/

ITK-SNAP. http://www.itksnap.org/

Kachelriess, M., Watzke, O. & Kalender, W.A. 2001. Generalized multi-dimensional adaptive filtering for conventional and spiral single-slice, multi-slice, and cone-beam CT, *Med. Phys.*, 28 (4), 475–490.

Lindstrom, P. 2000. Out-of-Core Simplification of Large Polygonal Models, *Proceedings of ACM SIGGRAPH*.

Lopes, D.S. 2006. *Geometric Modeling of Human Structures Based on CT Data – a Software Pipeline*. Final Course Project for the Biomedical Engineering Degree. Universidade Técnica de Lisboa. Instituto Superior Técnico. Lisboa, Portugal.

Lorensen, W.E. & Cline, H.E. 1987. Marching Cubes: a High Resolution 3D Surface Construction Algorithm. *Computer Graphics*, Vol. 21, No. 4.

MATLAB® – The Language of Technical Computing. http://www.mathworks.com/products/matlab/

McDuffee, L.A., Stover, S.M., Bach, J.M., Taylor, K.T. 2000. An in vitro biomechanical investigation of an equine interlocking nail. *Veterinary Surgery*, Vol. 29, p. 38–47.

McInerney, T. & Terzopoulos, D. 1996. Deformable Models in Medical Image Analysis: A Survey, *Medical Image Analysis*, 1 (2), 91–108.

ParaView. http://www.ParaView.org/HTML/Index.html

Perona, P. & Malik, J. 1990. Scale-Space and Edge Detection using Anisotropic Diffusion, *IEEE Transactions and Pattern Analysis and Machine Intelligence*, Vol. 12. No. 7.

Pham, D.L., Xu, C. & Prince, J.L. 2000. Current Methods in Medical Image Segmentation, *Annu. Rev. Biomed. Eng.*, 2, 315–337.

Schroeder, W.J., Zarge, J. A., & Lorensen, W.E. 1992. Decimation of Triangle Meshes, *Computer Graphics (SIGGRAPH92 Proceedings)*, 65–70.

Stiffler, K.S. 2004. Internal fracture fixation. *Clinical Techniques in Small Animal Practice*, 19 (3): 105–113.

Sullivan, J.M., Wu, Z. & Kulkarni, A. 2000. 3D Volumic Mesh Generation of Human Organs Using Surface Geometries Created from the Visible Human Data Set, *The Third Visible Human Project Conference*.

Sun, W., Starly, B., Nam, J. & Darling, A. 2005. Bio-CAD modeling and its applications in computer-aided tissue engineering. *Computer-Aided Design*, 37, 1097–1114.

Viceconti, M., Zannoni, C. & Pierotti, L. 1998. TRI2SOLID: an application of reverse engineering methods to the creation of CAD models of bone segments, *Computer Methods and Programs in Biomedicine*, 56, 211–220.

Vollmer, J., Mencl, R. & Müller, H. 1999. Improved Laplacian Smoothing of Noisy Surface Meshes, *Computer Graphics Forum, Eurographics '99*.

Watkins, J.P. 1990. Intramedullary interlocking nail fixation in foals. *Veterinary Surgery*, 19: 80.

Watson, D.F. 1981. Computing the n-dimensional Delaunay tessellation with application to Voronoi polytopes. *Comput J*, 24 (2), 167–172.

Westin, C.-F., Richolt, J., Moharir, V. & Kikinis, R. 2000. Affine adaptive filtering of CT data. *Medical Image Analysis*, 4, 161–177.

Yushkevich, P.A., Piven, J., Hazlett, H.C., Smith, R.G., Ho, S., Gee, J.C. & Gerig, G. 2006. User-guided 3D active contour segmentation of anatomical structures: Significantly improved efficiency and reliability. *Neuroimage*, Vol. 31, Issue 3, 1116–1128.

Zhu, K.H. & Yuille, A. 1996. Region competition: unifying snakes, region growing, and Bayes/mdl for multiband image segmentation. *IEEE Trans. Pattern Anal. Mach. Intell.*, 18 (9), 884–900.

Computational Vision and Medical Image Processing – João Tavares & Natal Jorge (eds)
© 2008 Taylor & Francis Group, London, ISBN 978-0-415-45777-4

Filtering ECG signals using Fractional Fourier Transform

Juan M. Vilardy, Cesar O. Torres & Lorenzo Mattos

Optics and Computer Science Laboratory, Universidad Popular Del Cesar, Valledupar, Colombia

ABSTRACT: The present paper shows the application of the Fractional Fourier Transform (FrFT) for the filtering Electrocardiographic signals (ECG). The noise classes that affect signals ECG, such as: Myolectrical noise, noise by interference of the electrical net and variations in the base line, are independent of signal ECG, but the Wigner distributions projections of such noise and signal ECG overlap so much in the time as frequency domains and therefore the conventional filtering techniques are not applicable to this type of distorted signals. This difficulty can be solved by a Wigner distribution rotation of the noisy signal using a FrFT to separate the Wigner distribution projections of the noise and the signal ECG nondistorted, obtaining this way that the noise can be blocked by a simple mask (multiplicative filter) and finally to recover signal ECG without noise an inverse FrFT is applied. At the moment that the distribution of Wigner is rotated by a FrFT, the new coordenates axes represent fractional domains with respect to the coordenates axes previous (time and frequency).

1 INTRODUCTION

The Fourier transform (FT) is one of the most frequently used tools in signal analysis. A generalization of the Fourier transform is the fractional Fourier transform (FrFT) and has become a powerful tool for time-varying signal analysis. In time-varying signal analysis, it is customary to use the time–frequency plane, with two orthogonal time and frequency axes. Because the successive two forward Fourier transform operations will result in the reflected version of the original signal, the FT can be interpreted as a rotation of signal by the angle $\pi/2$ in the time–frequency plane and represented as an orthogonal signal representation for sinusoidal signal. The FrFT performs a rotation of signal in the continuous time–frequency plane to any angle and serves as an orthonormal signal representation for the chirp signal. The fractional Fourier transform is also called *rotational Fourier transform* or *angular Fourier transform* in some documents.

Besides being a generalization of the FT, the FrFT has been proved to relate to other time-varying signal analysis tools, such as Wigner distribution, short-time Fourier transform (Almeida et al. 1994), wavelet transform, and so on. The applications of the FrFT include solving differential equations, quantum mechanics, optical signal processing, timevariant filtering and multiplexing, swept-frequency filters, pattern recognition, and time–frequency signal analysis. Several properties of the FRFT in signal analysis have been summarized in some papers (Almeida et al. 1994).

This paper is organized as follows. In Section 2, the preliminaries on Fractional Fourier transforms and Wigner distribution Function are reviewed. The Filtering ECG signals using fractional Fourier transform are described in Section 3. Finally, conclusions are made in Section 4.

2 PRELIMINARIES

2.1 *Fractional Fourier Transform (FrFT)*

The Fractional Fourier transform (FrFT) is a generalization of the identity transform and the conventional Fourier transform (FT) into fractional domains. The traditional Fourier transform decomposes signal by sinusoids whereas Fractional Fourier transform corresponds to expressing the signal in terms of an orthonormal basis formed by chirps. The Fractional Fourier transform can be understood as a Fourier transform to the *pth* power where *p* is not required to be an integer. The fractional Fourier transform (FrFT) of order *p*, is a linear integral operator that maps a given function $f(x)$ onto function $f_p(\xi)$, by: (Ozaktas et al. 2001):

$$f_p(\xi) = \Im^p[f(x)] = \int_{-\infty}^{+\infty} K_p(\xi, x) f(x) dx \qquad (1)$$

Where kernel this defined by:

$$K_p(\xi, x) = C_\alpha \exp\left\{-i\pi(2\frac{x\xi}{\sin\alpha} - (x^2 + \xi^2)\cot\alpha)\right\} \qquad (2)$$

with:

$$C_\alpha = \frac{\exp\{-i[(\pi\,\mathrm{sgn}(\sin\alpha)/4) - \alpha/2]\}}{\sqrt{|\sin\alpha|}} \qquad (3)$$

$$\alpha = \frac{\pi p}{2} \qquad (4)$$

Observing that α, when defined as a real number, only appears as argument of trigonometric functions in the equations (2) and (3), the definition (1) is periodic in p, with period 4.

The most important properties of the fractional Fourier transform are:

1) *Fractional order additivity:*

$$\mathfrak{I}^{p1}\{\mathfrak{I}^{p2}[f(x)]\} = \mathfrak{I}^{p1+p2}[f(x)] \qquad (5)$$

2) *Unitarity:*

$$\{\mathfrak{I}^{p}[f(x)]\}^* = \mathfrak{I}^{-p}[f^*(x)] \qquad (6)$$

Where: * it denotes the conjugated complex.

3) *Reduction to the Fourier transform when $p = 1$.*

2.2 Wigner distribution function

The Wigner distribution function is a powerful time-frequency analysis tool and it can be used to illustrate the time-frequency properties of a signal. The Wigner distribution of a function $f(t)$ is defined as (Pei et al. 2001):

$$W_f(t,v) = \int_{-\infty}^{+\infty} f(t+\frac{\tau}{2})f^*(t-\frac{\tau}{2})e^{-2\pi v\tau}d\tau \qquad (7)$$

And it can be interpreted as a function that indicates the distribution of the signal energy over the time-frequency space. The most significant properties of the Wigner distribution and the relationships between Wigner distribution and FrFT are stated in the following equations:

1) $\int W_f(t,v)dv = |f(t)|^2$ (8)

2) $\int W_f(t,v)dt = |F(v)|^2$ (9)

3) $\iint W_f(t,v)dtdv = \|f\|^2 = En[f]$ (10)

4) If $g(t) = h(t)*f(t)$,

     then $W_g(t,v) = \int W_h(t-\tau,v)\,W_f(\tau,v)d\tau$ (11)

5) If $g(t) = h(t)f(t)$,

     Then $W_g(t,v) = \int W_h(t,v-v')W_f(t,v')dv'$ (12)

6) *Wigner distribution of the Fourier transform is the Wigner Distribution of the original function rotated clockwise by a right angle.*

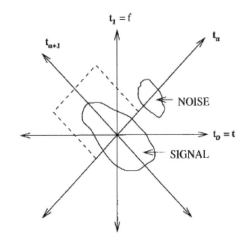

Figure 1. Wigner distribution of the FrFT of a function and noise separation on the Fractional Fourier domain (Kutay et al. 1997).

One of the most important properties of the FrFT states that the Wigner distribution of the FrFT of a function is a rotated version of the Wigner distribution of the original function:

$$W_{fp}(t,v) = W_f(t\cos\alpha - v\sin\alpha, t\cos\alpha + v\sin\alpha) \qquad (13)$$

The Wigner distribution is completely symmetric with respect to the time-frequency domains, it is always real but not always positive. The Wigner distribution exhibits advantages over the spectrogram (short-time Fourier transform): the conditional averages are exactly the instantaneous frequency and the group delay, whereas the spectrogram fails to achieve this result, no matter what window is chosen. The Wigner distribution is not a linear transformation, a fact that complicates the use of the Wigner distribution for time-frequency filtering.

3 FILTERING ECG SIGNALS WITH FrFT

The electrocardiographic signals are functions that represent the changes that happen in the electrical potentials of the heart in the dominion of the time, in where the voltage is indicated on the vertical axis, and the time on the horizontal axis.

The wave P is the deflection produced by the auricular depolarization, the complex QRS represents the time of ventricular depolarization, the wave T is the ventricular fast repolarización and the wave U is a deflection of low voltage, normally is positive.

The electrocardiographic signal was taken directly from the patient with the CARDIOBIP electrocardiograph (Martines et al. 2007), The following figure show the captured ECG signal of the patient with noise:

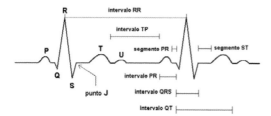

Figure 2. Waves of the ECG.

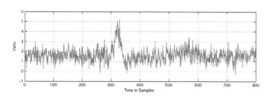

Figure 3. Noisy ECG signal.

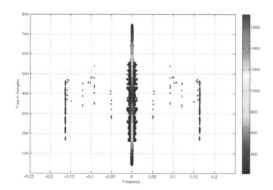

Figure 4. Wigner distribution of the noisy ECG signal.

The Wigner distribution of the noisy ECG Signal, is:

In order to make the filtering of noisy ECG signal, we used the FrFT for the optimal filtering in fractional Fourier domains (Kutay et al. 1997, & Erden et al. 1999), thus:

$$s = \mathfrak{I}^{-p}\left(h \bullet \mathfrak{I}^{p}(y)\right) \tag{14}$$

Where y is the noisy ECG signal, h is the multiplicative filter and s is the filtered ECG signal, its note that for $p = 1$, this estimate corresponds to filtering in the conventional Fourier domain.

Observing the Wigner distribution of the noisy ECG signal in the figure 4, and using the noise separation on the Fractional Fourier domain like in the figure 1, we used a multiplicative pass-stopband filter

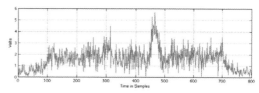

Figure 5. Magnitude of the fractional Fourier transfom of order $p = 1.5705$ of the noisy ECG signal.

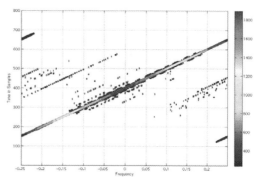

Figure 6. Wigner distribution of the noisy ECG signal transformed.

(Pei et al. 2001). The Fourier transform function of the pass-stopband filter is:

$$H(u) = 1, \quad when\ u_0 - B/2 < u < u_0 + B/2$$
$$H(u) = 0, \quad otherwise. \tag{15}$$

The order p of the farctional Fourier transform in equation 14 and the filter parameters (u_o and B), are (Pei et al. 2001):

$$p = \cot^{-1}\left(\frac{w_1}{t_1}\right) \tag{16}$$

$$u_0 = (t_0 + t_1)\cos(p)/2$$
$$B = \left|(t_1 - t_0)\cos(p)\right| \tag{17}$$

Using figure 4, it is had: $w_1 = 0.1257$, $t_0 = t_1 = 370$, $p = 1.5705$, $u_o = 0.1096$ y $B = 0$. It apply the farctional Fourier transform to noisy ECG Signal with the following result (Arikan et al. 1996):

The Wigner distribution of the noisy ECG Signal transformed, is:

The Filtered signal with the equation 14, is:

The Wigner distribution of the clean ECG Signal, is:

By simple observation it note that the filtering using FrFT can be enable effective noise elimination, for the following types noise: White Gaussian noise, chirp noise and others.

225

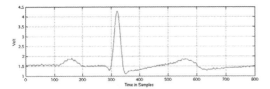

Figure 7.   Filtered ECG signal.

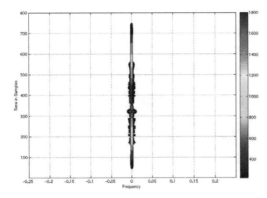

Figure 8.   Wigner distribution of the clean ECG signal.

4   CONCLUSIONS

In this paper, we applied the FrFT for the filtering of ECG signals and the signal quality can be improved after filtering the noise. The basic teory of the filtering is the relationship between the FrFT and the Wiger Distribution Function for to make the filtering process on the Fractional Fourier domains adequates, eliminating the noise in the ECG signals. We have used the FrFT to illustrate the effect of the la rotation on the Time-Frequency representations as The Wigner Distribution Function and its application to the noise filtering signal.

REFERENCES

Almeida L. 1994. The Fractional Fourier Transform and Time-Frequency Representations. In IEEE Transactions on Signal Processing, Vol. 42, No. 11.

Arikan O, Ozaktas H, Kutay M, & Bozdaki G. 1996. Digital Computation of the Fractional Fourier Transform. In IEEE Transactions on Signal Processing, Vol. 44, No. 9.

Kutay M, Ozaktas H, Arikan O, & Onural L. 1997. Optimal Filtering in Fractional Fourier Domains. In IEEE Transactions on Signal Processing, Vol. 45, No. 5.

Erden M, Kutay M, & Ozaktas H. 1999. Repeated Filtering in Consecutive Fractional Fourier Domains and Its Application to Signal Restoration. In IEEE Transactions on Signal Processing, Vol. 47, No. 5.

Pei S, & Ding J. 2001. Relations between Fractional Operations and Time-Frequency Distributions, and Their Applications. In IEEE Transactions on Signal Processing, Vol. 49, No. 8.

Ozaktas H, Zalevsky Z, & Kutay M. 2001. The Fractional Fourier Transform with Applications in Optics and Signal Processing. In Jhon Wiley & Sons (eds). New York.

Martines R, David R, Vilardy J, Torres C, & Mattos L. 2007. Acquisition Electronic System for Visualization, Registering and Filtering of ECG Signals. In Revision by the Scientific Committee of I ECCOMAS Thematic Conference on Computational Vision and Medical Image Processing.

Computational Vision and Medical Image Processing – João Tavares & Natal Jorge (eds)
© 2008 Taylor & Francis Group, London, ISBN 978-0-415-45777-4

Test of algorithms for z localization of pathology in optical micrographs stack of a cell

A.C. Pinho, A.J. Silva & A.R. Borges
Institute of Electronics and Telematics Engineering of Aveiro (IEETA), University of Aveiro, Portugal

B.S.K. Mendiratta
Department of Physics, University of Aveiro, Portugal

ABSTRACT: This work aims at testing some available deconvolution algorithms for constructing a 3-D view from a Z-stack of wide field optical micrographs. Specifically, we test different algorithms for providing useful contrast in z-direction in the demanding situation that the size of the object in the z-direction and that of the features (x–y plane) are of the order of probe wavelength. Here, the concept of Point Spread Function, based on geometrical optics, is at the limit of its validity. We have analyzed a Z-stack of images of erythrocytes with basophilic stippling associated with pyrimidine 5'-nucleotidase deficiency. The relative z-positions of dotted features can provide useful information to pathologist in understanding the causes of the disease. Our tests on four algorithms show that despite the limiting conditions mentioned above we can obtain an improvement in the vertical discrimination and general contrast of the figure.

1 INTRODUCTION

Localization and the extent of an anomaly in biological tissue is an all pervasive problem in medical sciences. Tomographic techniques are widely employed to visualize pathologies inside the body that are about a cm or larger. However, not many techniques are available to "see" the pathologies in a sample of linear size less than mm. For hematologists, on the other hand, the samples, the red blood cells are a few ($\sim$8)μm across and about 2 μm thick (Hoffbrand, A. & Pettit, J. 1993), and in some cases it is useful to know the relative depth of features that appear in the x–y image of a single erythrocyte.

A three dimensional image can be constructed by taking x–y images at different z positions of the focusing objective of the microscope (Conchello, J. A. 1994a, b, Vicidomini, G. 2005). Each image has, however, contributions from planes above the focusing plane as well as from the planes below (Vitri, J. et al. 1999a, b, McNally, J. et al. 1996). These out of plane contributions lead to blurring or loss of contrast in the image. Thus, the out of plane contribution must be carefully "subtracted" from the raw z-stack to construct a meaningful three dimensional image (Monvel, J. et al. 2001a, b, Vonesch, C. et al. 2006). The blurring of an out-of-focus image is mathematically described by Point Spread Function (PSF). This is

an instrumental function which can be experimentally determined or theoretically calculated (McNally, J. et al. 1996). The blurring effects occur on every process of image acquisition, but they are particularly important in biological samples of micron size where size of the features is close the optical resolution (Monvel, J. et al. 2001). The basic strategy of "deblurring" is to apply the "inverse" of PSF or the deconvolution algorithm to the raw images and thus obtain the true image (Schlecht, J. et al. 2006). The main aim of the our study is to show that the available algorithms can be applied to the optical images of a single erythrocyte and though the final 3-D image is noisy the relative z-position of the submicron features can be determined with a resolution close to the step size, 100 nm, of the z-stack.

The method, described below, consists of test of variable parameter PSFs on simulated 3-D image of spheres and optimization of the associated decovolution algorithms so that we can reproduce the simulated image. The optimized procedure is then applied to real images of blood smear of patients with the disorder associated with pyrimidine 5'-nucleotidase deficiency. The disorder shows up as dotted structures called basophilic stippling in the erythrocyte cell. The principal deconvolution algorithms tested were (A) non iterative methods: Wiener (WF) and Regularized Filters (RF) (Jianga, M. et al. 2003a, b, Holmes, T. J. 2002); and (B) iterative methods,

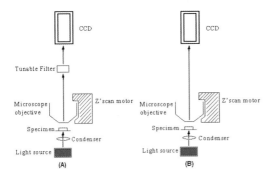

Figure 1. Schematic diagram of the experimental Setup. **A**: with the tunable filter **B**: without the tunable filter.

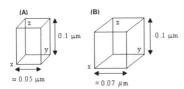

Figure 2. Schematic diagram of the size of the pixel volume of the two different setups, respectively **A** and **B**.

Richardson-Lucy (RL) and Blind Deconvolution (BD) (Kempen, G. M. P. 1999a, b, Holmes, T. 2002b, c, Jianga, M. et al. 2003c, d, Vicidomini, G. 2005d, e, Lai, X. et al. 2005).

2 DISCRIPTION OF THE METHOD

2.1 Description of image acquisition system

The experimental setup consists of a microscope (Zeiss Axioscop), a tunable liquid crystal filter (Varispec) and a peltier cooled high resolution (1024 × 1280) CCD camera (Sensicam, PCO, Germany). An oil immersion 100× objective (N.A.1.4) was used to obtain wide field images in the transmission mode. The Z-scan step motor coupled to the objective has span of 100 μm and minimum step size of 100 nm. The micrographs were taken with (mode A) and without (mode B) tunable filter inserted between the camera and the microscope output tube. The tunable filter allows taking pictures at a selectable wavelength with a spectral window of 10 nm.

The two different modes provide different pixel resolutions and the "voxel" size is indicated in Figure 2.

2.2 Three principal phases of the study

(1) Application of deconvolution algorithms to simulated 3-D images of ideal spheres. In this part, the ideal image was convoluted with some theoretical PSFs and the aim was to compare different deconvolution algorithms for convergence and accuracy in reproducing the original stimulated configuration of the spheres. This step of the study allows to us to better understand the different methods and to choose the optimized algorithm.

(2) Deconvolution of 3-D stack of digitalized images of biological samples with different algorithms assuming that the images are convoluted with theoretical PSFs; the choice of PSFs was restricted to those that showed good results in the first step described above.

(3) PSF can be directly extracted form the real images. Therefore in this phase we tested the performance of the deconvolution algorithms on the real images convoluted with real PSF.

Simulated phantom models that were analyzed in the first step of the study are schematically shown in Figure 3. In all the figures, henceforth, an image or intensity plot is referred to as a matrix. The PSF is characterized by an ideal point source at the focusing plane and becomes diffuse as we move away in the z direction; thus it is like a cone. In the vertical, x–z plane, the image I(x,z) of the ideal point source is approximately in the form of "X" (McNally, J. et al. 1996).

To reproduce the conditions of wide field transmission microscopy, the total intensity at each z-plane was normalized to the same constant value. Six PSFs were tested, with two variable parameters: the diameter of the radiating sphere **d**, and the step size **p** in the z direction. The latter parameter affects the opening of the emitting cone or of "X". Following values of the parameters were tested.

P1sint: (d = 2 px; p = 0.5); P2sint: (d = 2 px; p = 0.1);
P3sint: (d = 2 px; p = 0.01); P4sint: (d = 2px; p = 0.001);

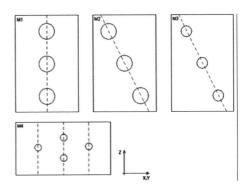

Figure 3. Schematic XZ plane view of stimulated spheres; M1 (d = 40px); M2 (d = 40px); M3 (d = 20px) and M4 (d = 12px). The sizes of the matrices are (X = 61, Y = 61, Z = 101) px for M1, M2 and M3, and (X = 101, Y = 101, Z = 51) for M4.

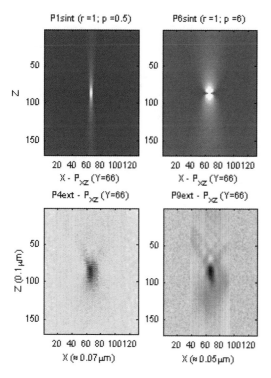

Figure 4. XZ view of 4 different PSFs. The upper figures are from the theoretical PSFs (P1sint and P6sint); and the lower figures are from the extracted PSFs (P4ext and P9ext). These PSFs were used for different deconvolution algorithms.

P5sint: (d = 10 px; p = 1.2); P6sint: (d = 2 px; p = 6);

The PSF from real images was extracted by isolating the smallest feature and following its image in different z-planes. This PSF was then used in conjunction with deconvolution algorithms to recover the 3-D view of the original object; twelve extracted PSFs were tested.

3 RESULTS AND DISCUSSION

3.1 Deconvolution of simulated data

Analyzing the results obtained by different deconvolution algorithms, applied to the simulated objects imaged through various PSFs, we observe that the opening of the "X" of the PSF directly affects the convergence of the deconvolution algorithms. A bigger opening of the PSF induces a faster convergence but also leads to a smaller visual accuracy.

In the iterative deconvolution methods, the size and the position of different spheres influences the number of iterations necessary to achieve a good convergence with the original data. The best results for the simulated data were obtained with theoretical PSFs P1sint, P2sint

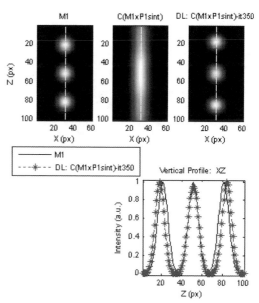

Figure 5. XZ view of the matrixes: M1, Convolution of M1 with P2sint {C(M1 × P1sint)}, and respective Deconvolution with RL algorithm for 350 iterations (upper panels). Lower panel shows variation of the intensity along the vertical line that passes through the centre of the three spheres. The full black line refers to the original matrix M1 and the dashed blue line to the deconvoluted matrix.

and P6sint, for range of iterations between 10 and 1500. One example this result is presented on Figure 5.

3.2 Results on real images of erythrocytes

While deconvoluting the images of the erythrocytes it was observed that z movement of objective created some vibrations in the x–y plane and CM of small features was shifted. All the results presented hereafter have been corrected for this registration shift in x–y plane. In different stacks of digitalized images of erythrocytes, by applying the synthetic PSFs, only non iterative methods (e.g. Wiener, WF) show an increase of discrimination in Z direction. Moreover, it was observed that acceptable results were obtained only with larger opening of the "X" form of the PSF.

The results showed a substantial increase in image contrast accompanied by mere than 40% increase in the overall intensity; (Figures 6–7).

The results obtained by deconvoluting with the different PSFs extracted from the smallest features of the images showed some artifacts due to the cut-off or smoothing parameters of the PSF that were used to give a continuous function. Despite this, two of the PSFs (P4ext and P9ext) showed improvement in the contrast in z-direction; (Figures 8–10). This result was obtained with the application of the iterative algorithm RL.

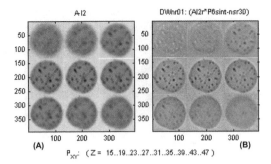

(A)

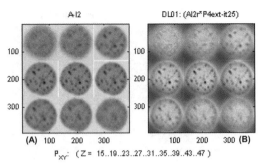

(B)

P_{xy}: (Z = 15..19..23..27..31..35..39..43..47)

Figure 6. XY images of images of an erythrocyte. All these figures have a Z step of 500 nm between each image. (A) raw image; (B) deconvolution of the raw image with the WF algorithm and a theoretical PSF, P6sint, for an associated noise to signal ratio (nsr) of 30.

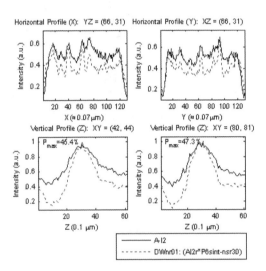

Figure 7. The upper figures show the intensity variation along the two perpendicular axis lines that pass trough the best z-focused x–y image of the cell. The lower figures show the intensity variation along the z-lines that pass trough two different dots of basophilic stippling on the cell. The black line refers to the raw image and the blue to the deconvoluted image (WF algorithm).

In away to reduce the burring of the extracted PSF, there were applied to it the deconvolution algorithm WF, that showed better results when using the theoretical PSF, P6sint. This deconvoluted PSF were then applied to the matrix of biological data, with the iterative algorithm RL. This method results in a more realistic final image, showing signatures of the characteristic "donut" form of the red blood cell; and gives approximately the same increase of intensity as obtained with the raw extracted PSF. One example of this result is presented in Figures 11–12.

(A)

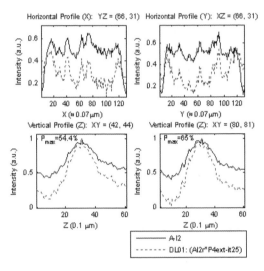

(B)

P_{xy}: (Z = 15..19..23..27..31..35..39..43..47)

Figure 8. (A) raw image; (B) deconvolution of the raw image with the RL algorithm and a extracted PSF, P4ext, for 25 iterations.

Figure 9. Results of intensity variations after deconvolution with RL algorithm and an extracted PSF (P4ext) for 25 iterations. The black line refers to the raw image and the blue to the deconvo-luted image.

4 CONCLUSIONS

From the analysis presented above we can draw the following conclusions:

1. Deconvolution of a stack of images with a PSF not only gives a quantitative measure of the relative z-positions of the objects but also improves the contrast of the objects in the x–y plane. A better resolution in the x–y plane and z direction is quite useful in doing a reliable image analysis and correlating the quantifying parameters with those obtained with biochemical assays. The parameters are also useful as statistical and dynamic descriptors of the pathology. A good 3-D image of the anomalies of the cell can be used to quantify with

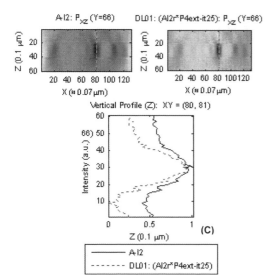

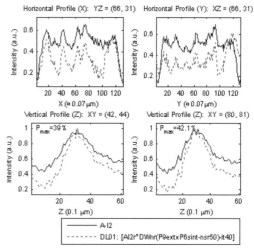

Figure 12. Results of intensity variations after deconvolution with RL algorithm and the restored extracted PSF (P9ext) for 40 it-erations. The black line refers to the raw image and the blue to the deconvoluted image.

Figure 10. The figures (A) and (B) are XZ views of the image of basophilic stippling; (A) raw image and (B) the deconvoluted image with RL and PSF P4ext; (25 iterations). The figure (C) shows the intensity variation along z-line that passes through the centre of the basophilic stippling dot. The black line corresponds to the raw image and the blue line to the de-convoluted image.

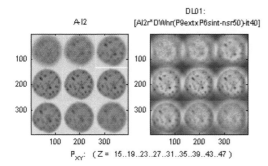

Figure 11. (A) raw image; (B) deconvolution of the raw image with the RL algorithm and the restored extracted PSF, P9ext; 40 itera-tions.

confidence geometrical parameters like surface area, volume, inter-particle distance.

2. Different PSFs and deconvolution algorithms were tried. For synthetic PSF, the best results were obtained using the non iterative Wiener Filter algorithm. For the PSF extracted from the point like features of the images itself, the best strategy is to use iterative Richardson-Lucy algorithm.

3. In the absence of availability of good quality data on real PSF of the microscope, theoretically reasonable PSFs can be optimized to obtain a good and useful resolution even when the objects are of the size of the probe wavelength, i.e. in the diffraction limit.

4. This work can be easily applied to the study of other blood disorders and, in principle, to other research problems in cytology.

REFERENCES

Conchello, J. A. 1994. Fluorescence photobleaching correction for expectation-maximization algorithm. Institute for Biomedical Computing. Washington University.

Hoffbrand, A. V. & Pettit, J. E. (3rd ed) 1993. Essencial Haematology.

Holmes, T. J. 2002. Background of Deconvolution. Auto-Quant Imaging.

Jianga, M. & Wang, G. 2003. Development of blind image deconvolution and its applications. Journal of X-Ray Science and Technology 11 (13–19).

Kempen, G. M. P. 1999. Image Restoration in Fluorescence Microscopy. Delft University. The Netherlands.

Lai, X., Lin, Z., Ward, E. S. & Ober, R. J 2005. Noise suppression of point spread functions and its influence on deconvolution of three-dimensional fluorescence microscopy image sets. Journal of Microscopy, Vol. 217. pp. 93–108.

McNally, J., Karpova, T., Cooper, J. & Conchello, J.A. 1999. Three-Dimensional Imaging by Deconvolution Microscopy. ID meth.1999.0873.

Monvel, J. B., Calvez, S. & Ulfendahl, M. 2001. Image Restoration for Confocal Microscopy: Improving the Limits of Deconvolution, with Application to the Visualization of the Mammalian Hearing Organ.

Sarder, P. & Nehorai, A. 2006. Deconvolution methods for 3-D fluorescence microscopy images. IEEE, Signal Processing Magazine.

Schlecht, J., Barnard, K. & Pryor, B. 2006. Statistical Inference of Biological Structure and Point Spread Functions in 3D Microscopy.Third International Symposium on 3D Data Processing, Visualization and Transmission. IEEE.

Vicidomini, G. 2005. 3D Image Restoration in Fluorescence Microscopy. University of Genoa: Italy.

Vitri, J. & Llacer, J. 1996. Reconstructing 3D light microscopic images using the EM algorithm. Elsevier Science. 0167-8655.

Vonesch, C., Aguet, F. Vonesch, J. L. & Unser, M. 2006. The Colored Revolution of Bioimaging. IEEE, Signal Processing Magazine.

Computational Vision and Medical Image Processing – João Tavares & Natal Jorge (eds)
© 2008 Taylor & Francis Group, London, ISBN 978-0-415-45777-4

Clustering of BOLD signals recorded during rest reveals more inter-subject constancy than EEG-fMRI correlation maps

S.I. Gonçalves, P.J.W. Pouwels, J.P.A. Kuijer, R.M. Heethaar & J.C. de Munck

VU University Medical Center, Amsterdam, The Netherlands

ABSTRACT: In this paper we apply an hierarchical clustering algorithm to resting state BOLD signals recorded during simultaneous measurement of Electroencephalogram (EEG) and functional magnetic resonance imaging (fMRI). Results of 15 subjects showed that in all cases clusters containing primarily occipital, parietal and frontal lobes were obtained. Furthermore, we found that in all cases, visual and somatosensory cortices were separated into different clusters. Contrary to the inter-subject constancy of the BOLD signal clusters, the statistical parametric maps (SPM's) resulting from correlating the alpha power time series with BOLD showed a much larger variability, both in terms of spatial distribution and statistical significance. Also the individual EEG spectrograms varied considerably from subject to subject. These results suggest that the BOLD signals have a larger inter-subject constancy and that the variability in the EEG-fMRI SPM's appears to be due to the variability associated with the EEG alone.

Keywords: Co-registered EEG-fMRI, clustering, resting state, correlation

1 INTRODUCTION

The simultaneous recording of Electroencephalogram (EEG) and functional magnetic resonance imaging (fMRI) is a recent technique that provides new insights into the mechanisms of spontaneous brain activity. In fact, since spontaneous brain activity is non-reproducible, it cannot be studied consistently with EEG and fMRI separately. Only with the simultaneous recording of the two techniques is it possible to establish the BOLD correlates of spontaneous EEG. In this context, the EEG is used to detect the events of interest and the fMRI is used to make contrasts between periods with and without events. The first applications of simultaneous EEG-fMRI were in the field of epilepsy (Krakow et al. 2001, Lemieux et al. 2001, Salek-Haddadi et al. 2002), where the BOLD correlates of ictal and interictal activity were studied. This was followed by the application of simultaneous EEG-fMRI to study the spontaneous variations of the alpha rhythm. The first studies (Goldman et al. 2002, Laufs et al. 2003a, Laufs et al. 2003b, Moosmann et al. 2003) had a focus on group results whereas the study of Gonçalves et al. (2006) had the focus on individual results. The group results showed that the spontaneous power variations of the alpha rhythm are negatively correlated to the BOLD signal in occipital, parietal and frontal areas whereas they are positively correlated to BOLD signal in the thalamus. However, the individual results presented in Gonçalves et al. (2006)

show a striking inter-subject variability in the correlations between spontaneous power variations of the alpha rhythm and the BOLD signal as well as in the individual spectrograms derived from the EEG data.

In this paper, we apply simultaneous EEG-fMRI to acquire resting state in order to investigate the sources of inter-subject variability in the correlation between alpha power spontaneous variations and BOLD. In particular, we investigate whether the variability is also associated to the BOLD signal. For that, we use the clustering algorithm described in Van't Ent et al. (2003), based on Ward's hierarchical clustering method (Ward 1963) to group voxels such that within each cluster, the BOLD signal time courses are as similar as possible. The clustering results are compared to the SPM's resulting from correlating the spontaneous variations of the alpha rhythm with BOLD.

2 MATERIALS AND METHODS

2.1 Subjects

Simultaneous EEG-fMRI data were acquired from 15 healthy subjects (7 males, mean age 27, ±9 years) while they lied in the scanner with eyes closed, resting without falling asleep, in a room that was kept in the dark. The goal was to record resting state data with the focus on the spontaneous variations of the alpha rhythm. For all subjects, EEG-fMRI data were acquired in a continuous way.

2.2 Acquisition of EEG data

The EEG was acquired using an MR compatible EEG amplifier (SD MRI 64, Micromed, Treviso, Italy) and a cap providing 64 Ag/AgCl electrodes positioned according to the extended 10–20 system. The reference electrode was positioned between Pz and POz. For subject safety reasons, the wires were carefully arranged such that loops and physical contact with the subject were avoided. Later while processing the data, all channels were re-referenced to average reference.

EEG data were acquired at a rate of 2048 Hz using the Clinic-Acquisition software package (Micromed, Treviso, Italy). An anti-aliasing hardware low-pass filter at 537.6 Hz is applied. The EEG amplifier has a resolution of 22 bits, an input impedance larger than 10 MΩ and a CMRR = 105 dB at 50 Hz directly on the inputs. Each channel has differential inputs and uses one sigma-delta AD converter. For RF protection, a low-pass filter at 600 Hz (20 dB/decade) is applied and a 10 kΩ current limiting resistor is attached to each electrode lead.

2.3 Acquisition of fMRI data

Functional images were acquired on a 1.5 T MR scanner (Magnetom Sonata, Siemens, Erlangen, Germany) using a T2* weighted EPI sequence (TR=3000 ms, TE = 60 ms, 64×64 matrix, FOV = 211×211 mm, slice thickness = 3 mm (10% gap), voxel size = $3.3 \times 3.3 \times 3$ mm^3) with 24 transversal slices covering the complete occipital lobe and most of the parietal and frontal lobes. In the protocol, 600 volumes (i.e. 30 minutes of data) were acquired for each subject except for subject 2 for whom 400 volumes were acquired (i.e. 20 minutes of data). The pulse and respiration signal as well as the electrocardiogram (ECG) were recorded during scanning. The ECG was recorded using the electrodes provided with the scanner which consisted of Ag/AgCl electrodes with carbon leads.

2.4 Analysis of EEG data

The EEG artefact was corrected by applying an in-house developed algorithm which is described in detail in Gonçalves et al. (2007).

In order to determine the spectral characteristics of interest of the corrected EEG data the fluctuations in power of the alpha band were computed. An advanced FFT algorithm (Frigo and Johnson, 2005) was applied in order to compute a spectrogram sampled at the same frequency as the fMRI data, i.e. 0.33 Hz. Thus, the FFT was computed using a rectangular window of 3 s without overlap. Subsequently, the power time series were averaged over all alpha band frequencies (8–12 Hz), thus obtaining a single power time series per channel.

In this study, a classic bipolar montage was used in order to emphasize local variations. For the analysis,

the following derivations were considered: O1, O2, Oz, _PO3, _PO4, _PO7 and POz _PO8. The power time series, averaged over these channels, was considered for further calculations.

2.5 Analysis of fMRI data

The MR data were motion corrected and spatially smoothed using a 5 mm radius Gaussian kernel. Next, the BOLD signal in each voxel was correlated to the average power time series using the General Linear Model (GLM) (Cox et al., 1995; Kherif et al., 2002; Worsley et al., 2002). In the context of the GLM, the average power time series was taken as the regressor and both an offset and a linear trend were considered as covariates. Furthermore, the hemodynamic response function (HRF) was determined from the model according to De Munck et al. (2007).

The correction for multiple comparisons was made by means of controlling the false-discovery rate (FDR) according to Benjamini and Hochberg (1995).

The clustering algorithm was applied to smoothed, normalized data which was downsampled by a factor of 2 in order to reduce the computation time.

We used software developed in-house and based on C++ and the GUI toolkit (QT version 2.3.3).

3 RESULTS

Typical results obtained for the BOLD signal clustering are shown in fig. 1 (left-hand side) for 3 subjects together with the corresponding SPM's showing areas where the correlation between the BOLD signal and the spontaneous variations of the alpha rhythm is significant. For all subjects, clusters containing primarily the frontal, parietal and occipital cortices respectively were found. Furthermore in some subjects, clusters containing the thalamus were also found, although this finding was less consistent. However, the SPM's (right-hand side of fig. 1) containing regions where the correlation between the BOLD signal and the spontaneous variations of the alpha rhythm is significant show a much larger inter subject variability, as can be seen by comparing the results of subject 6 (first row in fig. 1) with those of subject 13 (third row in fig. 1). The corresponding spectrograms derived from the EEG data are also quite different, as can be seen in fig. 2. In fact, the spectrogram corresponding to subject 6 shows a highly modulated alpha rhythm (quite prominent at 10 Hz) whereas the spectrogram corresponding to subject 13 shows an almost constant alpha rhythm.

An overview of the results obtained for all subjects is presented in Table 1. For each subject the following information is given:

– Nr. Clusters – Number of clusters that must be considered in order to separate the visual and somatosensory areas;

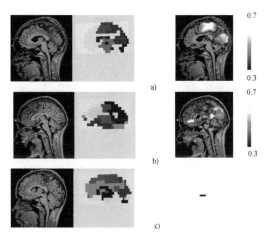

a)

b)

c)

0.7

0.3

0.7

0.3

Figure 1. Clustering results (left hand-side column) corresponding to subjects 6, 12 and 13. The right hand-side column shows the SPM's containing the areas showing significant correlations between alpha power and BOLD at an FDR of 5.6×10^{-6}. The color coded bar shows the correlation range. a) Subject 6. b) Subject 12. c) Subject 13. For this subject no significant correlations were found at the chosen FDR level.

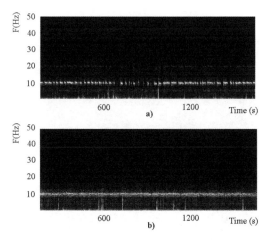

Figure 2. Spectrograms derived from the EEG data recorded during the MR scanning. a) Subject 6. b) Subject 13.

– *Frontal* - Existence of a cluster containing mainly the frontal lobe (given by a + sign);
– *Occipital* - Existence of a cluster containing mainly the occipital lobe (given by a + sign);
– *Parietal* - Existence of a cluster containing mainly the parietal lobe (given by a + sign);
– *Thalamus* - Existence (given by a + sign) or non-existence (given by a - sign) containing mainly the thalamus;

Table 1. Overview of the results obtained for all subjects (see text for explanation).

Subject	Nr. Clusters	Frontal	Occipital	Parietal	Thalamus	EEG/ fMRI SPM's
1	7	+	+	+	-	-
2	7	+	+	+	+	++
3	7	+	+	+	+	+
4	6	+	-	+	-	-
5	7	+	+	+	+	+
6	4	+	+	+	+	++
7	7	+	+	+	+	++
8	7	+(p)	+	+(f)	+	++
9	12	+	+	+	+	++
10	7	+(p)	+	+(f)	+	-
11	12	+	+	+	-	++
12	7	+	+	+	+	++
13	7	+(o)	+(f)	+	-	-
14	7	+	+	+	-	+
15	7	+	+	+	-	++

– *EEG/fMRI SPM's* - Characteristics of the SPM's showing the areas where the BOLD signal is correlated to the alpha power. The "++" sign means that the SPM is significant and quite wide spread; the "+" means that the SPM is significant but more limited to occipital areas; the "−" sign means that the SPM is not significant.

As can be observed from the table, the variability associated to the SPM's is much larger than the variability in the clusters that are obtained from BOLD resting state data.

4 DISCUSSION AND CONCLUSIONS

The results presented in this paper show that the resting state BOLD signal varies much less from subject to subject than the SPM's showing the areas where the correlation between the spontaneous variations of the alpha power and the BOLD signal are significant. Furthermore, the comparison of the spectrograms corresponding to different subjects also shows large variability. This suggests that the inter-subjects variability of the EEG-fMRI SPM's is due to the EEG alone. This maybe because the EEG indeed changes more from subject to subject or simply because the model that is used to model the BOLD signal from the EEG is not accurate enough.

REFERENCES

De Munck, J. C., Gonçalves, S. I., Huijboom, L., Pouwels, P. J. W., Kuijer, J. P. A., Heethaar, R.M. & Lopes da Silva, F. H., 2007. The hemodynamic response of the alpha rhythm: an EEG/fMRI study, NeuroImage, accepted.

Goldman, R. I., Stern, J. M., Engel Jr., J. & Cohen, M., 2002. Simultaneous EEG and fMRI of the alpha rhythm. NeuroReport, **13(18)**, 2487–2492.

Gonçalves, S. I., de Munck, J. C., Pouwels, P. J. W., Schoonhoven, R., Kuijer, J. P. A., Maurits, N. M., Hoogduin, J. M., Van Someren, E. J. W., Heethaar, R. M. & Lopes da Silva, F. H., 2006a. Correlating the alpha rhythm to BOLD using simultaneous EEG/fMRI: inter-subject variability. NeuroImage, 30(1), 203–213.

Gonçalves, S.I., Pouwels, P. J. W., Kuijer, J. P. A., Heethaar, R. M. & De Munck, J. C., 2007, Artifact removal in co-registered EEG/fMRI by selective average subtraction. Clin. Neuroph., under review.

Krakow, K., Allen, P. J., Symms, M. R., Lemieux, L., Josephs, O. & Fish, D. R., 2000. EEG recording during fMRI experiments: image quality. HBM, **10**, 10–15.

Laufs, H., Krakow, K., Sterzer, P., Eger, E., Beyerle, A., Salek-Haddadi, A. & Kleinschmidt, A., 2003a. Electroencephalographic signatures of attentional and cognitive default modes in spontaneous brain fluctuations at rest. PNAS, **100(19)**, 11053–11058.

Laufs, H., Kleinschmidt, A., Beyerle, A., Eger, E., Salek-Haddadi, A., Preibisch, C. & Krakow, K., 2003b. EEG-correlated fMRI of human alpha activity. NeuroImage, **19**, 1463–1476.

Lemieux, L., Krakow, K. & Fish, D.R., 2001, Comparison of spike-triggered functional MRI BOLD activation and EEG dipole model localization, NeuroImage, **14(5)**, 1097–1104.

Moosmann, M., Ritter, P., Krastel, I., Brink, A., Thees, S., Blankenburg, F., Taskin, B., Obrig, H. & Villringer, A., 2003. Correlates of alpha rhythm in functional magnetic resonance imaging and near infrared spectroscopy. NeuroImage, **20**, 145–158.

Salek-Haddadi, A., Merschhemke, M., Lemieux, L. & Fish, D.R., 2002, Simultaneous EEG-Correlated Ictal fMRI, NeuroImage, **16**(1): 32–40.

Computational Vision and Medical Image Processing – João Tavares & Natal Jorge (eds)
© 2008 Taylor & Francis Group, London, ISBN 978-0-415-45777-4

New method of segmentation using color component

Jose C. Peña, Yamelys Navarro & Cesar O. Torres

Optics and Computer Science Laboratory, University Popular of the Cesar, Valledupar, Cesar, Colombia

ABSTRACT: Digital Image Processing allows to manipulate color images as conveniently and easily as monochrome images; color images may be represented in three different color formats, allowing to select the most appropriate format for any color processing application. In this work we based on the color system named C1, C2 and C3 the modified components, developed by Baez Rojas, et al. These components were digitally obtained, with this technique, it is possible to calculate the modified components skeleton, during the segmentation process for arbitrary images; we named CR1 CR2 and CR3 these components. Precision of segmentation was improved by processing individual components with each skeleton component separately and correcting within the software the noise aroused during the thresholding segmentation process and for the obtain the information retrieval and the improvement of the images a brightness and contrast filters are used; the filters were applied to the region of the image that at first did not show information, for this procedure we used one mask. The aim of the System described in this paper is to segment and to recover information in images. The proposed algorithm was applied to the coal macerals images. The technique is able to separate the maceral from the background resin without loss of information. The method is possibly more effective than other used in previous articles.

1 INTRODUCTION

Partitioning of an image into several constituent components is called segmentation. Segmentation is an important part of practically any automated image recognition system, because it is at this moment that one extracts the interesting objects, for further processing such as description or recognition.

In the analysis of the objects in images it is essential that we can distinguish between the objects of interest and "the rest." This latter group is also referred to as the background. The techniques that are used to find the objects of interest are usually referred to as segmentation techniques.

The state of the art for the segmentation to color this based a great amount of works that present techniques, models and algorithms for the segmentation of these images of color. These techniques are divided in four great groups, the first group conform the techniques of segmentation based on the values of pixel. The second group contains techniques of segmentation based on the area. The techniques based on borders belong to the third group and finally the techniques based on the physics are the fourth group.

The segmentation based on the value of pixel includes the techniques based on the histogram. In other words, the histogram of the image is obtained, some maximums are identified and intervals are analyzed that surround to this maximum one during the segmentation process. The segmentation by means of the group of pixels in some space of color according to one or more characteristic. Finally, another great set of algorithms that use the value of pixel as reference for the segmentation composes those algorithms that use diffuse group in the color spaces.

The proposed method is based on algorithm of segmentation for color images, based on a system of color proposed by Baez et al. The final application of the algorithm is the automatic segmentation of coal macerales in digital images color.

2 SYSTEM OF COLOR PROPOSED BY BAEZ et al.

The method involves three channels and they imagine matrix with the following equation:

$$\begin{bmatrix} C1 \\ C2 \\ C3 \end{bmatrix} = \begin{bmatrix} 1/2 & -1/2 & 0 \\ 0 & 1/2 & -1/2 \\ -1/2 & 1/2 & 0 \end{bmatrix} * \begin{bmatrix} R \\ G \\ B \end{bmatrix} \qquad (1)$$

The components were written of the following form:

$$C1 = \left(\frac{R-G}{2} \right); C2 = \left(\frac{G-B}{2} \right); C3 = \left(\frac{B-R}{2} \right) \qquad (2)$$

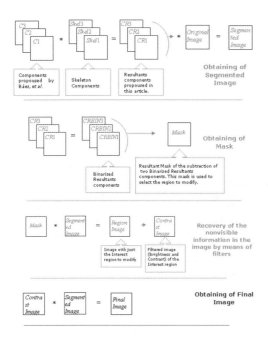

Figure 1.   Segmentation process diagram.

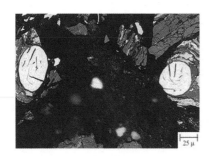

Figure 2.   Original image.

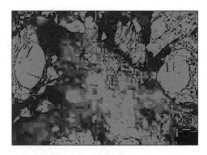

Figure 3.   Resultant component CR1.

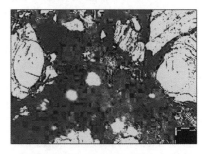

Figure 4.   Resultant component CR2.

Figure 5.   Resultant component CR3.

3   NEW SEGMENTATION PROCESS

By means of the images digital processing, we developed the theoretical and algorithms bases by means of which we extract the information from the studied images, provided by the National Institute of Coal INCAR of the CSIC (Oviedo – Spain) and taken from the Petrografic Atlas as shown in Fig. 2, All the photomicrographies were taken in reflected white light with and without the use of a polarizer, an analyzer, and a plate of the retarder. In some cases the photomicrographies were taken with fluorescent light using the ultraviolet illumination; these images are previously characterized by each Institution and from that information we were based for the study of the images.

Skeletonization is useful when we are interested in the recuperecion of informacion in images there are several algorithms which were designed for this aim. In this project we are concerned with one of them Algorithm developed in Matlab 7.4 as shown in Fig. 9, in that the is obtained initially the separation of each component, C1, C2 and C3, of the original image, proposed by Báez, these same images are skeletonizated and multiplying point to point each one of these resulting components with the skeleton component we obtain the propose components in this investigation, CR1, CR2 and CR3 as shown in Fig. 3, 4 and 5, and finally the segmented image as shown in Fig. 1.

4   RESULTS

4.1   Software description

By using this software, users can choose between making segmentation or mapping the component which

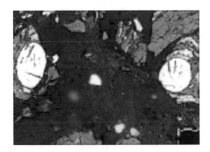

Figure 6. Segmented image.

Figure 7. Mask.

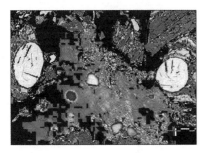

Figure 8. Final image.

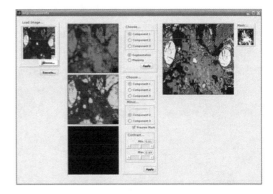

Figure 9. Segmentation software by color components.

5 CONCLUSIONS

A new method for the color images segmentation has been developed. The algorithm has been proven successfully in coal macerals images, segmenting from digital photographies.

With the new proposed method was obtained the information retrieval in certain region demarcated by one mask in color images. The identification of the nonvisible information was obtained by means of the skeletonization of each component. By our method we achieve to separate single the resin of the maceral without eliminating the interest information that by other methods was taken as resin. The recovery of this information was achieved applying brightness and contrasts filters. The final or resultant image is a segmented image to which we added the recovered and contrasted region.

REFERENCES

1. J.J. Báez Rojas, M.L. Guerrero *b*, J. Conde Acevedo*c*, A. Padilla Vivanco y G. Urcid Serrano Revista Mexicana de Física (2004).
2. Lim, Jae S., *Two-dimensional signal and image processing*, (USA, Ed. Prenctice-Hall, 1996).
3. Yu-Ichi Ohta, Takeo Kanade, and Toshiyuki Sakai, *Computer Graphics and Image Processing* (1980) 222.
4. R. Ohlander, K. Price, and D.R. Reddy, *Computer Graphics and Image Processing* (1978) 313.
5. Kenneth R. Castleman, *Digital Image Processing* (USA, Ed.Prentice Hall, 1996).
6. L.M. Lifshitz, S.M. Pizer, "A Multiresolution Hierarchical Approach to Image Segmentation Based on Intensity Extrema", *IEEE Trans. on Pattern Analysis and Machine Intelligence*, vol. 12, no. 6, pp 529–539, june 1990.
7. MOLINA, R. Introducción al Procesamiento y Análisis de Imágenes digitales, Universidad de Granada.

contain more information, for first process, takes the original image and multiplique it with choosed component as shown in Fig. 6, and for procedure of mapping, takes the segmented image and the component used for segmentation. In order to recover information, we applied one mask as shown in Fig. 7 (it allows to select the region that contains information which is not visible in the image) that results of substract two components selected by the user (the mask which shows the awaited results by user is selected, by having six possible options). To the Image produced from substraction, we apply an improvement of contrast and brightness by using some filters. By adding segmented image with filtered image, we obtain final image where the difference between maceral and resine is founded shown in Fig. 8.

Computational Vision and Medical Image Processing – João Tavares & Natal Jorge (eds)
© 2008 Taylor & Francis Group, London, ISBN 978-0-415-45777-4

Analysis of pupil fluctuations for detection of deception

D. Iacoviello

Dept. of Computer and Systems Science "A.Ruberti", "Sapienza" University of Rome

ABSTRACT: The relation between pupil fluctuation and deception is considered. The response of pupil to light or noise stimuli has been extensively studied, together with the relation between anomalous fluctuations and drug or alcohol abduction. Moreover it is known that cognitive process, fatigue and emotions have influence in pupil fluctuation and in its reaction to stimuli. Since 1940 the idea of studying the pupil movement and fluctuations as index of deception has been considered; nevertheless, since many different cognitive processes can influence pupillary dilatation, it is necessary to improve the technique and to assess the accuracy of pupil movement as such a possible parameter. The present paper analyses the pupil fluctuation, the movements of its center and the increase of the number of eye blinking in order to propose possible indicators of deception. The first results proposed appear to be encouraging; of course in all laboratory studies of deception the simulation context is different from real-life investigations; this limitation will require further development.

1 INTRODUCTION

In this paper the analysis and evaluation of correlation between pupil fluctuation and deception are considered. In general, the analysis of pupil's fluctuations is useful for non-invasive diagnosis; the response of pupil to light or noise stimulus has been extensively studied since it may be assumed as indicator of many different diseases. In normal conditions, the pupil of human eyes fluctuates to adapt the amount of light to the retina and when gazing at fixed object (pupil noise). When a light or noise stimulus is presented to a subject, the pupil response is not instantaneous because of the action of the sphincter muscle. The delay in the pupil response to a stimulus is the latency; typical parameters in the study of pupil fluctuation are described in Wilhelm 2003.

Some correlations between the human fatigue during visual display terminal task and the variation in the pupil diameter are discussed in Higuchi 2003. The pupil fluctuations seem to be sensitive indicator of mental activity; in O'Neill 2001 the short term memory has been studied by the pupillometry analysis with respect to narcoleptic patients. In O'Neill 2000 acoustic impulse test is described, trying to quantify the parasympathetic and sympathetic pupillary dilation components. Also sleep disorder can give rise to anomalous pupil fluctuations, Oroujeh 1995, Agarwal 2005 and Merritt 2004; changes in pupil reaction to light stimulus are studied in Alzheimer's and Parkinson's disease patients, Fotiou 2000, Granholm 2003: the aim is to obtain early diagnosis. The effects of consuming drugs and alcohol or the consequence of specific pharmacologic treatment may be observed analysing the variations of pupil size and pupillary light reflexes, Oswald 2004 . Pupil analysis can be useful as an early sign of development of systemic autonomic neuropathy in patients with type 1 and 2 diabetes, Pittasch 2002, Cahill 2001; for these diseases also relations between pupillary and cardiovascular autonomic function have been assessed. Changes in the pupillary responses have been studied with respect to: melancholic (Fountoulakis 1999), schizophrenic (Granholm 2004) and stressed (Jomier 2004) subjects; in all these situations the idea is to study the response of the autonomic nervous system by quantifying the variation of the pupil size after stimuli of different nature. The variation of pupil diameter is studied also with respect to dark adaptation, Brown 2004, with particular attention to Horner syndrome. An interesting review on the state of art of clinical application of pupillography can be found in Wilhelm 2003; in particular the characteristics of pupillographic devices are extensively described.

As can be noted, great effort has been devoted to the study of the response of the pupil to various stimuli, aiming at non invasive diagnosis of different diseases. The aim of this paper is to show and evaluate possible correlation between the fluctuation of the pupil, its movement in the sclera, the increase of the number of eye blinking and deception. It is well known that emotions, cognitive processing and mental activity are correlated to anomalous fluctuations of the pupil, Dioniso 2001; in particular in Vendemia

2006 the pupillometry is considered as an indicator of deception along with other techniques, such as voice analysis, brain wave, thermal imaging, and its reliable is considered also in view of court admissibility, Stern 2006. In this paper the possibility of considering pupil fluctuations and its movement as index of deception is considered. This argument is of great interest in psychophysiological field, but, it is necessary to improve the technique and to assess the accuracy of this possible indicator, as noted in Vendemia 2006. Two different kinds of problems have to be considered. The first aspect in considering the pupil anomalous fluctuation as an index of deception is the procedure in which the questions should be presented to the subject; in particular, in Dioniso 2001 the subjects were required to delay their responses to questions by 8–10 s. Moreover, as noted in Dionisio 2001, the laboratory context is intrinsically different from real life investigations. In Wang the interesting aspect considered is about the "amount of deception" and its predictability using pupil dilation; important variables to be measured are the pupil size between the beginning and the end of the processing. This latter consideration introduces the second problem to be taken into account; it is related to the quality of the images. Despite the images captured by the pupillometer may appear of good qualities, especially with clear eyes, it is not easy to process this signal, not only because of the degradation introduced by instrumental devices, but also for the natural movement of the subjects, for eye blinking, eyelids, for the texture of the iris, or for a particularly teary eye. Attention should be devoted to illumination sources. In this paper every image of the eye will be binarized to enhance the pupil and evaluate its measure, and the coordinates of its center. The binarized image may contain other black objects besides the pupil; it may be easily identified taking into account that it is always the darkest elliptic object, not ragged and without holes. The fluctuations of the diameter will be considered, but no significant difference could be observed with respect to the properties of fluctuations of the area. Another significant aspect that will be considered is the movement of the pupil in the eyes; in particular the coordinate of its center of mass will be estimated and it will be noted that in case of deception there is an increase in the horizontal movement of the pupil.

Also the number of eye blinking is considered; nevertheless, especially for long acquisition, the subject may get tired and the number of eye blinking increases; to avoid these false detections an acquisition period of 20–25 seconds has been considered.

In Section 2 the experimental set up and the image analysis are described; in Section 3 the parameter of interest for the problem are described along with the experiments performed. In Section 4 some conclusions and discussion on future work are presented.

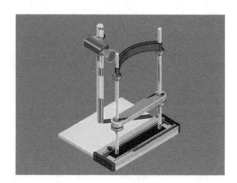

Figure 1. Experimental set up.

2 EXPERIMENTAL SET UP AND IMAGE ANALYSIS

In the analysis of pupil fluctuation and movement in the eyes for detection of deception it is particularly important the experimental set up and the procedure of asking questions to the subject.

2.1 Experimental set up

As far as the experimental set up is concerned the pupillometer shown in Fig.1 was considered.

There is a chin and forehead rest, and a micro infrared ccd camera is positioned at 20 cm from the eyes. The subject was asked to look into the camera lens. The camera was connected to a PC by a commercial video capture board with a frame rate of 30 frames/sec at a resolution of 352×288, with a 8-bit intensity level.

2.2 Image analysis

It is not easy to analyse pupil images, since many different kind of degradations are present, such as eye blinking, eyelids, the movements of the subject, or the reflection of light sources. It may be useful to preprocess the image by filtering operation; the contour of the pupil becomes less sharp but also the acquisition noise and the iris texture will be smoothed.

In order to analyse the sequence of pupil images it is convenient to segment each image. To segment the image a thresholding operation was considered, taking into account that the pupil is always the darkest element of the eye. The threshold has been chosen by histogram arguments, considering the value that best separates the dark elements from the others.

Nevertheless simple thresholding operation may not be sufficient to identify the pupil as the unique black element over a white background. Once the image is segmented morphological properties may be calculated for each black object isolated, for example

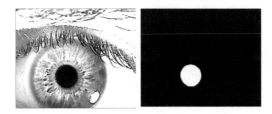

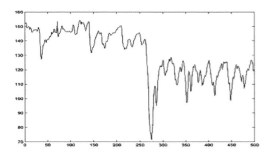

Figure 2. a) Frame from a sequence of pupil captured by the pupillometer; b) binarization of the image of Figure 2a)

Figure 3. Pupil fluctuation in presence of light stimulus.

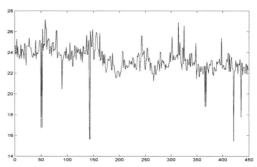

Figure 4. Fluctuation of the Major Diameter in absence of stimuli; on the abscissa the number of the frame of the sequence. The length of the Major Diameter is in pixel unit. The standard deviation of this signal is 1.02.

the area, the eccentricity, the lengths of the Major and Minor axis, the coordinates of its center of mass. The pupil can be univocally identified considering that it is always an elliptic object with a low value of eccentricity. In Figure 2a a single frame of a sequence of pupil is reported, whereas in Figure 2b the pupil is enhanced over the background; some of its morphological properties have been calculated, for example, its area is equal to 3367 pixels, its Major and Minor Axis are respectively 67 and 63 pixels, the coordinates of its center of mass are 142 and 178.

The major axis and the coordinates of the center of mass will be considered for detection of deception, whereas other morphological properties, such as the area and the minor axis, are considered to enhance, if necessary, the pupil over other black objects.

3 EXPERIMENTS

The analysis of pupil fluctuation has been extensively performed in case of light or noise stimulus to calculate the latency time. In this case the signal, even if affected by pupil noise, has a known profile and different kind of modelization are possible, for example by arcotangent function. Moreover the instant in which the stimulus is presented to the subject is well known; see Fig.3 for an example of pupil fluctuation in presence of light stimulus. In the case considered in this paper the fluctuation of pupil major diameter presents anomalous profile in unknown instants.

The effects of deception that will be considered are the following: the variations of the fluctuation of pupil Major diameter, the movements of the center of mass of the pupil and the eventual increase of the eyes blinking.

3.1 Stimuli

It is important to stress that the major difficulty in the analysis of the effects of deception on pupil movements, or also on physiological parameters variations, is first of all in preparing ad hoc experiments to simulate the same conditions of a person who is lying. To this aim we considered ten informed and consenting individuals with ages between 35 and 65 years old. No participants smoked cigarettes or drank caffeinates beverages within 2hr prior to testing. In order to avoid non significant signals due to the effort for the subject to look in the camera, the acquisition time was set equal to 25 seconds. Three different experiments were carried out for each subject:

1) a sequence in absence of any stimulus has been captured; this is important to have a reference signal;
2) a sequence in presence of a question has been captured: after few seconds from the beginning of the acquisition period a question is asked to the subject; the subject is invited to tell the truth;
3) the experiment b) is repeated with different question inviting the subject to tell a lie.

An important aspect is that the subject does not know the questions in advance; this implies that also in the case in which the subject is telling the truth there is a non negligible mental effort.

For each patients three avi files were available. From these files were analyzed: the fluctuation of the Major diameters, the movements of the center of mass of the pupil and the number of eyes blinking. In Fig. 4 is

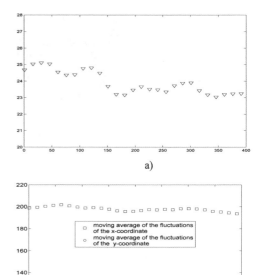

a)

b)

Figure 5. Behaviour of the coordinates of the center of pupil; on the abscissa the number of the frame of the sequence. The standard deviation of the x-coordinate is 6, the standard deviation of the y coordinate is 1.2.

reported an example of the fluctuation of the Major Diameter, whereas the movement of the coordinate of the pupil center is plotted in Fig. 5, both in case of no stimulus. Also the standard deviation of these signals are reported. Note that in case 1) of absence of stimulus the profile of the major diameter fluctuation is quite regular, often showing a sinusoidal behaviour with pupil noise superimposed; the mean value of the signal is quite constant. Nevertheless pupil noise could hide eventual bias due to mental effort or deception. The same considerations hold also referring to the signals describing the variations of the coordinates of the center of pupil.

To enhance the change in the mean value of a signal s it is useful to analyze the variation of its mean value on intervals of length T with moving windows of length h, $1 \leq h \leq T$. In all the experiments it has been set $T = 50$ frames, corresponding to about 2 seconds of signals and $h = 15$. The function considered as an indicator for deception is the following:

$$f(i) = \frac{\sum_{j=0}^{T-1} s(i+j)}{T}, \quad i = 1, ..., length(s), \, mod(h)$$

This function is applied to the three signals of interest, the Major Diameter and the two coordinates of the center of pupil, showing a more evident behaviour of the signal, as can be noted in Figs. 2a-2b in the case 1) of absence of stimulus. It is evident that in the case 1) the moving averages of the fluctuations of the major diameter and of the coordinate of the center of pupil are almost constant.

The same function f has been applied to the fluctuation of the Major Diameter and of the x and y-coordinate of the center of pupil in the case 2) of a question to which the subject must answer the truth and in the case 3) in which the answer must be a lie. The question has been asked to the subject at instant $t = 2$, corresponding to the 60th frame. In Figs. 7–9 the

Figure 6. Moving average of the fluctuations of the Major Diameter (a) and of the fluctuations of the coordinate of the center of pupil (b) in the case of experiment 1).

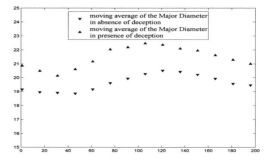

Figure 7. Moving averages of the Major Diameter in absence and presence of deception; the standard deviations of the Major Diameter in absence and presence of deception are 0.8 and 1.37 respectively.

moving averages relative to the fluctuation of the Major Diameter, and the coordinates of the center of pupil of the same subject of Figs.1–2 are plotted together in the two situations 2) and 3), in order to show the different behaviour of the function f.

It can be observed that when a deception is present there is a change in the function f, both when the signal considered is the Major Diameter and when the signal is the horizontal coordinate of the pupil center. The increase of the movement of the pupil along the vertical axis is less evident. For each signal the standard

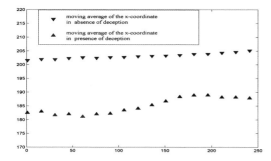

Figure 8. Moving averages of the x-coordinate of the center of pupil in absence and presence of deception; the standard deviations of the x-coordinate in absence and presence of deception are 1.4 and 3.32 respectively.

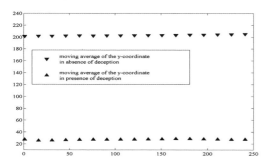

Figure 9. Moving averages of the y-coordinate of the center of pupil in absence and presence of deception; the standard deviations of the x-coordinate in absence and presence of deception are 1.4 and 3.38 respectively.

deviation has been evaluated, as indicated in the figures caption; in case of deception the increase of the standard deviation of the signal is evident.

It is useful to note that the fluctuations of the first experiment (no questions) and the fluctuations of the second experiment (question with true answer) are different, since the individual in the case 2), even in absence of deception, is subject to a mental effort. Of course, as shown in Figs. 7–9, the behaviour of the considered signals in the case 2) are more regular than the corresponding ones of the case 3).

The same experiments have been repeated for all the subjects, obtaining the same kind of results.

It is possible by alarm procedures to automatically detect the instants in which there is a significant probability that a subject is lying. For example a conservative simple procedure may consider as critical situation of possible deception the case in which one of the signals f considered above leaves a confidence region related to its standard deviation.

On the same sequences analysed the number of eye blinking has been evaluated. As already noted in the Introduction, the number of eye blinking may increase in case of too long acquisition, even in absence of

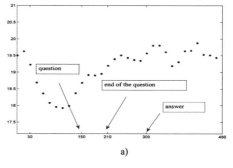

a)

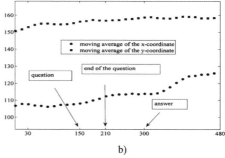

b)

Figure 10. Data analysis in case of a subject who can answer as he prefers, with a truth or with a lie; a) Moving averages of the major diameter fluctuation; b) Moving averages of the x and y coordinates of the pupil center. It is evident the changes in the mean values of the considered signals. Moreover there is an increase of 40% of eye blinking. In this case there is a correct detection of a deception occurring at instant 10 sec. corresponding at frame 300.

deception. Comparing the number of eye blinking in absence or presence of deception it can be observed an increase of almost 45%.

Note that when a subject is lying there are three possible evidence based on pupil movements analysis: the fluctuation of the major diameter, the fluctuation of the coordinate of the center of pupil (the y-coordinate appeared to be less informative than the fluctuation of the x- coordinate) and the number of eye blinking.

As a final test the subjects involved in the experiments 1)-2)-3) were invited to answer to a question as they prefer, with a truth or with a lie. In this preliminary study all the above three informative elements were considered for detection of deception. When all these elements were significantly different from the corresponding reference situation of no stimulus a deception is supposed, see Figs.10a)–10b); it is of course a conservative rule. The percentages of false positive (i.e. detect a lie when there is no lie) and the percentage of false negative (i.e. do not detect a lie when there is a lie) were less than 10%. Nevertheless it is worth noting that experiments in simulated context without deep anxiety are of course different from real situations.

245

3.2 Discussion

From the experiments considered some observations are in order:

- the light conditions are important to avoid albedos and shadows and improve the pupil segmentation;
- each subject has his own response time and his own mental effort;
- the indicators proposed for detection of deceptions appear to be reliable; the most important seems the one related to the movements of the eyes along the x-coordinate;
- after the subject says a lie the frequency of eye blinking increases;
- it is not easy to simulate the tension of a situation in which an individual really wants to lie and, most of all, does not want to be recognized as a liar.

As a concluding remark this preliminary study on correlation between pupil fluctuation/movements and deception shows that the three parameters considered, the major diameter, the center of pupil and the number of eye blinking seem to be useful for detection of deception; to avoid false alarm it is useful to consider them together.

4   CONCLUSIONS AND FUTURE WORK

The use of pupil movement for non invasive analysis has been extensively considered. In this paper the correlation between pupil movements/fluctuations and deception is considered. This intriguing problem presents many difficulties, due mainly in defining a function able to univocally identify deception. Some experiments are presented, showing that a moving average of the signal representing the Major Diameter, and the coordinates of the center of pupil appears sensitive to deception. Also the frequency of eye blinking increases after the subject says a lie. Future effort will be devoted in comparing the reliability of the pupil for detection of deception with other techniques, such as voice analysis, brain wave, thermal imaging. A greater number of participants is advisable to a more detailed analysis; moreover a limitation, that will require further development and that is inherent in all laboratory studies of deception, is that the simulated interrogative context is different from real investigation.

REFERENCES

H. Wilhelm, B. Wilhelm, Clinical applications of pupillography, *J.Neuro Ophthalmol*, vol.23, n.1, 2003

S. Higuchi, Y. Motohashi, Y. Liu, M. Ahara, Y. Kaneko, Effects of VDT tasks with a bright display at night on melatonin, core temperature, heart rate, and sleepiness, *Journal of Applied Physiology*, **94** (5), 1773–1776, (2003).

W. O'Neill, K. Trick, The narcoleptic cognitive pupillary response, *IEEE Trans. on Biomedical Engineering*, **48** (9) 963–968, (2001).

W. O'Neill, S. Zimmerman, Neurological interpretation and the information in the cognitive pupillary response, *Methods Inf Med*, **39** (2), 122–124, (2000).

A.M. Oroujeh, W. O'Neill, A.P. Keegan, S.L. Merritt, Using recursive parameter estimation for sleep disorder discrimination, *International Conference on Acoustic, Speech and Signal Processing*, 3, 1928–1931, (1995).

R. Agarwal, Detection of rapid-eye movements in sleep studies, *IEEE Trans.on Biomedical Engineering*, **52** (8), 1390–1396, (2005).

S.L. Merritt, H.C. Schnyders, M. Patel, R.C. Basner, W. O'Neill, Pupil staging and EEG measurements of sleepness, *International Journal of Psychophysiology*, **52**, 97–112, (2004).

F. Fotiou, K.N. Fountoulakis, M. Tsolaki, A. Goulas, A. Palikaras, Changes in pupil reaction to light in Alzheimer's disease patients: a preliminary report, *International Journal of Psychophysiology*, 37, 111–120, (2000).

E. Granholm, S. Morris, D. Galasko, C. Shults, E. Rogers, B. Vukov, Tropicamide effects on pupil size and pupillary light reflexes in Alzheimer's and Parkinson's disease, *International Journal of Psychophysiology*, 47, 95–115, (2003).

L.M. Oswald, G.S. Wand, Opioids and alcoholism, *Physiology and Behavior*, **81**, 339–358, (2004).

D. Pittasch, R. Lobmann, W. Behrens-Baumann, H. Lehnert, Pupil signs of sympathetic autonomic neuropathy in patients with type 1 diabetes, *Diabetes Care*, **25** (9), 1545–1551, (2002).

M. Cahill, P. Eustace, V. De Jesus, Pupillary autonomic denervation with increasing duration of diabetes mellitus, *Br J Ophthalmol.* **85** (10), 1225–1230, (2001).

K. Fountoulakis, F. Fotiou, A. Iacovides, J. Tsiptsios, A. Goulas, M. Tsolaki, C. Ierodiakonou, Changes in pupil reaction to light melanchonic patients, *International Journal of Psychophysiology*, 31, 121–128, (1999).

E. Granholm, S.P. Verney, Pupillary responses and attentional allocation problems on the backward masking task in schizophrenia, *International Journal of Psychophysiology*, **52**, 37–51, (2004).

J. Jomier, E. Rault, S.R. Aylward, Automatic quantification of pupil dilatation under stress, *IEEE*, 249–252, (2004).

S.M. Brown, A.M. Khanani, K.T. Xu, Day to day variability of the dark-adapted pupil diameter, *Journal Cataract Refract Surg.*, **30**, 639–644, (2004).

D.P. Dioniso, E. Granholm, W.A. Hillix, W.F. Perrine, Differentiation of deception using pupillary responses as an index of cognitive processing, Psychophysiology, vol.38, 205–211, 2001.

J.M.C. Vendemia, M.J. Schillaci, R.F. Buzan, E.P. Green, S.W. Meek, Credibility assessment: psychophysiology and policy in the detection of deception, American Journal of Forensic Psychology, vol.24, 4, 53–85, 2006.

J.T. Wang, M. Spezio, C.F. Camerer, Pinocchio's pupil: using eyetracking and pupil dilation to understand truth telling and deception in games, *American Economic Review*, to appear.

J.A. Stern, The gaze control System and detection of deception, The Journal of Credibility Assessment and Witness Psychology, vol.7,2,146–148, 2006

Computational Vision and Medical Image Processing – João Tavares & Natal Jorge (eds)
© 2008 Taylor & Francis Group, London, ISBN 978-0-415-45777-4

Geometry calibration for X-ray equipment in radiation treatment devices

B.P. Selby & G. Sakas
Cognitive Computing and Medical Imaging, Fraunhofer IGD, Darmstadt, Germany

S. Walter
Medical Imaging, Medcom GmbH, Darmstadt, Germany

U. Stilla
Photogrammetry and Remote Sensing, Technische Universitaet Muenchen, Muenchen, Germany

ABSTRACT: It is essential for radiological tumor treatment to position a patient very accurate in the treatment device. To avoid sub-millimetre misalignments for a patient in the treatment facility, X-ray images acquired from within the device can be compared to the planning CT. But as slight displacements of the treatment beam nozzle, the digital X-ray panels and the X-ray tubes may arise over time and with movements of the dynamic parts of the treatment device, the underlying geometry model for the patient set-up correction may become inaccurate. To solve this problem, an automatic calibration routine is proposed, which bases on the known geometry of a calibration phantom and X-ray images acquired during the calibration routine. The result from the registration of the X-ray projections of the phantom and its known geometry is used to update the geometric model of the respective beamlines to enable accurate alignment correction.

1   INTRODUCTION

A great percentage of tumor diseases can be treated by the application of radiological doses onto the diseased tissue. To destroy carcinogen cells and to save surrounding tissue it is necessary to direct the treatment beam exactly onto the radiation target. Today, especially particle beam treatment techniques allow a very high accuracy of less than 1 mm and patient fixation equipment even assures set-up preciseness for the patient of less than 0.5 mm (Verhey et al. 1982). To avoid misalignments of the patient in the treatment device, a positioning fine-tuning is done by manual step-by-step correction, reducing the remaining offsets between a reference CT and X-ray images acquired from within the treatment device (Heufelder et al. 2004).

This procedure causes inaccuracies, which cannot be eliminated if the geometric model for the X-ray beamlines is not defined properly or if it is not up-to-date. Especially common systems that use X-ray flat panels mounted in a rotating gantry that delivers the particle beam suffer from this problem, because distortions of the gantry from the weight of the beam nozzle or misplacements of the flat panels from a flap-in and flap-out procedure cause geometric changes that affect the projection of the X-ray images

and therewith lead to incorrect patient alignment corrections.

To overcome this problem, the geometric model for the imaging equipment of the treatment device should be re-calibrated frequently. In this contribution an approach is presented, that allows an automatic image based calibration of the geometry model of the X-ray imaging devices. The modified model parameters are then taken into account for the treatment position set-up correction procedure.

2   STATE OF THE ART

During assembly of the treatment facility, the X-ray equipment is installed inside the rotatable gantry. The flat panels are either located on a special mounting or directly mounted to the beam delivery nozzle. In either case the geometric set-up can be determined via laser tracking systems, which gives an acceptable initial geometric definition of the system, but does not take any displacements into account that occur during lifetime of the machine.

Today, the periodic quality control for the geometric set-up of the imaging devices inside the treatment facility does not include a phantom based verification of the X-ray equipment (Dunscombe et al. 2007).

To track the displacement of the isocenter of an X-ray beamline, it is common practice to place a crosswire in front of the respective X-ray tube. With each acquisition, the wire is projected onto the X-ray panel and becomes visible in the image. An offset of the intersection point of the wires and the centre of the image indicates a displacement of the isocenter and can be evaluated after each image acquisition. The disadvantage of this procedure is, that it is not possible to identify image distortions resulting from a flat panel rotation or a tube movement. Despite that, the crosswire leads to unwanted artefacts in the X-ray images.

To overcome these problems, we propose an automatic calibration on a regular basis that allows adjusting the geometric model of the X-ray systems and therewith increases the accuracy of the patient alignment. The procedure does not lead to any modifications of the X-ray images themselves.

3 METHODS

The refinement of the X-ray geometry definition involves several working steps during a calibration session:

- Manual alignment of the phantom on the treatment table
- Acquisition of X-ray images for several different gantry angles and snout positions
- Automatic calculation of corrections for the X- ray beamline geometry.

During a treatment session the corrections for the X-ray beamline geometry are applied to the initial beamline model M_0.

3.1 Description of parameters for the X-ray geometry

The X-ray tube and the X-ray flat panel define an X-ray beamline. The geometric set-up of the X-ray beamline is described by Model M_0, which contains:

- Position of the X-ray tube
- Position of the X-ray panel
- Normal Vector of the panel plane
- Vector in Panel Y-direction.

Depending on if the X-ray equipment is mounted to the rotating gantry or if it is positioned at a static position in the treatment room, the respective coordinate systems of the tube and the panel are defined in a gantry coordinate system or a fixed reference system.

3.2 Layout and positioning of the phantom

The phantom used for the calibration is a glass cylinder containing 26 gold spheres each with a diameter of 1.5

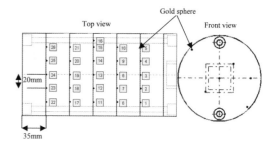

Figure 1.   Schematic view of a calibration phantom. Seen from top (left) and from front (right).

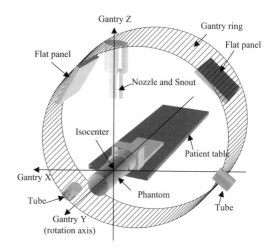

Figure 2.   Placement of the phantom.

mm that have a high absorption coefficient for X-ray beams (see Fig. 1).

The spheres are numbered from 1 to 26, whereas sphere 13 is the central sphere.

The phantom is positioned manually on the treatment table, so that the central sphere is located at the isocenter of the treatment device, which is, in the optimal case, the rotation centre of the gantry. To align the phantom properly on the tabletop in the device, a laser system is used (see Fig. 2).

3.3 Image acquisition for the calibration process

If the X-ray equipment is mounted to the rotating gantry, one image per beamline has to be acquired for several gantry angles, for example in steps of 10° degrees of gantry rotation. As the alignment of the equipment can vary in dependence of the rotation direction, each angle is calibrated for a clockwise and a counter-clockwise rotation.

If the system has a movable snout for the treatment beam, the calibration should be done for several snout positions, as the treatment table may be

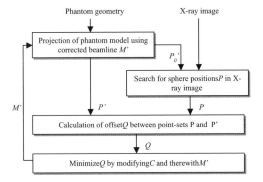

Figure 3.   Calculation of a correction.

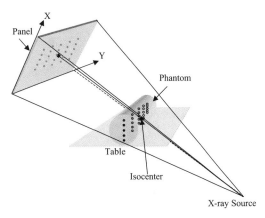

Figure 4.   Projection of the phantom.

realigned automatically for the treatment beam isocenter and herewith affects the alignment of the calibration phantom.

In total, the calibration of one beamline affords one image for one set of:

- Gantry angles with clockwise rotation,
- gantry angles with counter-clockwise rotation,
- snout positions.

For each image one correction can be computed which modifies the initial beamline model according to the mapping in 1:

$$M'_{\alpha,d,sn} \xleftarrow{C\alpha,d,sn} M_0 \qquad (1)$$

where M' is the corrected beamline model, C is the correction, α is the respective gantry angle, d is the rotation direction and sn is the snout position.

3.4 *Automatic calculation of corrections for a beamline model*

The correction $C_{\alpha,d,sn}$ for one beamline is determined by projecting a model of the known phantom geometry onto the respective panel plane and then calculating the modifications to the initial beamline model that allow a mapping of the projected phantom model spheres P' onto the real X-ray image spheres P for the device configuration (α, d, sn). Therefore the spheres P are detected in the X-ray images and are registered with the projected positions P' (see Fig. 3).

3.4.1 *Projection of the model of a phantom*
As the locations of the spheres in the phantom are known, they can be projected into an X-ray panel plane, according to a corrected beamline model $M_{\alpha,d,sn}$ (see Fig. 4).

The geometric correction C applied to the initial beamline model M_0 consists of:

- Tilt around the X- and Y-axes of the panel
- Rotation around the normal vector of the panel

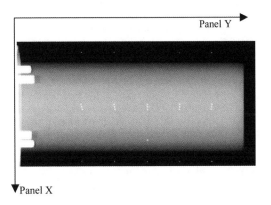

Figure 5.   X-ray image of phantom.

- Movement of the panel in X-, Y-axis and normal direction
- Movement of the X-ray tube in X-, Y- and Z- direction.

The resulting coordinates of spheres in the plane of the flat panel depend on the position of the X-ray tube and the position and alignment of the panel. The resulting coordinates are given by the intersection point of a ray from the X-ray source through the respective sphere and the panel plane. They are defined in 2D space.

3.4.2 *Detection of sphere positions in X-rays*
The spheres appear as disks of high grey value intensity in the X-ray image (see Fig. 5).

Equation 2 gives the radius R of such a disk in the X-ray image:

$$R_{disk} = R_{sphere} * \frac{\left|\vec{V}_{disk}\right|}{\left|\vec{V}_{sphere}\right|} \qquad (2)$$

249

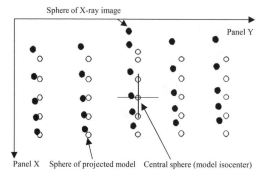

Sphere of X-ray image

Panel Y

Panel X   Sphere of projected model   Central sphere (model isocenter)

Figure 6. Spheres of phantom model and real X-ray projected into plane of the X-ray flat panel.

where V_{disk} is the vector from the X-ray source to a projected disk and V_{sphere} is the vector to a sphere.

The goal of the detection procedure is to find the centres P of the spheres in an X-ray image. Detection of a single disk is hardly possible because of artefacts in the image, which can result for example from the mounting of the phantom. For this reason, the detection procedure for each disk is performed nearby the positions of the projected phantom model P'.

For each pixel in a distance of ±15 mm nearby a point of P', a value S is calculated by equation 3:

$$S = \sum_{i=-R-1}^{R+1} \sum_{j=-R-1}^{R+1} \left(I(x+j, y+i) * k(x+j, y+i) \right)$$

$$k = \begin{cases} +1 & for \quad j^2 + i^2 \leq R^2 \\ -1 & else \end{cases}$$

(3)

where R is the radius of a disk in pixels and I the intensity of a pixel. For each point of P' a maximum for S can be found, which determines the location x and y of the respective disk centre of P.

3.4.3 Registration of the phantom model and the X-ray image

After the spheres of the phantom model have been projected computationally into the panel plane and the disk centres in the X-ray image have been found, the next step is to find the correction C for the initial beamline model that minimizes the offset between the spheres in an X-ray image of the phantom and the projected model of the phantom (see Fig. 6).

The transformation between the 2D point-sets is a non-rigid transformation. It is caused by the misalignment of the X-ray equipment in 3D space.

To minimize the offset between the positions, a downhill simplex algorithm is used (Teukolsky et al. 1992), which minimizes the squared distances between

the projected points of the model and the X-ray image as shown in equation 4:

$$Q = \sum_{i=1}^{26} \left((x_i' - x_i)^2 + (y_i' - y_i)^2 \right)$$

(4)

where Q denotes the value to minimize, i is the index for a sphere, x' and y' are the coordinates of the projected phantom spheres P' and x and y are the centres of the disks P in the X-ray image.

The optimisation is done in 9 dimensions, to find all 3 panel rotations, 3 panel shifts and 3 tube shifts at once.

After one pass of the optimisation procedure, a remaining error Q indicates how good the beamline model M_0 could be corrected. The impact of the largest error Q_{imp} on the patient alignment is estimated by equation (5):

$$Q_{imp} = \sqrt{\max_i \left((x_i' - x_i)^2 + (y_i' - y_i)^2 \right)} * \frac{|\vec{V}_{Ri'}|}{|\vec{V}_{Pi'}|}$$

(5)

where $V_{Pi'}$ is the vector from the beam source to the projected point with the maximal error and $V_{Ri'}$ is the vector to the corresponding point of the phantom in 3D space of the treatment room. If the error value exceeds a certain threshold of for example 1 mm, the algorithm tries to redetect the respective disk in the X-ray image, assuming, that starting detection at a refined position, generated through beamline model M', the result becomes more accurate. After a new detection, the registration is resumed.

The final result is obtained, as soon as the maximal error value falls below the threshold, or no improvement could be achieved by repetition.

3.5 Application of the adjusted geometry model in treatment sessions

During the calibration, a number of corrections have been calculated, that transform an initial beamline M_0 into a corrected beamline $M'_{\alpha,d,sn}$. During the treatment of a patient, the relevant values to select the correct beamline model for the patient alignment correction are gantry angle α, rotation direction d and snout position sn. The gantry angle and the snout position are provided by the treatment plan (DICOM 2006). This is not the case for the rotation direction of the gantry, which may rotate in any direction to arrive at the proper angle.

In order to obtain the best model M' for the given treatment situation, first the nearest gantry angles α_0 and α_1 for a current angle α are determined from the list of calibrated angles, so that $\alpha_0 \leq \alpha \leq \alpha_1$.

Then the following sets of corrections are chosen: $C_{\alpha0,CW}(sn)$, $C_{\alpha0,CCW}(sn)$, $C_{\alpha1,CW}(sn)$ and $C_{\alpha1,CCW}(sn)$,

Figure 7. Photography of a calibration phantom.

Figure 8. Projection of phantom model onto misaligned X-ray (left) and projection after the alignment correction (right).

where C is a set of corrections for all snout positions sn and CW or CCW denotes the rotation direction clockwise or counter-clockwise.

From each set of corrections, the two existing corrections for a snout position sn_0 and sn_1 near to the current snout position are selected so that $sn_0 \leq sn \leq sn_1$. Now the correction values of each of the 8 corrections $C_{\alpha 0,CW,sn0}$, $C_{\alpha 0,CW,sn1}$, $C_{\alpha 0,CCW,sn0},\ldots$, $C_{\alpha 1,CCW,sn1}$ are interpolated.

For the gantry angles and the snout positions we use bilinear interpolation as in equation 6:

$$C_{\alpha,sn} = \sum_{i=0}^{1}\sum_{j=0}^{1} C_{\alpha i,snj} * w_{\alpha i,snj} \qquad (6)$$

with

$$w_{\alpha i,snj} = \frac{|\alpha - \alpha_i|}{\alpha_1 - \alpha_0} * \frac{|sn - sn_j|}{sn_1 - sn_0} \text{ and } \sum_{i=0}^{1}\sum_{j=0}^{1} w_{\alpha i,snj} = 1$$

We obtain two corrections, one for each rotational direction of the gantry. Our final correction that is applied to the initial beamline model M_0 is the average of these.

4 RESULTS

Tests have been performed in three different gantry constructions:

1. System with rotatable gantry but static panels mounted in treatment room
2. System with flat panels mounted to a nozzle in a rotatable gantry
3. System with panels in a rotatable gantry with movable snout

A phantom with 26 gold spheres was used for the tests (see Fig. 7).

In the first test scenario the calibration has been performed for only one gantry angle, as the flat panels were fixed in the room. Nevertheless, aberrations of the initial beamline geometry could be determined during the calibration session. The maximum aberration was a tube shift of 12.5 mm. After the calibration we simulated a treatment situation, using a patient phantom. Before the calibration, the remaining alignment

error for the patient was 3.1 mm. After the calibration we could achieve alignments with accuracy better than 0.5 mm using the X-ray equipment.

The set-ups with moving X-ray devices have been calibrated for gantry angles in 45° steps clockwise and counter-clockwise. In the case of the movable snout, we used snout movements in steps of 25 mm.

In both cases, the initial geometric set-up of the systems were misaligned by more than 25 mm for the tubes and about 18 mm shift and 0.8° normal rotation for the panels (see Fig. 8).

Movement of the snout did affect the beamline alignment in less than 1.0 mm, but gantry rotations of 45° caused alignment deviations of maximal 7 mm and 0.8° for the panels and 10 mm for the X-ray tubes. Differences between clockwise and counter-clockwise rotations caused variations in the panel shift of a maximum of 2.0 mm.

Before the calibration the maximal error for the best patient set-up was 8.7 mm. After the calibration routine the maximal alignment error was 0.9 mm.

The calculation time for the calibration of one beamline at one specific gantry angle and snout position varies between 10 and 20 seconds on a standard PC, depending on the offset between the initial beamline geometry and the real geometry. This results in a total calibration time of up to 1:45 h for 8 gantry angles in both directions, two beamlines and 10 different snout positions, without the time needed for image acquisition.

5 CONCLUSIONS

Through the geometry calibration of the beamlines the accuracy for the patient alignment could be raised dramatically, whereas the initial set-up of the geometry is often unacceptable if alignments with less that 0.5 mm error have to be achieved. The automatic alignment procedure proposed in this contribution could reduce the patient alignment errors and can additionally serve

251

as an indicator for the initial accuracy of the beamline installation.

The relative high time exposure for the calibration is noncritical because the calibration is not done during the treatment session and can be conducted during a regular quality validation of the treatment device.

It would be possible to reduce the remaining errors after the calibration (maximum 0.9 mm) if:

- More gantry angles and snout positions would be used
- Information about the rotation direction of the gantry for a specific treatment set-up could be provided to the application
- A larger phantom would be used, as distortions are mainly visible as offsets of the outer spheres. A candy cane type phantom could increase the accuracy (Claus 2006)
- A method more suitable for the initial alignment of the calibration phantom on the treatment table than the laser alignment system could increase the accuracy of the final calibration result.

REFERENCES

Claus, B. E. H. 2006. Geometry calibration phantom design for 3D imaging. Physics of Medical Imaging 6142: 823–834.

DICOM 2006. Digital Imaging and Communications in Medicine Part 3. Rosslyn: National Electrical Manufacturers Association.

Dunscombe, P. et al. 2007. Development of quality control standards for radiation therapy equipment in Canada. Journal of applied clinical medical physics 8(1): 108–118.

Heufelder, J. et al. 2004. Fuenf Jahre Protonentherapie von Augentumoren am Hahn-Meitner-Institut Berlin. Zeitschrift fuer Medizinische Physik 14: 64–71.

Teukolsky, S. A. et al. 1992. Numerical Recipes in C. Cambridge: Cambridge University Press.

Verhey, L. J. et al. 1982. Precise positioning of patients for radiation therapy. International Journal of Radiation Oncology 8(2): 289–294.

Computational Vision and Medical Image Processing – João Tavares & Natal Jorge (eds)
© 2008 Taylor & Francis Group, London, ISBN 978-0-415-45777-4

Predicting muscle fascicule paths for use in customized biomechanical models

E.A. Audenaert
Department of Orthopedic Surgery and Traumatology, Ghent University, Belgium

G.T. Gomes
Department of Anatomy, Embryology, Histology and Medical Physics, Ghent University, Belgium

A. Audenaert
Department of Environment, Technology and Technology Management, University of Antwerp, Belgium

ABSTRACT: Patient specific biomechanical models can be applied in the design and analysis of orthopaedic implants, preoperative planning and accurate interpretation of post operative outcomes. The first step to obtain reliable customized models is by having a precise estimate of the patient specific morphology. In this study a method is presented to create patient specific models from an idealized generic musculoskeletal model. Bony morphology, wrapping structures and muscle attachment sites are derived by non-uniform scaling of a generic musculoskeletal model. This data is used as an input for a wrapping algorithm that predicts muscle fascicule paths.

1   INTRODUCTION

The prevalence of omarthrosis, possible necessitating joint replacement, varies from 2.5% at the age of 50 up to more than 10% at the age of 80. Due to individual variations in shoulder anatomy, the outcome of such surgery is often difficult to predict. To date optimum prosthetic configuration and technique is not yet defined, especially when non-anatomical reconstruction or arthroplasty is performed. Moreover it is known that an even small difference in operative prosthetic positioning importantly affects the final functionality.

Customized shoulder models should therefore be used for simulations and preoperative planning in total shoulder arthroplasty. For such models to be realistic, a reliable method needs to be developed to define the origin, insertion, location and lapses of the muscle fascicules surrounding the shoulder as well as acquiring detailed 3D reconstructions of all bones involved in shoulder movement. (Kaptein & van der Helm 2004).

2   MATERIAL AND METHODS

2.1   *Describing muscle-tendon paths*

In musculoskeletal modeling, most muscles cannot be represented as straight lines from origin to insertion because bony and musculotendinous morphology of neighboring structures causes them to wrap. (Charlton & Johnson 2001).

The majority of these passive structures can adequately be described as simple geometric shapes like spheres and cylinders. A technique for describing smooth muscle paths over multiple obstacles was developed by minimizing total path length. We assumed no frictions to exist between muscle and bony surface. From the muscle origin to insertion all possible smooth paths, without intersection of the obstacles, are calculated and the shortest smooth path will be selected as the real shortest muscle path between these two points. (Fig. 1)

Comparing predicted muscle paths with 3D cadaveric measurements of muscle fascicule paths validated the method.

2.2   *Customizing bone-muscle models*

We hypothesized that muscle fascicule position and orientation could be reliably defined by carefully suturing very thin flexible metal wires to the muscles according to the fibres. First, a computed tomography analysis of a porcine bone-muscle model containing different copper and iron wires with a thickness ranging from 0.7 to 1.3 mm was performed. The Mimics® (Materialise NV, Heverlee, Belgium) software was used for the 3D reconstructions and further analysis. The properties –scattering and visibility versus flexibility- of the different reconstructed metal wires were compared in relation to the porcine muscle-bone model. The method was then applied on different

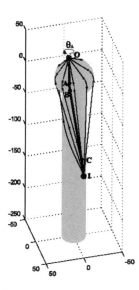

Figure 1. Optimization of a spherical-cylindrical wrapping path.

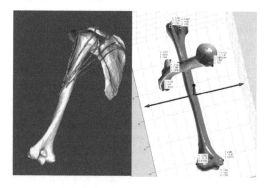

Figure 2. 3D reconstructions of bone and muscle components can be used for scaling to a customized model.

human cadaver models. In these models the position of the muscles origins and insertions were defined relative to the bony reconstructions. A spherical-cylindrical wrapping algorithm was applied to define the different muscle paths in-between. By applying a scaling algorithm on a specific bone we tried to predict the position of muscle origins and insertions on a second one in order to obtain a complete customized model. (Fig. 2).

3 RESULTS

The thinnest (0.7 mm) copper wire, was reliably identified as having the least scattering on computed tomography images and could still be used for automated segmentation. It was therefore selected for application to all shoulder girdle muscle in 12 human

cadaver specimens. A spiral computed tomography scan with a 1-mm slice thickness was obtained of all specimens and 3D reconstructed. The transformation from one cadaver model to a specific other one is achieved by a volumetric morphing technique using non uniform scaling, translations, rotations and shear matrices combined with the optimum fitting around bony and muscle structures. Analysis of the reconstructions of the muscle-bone model of the cadaver specimen confirmed that muscle fascicule positions can be adequately estimated with this method.

4 DISCUSSION

The necessity for the generation of subject specific biomechanical models that are aimed at clinical applications, in particular in the design and analysis of orthopaedic implants, preoperative planning and accurate interpretation of post operative outcomes, is becoming increasingly apparent. Currently, musculoskeletal imaging techniques like magnetic resonance imaging and computed tomography scans can potentially provide a source for complete individualized models, however, the constraints of clinical reality e.g. cost, exposure to unacceptable levels of radiation and time, preclude at the moment the creation of full complex models of a region of interest. This gap can be bridged by creating an idealized generic musculoskeletal model that can be morphed into a specific patient model using limited computed tomography and morphometric data obtained pre-operatively and a database being created from a cadaver study where bone geometry and muscle fascicule paths are measured in detail.

5 CONCLUSION

Within the context of musculoskeletal modeling a precise estimation of the musculotendinous path is highly crucial because muscle moment arms en forces directly depend on it. As any model is based on a variety of measurements and estimates e.g., centroids, kinematics, musculotendinous properties, anatomy etc, each estimate needs to be as precise as possible. The first step to obtain reliable customized models is by defining a good estimate of the patient specific morphology.

REFERENCES

Charlton IW & Johnson GR. 2001. Application of spherical and cylindrical wrapping algorithms in a musculoskeletal model of the upper limb. J Biomechanics 34: 1209–1216.
Kaptein BL & van der Helm FC. 2004. Estimating muscle attachment contours by transforming geometrical bone models. J Biomechanics 37(3):263–73.

Computational Vision and Medical Image Processing – João Tavares & Natal Jorge (eds)
© 2008 Taylor & Francis Group, London, ISBN 978-0-415-45777-4

A novel model of bottom-up visual attention using local energy

A. García-Díaz, X.R. Fdez-Vidal, R. Dosil & X.M. Pardo
Universidade de Santiago de Compostela, Santiago de Compostela, Spain

ABSTRACT: In this paper we propose a novel model for the implementation of the Koch & Ullman architecture of bottom-up visual attention. We use two features to measure saliency: local energy and color. Local energy is extracted as the envelope of a bank of complex-valued log Gabor filters (resembling the receptive fields of complex cells from visual cortex), giving regions of maximum phase congruency (PC) in the image, which have proven perceptually relevant. Previous PC measures have been employed in edge extraction but they aren't suitable in attention applications. In our approach, across the scales of each orientation, we compute the T^2 Hotelling statistic, as a multivariate measure of variance, that is of maximum PC, and we take it as a conspicuity map. We forward normalize and integrate these maps with color maps to form a unique final saliency measure, taking the highest values of variance at each point. With this model we reach improved or similar performance in comparison to a state of art model, when we reproduce the same test experiments.

Keywords: (Attention, bottom-up, saliency, feature integration, principal component analysis)

1 INTRODUCTION

Visual attention plays a crucial role in human visual system (HVS), avoiding for the overloading which would imply the processing of all the data entering from the visual field ($\sim 10^8$ bits). It reduces the amount of information to be processed, and therefore which reach consciousness. And in short, It reduces the complexity of the problem of viewing, allowing a particular solution according to the available resources (Tsotsos 2005).

The problem of complexity and limitation of resources is shared by computing processing of images or video sequences in real time. Hence attention seems to be an adequate biologically inspired solution which we can apply in a wide variety of computing problems, like vision system for robots (Witkowski & Randell 2006); visual behavior generation in virtual human animation (Peters 2003); content-based image retrieval (Marques et al. 2006); etc.

The development of computational models of visual attention started with the publication of the feature integration theory by Treisman & Gelade (1980). This idea was after gathered by Koch & Ullman (1985), to conceive a computational architecture for a saliency-based model of visual attention, which assumes the existence of two different types of attention: one image-based, called bottom-up and other guided by knowledge, the so called top-down attention. These two kinds of attention interact each other to deliver a global measure of saliency to drive visual selection.

Any way, it can be argued that understanding of bottom-up saliency should definitely help to elucidate the mechanisms of attention (Zhaoping 2005).

Many different models of visual attention have been hold to date, with different approaches based mainly in psychophysical and neurophysiological knowledge of the human visual system (HVS). Well known examples are the Guided Search (Wolfe 1994), the SMT proposed by Tsotsos (1995), the connectionist approach made by Grossberg (2005), the particular implementation of the Koch and Ullman architecture made by Milanese (1995) using color, orientation and edge magnitude as low level relevant features, or the alternative implementation conceived by Itti et al. (1998) making use of contrast, color and orientation as separated features, in a center-surround multi-scale approach. This last can be regarded as the most developed model of visual attention, considering the fact that it has been compared with human performance (Itti & Koch 2000; Itti 2006; Ouerhani et al. 2004) and tested in a variety of applications (Walther 2006; Ouerhani & Hugli 2003).

Other studies have taken advantage of information theory and have oriented their work to discover the statistical structure of what we see and link it to the attentional processes. The intrinsic sparseness of natural images has been pointed out by Olshausen & Field (1996), who have proposed that an efficient coding which maximizes sparseness is sufficient to account for cellular receptive fields. Hoyer & Hyvärinen (2000) have applied the independent component analysis to the feature extraction on colour and stereo images, obtaining features resembling simple cells receptive fields. Rajashekhar et al. (2006) have studied the statistical structure of the points that attract the eye fixations, modeling a set of low level gaze attractors.

In this paper we assume the Koch and Ullman architecture and we restrict our study to the extraction of an image-based saliency map, leaving apart the top-down influences for a future work. The main difficulties that face a saliency-based bottom-up attention system are the election of an adequate set of low level features, which efficiently serve to measure the relevancies of a region of the image, and their integration in a final saliency measure. With this aim, we propose a novel model of bottom-up attention consistent with the pointed sources of knowledge -psychophysics, physiology and information theory- at the moment, on the HVS and the statistical structure of natural images. In this model we adopt the detection of regions with maximum local energy (Kovesi 1996; Morrone & Burr 1988).

The integration process is addressed in a statistical way which assures for both sparseness population increase (Weliky et al. 2003) and pop-out effects of orientation, size and colour singletons as have been extensively observed in the mammalian visual system (Zhaoping 2005; Wolfe 2004). Our approach is based on the local energy measure with a bank of complex-valued log Gabor filters resembling receptive fields of cortex cells, and the statistical extraction of conspicuity maps with the T^2 Hotelling transform, obtained from principal component analysis performed on the sets of scaled responses.

We have succeeded improving or equalling the performance achieved by Itti & Koch (2001) in identical experiments on target detection within cluttered natural scenes.

In the section 2 we describe the model proposed; in section 3 we show some comparative results with previous models; in section 4 we make a brief discussion of the results; and finally section 5 deals with conclusions.

2 MODEL OVERVIEW

2.1 Feature maps extraction

First, we extract three initial channels from the image: the intensity and two color double-opponent components, r-g and b-y.

2.1.1 Color maps
To perform the extraction of color conspicuity, we use two components of opponencies r-g and b-y, and by means of center-surround operations, as described in Walther (2006; appendix A.2), we obtain a pyramid of Gaussian subsampled maps for each color component.

2.1.2 Local energy maps
From the intensity channel we obtain four oriented sets of scales $(0°, 45°, 90°, 135°)$. We use a bank of complex log Gabor filters in phase quadrature, constructed in the frequency domain, and we extract the local energy as the modulus of this complex vector. The log Gabor filters have been proposed by Field (1987) to overcome some disadvantages of Gabor filters related to the fact that these are non-zero for negative frequencies and present a non-zero DC component, giving rise to artifacts. The log Gabor filters present a symmetric profile in a logarithmic frequency scale, and thus, one additional advantage relies in their long tail towards the high frequencies; since natural images present scale invariance, this is, they present amplitude profiles that decays with the inverse of the frequency (Field 1993), then a filter that presents a similar behavior, should be able to properly encode those images (Kovesi 1996).

Since these logarithmic scaled filters have no analytic representation in the spatial domain, we construct the banks of filters in the frequency domain, and we perform the convolution by means of a product with the FFT transform of the intensity of the image. In the frequency domain, the applied log Gabor filter takes the expression:

$$\log Gabor(f, \alpha; f_i, \alpha_i) = e^{-\frac{(\log(f/f_i))^2}{2(\log(\sigma_{fi}/f_i))^2}} e^{-\frac{(\alpha-\alpha_i)^2}{2(\sigma_\alpha)^2}} \quad (1)$$

In our implementation we have used 5 scales and the central frequencies of the filters were spaced by one octave; other parameters are the minimum wavelength $(\lambda_{min} = 1)$, the angular standard deviation $(\sigma_\alpha = 2)$ or the frequency bandwidth (two octaves).

2.2 Integration procedure and local energy saliency

2.2.1 T^2 statistic as a measure of maximum variance in phase congruency
The simplest way to perform integration of feature maps would be to compute their average value, but this provides noisy and ambiguous results, and don't take into account the competition process involved in visual attention, which have been observed in psychophysical experiments. For the integration of the early feature maps in bottom-up attention, they have been proposed several approaches. In the frame of Koch and Ullman architecture, Milanese (1995) implemented a relaxation process by means of a non-linear updating rule which updates all the feature maps to satisfy a convergence criterion, and defining a heuristic energy function to minimize. Itti & Koch (2000) have proposed an integration process based on the averaging after the filtering of maps with iterative DoG

filters, providing local within-feature and inter-feature competition.

But in our case, as can be seen in Figure 1, we perform a principal component analysis (PCA) across the local energy maps for each orientation, using each scale response as an initial component vector; and then we take the resulting T^2 Hotelling statistic as a measure of relevance, giving rise to a unique conspicuity map for each orientation. This is the key point of our approach to the integration process, because of the fact that it provides an efficient mechanism for on/off detection and orientation and size pop-out effects, widely observed in psychophysics experiments, by mean of a multivariate measure of the phase congruency variance. Furthermore, this procedure removes most noise and irrelevant regions, getting an important gain in population sparseness. And hence, it has the benefit of introduce a more efficient within feature competition.

2.2.2 Spatial competition and normalization saliency

Consecutively, we perform a spatial competition and normalization which lies on a center surround contrast operation by applying three iterations of Difference of Gaussians (DoG) filtering. These filters don't present the disadvantages pointed for the Gaussians. The precise number of iterations can vary with the size of the image to normalize.

Our procedure employs DoG filters in a very close manner to that implemented by Itti & Koch (2000), except that in the last iteration, we take the excitatory signal instead of the difference, avoiding the influence of the inhibitory Gaussian; and hence, achieving a strengthening of the regions with high variance.

So, the obtained orientation maps are involved in spatial competition and normalization, with three iterations of DoG filtering.

After this normalization, the next integration step is performed taking the maximum of the variance at each point for each orientation, reinforcing our strategy which aims to maximize the variance; and therefore giving rise to a local energy saliency map, which we normalize with other three iterations of DoG filtering, as well as the color map, before of their averaging in the final saliency map.

2.3 Fixations and target detection

To sequentially perform the fixations on an image using the saliency information already obtained, we have implemented a simplified version of the WTA neural network used by Itti & Koch (2001) in their experiment, but maintaining the basic assumptions for the focus of attention (FOA) size and considering the target detected when the FOA intersect its mask.

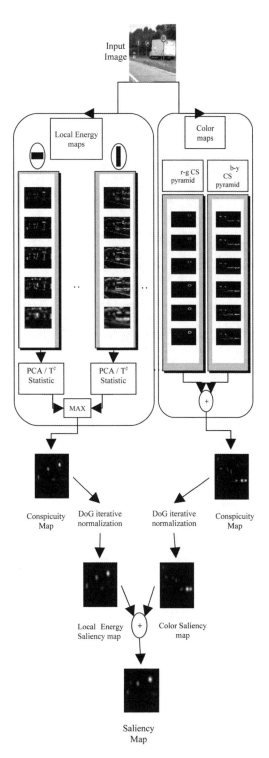

Figure 1. Computing of Saliency in the model of bottom-up attention, based on local energy.

257

Table 1. Average number of false detections before target found.

Red can	1.67 ± 1.81
Triangle	1.09 ± 1.53
Traffic[a]	0.37 ± 0.86
Traffic[b]	0.77 ± 1.32

[a] Before first sign found.
[b] Before all signs found.

Figure 2. Results obtained with some images with traffic signs, showing the original image (left), the saliency map (center) and the fixations performed (right).

Figure 3. Results obtained with two images with triangles, showing the original image (left), the saliency map (center) and the fixations performed (right).

Figure 4. Results obtained with two images with a red can, showing the original image (left), the saliency map (center) and the fixations performed (right).

3 EXPERIMENTAL RESULTS

In this section we show some detection results with two sets of images used by Itti & Koch (2001), and some images with the respective saliency maps and the consecutive FOA performed before target detection. One set contains traffic signs, other an emergency triangle, and last, another one with a red can. In all cases we take account of the false fixations before the first target detection, and in the case of traffic signs we also compute the number of false fixations until all the signs are detected.

All the images are public and can be found in (http://ittilab.usc.edu).

3.1 Images sets

Each set of images contains the respective binary images with the masks for target detection. The set of images with traffic signs consists of 45 images of 512x384 with 24-bit color, containing one or more signs: 17 containing one sign, 19 with 2, 6 with 3, 2 with 4, and 1 with 5. The FOA for these images is a disk of radius 64.

The Triangle's set have 32 images and the red can set have 59, both containing one target per image. All the images are sized 640x480, 24-bit color, and the FOA used with them was a disk of radius 80.

3.2 Results

Results are summarized in table 1 where the average of false fixations before target detection is showed as well as its standard deviation.

In Figures 2, 3 and 4 we also show some images with their respective saliency maps and the subsequent fixations performed before target(s) detection with the three sets of images employed. We can observe how false fixations are directed to relevant objects which act as distractors capturing the attention before one of the targets, as is the case of the car and the bottom of the lorry (figure 2), or the colored flowers surrounded by uniform textures (figure 4).

4 DISCUSSION

In table 2 we show the results obtained on the same images by Itti & Koch (2001). As can be seen, there has been a clear improvement in the first sign detection rate and in the triangle detection rate. The detection of all signs has reached a similar rate than Itti & Koch result, and the red can a slightly poorer one.

Table 2. Results by Itti & Koch (2001).

	Iterative
Red can	1.24 ± 1.42
Triangle	1.42 ± 1.67
Traffic[a]	0.52 ± 1.05
Traffic[b]	0.70 ± 1.18

[a] Before first sign found.
[b] Before all sign found.

We found that the saliency maps detect in a well suited manner regions containing other relevant objects for general purpose like cars, steps, etc. Hence, our model has attained better or similar performance with fewer features, less maps and less iterative process, reducing the processing load.

5 CONCLUSIONS

As we have shown, expectancies on the performance of our approach to saliency-based bottom-up attention have been widely satisfied in these initial tests. Local energy and information decorrelation have revealed suitable providing statistically independent image descriptors, giving rise to a simple model of attention with slightly improved performance respect to the model of bottom-up attention hold by Itti et al. (1998), which can be at present considered the state of art. Furthermore, we use less features and a lower number of maps when computing bottom-up saliency than Itti and colleagues do.

We can conclude that the results strength the starting hypothesis, revealing local energy as a suitable measure in the extraction of regions perceptually relevant, which attract human attention.

In progress and future work will include further improvements of the model (mainly, a new approach to color feature extraction), and experiments to test particular aspects (pop-out effects, human-like performance, etc.), being the aim the development of the model in a complete attention system, introducing motion as a relevant additional feature, as well as top-down guidance to bias the bottom-up saliency, for its application in active vision tasks for general purpose.

ACKNOWLEDGEMENTS

This work has been financially supported by the Ministry of Education and Science of the Spanish Government through the research project TIN2006-08447.

REFERENCES

Field, D.J. 1987. Relations Between the Statistics of Natural Images and the ResponseProperties of Cortical Cells. *Journal of the Optical Society of America A*, 4(12): 2379–2394.

Field, D.J. 1993. Scale-invariance and self-similar 'wavelet' transforms: an analysis of natural scenes and mammalian visual systems. In M. Farge, J. Hunt and J. Vassilicos(eds.), *Wavelets, Fractals and Fourier Transforms*, 151–193. Oxford: Clarendon Press.

Grossberg, S. 2005 Linking attention to learning, expectation, competition and consciousness. In L. Itti, G. Rees, J. Tsotsos (eds.), *Neurobiology of attention*. 652–662. Elsevier.

Hoyer, P.O. & Hyvärinen A. 2000. Independent component analysis applied to feature extraction from color and stereo images. *Network: Computation in Neural Systems,* 11: 191–210.

Itti, L., Koch C. & Niebur, E. 1998. A model of saliency-based visual atttention for rapid scene analysis. *IEEE Transactions on Pattern Analysis and Machine Intelligence*, 20 (11): 1254–59.

Itti, L. & Koch, C. 2000. A saliency-based search mechanism for overt and covert shifts of visual attention. *Vision Research*, 40: 1489–1506.

Itti, L. & Koch, C. 2001. Feature combination strategies for saliency-based visual attention systems. *Journal of Electronic Imaging,* 10 (1): 161–169.

Itti, L. 2006. Quantitative modeling of perceptual salience at human eye position. *Visual Cognition*, 14(4/5/6/7/8) :959–984.

Koch, C. & Ullman, S. 1985. Shifts in selective visual attention: towards the underlying neural circuitry. *Human neurobiology*, 4(4): 219–227.

Kovesi, P. 1996. *Invariant Measures of Image Features from Phase Information.* Ph.D., The University or Western Australia.

Marques, O., Mayron, L.M., Borba, G.B. & Gamba, H.R. 2006 Using visual attention to extract regions of interest in the context of image retrieval *Proceedings of The 2006 ACM Multimedia Conference.*

Milanese, R., Gil, S. & Pun, T. 1995. Attentive mechanisms for dynamic and static scene analysis. *Optical engineering*, 34(8): 2428–34.

Morrone & Burr 1988. Feature Detection in Human Vision: A Phase-Dependent Energy Model. *Proceedings of the Royal Society of London B*, 235: 221–245.

Olshausen, B.A. & Field, D.J. 1996. Emergence of Simple-Cell Receptive Field Properties by Learning a Sparse Code for Natural Images. *Nature*, 381: 607–609.

Ouerhani, N. & Hugli, H. 2003. Maps: Multiscale attention-based presegmentation of color images. Lecture Notes in Computer Science, 2695: 537–549

Ouerhani, N., Wartburg, R., Hugli, H. & Mueri R. 2004. Empirical validation of the saliency-based model of visual attention. Electronic letters on computer vision and image analysis, 3(1): 13–24.

Peters, C. & O' Sullivan, C. 2003, Bottom-Up Visual Attention for Virtual Human Animation. *CASA*: 111–117,

Rajashekhar, U., Bovik, A.C., Cormack, L.K. 2006. Visual search in noise. *Journal of Vision*, 6: 379–386

Treisman, A. & Gelade, G. 1980. A Feature-Integration Theory of Attention. *Cognitive Psychology*, 12: 97–136.

Tsotsos, J.K., Culhane, S., Wai, W., Lai, Y., Davis, N. & Nuflo, F. 1995. Modeling visual attention via selective tuning, Artificial Intelligence, 8(1-2): 507–547.

Tsotsos, J.K. 2005 Computacional foundations for attentive Processes. In L. Itti, G. Rees, J.K. Tsotsos (eds) *Neurobiology of Attention*, 3–7. Elsevier Academia Press.

Walther, D. 2006 *Interactions of visual attention and object recognition: computational modeling, algorithms, and psychophysics*. PhD., California Institute of Technology.

Weliky, M., Fiser J., Hunt, R.H. & Wagner, D.N. 2003. Coding of natural scenes in primary visual cortex. *Neuron* 37: 703–718.

Witkowski, M. & Randell, D. 2006. Modes of Attention and Inattention for a Model of Robot Perception. *Proc. Towards Autonomous Robotic Systems (TAROS-06)*: 246–253.

Wolfe, J.M. 1994 Guided Search 2.0 A Revised model of visual search. Psychonomic Bulletin, 1(2): 202–238.

Wolfe, J. M. and Horowitz, T. S. 2004 What attributes guide the deployment of visual attention and how do they do it?. Nature Reviews. Neuroscience, 5(6): 495–501.

Zhaoping, L. 2005. The primary visual cortex creates a bottom-up saliency map In L. Itti, G. Rees, J.K. Tsotsos (eds) Neurobiology of Attention, 570–575. Elsevier Academia Press.

Computational Vision and Medical Image Processing – João Tavares & Natal Jorge (eds)
© *2008 Taylor & Francis Group, London, ISBN 978-0-415-45777-4*

Coarse comparison of outlines shapes descriptors

Saliha Aouat & Slimane Larabi
*Computer Science Department, Faculty of Electronic and Computer Science,
University of Sciences and Technologies HOUARI BOUMEDIENNE, El Alia, Algiers, Algeria*

ABSTRACT: In order to retrieve shape in images database, we follow the strategy that achieves this process in two stages: coarse matching and full matching. In this paper, we assume that we have a database of descriptors of shapes written according the XML language XLWDOS [6, 7]. We propose a new method for coarse matching of silhouettes represented by their XLWDOS descriptors that may be used for shape indexing. In order to take in account the deformation of shapes due to noise or to varying camera position, a reducing technique is proposed that deduces the same index from the two slightly different descriptors. Experiments over real images are conducted and explained.

1 INTRODUCTION

Various methods have been developed in order to represent shape in an abstract and efficient way. The most interesting methods are the part-based methods where silhouette is decomposed into parts [4, 9, 11, 12, 14, 15], the aspect-graph methods that are viewer-centered representations of a three-dimensional object [2, 5], methods that use the medial axis of silhouettes. [3, 13, 16] and appearance-based methods [10].

The recognition of object using database of shapes is a current and difficult problem. Different methods based on different representations of shapes have been proposed. All of these methods proceed on descriptors comparison and a similarity measure is then computed.

Indexing database of images facilitates the retrieval problem of any image. Many solutions have been proposed but depending on the shape descriptors.

In order to retrieve shape in images database, we propose to achieve the comparison in two stages: a coarse matching (indexing) and a full matching (similarity measure).

In this work, we use the part-based method published in [6, 7] for object representation from its silhouettes. We propose a new method for coarse matching of silhouettes represented by their XLWDOS descriptors that may be used for shape indexing. In order to take in account the deformation of shapes due to noise or to varying camera position, a reducing technique is proposed that deduces the same index from the two slightly different descriptors.

In this paper, we firstly introduce the silhouette description according the language XLWDOS. In the second section, we present our method for coarse comparison of descriptors. In the last section, we present and discuss the obtained experimental results.

2 SILHOUETTES DESCRIPTION ACCORDING TO *XLWDOS* LANGUAGE: A SUMMARY [6, 7]

The first step of shape description is its decomposition into parts and separating lines. For this, we sweep the outline shape from the top to bottom, and we locate concave points for which the outer contour changes the direction top-bottom-top or bottom-top-bottom (see fig. 1).

The outline shape is decomposed at these points into parts, and separating lines: either, two parts or more are joined with a third part through a junction line, or a part is joined with two parts or more through a

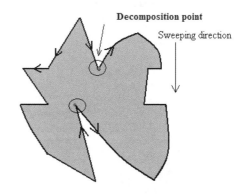

Figure 1.   Location of decomposition points.

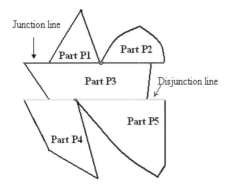

Figure 2.   Parts, junction and disjunction lines.

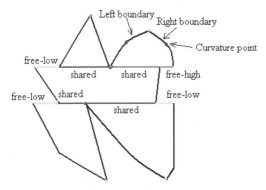

Figure 3.   Description of parts and separating lines.

disjunction line. This process applied for example to silhouette of figure 1 produces five parts, one junction and one disjunction line (see fig. 2). The part and separating lines are numbered from the top to bottom and left to right.

The second step is the representing of these elements with a graph structure where nodes are the different parts and separating lines.

The third step is the description of each element in order to guaranty the uniqueness of the outline shape representation.

Part is defined by its two boundaries (left and right) (see fig. 3). Using the curvature points located by one of the known algorithms [1], the two boundaries of each part are segmented into set of elementary contours (line, convex and concave contours) and described by the parameters: type (line, convex or concave curve), degree of concavity or convexity, angle of inclination and length.

Separating lines are decomposed into segments. Each segment is described with three parameters: type, the reference numbers of linked parts and its length. Three types for segment are possible: Shared if it is common for two parts, Free-High if it is neighbor only

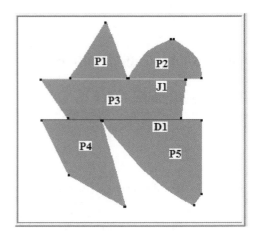

Figure 4.   The output of XLWDOS software.

to the high part and Free-Low if it is neighbor only to the low part (see fig. 3).

In order to guaranty the uniqueness of the proposed description, the sweeping of the outline shape must be done following one of the two direction of the rectangle of minimum area encompassing it.

The graph structure of the outline shape is translated into an XML descriptor written following the language XLWDOS. The coarse descriptor of silhouette written with the XLWDOS language is the descriptor for which only appear parts, composed parts, the junction and disjunction lines without their geometrical descriptions. For example, the coarse descriptor of the silhouette of figure 1 is:

$<CP><CP>P_1 P_2 J_1 P_3 </CP> D_1 P_4 P_5</CP>$

Where: $<CP>$ and $</CP>$ are marks for composed part, P_i designates the part ith part, J_k designates the kth junction line and D_n designates the nth disjunction line.

To obtain the full XLWDOS descriptor, we replace in the coarse descriptor each part or separating line by its description. We give in follow the XLWDOS descriptions of parts and separating lines obtained from the XLWDOS software (see fig. 4):

$<P1><L>$ r 57 75 $</L><R>$ r 111 75 $</R></P1>$
$<P2><L>$ cv 11 42 53 $</L>$
     $<R>$ r 180 4 cv 16 124 53$</R></P2>$
$<P3><L>$ cv 3 125 51$</L><R>$r 83 51$</R><P3>$
$<P4><L>$ r 116 74 r 150 42 r 180 1$</L>$
     $<R>$ cv 50 105 116 $</R></P4>$
$<P5><L>$ cv 6 137 113 $</L>$
     $<R>$ r 90 99 r 56 15 $</R></P5>$
$<J1>$w P3 40 s P1 P3 76 w P3 2 s
P2 P3 77 h P2 21$</J1>$
$<D1>$ w P4 36 s P3 P4 85 h P3 2 s
P3 P5 106 w P5 27$</D1>$

where:

- **L, R** designate the left and right boundary
- **r** designates a right contour followed by its attributes: angle of orientation and length
- **CV** and **CC** designate respectively a convex and concave contour followed by its attributes: concavity degree, angle of orientation and length
- **S** designates a shared segment followed by its attributes: numbers of linked parts, its length
- **W, H** designates respectively a Low and High segment followed by its attributes: the number of linked part, its length.

3 COARSE COMPARISON OF SILHOUETTE DESCRIPTORS

3.1 Sensitivity to noise of XLWDOS descriptors

The XLWDOS descriptor of silhouette is sensitive to noise.

Figure 5 represents the same shape as figure 1 with some differences due to additional pixels in the outline. Unfortunately their coarse and full XLWDOS descriptors are different (see fig. 6).

Indeed the coarse descriptor silhouette of figure 5 is:

<CP><CP>P₁ P₂ J1 P₃</CP>
    D₁ <CP>P₄ D₂ P₈ P₉</CP>
        <CP> <CP>P₅ P₆ J₂ P₇</CP>
            D₃ P₁₀ P₁₁
        </CP>
</CP>

3.2 Basic principle of coarse matching of XLWDOS descriptors

First stage: Matching of separating lines.

This step consists to explore descriptors of two different silhouettes and to match their separating lines. Two separating lines are matched if they have the same descriptions: the different segments have the same type (shared, free low, free high) and the same length relatively to the length of the minimum rectangle encompassing the silhouette.

The second step is to match composed parts associated to the matched separating lines. This implies that a set of pairs of elements (part or composed part) having the same position relatively to matched separating lines will be grouped and will be matched.

For example, we give the coarse descriptors of the two silhouettes of figure 1 and 5:
Silhouette S1 (fig.1):

<CP><CP>P₁ P₂ **J₁** P₃ </CP> **D₁** P₄ P₅</CP>

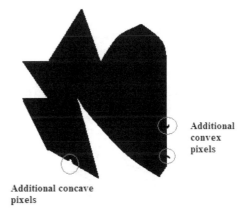

Additional convex pixels

Additional concave pixels

Figure 5. Noisy silhouette.

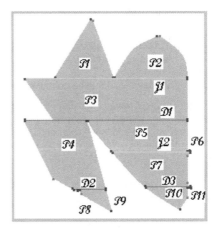

Figure 6. XLWDOS decomposition for noisy shape.

Silhouette S2 (fig.5):

<<CP><CP>𝒫₁ 𝒫₂ 𝒥1 𝒫₃</CP> 𝒟₁ <CP>𝒫₄ 𝒟₂ 𝒫₈ 𝒫₉</CP> <CP><CP>𝒫₅ 𝒫₆ 𝒥₂ 𝒫₇</CP>𝒟₃ 𝒫₁₀ 𝒫₁₁</CP></CP>

After the matching of J1 of (S1) and 𝒥1 of (S2), the following pairs are constituted and will be matched:

(P₁, 𝒫₁), (P₂, 𝒫₂), (P₃, 𝒫₃).

Also, as D₁ is matched to 𝒟₁, the following pairs are constituted:

(P₄, <CP>𝒫₄ 𝒟₂ 𝒫₈ 𝒫₉</CP>) and
(P5, <CP><CP>𝒫₅ 𝒫₆ 𝒥₂ 𝒫₇</CP>𝒟₃ 𝒫₁₀ 𝒫₁₁</CP>)

Second stage: Reducing composed pars to parts

In order to match the different elements of constituted pairs, we must find if we can reduce any composed

263

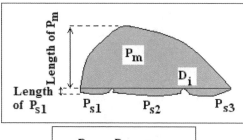

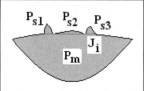

Figure 7.   Composed part with noised parts.

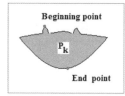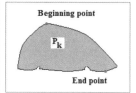

Figure 8.   Applying the reducing process.

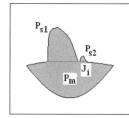

Figure 9.   Applying the reducing process over another example.

part member of a constituted pair to a single part. This case occurs when noise produces a separating line thus an additional composed part.

The noise altering the silhouette boundary may be:

– Additional pixels producing composed part with disjunction line instead of a single part (see fig. 7).
– Additional pixels producing composed part with junction line instead of a single part (see fig. 7).

We define:

– A *secondary part* P_{si} the part appearing at the left of the junction line or at the right of the disjunction line.
– A *main part* P_m the part appearing at the right of the junction line or at the left of the disjunction line.
– the length of any part is the number of rows that it contains.

When the lengths of many secondary parts P_{si} are very short compared to the length of the main part, we apply the reducing process that translates the composed part $<CP> P_{s1} P_{s2} ... P_{sn} J_i P_m </CP>$ or $<CP> P_m D_i P_{s1} P_{s2} ... P_{sn} </CP>$ to a single part P_k.

In this case, a new part P_k is then created replacing the set of main and secondary parts.

$P_{s1} P_{s2} ... P_{sn} J_i P_m \rightarrow P_k$ or
$P_m D_i P_{s1} P_{s2} ... P_{sn} \rightarrow P_k$.

where the boundary of P_k is obtained as the boundaries of all parts and the segments of separating line having Free_High or Free_Low as attributes. The left and right boundaries of P_k begin at the highest left point and terminates at the lowest right point.

Figure 8 illustrates the new part obtained knowing that the lengths of the three parts $P_{s1} P_{s2} P_{s3}$ are very short relatively to the length of the Part P_m.

Figure 9 illustrates the application of the reducing process in case when there is only one noised part (P_{s2}).

The reduction process is recursive. It is applied while there is in the XLWDOS descriptor composed parts that verify the reducing condition beginning by the more internal composed part in the descriptor.

Third stage: Coarse comparison of XLWDOS descriptors

Let S_R (shape request), S_M (shape model) are two shapes. After the computation of the XLWDOS descriptors $Dt(S_R)$ and $Dt(S_M)$, we may compare the two shape descriptors as follow:

– Matching separating lines of $Dt(S_R)$ and $Dt(S_M)$
– Application of the reducing process over the constituted pairs.
– After this, if all obtained pairs are constituted by only single parts, we deduce that the two shapes are coarsely matched and the coarse descriptor constitutes the shape index.

In this favorable case, we may match the parts of the two shapes and compute the similarity between them.

4   EXPERIMENTATION

In this section we show how we apply the proposed method over real images.

– The first application is the matching of two images of a moving object where the extracted shapes are little different.

Figure 10 illustrates two images of a moving object. After silhouettes extraction using background

264

Figure 10. Images of moving car (toy).

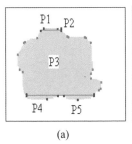
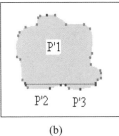

(a)   (b)

Figure 11. Extracted silhouettes and results of XLWDOS decomposition.

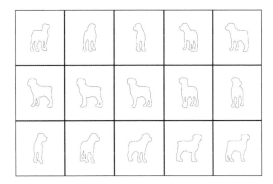

Figure 12. A set of shapes of the database of Leibe and Schiele[8].

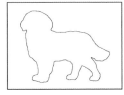
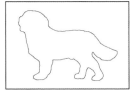

Figure 13. Two successive images of a dog.

subtraction technique, we show in Figure 11 the obtained results of their XLWDOS decomposition into parts, separating lines and elementary contours by means of inflexion points [1].

We can see that there is an additional noised part P2 in the XLWDOS of the shape (a).

We give the coarse XLWDOS descriptor for each silhouette:

(a) <CP> <CP>P_1 P_2 J_1 P_3</CP>
               D_1 P_4 P_5</CP>
(b) </CP> P'$_1$ D'$_1$ P'$_2$ P'$_3$</CP>

The coarse matching of (a) and (b) is possible because D_1 and D'_1 may be matched. The set of following of pairs are constituted:

(<CP>P_1 P_2 J_1 P_3</CP>, P'$_1$), (P_4, P'$_2$), (P_5, P'$_3$)

The following reduction is applied:

<CP>P_1 P_2 J_1 P_3</CP> → P''$_1$, where P''$_1$=$P_1 \cup P_2 \cup P_3$

The list of pairs of parts that will be considered for full matching are: (P''$_1$, P'$_1$), (P_4, P'$_2$), (P_5, P'$_3$).

From this coarse matching, the descriptors of the two silhouettes (a) and (b) are candidates for similarity measure.

– The second application of our method is the indexing of database of shapes. We used some images of database of shapes built by B. Leibe and B. Schiele [8] where each object is represented by 41 views

spaced evenly over the upper viewing hemisphere. Figure 12 illustrates a sequence of shapes corresponding to images of a dog taken by rotating the camera with an angle of 24°.

Figure 13 illustrates two successive images (S1) and (S2) of a dog. We give in figure 14 the result of their decomposition into parts and separating lines.

The two computed XLWDOS descriptors are:

DS(S1)=
<CP><CP><CP>P_1 D_1 P_2 P_3</CP> <CP>P_4 P_5 J_1 P_6</CP> J_2 P_7</CP> D_2 <CP>P_8 D_3 P_{10} <CP>P_{11} D_4 <CP>P_{12} D_5 P_{14} P_{15}</CP> P_{13}</CP></CP> P_9</CP>

$\mathcal{DS(S2)}$=<$\mathcal{CP}$><$\mathcal{CP}$>$\mathcal{P}_1$ <$\mathcal{CP}$> $\mathcal{P}_2$ $\mathcal{P}_3$ $\mathcal{J}_1$ $\mathcal{P}_4$ </$\mathcal{CP}$> $\mathcal{J}_2$ $\mathcal{P}_5$ </$\mathcal{CP}$>$\mathcal{D}_1$ <$\mathcal{CP}$>$\mathcal{P}_6$ $\mathcal{D}_2$ $\mathcal{P}_8$<$\mathcal{CP}$>$\mathcal{P}_9$ $\mathcal{D}_3$ $\mathcal{P}_{10}$ <$\mathcal{CP}$>$\mathcal{P}_{11}$ $\mathcal{D}_4$ <$\mathcal{CP}$>$\mathcal{P}_{12}$ $\mathcal{D}_5$ $\mathcal{P}_{14}$ $\mathcal{P}_{15}$ </$\mathcal{CP}$> $\mathcal{P}_{13}$ </$\mathcal{CP}$></$\mathcal{CP}$> </$\mathcal{CP}$> $\mathcal{P}_7$ </$\mathcal{CP}$>

Firstly we match the separating lines and we find the following matches:

(J_1,$\mathcal{J}_1$), (J_2,$\mathcal{J}_2$), (D_2,$\mathcal{D}_1$),(D_3,$\mathcal{D}_2$),(D_4,$\mathcal{D}_4$),(D_5,$\mathcal{D}_5$)

Thus, the following pairs of elements are constituted:

(J_1,$\mathcal{J}_1$)→ (P_4, $\mathcal{P}_2$) (P_5, $\mathcal{P}_3$), (P_6, $\mathcal{P}_4$)
(J_2,$\mathcal{J}_2$)→(<CP>P_1 D_1 P_2 P_3</CP>, $\mathcal{P}_1$), (P_7, $\mathcal{P}_5$)
(D_2,$\mathcal{D}_1$)→ (P_9,P_7), (P_8, $\mathcal{P}_6$)
(D_3,$\mathcal{D}_2$)→(P_{10},$\mathcal{P}_8$),(P_{11}, <$\mathcal{CP}$> $\mathcal{P}_9$ $\mathcal{D}_{10}$ $\mathcal{P}_{10}$ $\mathcal{P}_{11}$</$\mathcal{CP}$>)
(D_4,$\mathcal{D}_4$)→ (P_{12}, $\mathcal{P}_{12}$), (P_{13}, $\mathcal{P}_{13}$)
(D_5, $\mathcal{D}_5$)→ (P_{14}, $\mathcal{P}_{14}$), (P_{15},$\mathcal{P}_{15}$)

265

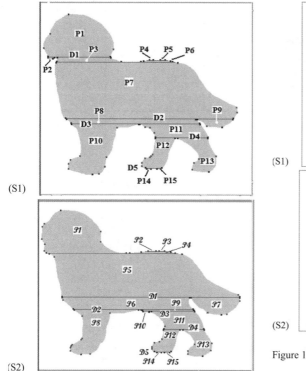

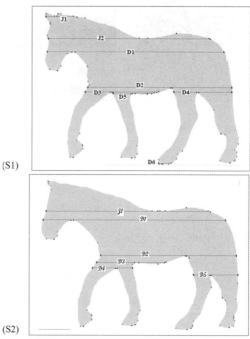

(S1)

(S2)

Figure 14. The result of XLWDOS decomposition.

Figure 16. Decomposition of the two shapes.

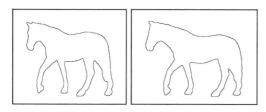

Figure 15. Two images of the same object (horse).

We can see that:

– P2 is noised part; we can then reduce the composed part $<CP>P_1\ D_1\ P_2\ P_3</CP>$ to a single part P_2' whose boundary is the union of the boundary of the three parts.

– $\mathcal{P}_{10}$ is noised part; we can also reduce the composed part $<\mathcal{CP}>\ \mathcal{P}_9\ \mathcal{D}_3\ \mathcal{P}_{10}\ \mathcal{P}_{11}</\mathcal{CP}>$ to a single part $\mathcal{P}_{11}'$ whose boundary is the union of the boundary of the three parts.

The two shapes (S1) and (S2) are coarsely matched and we can use for their indexing the coarse XLWDOS descriptor.

Figure 15 illustrates two different images of a horse but not successive. The computation of their XLWDOS descriptors and the matching of the separating lines gives the following results (see figure 16):

The pairs $(J2,\mathcal{J}1)$, $(D1,\mathcal{D}1)$, $(D2,\mathcal{D}2)$, $(D4,\mathcal{D}5)$ are matched, but the separating lines D3, D5 aren't matched respectively with $\mathcal{D}3$ and $\mathcal{D}4$.

We deduce then the two shapes are very different, and they must have different indexes.

5 CONCLUSION

We presented in this paper a new method for coarse silhouettes comparison that may be used for shape indexing. The silhouettes are described according to XLWDOS language. We have proposed a reducing technique allowing the descriptors comparison taking in account the noise. To validate the proposed approach, real image silhouettes have been used. The next step of our work is the full matching and the computation of the similarity measure between two coarsely matched descriptors.

REFERENCES

[1] D. Chetverikov. A Simple and Efficient Algorithm for Detection of High Curvature Points in Planar Curves, 10th International Conference, CAIP.

[2] C. M. Cyr and B. B. Kimia, A similarity-based aspect-graph approach to 3D object recognition, International Journal of Computer Vision, 57 (1), 5–22, 2004.

[3] D. Geiger, T. Liu, and R. V. Kohn, Representation and Self-Similarity of shapes, IEEE Transactions on Pattern Analysis and Machine Intelligence, Vol. 25, No 1, January 2003

[4] D. H. Kim, I. D. Yun and S. U. Lee, A new shape decomposition scheme for graph-based representation, Pattern Recognition, 38(2005).

[5] J.J. Koenderink and V. Doorn, the internal representation of solid shape with respect to vision, Biol. Cyber., 32, 1976.

[6] S. Larabi, S. Bouagar, F. M. Trespaderne and E. F. Lopez, LWDOS: Language for Writing Descriptors of Outline Shapes, In the LNCS proceeding of Scandinavian Conference on Image Analysis, June 29–July 02, Gotborg, Sweden, 2003.

[7] S. Larabi, Textual description of images, In the proceeding of International Conference CompImage (Computational Modelling of objects represented in images, Fundamentals and Applications), Coïmbra, Portugal, October 2006.

[8] B. Leibe and B. Schiele, Analyzing Appearance and Contour Based Methods for Object Categorization. In International Conference on Computer Vision and Pattern Recognition (CVPR'03), Madison, Wisconsin, June 2003.

[9] F. Mokhtarian, Silhouette-Based isolated object recognition through curvature scale space, IEEE Transactions on Pattern Analysis and Machine Intelligence, 17(5): 539–544, 1995.

[10] H. Murase and S. K. Nayer, Visual learning and recognition of 3-D objects from appearance, International Journal of Computer Vision, vol. 14, Issue 1, January 1995.

[11] I. Pitas and A. N. Venetsanopoulos, Morphological shape decomposition, IEEE Transactions on Pattern Analysis and Machine Intelligence, vol. 12(1) pp38–45, 1990.

[12] P. L. Rosin, Shape partitioning by convexity, British Machine Vision Conference 1999.

[13] C. Ruberto, Recognition of shapes by attributed skeletal graphs, Pattern Recognition, 37(2004):21–31.

[14] K. Siddiqi and B. B. Kimia, Parts of visual form computational aspects, IEEE Transactions on Pattern Analysis and Machine Intelligence, vol. 17(3) pp239–251, 1995

[15] L. Yu and R. Wang, Shape representation based on mathematical morphology, Pattern Recognition Letters, 2004.

[16] S.C. Zhu and A. L. Yuille, FORMS: A flexible object recognition and modeling system, Fifth International Conference on Computer Vision, June 20–23, M.I.T. Cambridge, 1995.

Brain volumetry estimation based on a suitable combination of statistic and morphological techniques

J.M. Molina
Computer Sciences Faculty, University of Ciego de Ávila, Ciego de Ávila, Cuba

A.J. Silva
Department of Electronics and Telecommunications, University of Aveiro, Aveiro, Portugal

M.A. Guevara
Computer Sciences Faculty, University of Ciego de Ávila, Ciego de Ávila, Cuba

J.P. Cunha
Department of Electronics and Telecommunications, University of Aveiro, Aveiro, Portugal

ABSTRACT: Magnetic Resonance Imaging has become one of the most important tools for anatomic and functional assessment of the complex brain entities. In neurology are present different indicators related with the brain volume measurements, which have a direct impact on several fields such as diagnosis, surgical planning, study of pathologies, disease preventing and tracking the evolution of diseases under (or not) medical treatments, etc. In this work we propose a new automatic brain volumetry estimation method based on the suitable combination of histogram analysis, optimal thresholding, prior geometric information and mathematical morphology techniques. To validate our method we compare our results with three different well established methods in the neuroscience community: Brain Extraction Tool, Brain Suite and Statistical Parametric Mapping. A dataset of 25 patient studies were evaluated concerning to precision, resolution as well as inter-examination features and statistically we demonstrated that our method present competitive results in relation to the others.

1 INTRODUCTION

Magnetic Resonance Imaging (MRI) has become one of the most important tools for anatomic and functional assessment of the complex brain entities. In neurology are present different indicators related with the brain volume measurements, which impacts to several fields such as diagnosis, surgical planning, study of pathologies, disease preventing and tracking the evolution of diseases under (or not) medical treatments, etc. The foremost significant neurodegenerative diseases (ND) are Multiple Sclerosis (MS), Frontotemporal Dementia, (Pick's Disease), Alzheimer, Amyotrophic Lateral Sclerosis (ALS or Lou Gehrig's Disease), Parkinson, Huntington, and Prion diseases, which ones debuts pathologically with brain volume variations in most of the cases.

In this work we propose a new method for automatic brain volumetry, which is based on the application of histogram analysis, optimal thresholding, prior geometric information and mathematical morphology techniques. We compare our results with three different well established methods in the neuroscience community for brain volumetry: Brain Extraction Tool (BET), Brain Suite (BS) and Statistical Parametric Mapping (SPM) respect to accuracy and precision. A dataset of 25 patient studies were evaluated with all methods, including our method that we named "Brain Volumetry Estimation" (BVE). Features under analysis were precision, resolution and inter-examinations. Despite the conceptual similarity between the reference patterns methods, important differences are revealed. Finally, it was demonstrated statistically that our method has a competitive results respect to the others.

This work is divided as follows: section 2, outlines aspects related with materials and methods. In section 3 we describe results obtained and section 4 conclusions are presented.

1.1 Study parameters

We used MRI images obtained on a 1.5T Sigma CV/I NV/i (GE). Volumetric data was acquired using a coronal 3D SPGR T1 sequence (TE: Min. full, TR: 20–30 ms, TI: 450 ms, 2 NEX) with a high spatial resolution (Matrix: 512×512, thickness: 1.5 mm) and optimized T1-contrast (FOV: 24). The volumetric assessment included the entire head, producing 120–126 partitions.

Discrete MRI patient volumes under study are defined as:

$$St = \{(x, y, z) \in N^3 : 0 < x < R, 0 < y < C, 0 < z < S\} \quad (1)$$

$$i = \{i_1, i_2, i_3 ... i_n\} \qquad n = 255 \quad (2)$$

Where, R and C mean row and column size respectively (512 pixels). S represents the numbers of slices between 120 and 126. The studies intensities are defined as one-dimensional array i with n intensity levels ($n = 256$).

1.2 Background correction

MRI capture is a very susceptible process that includes physical variables which can introduce noise and artifacts that affects directly the image quality of the image. The most common types of errors are divided in four types: electronic random noise (ERN) related to thermal effects, magnetic field inhomogeneities (MFI), biological tissue variations, and partial volume effects. Most of the failures and the excessive time consuming in the MRI brain volumetry estimation process occurs on preprocessing stage, due to algorithm complexity used to correct the MFI and ERN [1].

Several filter techniques [2] has been applied to remove the ERN in brain MRI. The T1-wighted MRI has a common statistical noise behavior, mostly white, additive Gaussian, particularly skewed Gaussian in some cases [3]. We successfully overcome this drawback with a correct background correction step. Background correction is achieved by mean of a suitable combination of mathematical morphology operators and prior geometric information, without to affect the foreground features. It also avoids a wrong answer in the image contrast enhancement process because excessively magnifying image noise effect [4, 5].

Background correction is carry out by mean of the image threshold and mathematic morphology. First is taken a background subvolume from the total volume by means of the (algorithm. 1, fig. 1) and the global maximum is computed. Next, this volume is thresholded to obtain a initial mask (eq. 4, fig. 2.a)) and afterwards is applied a morphological close operation (eq. 5, fig. 2.b), to avoid the hole containing inside the obtained mask. Finally the resulted mask (*cmask*) is multiplied element by element with the original study (eq. 6, fig. 2.c); this process allowed a fast and accurate background correction that not affects foreground features.

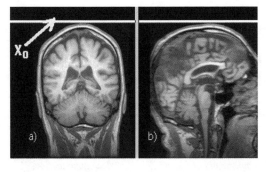

Figure 1.   Coronal and axial view of study with X_0 profile in the upper part of the head.

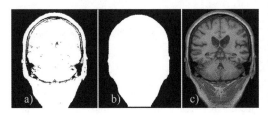

Figure 2.   a) Coronal slice after apply eq3 (*tmask* = 57) b) Morphological close of a). c) Resulting slice after multiply 1a) by 2c).

Algorithm 1: *Background correction.*
%: Heuristically was demonstrated that a threshold at level 100 is a good choice (skull_threshold = 100)
for i = 1 to R
　　　　　prof = St(i,∀y, ∀z)
　　　　　prof_sum = prof
　　　　　If St(i,∀y, ∀z) > prof_sum
　　　　　　　$X_0 = i$
　　　　　　　% X_0 means the higher value
　　　　　　　% out of the skull
　　　　　end
end

$$nMax = gmax \, (St(\, 1 < x < X_0 ; 1 < y < C; 1 < z < d)) \quad (3)$$

$$tmask_{(x,y,z)} = \begin{cases} 0 & \forall \, St_{(x,y,z)} < nMax \\ 1 & \forall \, St_{(x,y,z)} \geq nMax \end{cases} \quad (4)$$

$$cmask = (tmask \bullet se) \quad (5)$$

$$st\_bgc = mult(cmask_{(x,y,z)}, St_{(x,y,z)}) \quad (6)$$

where *gmax* returns the maximum global value of his argument. *tmask* is the resulting volumetric mask obtained by thresholding (fig. 2a), *cmask* is the volumetric mask obtained by the application of a morphological close operator ("•") (fig. 2.b) with an spherical structural element "se" with radius equal 11. "*mult*"

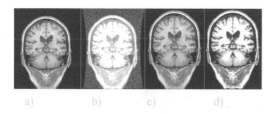

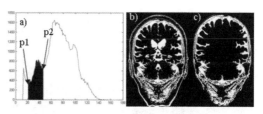

Figure 3. a) Original image, b) Result of applied CLAHE to (a), c) Background corrected of a, d) result after apply CLAHE to (c).

Figure 4. Intensities frequency histogram, a) dark zone in the graph corresponds to the intensities of the CSF. b) CSF thresholding. c) Image (b) filtered with mccs.

is the resulting image obtained after multiply *cmask* by the original image. st_bgc is the study volume with corrected background (fig. 2c).

The background correction is a fundamental step before contrast enhancing stage because it avoids wrong answer due to excessively magnifying image background noise effect (fig. 3b) [6, 7].

1.3 *Image enhancement*

Several image contrast enhancement techniques reported in the literature were evaluated [8–17] but in our work the better result was obtained with a combination of median filter and the contrast-limited adaptive histogram equalization (CLAHE) [16] (fig. 3d). Initially is applied a median filter with a $3 \times 3 \times 3$ window to smooth lightly the study image and eliminate impulses of short duration, which do not affect the features of objects of interest.

$$st\_clahe = CLAHE(st\_bgc) \qquad (7)$$

1.4 *Thresholding*

In our approach contrary to others methods we consider more convenient to start the thresholding process from the analysis of the CSF intensities spectrum instead the own brain. The CSF intensities have a thinner spectrum and more regular intensities than the brain what implies more stability in the system response.

To perform the automatic brain detection first is produced IFH of the MRI voxels to know intensities behaviors. [17–19], which convenient for optimal thresholding. Hereafter, local threshold values are selected based on the fact that most of the relatively low intensity voxels in the MRI belong to the CSF, due to the background noise has already been removed. Then a local thresholding is performed over the study volume to produce a 3-D CSF binary mask (eq. 8, fig. 4.a).

$$CSF_{(x,y,z)} = \begin{cases} 1 & \forall P_1 \leq St\_clahe_{(x,y,z)} < P_2 \\ 0 & otherwise \end{cases} \qquad (8)$$

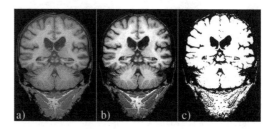

Figure 5. a) It is the first approach in the VBE, b) resulting image after to apply the clahe. c) Resulting binary volumetric mask after threshold 4b.

$$CSFEnh = mcss(CSF) \qquad (9)$$

Where *mcss* means the maximum foreground connected subset.

The selection of the largest foreground connected component (fig. 4.) is the previous step to brain auto-detection.

1.5 *Brain detection*

Morphological approaches [20–22] have already been widely used in nonlinear image filtering, noise suppression, smoothing, shape recognition and edge detector and segmentation in general. We apply mathematical morphology in two moments at this stage: first enhancing the binary mask to brain auto-detection (fig. 5.b) and second the automatic brain detection.

In fig. 5.c we have been selected the largest foreground connected component in foreground. Here is possible to observe how still reminds undesirables object that are not CSF.

$$fmask = (CSFEnh * se) \qquad (10)$$

$$fa = mult(fmask_{(x,y,z)}, St_{(x,y,z)}) \qquad (11)$$

Where, *fa* means a first approach to the brain segmentation, composed by *CSF* in the subarachnoid space and other structures (Fig. 4). "*se*" means a spherical structural element with radius equal 5.

$$facl = claheop(fa) \qquad (12)$$

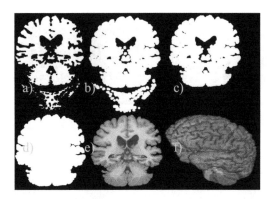

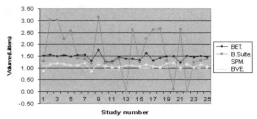

Figure 6a. Graphical methods representations.

Figure 6. a) Means the sbm erode (eq16), afterwards 5b) is a morphological close. C) Is the max foreground connected object, d) is morphological close, e) is the fill holes operator and f) is a coronal slice of the brain segmentation. g) 3-D view of the brain segmentation.

First approach to the VBS (fig. 5), observe how still reminds undesirable connected structures mostly compound for brain tissue with spurious structures such as the major venous sinuses, other blood vessels, dura, marrow, scalp, and soft tissue of the neck.

$$thv = max(facl)*77\% \qquad (13)$$

$$fbm = \begin{cases} 0 & \forall facl \;\; thv \leq facl_{(x,y,z)} \\ 1 & otherwise \end{cases} \qquad (14)$$

$$mer = fbm \ominus se1 \qquad (15)$$
$$mcl = mer \oplus se2 \qquad (16)$$
$$bobjt = mcss(mcl) \qquad (17)$$
$$bhl = fh(bobjt) \qquad (18)$$
$$fbm = (bhl \bullet se3) = (bhl \oplus se3) \ominus se3 \qquad (19)$$
$$ffbma = mult(fbm_{(x,y,z)} * St_{(x,y,z)}) \qquad (20)$$

$$Stvol = V_{VOX} \sum_{z=1}^{D} \sum_{y=1}^{C} \sum_{x=1}^{F} \begin{cases} 1 & if \;\; ffbm_{(x,y,z)} > 0 \\ 0 & otherwise \end{cases} \qquad (21)$$

Where: In (eq.12) *claopop* is an operator that performs the volume CLAHE; *facl* is the contrast enhanced volume. In *thv* is the cut value for threshold process, *max* represents an operator that computes the global maximum intensity in the volume. In (eq. 15) *mer* is the morphologic erosion of *fbm* with a spherical structural element "*se1*" of radius equal 3 (fig. 7a). *mcl* is the morphologic dilatation of mer with a spherical structural element "*se2*" of radius equal 3 (fig. 7b). *bobjt* is maximum connected subset in foreground (*mcl*), (fig. 7c). In (eq. 18) *bhl* is mask without holes after applying the *fh* operator,(fig. 7d). In (eq. 19) *fbm* is the morphologic close *bhl* with a spherical structural element "se2" of radius equal 3 (fig. 7e). In (eq. 20) *ffbma* is a skull strip of study (*st*),(fig. 7f). In (eq. 21) *Stvol* is the volume of the study, *Vvox* means the volume of a *voxel*.

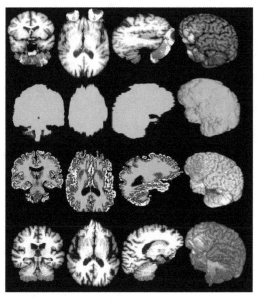

Figure 7. Segmentation and volumetry estimation resulted on the same patient volume. Up to bottom , first row are BET results, second row BS, third SPM and fourth BE.

2 RESULTS

To evaluate the robustness of our BE method we used an aleatory (not selected) dataset formed by 25 MRI studies of patients belonging to Coimbra region, Portugal. The MRI images were produced by the same scanner under the same protocol.

Our results were compared with three different well established methods in the neuroscience community for brain volumetry: Brain Extraction Tool (BET), Brain Suite (BS) and Statistical Parametric Mapping (SPM). For this purpose we computed the volumetric data of the all patient studies for each method in automatic way without user intervention. Regardless of the correspondence between the obtained results, important differences were revealed (see Table 1 and fig. 6). Table 1 shows the volumetric data of all studies (rows) performed by each method (columns), also

272

Table 1. Mean and deviation (SD) values for each method through the dataset.

Method	Mean	S.D.
BET	1,46	0,12
Bsuite	1,69	0,93
SPM	1,10	0,09
BVE	1,04	0,09

Table 3. Absolute value of the difference among the mean values of the brain volumes studies obtained by each method.

Meth.	BET	B.S.	SPM	BVE
BET	0,00	0,24	0,35	0,41
BS.	0,24	0,00	0,59	0,65
SPM	0,35	0,59	0,00	0,06
BVE	0,41	0,65	0,06	0,00

Table 2. Volumetry data results.

St #	BET	B.S.	SPM	BVE	Mean	Strd.Dev.
1	1.52	1.28	0.89	1.10	1.20	0.27
2	1.57	3.02	1.17	1.11	1.72	0.89
3	1.49	3.02	1.20	1.20	1.73	0.87
4	1.55	2.22	1.18	1.13	1.52	0.50
5	1.47	2.56	1.13	1.11	1.57	0.68
6	1.54	1.40	1.11	1.15	1.30	0.21
7	1.56	1.39	1.23	0.93	1.28	0.27
8	1.30	1.19	0.87	0.92	1.07	0.21
9	1.77	3.14	1.15	1.26	1.83	0.91
10	1.27	1.20	1.08	0.95	1.12	0.14
11	1.27	1.23	1.05	0.97	1.13	0.14
12	1.47	1.38	1.04	0.92	1.20	0.26
13	1.38	0.00	1.11	1.09	0.90	0.61
14	1.38	2.60	1.11	1.10	1.55	0.71
15	1.33	1.24	1.07	0.99	1.16	0.16
16	1.63	2.24	1.15	1.00	1.50	0.56
17	1.32	2.61	1.07	1.01	1.50	0.75
18	1.43	2.67	1.05	1.01	1.54	0.78
19	1.50	1.31	1.19	1.03	1.26	0.20
20	1.49	0.11	1.22	1.11	0.98	0.61
21	1.24	2.64	1.03	0.93	1.46	0.80
22	1.51	0.00	1.18	1.11	0.95	0.66
23	1.46	1.23	1.05	1.05	1.20	0.19
24	1.50	1.31	1.17	0.96	1.24	0.23
25	1.47	1.37	1.07	0.99	1.23	0.23

was include the computed mean and standard deviation values of each particular study. Mean and standard deviation demonstrated that the results obtained by our method are in the vicinity defined by the results of the other methods. Figure 6 delineate a graphical line chart representation of each method by patient studies, revealing that our approach performance has a very close behavior in relation with the others methods and still more with the SPM method.

Table 2 shows computed mean and standard deviation (SD) values for each method through the dataset. Here is possible to observe that SD in our approach (as in SPM) is closely clustered around the mean than BET and BS, which is a more correct answer to the dataset. Table 3 outline the absolute value of the difference between the mean values of the brain volumes studies obtained by each method, where again is demonstrated that the performance of our method is closer to the others.

Figure 7 shows segmentation results of slices of the same volume after apply different volumetry methods: a) BET, b) BS, c) SPM and d) our method., from left to right appears the coronal, axial, sagital slices and 3D volumes, where is possible to observe visually that our results are similar, but better that the results obtained by BET and BS methods.

3 CONCLUSIONS

In this paper we have outlined for novel ideal a new and simple fully automatic method for brain volumetry estimation inspired in two fundamental steps: first to produce a background noise correction algorithm process based on mathematical morphology and prior geometric information. Second very important step is the fact recognizes the cerebrospinal fluid rim in subarachnoid space as starting point to automatic brain detection. This approach represents a new suitable combination of morphological and statistical techniques, in which are including histogram analysis, optimal thresholding, prior geometric information and mathematical morphology. Statistically was demonstrated with a representative dataset of 25 patient studies, that our method (BVE) presents competitive results in relation with others well established methods in the neuroscience community. Finally a complete algorithm prototype was implemented and tested successfully in MATLAB version 7.3.

ACKNOWLEDGMENT

The authors wish to thank to IEETA institute of Aveiro University, Project CTS 2004/14 for financial support.

REFERENCES

1. Agath, C., Rajapakse, Member, IEEE, Jay, N. Giedd, and Judith, L., *Statistical Approach to Segmentation of Single-Channel Cerebral MR Images.* IEEE Transaction on Medical Imaging, 1997. **16**(2): p. 176–186.
2. Gerig, G., O. Kubler, R. Kikinis, and F. A. Jolesz, *Non-linear anisotropic filtering of MRI data.* IEEE Trans Med. Imag, 1992. **11**(2): p. 221–232.
3. DeCarli, C., J. Maisog, D.G.M. Murphy, D. Teichberg, S.I. Rapoport, and B. Horwitz, *Method for quantification*

of brain, ventricular and subarachnoid CSF volumes from MR images. J. Comput. Assist. Tomogr, 1992. **16**(2): p. 274–284.

4. Jin, Y., et al. *Contrast Enhancement by Multi-Scale Histogram Equalization*. 2001.

5. Pizer, S.M., et al. *Adaptive Histogram Equalization and Its Variations*. in *CVGIP*. 1987.

6. Zimmerman, J.B. and S.M., Pizer., *An evaluation of the effectiveness of adaptive histogram equalization for contrast enhancement*. IEEE Transactions on Medical Imaging, 1988. **7**(4): p. 304–312.

7. Lau, S.S.Y., *Global image enhancement using local information*. Electronics Letters, 1994. **30**(2): p. 122–123.

8. Kim, J.Y. and Kim, L.S., *An advanced contrast enhancement using partially overlapped sub-block histogram equalization*. IEEE Transactions on Consumer Electronics, 2001. **11**(4): p. 475–484.

9. Kim, T.K. and Paik, J.K., *Contrast enhancement system using spatially adaptive histogram equalization with temporal filtering*. IEEE Transactions on Consumer Electronics, 1998. **44**(1): p. 82–87.

10. Kim, Y.T., *Contrast enhancement using brightness preserving bi-histogram equalization*. IEEE Transactions on Consumer Electronics, 1997. **43**(1): p. 1–8.

11. Chen, S.-D. and Ramli, A.R., *Contrast enhancement using recursive mean-separate histogram equalization for scalable brightness preservation*. IEEE Transactions on Consumer Electronics 2003. **49**(4): p. 1301–1309.

12. Wang, Y. and Chen, Q., *Image enhancement based on equal area dualistic sub-image histogram equalization method*. IEEE Transactions on Consumer Electronics, 1999. **45**(1): p. 68–75.

13. Pisano, E.D. and Cole, E. B., B. M. H., *Image Processing Algorithms for Digital Mammography: A Pictorial Essay*. Radiographics, 2000. **20**: p. 1479–1491.

14. Chen, S.-D. and Ramli, A.R., *Minimum mean brightness error bi-histogram equalization in contrast enhancement*. IEEE Transactions on Consumer Electronics, 2003. **49**(4): p. 1310–1319.

15. Caselles, V., Lisani, J.L., *Shape preserving local histogram modification*. IEEE Transactions on Image Processing, 1999. **8**(2): p. 220–230.

16. Pisano, E., Zong, S., Hemminger, M., De Luca M., Johnsoton, R., Muller, K., Braeuning, M. and Pizer, S., *Contrast Limited Adaptive Histogram Equalization Image Processing to Improve the Detection of Simulated Spiculations in Dense Mammograms*. Digital Imaging, 1998. **11**(4): p. 193–200.

17. Yutaka Hata, S.K., Shoji Hirano, Hajime Kitagaki, and Etsuro Mori, *Automated Segmentation of Human Brain MR Images Aided by Fuzzy Information Granulation and Fuzzy Inference*. IEEE Transactions on Systems, Man and Cybernetics., 2000. **30**(3).

18. Marjin, E. Brummer, R., M., Merseareu., Robert L. Eisner., and Richard R. J. Lewine, *Automatic Detection of Brain Contours in MRI Data Sets*. IEEE Trans Med. Imag, 1993. **12**(2): p. 153–166.

19. Friedlinger, M., Männer, R., Schröder, J. and Schad, L.R. *Ultra-fast Automated Brain Volumetry Based on Bispectral MR Imaging Data*. in *S. (Tagungsort: nnb; Datum: 1999)*. 1999.

20. Serra, J. *Image Analysis and Mathematical Morphology*. 1982. New York: Academic.

21. Maragos, P., *Tutorial on advances in morphological image processing and analysis*. Optical Eng., 1987. **26**.

22. Takahashi, Y., Shio, A., Ishii, K., *Morphology based thresholding for character extraction*. IEICE Trans. Inform. Syst., 1997. **E76-D**(7): p. 1208–1215.

Computational Vision and Medical Image Processing – João Tavares & Natal Jorge (eds)
© 2008 Taylor & Francis Group, London, ISBN 978-0-415-45777-4

Carotid artery ultrasound image segmentation using graph cuts

Amr R. Abdel-Dayem & Mahmoud R. El-Sakka

Computer Science Department, The University of Western Ontario, London, Ontario, Canada

ABSTRACT: This paper proposes a scheme for segmenting carotid artery ultrasound images using graph cuts segmentation approach. A weighted graph is constructed, where each image pixel is represented by a graph node. Edge weights are assigned based on both region homogeneity and domain specific information. Then, the segmentation problem is solved by finding the minimum cut through the constructed graph edges. Experimental results demonstrated the efficiency of the proposed scheme in segmenting carotid artery ultrasound images.

1 INTRODUCTION

Vascular plaque, a consequence of atherosclerosis, results in an accumulation of lipids, cholesterol, smooth muscle cells, calcifications and other tissues within the arterial wall. It reduces the blood flow within the artery and may completely block it. As plaque layers build up, it can become either stable or unstable. Unstable plaque layers in a carotid artery can be a life-threatening condition. If a plaque ruptures, small solid components (emboli) from the plaque may drift with the blood stream into the brain. This may cause a stroke. Early detection of unstable plaque plays an important role in preventing serious strokes.

Currently, carotid angiography is the standard diagnostic technique to detect carotid artery stenosis and the plaque morphology on artery walls. This technique involves injecting patients with an X-ray dye. Then, the carotid artery is examined using X-ray imaging. However, carotid angiography is an invasive technique. It is uncomfortable for patients and has some risk factors, including allergic reaction to the injected dye, renal failure, the exposure to ionic radiation, as well as arterial puncture site complications, e.g., pseudoaneurysm and arteriovenous fistula formation.

Ultrasound imaging provides an attractive tool for carotid artery examination. The main drawback of ultrasound imaging is the poor quality of the produced images. It takes considerable effort from clinicians to extract significant information about carotid artery contours and the possible existence of plaque layers that may exist. Furthermore, manual extraction of carotid artery contours generates a result that is not reproducible. Hence, a computer aided diagnostic (CAD) technique for segmenting carotid artery contours is highly needed.

(Mao *et al.* 2000) proposed a scheme for extracting the carotid artery walls from ultrasound images. The scheme uses a deformable model to approximate the artery wall. However, the result accuracy depends, to a large extent, on the appropriate estimation of the initial contour. Furthermore, the deformable model takes a considerable amount of time to approach the equilibrium state. It is worth mentioning that the equilibrium state of a deformable model does not guarantee the optimal state or contour shape.

(Abolmaesumi *et al.* 2000) proposed a scheme for tracking the center and the walls of the carotid artery in real-time. The scheme uses an improved star algorithm with temporal and spatial Kalman filters. The major drawback of this scheme is the estimation of the weight factors used by Kalman filters. In the proposed scheme, these factors are estimated from the probability distribution function of the boundary points. In practice, this distribution is usually unknown.

(Da-chuan *et al.* 1999) proposed a method for automatic detection of intimal and adventitial layers of the common carotid artery wall in ultrasound images using a snake model. The proposed method modified the Cohen's snake (Cohen *et al.* 1991) by adding spatial criteria to obtain the contour with a global maximum cost function. The proposed snake model was compared with the ziplock snake model (Neuenschwander *et al.* 1997) and was found to give superior performance. However, the computational time for the proposed model was significantly high. It took a long amount of time for the snake to reach the optimum shape.

(Hamou *et al.* 2004) proposed a segmentation scheme for carotid artery ultrasound images. The scheme is based on Canny edge detector (Canny *et al.* 1986). The scheme requires three parameters. The first parameter is the standard deviation of the Gaussian smoothing kernel used to smooth the image before applying edge detection process. The second and the third parameters are upper and lower bound thresholds to mask out the insignificant edges from the generated edge map. The authors empirically tuned these parameters, based on their own database of images. This makes the proposed scheme cumbersome when used with images from different databases.

(Abdel-Dayem *et al.* 2004) proposed a scheme for carotid artery contour extraction. The scheme uses a uniform quantizer to cluster image pixels into three major classes. These classes approximate the area inside the artery, the artery wall and the surrounding tissues. A morphological edge extractor is used to extract the edges between these three classes. The system incorporates a pre-processing stage to enhance the image quality and to reduce the effect of the speckle noise in ultrasound images. A post-processing stage is used to enhance the extracted contours. This scheme can accurately outline the carotid artery walls. However, it cannot differentiate between relevant objects with small intensity variations within the artery tissues. Moreover, it is more sensitive to noise.

(Abdel-Dayem *et al.* 2005a) used the watershed segmentation scheme (Vincent *et al.* 1991) to segment the carotid artery ultrasound images. Watershed segmentation schemes usually produce over-segmented images. Hence, a region merging stage is used to merge neighbouring regions based on the difference on their average pixel intensity. A single global threshold is needed during the region merging process. If this threshold is properly tuned, the scheme produces accurate segmentation results.

(Abdel-Dayem *et al.* 2005b) integrated multi-resolution-analysis with watershed-based segmentation scheme (Abdel-Dayem *et al.* 2005a) to reduce the computational cost of the segmentation process and at the same time reduce the sensitivity of the results with respect to noise. In this scheme, the image is decomposed into a pyramid of images at different resolutions using wavelet transform. Then, the lowest resolution image is segmented using (Abdel-dayem *et al.* 2005a) segmentation scheme. Finally, the segmented image is projected back to produce the full resolution image.

(Abdel-Dayem *et al.* 2005c) proposed a scheme for segmenting carotid artery ultrasound images uses a fuzzy region growing technique. Starting from a user defined seed point within the artery, the scheme creates a fuzzy connectedness map for the image. Then, the fuzzy connectedness map is thresholded using an automatic threshold selection mechanism to segment the area inside the artery. The scheme is a region-based scheme. Hence, it is resilient to noise. It produces accurate contours. This gain can be contributed to the fuzzy nature of objects within ultrasound images. Moreover, it is insensitive to the seed point location, as long as it is located inside the artery. However, the calculation of the fuzzy connectedness map is a computationally expensive process.

The previous segmentation schemes are either region-based or edge-based. By integrating both region and edge information with domain specific information during the segmentation process, we hope that better segmentation results would be achieved. Graph-based segmentation schemes will be our vehicle to achieve such integration.

The rest of this paper is organized as follows. Section 2 describes the graph cuts segmentation approach. Section 3 describes the proposed scheme in details. Section 4 presents the results. Finally, Section 5 offers the conclusions of this paper.

2 THE PROPOSED SOLUTION

The proposed scheme consists of four major stages. These stages are: 1) pre-processing, 2) graph construction and minimum cut finding, 3) post-processing, and 4) boundary extraction. Figure 1 shows the block diagram of the proposed method. In the following subsections, a detailed description of each stage is introduced.

2.1 Pre-processing stage

Ultrasound images suffer from several drawbacks. One of these drawbacks is that ultrasound images have relatively low contrast. Another sever problem is the presence of random speckle noises, caused by the interference of the reflected ultrasound waves. These factors severely degrade any automated processing and analysis of the images. Hence, it is crucial to enhance the image quality prior to any further processing. In this stage we try to overcome these problems by performing two pre-processing steps. The first is a histogram equalization step (Gonzalez *et al.* 2002) to increase the dynamic range of the image gray levels. In the second step, the histogram equalized image is filtered using a median filter to reduce the amount of the speckle noise in the image. It was empirically found that a 3×3 median filter is suitable for the removal of most of the speckle noise without affecting the quality of the edges in the image.

2.2 Graph construction and minimum cut finding

In this stage, the pre-processed image is segmented using graph cuts-based segmentation approach. First,

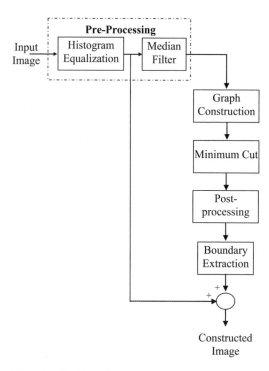

Figure 1. The block diagram of the proposed scheme.

a two terminal weighted graph is constructed for the image under consideration. Second, the weights of terminal-links and neighbour-links are set. Finally, the minimum cut through the graph is generated using (Boykov et al. 2004) algorithm. Graph nodes that remain connected to the source node represent object pixels, whereas nodes connected to the sink node represent background pixels.

The weight of a terminal-link is set to a value that reflects our confidence that the given pixel belongs to either the object or the background. Due to the nature of the carotid artery ultrasound images, the area inside the artery (which is the object of interest) is darker than the rest of the image. Hence, pixels with intensities less than a certain object threshold μ_{object} are connected by terminal links to both source and sink nodes with weights equal to one and zero, respectively. Meanwhile, pixels with intensities greater than a background threshold $\mu_{background}$ are connected by terminal links to the source and sink nodes with weights equal to zero and one, respectively. This way, the domain specific information is considered. All other nodes are connected to source and sink nodes with links that have certain weights. A node's weight is calculated by a non-negative decreasing function of the absolute differences between node's intensity and both of the object and the background thresholds, μ_{object}, $\mu_{background}$, respectively (this represents a region homogeneity

constraint). In the proposed scheme we used an exponential function to calculate the terminal-link weights. We can describe the assignment of the terminal-link weights by the following equations:

$$
W_{P,Source} = \begin{cases} 1 & \text{if } I_p \leq \mu_{object} \\ 0 & \text{if } I_p \geq \mu_{background} \\ e^{-\frac{|I_p - \mu_{object}|^2}{\alpha^2}} & \text{otherwise} \end{cases} \tag{1}
$$

and,

$$
W_{P,Sink} = \begin{cases} 0 & \text{if } I_p \leq \mu_{object} \\ 1 & \text{if } I_p \geq \mu_{background} \\ e^{-\frac{|I_p - \mu_{background}|^2}{\alpha^2}} & \text{otherwise} \end{cases} \tag{2}
$$

where, $W_{P,Source}$ and $W_{P,Sink}$ are the weights of the *terminal-link* connecting node P to the *source* and the *sink* nodes, respectively. I_P is the intensity of pixel P. μ_{object} and $\mu_{background}$ are the object and the background thresholds, respectively. α is a regulating term, used to control the rate of decay for the exponential weight function. This regulating term allows the weight function to cope up with the fuzzy (or defused) boundaries of the objects within the ultrasound images. We empirically set μ_{object} and $\mu_{background}$ to 10% of the lower and higher intensity ranges, respectively. Whereas α is set to 2% of the total intensity range. Hence, for 8-bit images, we set μ_{object} to 25, $\mu_{background}$ to 230 and α to 5. Note that, in ultrasound images, the object of interest appears darker than the background.

We used the 8-connectivity neighbourhood system to assign the *neighbour-link* weights. These weights are set based on local gradients according to the following equation:

$$
W_{P,Q} = e^{-\frac{|I_p - I_Q|^2}{\sigma^2}} \tag{3}
$$

where, $W_{P,Q}$ is the weight of the *neighbour-link* connecting nodes P and Q, I_p and I_Q are the intensities of pixels P and Q, respectively, and σ is the standard deviation of the gradient magnitude through the image. Note that *neighbour-link* weights represent the edge information.

By finding the minimum cut through the graph edges, a binary image, that separates the object from the background, is formed. The extracted object contains the area inside the carotid artery and some dark objects that usually exist in a given ultrasound image. The user can specify a seed point within the artery to extract the artery wall and neglect all other objects which are outside the region of interest.

2.3 Post-processing stage

The objective of this stage is to smooth the edges of the segmented area and to fill any gaps or holes that may present due to the presence of noise in ultrasound images. Hence, we used a morphological opening operation (Gonzalez *et al.* 2002), (Dargherty *et al.* 2003) with a rounded square structuring element of size W. The size of the structuring element can be adjusted, based on the maximum gap size in the segmented area, according to the following equation:

$$W = (h \times 2) + 1 \qquad (4)$$

where, W is the size of the structuring element and h is the maximum gap size that exists in the segmented image. We empirically found that generally, the maximum gap size does not exceed two pixels. Hence we used a 5×5 structuring element.

2.4 Boundary extraction stage

The objective of this stage is to extract the boundaries of the segmented regions. Various edge detection schemes can be used for this purpose (Gonzalez *et al.* 2002). In our system, we use a morphological-based contour extraction mechanism (Gonzalez *et al.* 2002), (Dargherty *et al.* 2003). First, the image produced by the previous stage is morphologically eroded using a 3×3 rounded square structuring element. Then, the eroded image is subtracted from the non-eroded image to obtain the boundary of the segmented region, which represents the artery wall. This operation can be described by the following equation:

$$\text{Boundary } (A) = A - (A \; \theta \; B) \qquad (5)$$

where, A is the post-processed image, B is the structuring element and θ is the erosion operator. Finally, the extracted contour is superimposed on the histogram equalized image to produce the final output of the proposed scheme.

3 RESULTS

The test images used in this research were obtained using an ultrasound acquisition system (Ultramark 9 HDI US machine and L10-5 linear array transducer). The output is digitized with a video frame grabber before saving it. The artery contours in these images were manually highlighted by an experienced

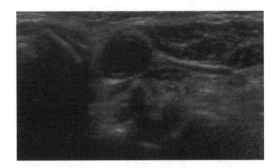

Figure 2. Original carotid artery ultrasound image.

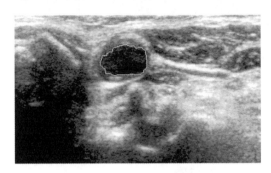

Figure 3. Final output using the proposed scheme.

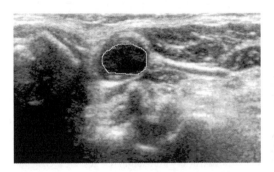

Figure 4. Gold standard image, where the artery contour is highlighted by an experienced clinician.

clinician. We consider these manually highlighted images as gold standards to validate our results.

Figure 2 shows one of the original ultrasound images that are used as an input to our system. Figure 3 shows the final output of the proposed scheme, where the extracted contour is superimposed on the histogram equalized image. Figure 4 shows the gold standard image (the artery contour is manually highlighted by an experienced clinician) for same test case, shown in Figure 2. The subjective comparison between Figure 3 and Figure 4 showed that the proposed scheme produces accurate artery contour. To demonstrate the contribution of the neighbour-link weights in the final

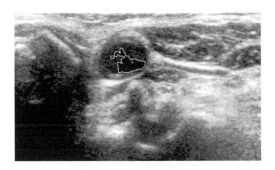

Figure 5. Final output using the proposed scheme when only the hard constraints are active.

segmentation results, Figure 5 shows the output of the proposed scheme when only the terminal-link weights are active (hard constraints). The comparison between Figure 3 and Figure 5 reveals that the incorporation of the neighbour-link weights improves the segmentation results, as they tend to hang neighbouring pixels together to produce meaningful objects.

4 CONCLUSIONS

In this paper, we proposed a novel scheme for highlighting the carotid artery contour in ultrasound images. The proposed scheme is based on using a graph cut approach to segment the image. The graph weights are formed in terms of both local intensity gradients (edge feature), as well as, penalty weights to assign every pixel to either object or background areas (region feature). Then, the image is segmented by finding the minimum cut through the graph.

Experimental results over a set of sample images showed that the proposed scheme provides a good estimation of carotid artery contours.

REFERENCES

Abdel-Dayem, A. & El-Sakka, M. 2004. A novel morphological-based carotid artery contour extraction. Proceedings of the Canadian Conference on Electrical and Computer Engineering 2: 1873–1876.

Abdel-Dayem, A., El-Sakka, M. & Fenster, A. 2005a. Watershed segmentation for carotid artery ultrasound images. Proceedings of the IEEE International Conference on Computer Systems and Applications: 131–138.

Abdel-Dayem, A. & El-Sakka, M. 2005b. Carotid Artery Contour Extraction from Ultrasound Images Using Multi-Resolution-Analysis and Watershed Segmentation Scheme. ICGST International Journal on Graphics, Vision and Image Processing 5(9): 1–10.

Abdel-Dayem, A. & El-Sakka, M. 2005c. Carotid Artery Ultrasound Image Segmentation Using Fuzzy Region Growing. Proceedings of the International Conference on Image Analysis and Recognition, ICIAR 2005, Springer-Verlag Berlin Heidelberg, LNCS 3656: 869–878.

Abolmaesumi, P., Sirouspour, M. & Salcudean, S. 2000. Real-time extraction of carotid artery contours from ultrasound images. Proceedings of the 13th IEEE Symposium on Computer-Based Medical Systems: 81–186.

Boykov, Y. & Kolmogorov, V. 2004. An Experimental Comparison of Min-Cut/Max-Flow Algorithms for Energy Minimization in Vision. IEEE transactions on Pattern Analysis and Machine Intelligence 26(9): 1124–1137.

Canny, J. 1986. Computational Approach To Edge Detection. IEEE Transactions on Pattern Analysis and Machine Intelligence 8(6): 679–698.

Cohen, L. 1991. On active contour models and balloons. Computer Vision, Graphics, and Image Processing: Image Understanding, 53(2): 211–218.

Dachuan, C., Schmidt-Trucksass, A., Kuo-Sheng, C., Sandrock, M., Qin, P. & Burkhardt, H. 1999. Automatic detection of the intimal and the adventitial layers of the common carotid artery wall in ultrasound B-mode images using snakes. Proceedings of the International Conference on Image Analysis and Processing: 452–457.

Dargherty, E. & Lotufo, R. 2003. Hands–on morphological image processing. The society of Photo-Optical Instrumentation Engineers.

Gonzalez, G. & Woods, E. 2002. Digital image processing. Second Edition, Prentice Hall.

Hamou, A. & El-Sakka, M. 2004. A novel segmentation technique for carotid ultrasound images. Proceedings of the IEEE International Conference on Acoustics, Speech and Signal Processing 3: 521–424.

Mao, F., Gill, J., Downey, D. & Fenster, A. 2000. Segmentation of carotid artery in ultrasound images. Proceedings of the 22nd IEEE Annual International Conference on Engineering in Medicine and Biology Society 3: 1734–1737.

Neuenschwander, W., Fua, P., Iverson, L., Szekely, G. & Kubler, O. 1997. Ziplock snake. International Journal of Computer Vision 25(3): 191–201.

Vincent, L. & Soille, P. 1991. Watersheds in digital spaces: an efficient algorithm based on immersion simulations. IEEE Transactions on Pattern Analysis and Machine Intelligence 13(6): 583–598.

Computational Vision and Medical Image Processing – João Tavares & Natal Jorge (eds)
© 2008 Taylor & Francis Group, London, ISBN 978-0-415-45777-4

A tissue relevance and meshing method for computing patient-specific anatomical models in endoscopic sinus surgery simulation

M.A. Audette
Innovation Center Computer Assisted Surgery (ICCAS), University of Leipzig, Leipzig, Germany

I. Hertel
Department of Otorhinolaryngology/Plastic Surgery, University of Leipzig, Leipzig, Germany

O. Burgert
Innovation Center Computer Assisted Surgery (ICCAS), University of Leipzig, Leipzig, Germany

G. Strauss
Department of Otorhinolaryngology/Plastic Surgery, University of Leipzig, Leipzig, Germany

ABSTRACT: This paper presents on-going work on a method for determining which subvolumes of a patient-specific tissue map, extracted from CT data of the head, are relevant to simulating endoscopic sinus surgery of that individual, and for decomposing these relevant tissues into triangles and tetrahedra whose mesh size is well controlled. The overall goal is to limit the complexity of the real-time biomechanical interaction while ensuring the clinical relevance of the simulation. Relevant tissues are determined as the union of the pathology present in the patient, of critical tissues deemed to be near the intended surgical path or pathology, and of bone and soft tissue near the intended path, pathology or critical tissues. The processing of tissues, prior to meshing, is based on the Fast Marching method applied under various guises, in a conditional manner that is related to tissue classes. The meshing is based on an adaptation of a meshing method of ours, which combines the Marching Cubes method and the discrete Simplex Mesh surface model to produce a topologically faithful surface mesh with well controlled edge and face size as a first stage, and Almost-regular Tetrahedralization of the same prescribed mesh size as a last stage.

1 INTRODUCTION

Functional endoscopic sinus surgery is an intervention whereby the surgeon inserts a surgical instrument in one of the patients nasal passages, in a manner guided by an endoscopic view, to get to as well as resect a pathology that occupies one or more sinuses in that patients cranium. This procedure also entails the avoidance of critical tissues, such as the optic nerve(s) and eyeball(s) near the pathology at hand. In order to train future ENT surgeons on this procedure, we are developing methods for computing patient-specific models from routine Computed Tomography data. These models express the tissues relevant to biomechanical interaction in terms of simple shapes, or elements, typically triangles and tetrahedra, in order to make its numerical simulation computationally tractable.

Currently, the anatomical models used in our simulation are obtained from manually labeled Tissue Maps, i.e.: a ENT surgeon manually identifies in CT relevant anatomical structures, namely the tumor, eyeball(s), optical nerve(s), and bone. Other, non-critical, soft tissues can be obtained by complementarity from CT, within a HU range of 500–1500. Together, these tissues comprise a Tissue Map, which is the starting point of the methods described herein. The objective of this paper is to propose a method for selecting the tissue subvolumes relevant to the procedure, and therefore to the simulation, and for tesselating these tissues so as to express them in terms of triangular and tetrahedral elements for subsequent real-time interaction. The method proposed here does not preclude a Tissue Map obtained through minimally supervised methods, the development of which is underway.

For the simulation, relevant tissues are determined as *the union of the pathology present in the patient, of critical tissues deemed to be near the planned surgical path or pathology, and of bone and soft tissue near these critical structures, or near a linearized, planned surgical path.* The relevant tissues must be tesselated and biomechanically modeled to allow a simulated interaction, where structures whose modeling is necessarily volumetric, such as the pathology to be resected, are decomposed into tetrahedral elements, while other structures are described in terms of triangular shell elements.

2 CONCEPTUAL MODEL

2.1 *Distance-based biomechanical relevance*

The first stage is the elaboration of a *Biomechanically Relevant Tissue Map*, which makes explicit those subvolumes of the Tissue Map that will be modeled for real-time interaction, with a view to excluding from consideration as much tissue as possible, for computational considerations, while keeping the simulation clinically meaningful. We make the assumption that non-pathological non-critical soft tissues that are relevant are relatively stiff and imbedded in bone, so that the restricting our attention to a subvolume and modeling the rest as rigid is close to constitutive reality.

The first step in producing the Biomechanically Relevant Tissue Map is a preprocessing of the CT data, as shown in Figure 1(a), to ensure that the image on which the Tissue Map is based is free of artefacts, namely blackout artefacts caused by tooth fillings. Currently, this is a simple cropping of the image volume: restricting the visible anatomical skin surface is viewed as preferable, from a simulation standpoint, to a slightly larger surface that is distorted by such an artefact. Furthermore, the subsequent surface meshing stage benefits from blurring binary volumes of relevant tissues, to make isosurface extraction well behaved, and this stage in turn benefits from zero-padding the Tissue Map at the boundaries coinciding with tissue, as shown in Figure 1(b). Critical tissues not relevant to the procedure, such as those in the orbital region opposite to the pathology, are not considered further.

In order to produce a final Biomechanically Relevant Tissue Volume, the processing of tissues, prior to meshing volumetric processing is achieved with a combination of two approaches: MINC-based 3D image processing software from the Montreal Neurological Institute (MNI 2007), and the well-known Fast Marching (FM) (Sethian 1999) method applied under various guises. First, *mincresample* and *mincreshape* are used to crop, zero-pad as well as orient the sampling of a volume along *xyz*. Second, the FM method is used to restrict tissues under consideration, based on a flexible notion of distance.

The Fast Marching method numerically simulates the Eikonal equation in an efficient manner:

$$F(x)\| \bigtriangledown T(x)\| = 1 \quad x \in \Omega \qquad (1)$$

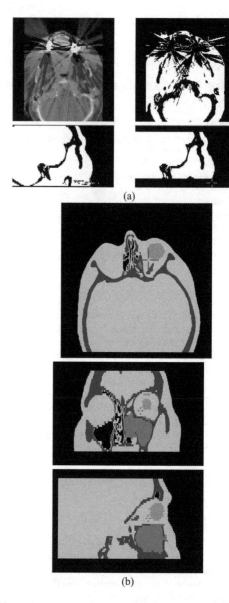

Figure 1. Preprocessing of CT data prior to determining biomechanically relevant tissues. (a) Elimination of blackout artefacts. Top: blackout artefact and distortive effect on soft tissue segmentation; bottom: sagittal illustration of cropping. (b) Tissue Map prior to determining biomechanically relevant subvolumes.

which in turn estimates the time T of arrival of a monotonic front $\partial \Omega$ over a domain Ω, whose speed of propagation is $F(x)$. When $F(x)$ is unity, this method computes a distance map. In our case, this front $\partial \Omega$ can be initialized (at $T = 0$) as either the boundary of a given tissue (identification of the boundary is based on a non-zero gradient of a binarization of the tissue class vs all other tissues), or a rasterization of a linearized path through user-provided points. Moreover, the method can deal with anisotropic spacing, as the upwind discretization of the gradient in equation 1 (Rouy and Tourin 1992),

$$F_{i,j,k} \begin{bmatrix} max(D_{i,j,k}^{-x}T, -D_{i,j,k}^{+x}, 0)^2 + \\ max(D_{i,j,k}^{-y}T, -D_{i,j,k}^{+y}, 0)^2 + \\ max(D_{i,j,k}^{-z}T, -D_{i,j,k}^{+z}, 0)^2 \end{bmatrix} = 1 \ , \qquad (2)$$

makes use of voxel spacing h_x, in computing $D_{i,j,k}^{-x} = \frac{T_{i,j,k} - T_{i-1,j,k}}{h_x}$, $D_{i,j,k}^{+x} = \frac{T_{i+1,j,k} - T_{i,j,k}}{h_x}$, as well as h_y and h_z respectively for the other difference operators.

Furthermore, we can place tissue-based constraints on the motion of the front. For example, if we are interested only in the subset of non-critical non-pathological ("generic") soft tissue that is within a distance ε_{csp} of critical structures and pathology, but not beyond bone tissue modeled as rigid, we binarize and combine ("OR") together critical structures and pathology, determine their joint boundary, and use this boundary to initialize a front enabled strictly on non-critical soft tissue. In other words, we constrain the front to halt at bone or air tissue, under the assumption that most bone tissue will be modeled as rigid, and that therefore soft tissue on the other side of it is unlikely to be biomechanically relevant. Similarly, if we are interested in soft tissue near the planned surgical path, in order to model it as elastic, in contrast with the rest of the soft-tissue boundary, we can consider a *linearized intended surgical path S* determined by user points $P = \{\mathbf{p}_i\}$: $S = \{E_i(\mathbf{p}_i, \mathbf{p}_{i+1})\}$, where E_i is an edge linking successive points. The distance of voxel $\mathbf{x}$, $d_{SP}(\mathbf{x})$, to this surgical path is given by:

$$d_{SP}(\mathbf{x}) = min_{E_i \in S} d_{edge}(\mathbf{x}, E_i) \ \text{where}$$

$$d_{edge}(\mathbf{x}, E_i) = min_{u+v=1} \|(u\mathbf{p}_i + v\mathbf{p}_{i+1} - \mathbf{x}\| \quad (3)$$

in practice, this distance is just a FM-based computation using an initial front coinciding with the voxels overlapping the edges E_i.

The end result is a refinement of the Tissue Map that discriminates between *biomechanically relevant* generic soft tissue, based on proximity to the planned surgical path or to critical or pathological tissue, and the *rest* of the generic soft tissue, which along with bone can be modeled as rigid in the simulation.

2.2 Tissue-guided surface and volume meshing of controlled mesh resolution

Once we have our Biomechanically Relevant Tissue Map, featuring critical, pathological, as well as generic elastic and rigid tissues, our next step is to express them

in terms of simple shapes, or elements, for biomechanical simulation. Our philosophy is to use as simple elements as required by the simulation. For example, the rigid tissue subvolume can be modeled exclusively as a collection of surface elements, since by definition no force will modify its interior. Likewise, we are not interested in how critical tissues, like the optic nerve, interact volumetrically with a surgical tool: we merely want the simulation to appropriately penalize a gesture that compromises its boundary, whereby surface meshing is once again sufficient. Other tissues, in particular the tumor that must be resected, and in our current implementation, elastic soft tissue in the nasal passage that is likely to interact with a surgical tool, are modeled as volumetric.

As a result of the requirement for flexibility of choosing between surface and volumetric elements, the meshing is a topologically faithful surface meshing stage followed by a tetrahedralization stage. The surface meshing stage combines Marching Cubes-based anatomical boundary identification and Simplex mesh-based decimation. The former provides our surface mesh with faithfulness to boundary and to topology, while the latter affords explicit control over mesh face area and edge size through the action of edge deletion and insertion. The Simplex Mesh is a discrete active surface model (Delingette 1999), characterized by each vertex being linked to each of 3 neighbours by an edge, and it is the dual of a triangulated surface. An image force can bind the model to the boundary of interest, even halting a model subject to a balloon inflation force, while other internal forces that improve face quality and continuity.

This Simplex model had been limited in the past by its topological equivalence with a sphere. However, to alleviate this issue, and in a manner comparable to our prior work (Audette, Delingette, Fuchs, Burgert, and Chinzei 2007), we instead initialize the simplex model with a dense surface mesh of high accuracy and topological fidelity, resulting from Marching Cubes (Lorensen and Cline 1987), based on the duality between a triangular surface and a 2-simplex mesh. We then decimate the simplex with topological T_1 and T_2 operators that act on each simplex face, where the former deletes an edge in order to fuse two faces into one, while the latter adds an edge to a face, subdividing it into two faces. Our strategy involves a penalty function computed for every face, as well as the sorting of all these faces by minheap or maxheap according to this penalty value, in order to "fix" the worst faces first. T_1 is applied to faces that are considerably smaller than desired and have a small (<6) number of edges, while we subdivide with T_2 those faces with a large number of edges (>7) and whose size is in excess of the desired resolution. The final simplex boundaries can be converted to triangulated surfaces by duality.

The last stage in our procedure partitions into tetrahedral elements those tissue volumes requiring volumetric interaction, on the basis of their triangulated mesh boundary. The volumetric meshing stage is a published technique, Almost-regular Tetrahedralization (Fuchs 2001), which automatically produces an optimal tetrahedralization from a given polygonal boundary, such as a triangulated surface. Optimality is defined as near-equal length of the tetrahedral edges, along with a sharing of each inner vertex by a nearly consistent number of edges and tetrahedra. This method features the optimal positioning of inner vertices, expressed as a minimization of a penalty functional, followed by a Delaunay tetrahedralization. We modify this technique by integrating into the penalty functional an edge length objective for each tetrahedron.

3 ANALYSIS

Our tissue-discriminating FM-based approach offers advantages over a combination of logical and traditional morphological operators acting only on soft tissue, which could include soft tissues within a zone of influence based on the morphological structuring element's size, but on the other side of a thin bone or air duct. Our approach is also more useful than a generic region of influence approach, being more restrictive of the volume considered elastic. Proceeding this way gives us the flexibility to eventually model curviplanar tissues, such as membranes in the nasal passage, as relatively sparse "thick shell" elements. The combination of Marching Cubes, Simplex and Almost-regular tetrahedralization provides topologically faithful meshing with controlled mesh size. The parameters, such as distances and edge lengths, are currently set arbitrarily, but will be optimized based on future simulations and clinical interaction.

4 EXAMPLES

Figure 2 illustrates the results of distance-based tissue relevance strategy. First, figure 2 (a) illustrates the consideration of a path-based distance to identify air voxels within $\varepsilon_{SP} = 10\,\text{mm}$ of the linearized path S.

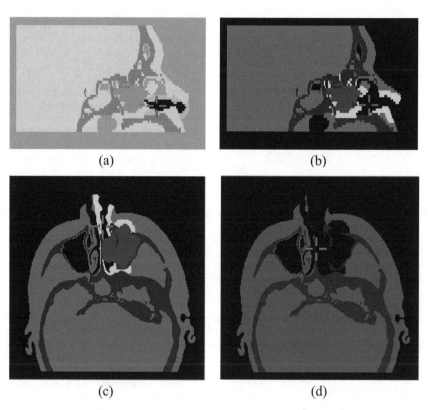

(a)

(b)

(c)

(d)

Figure 2. Use of Fast Marching method for restricting tissues. (a) Identification of air voxels near path (black: $d_{SP} < 5\,\text{mm}$; orange: $5\,\text{mm} < = d_{SP} < 10\,\text{mm}$; (b) sagittal and (c) axial plane images of identification of soft tissues near surgical path, critical tissues and tumor, based on air- and bone-inhibited distance; (d) soft tissue and bone modeled as rigid.

In figure 2(b) and (c), we also consider soft tissue voxels within $\varepsilon_{SP-air} = 4$ mm of this subset of air voxels in the nasal passage, as well as close to tumor (shown) and critical tissues ($\varepsilon_{csp} = 8$ mm in this case), with the subvolume of non-critical soft tissues modeled as elastic shown in light green. At all times, the algorithm processes information in 3D, not on a slice-by-slice basis, so it may appear that elastic soft tissue lies beyond bone in 2D in figure 2, when in fact there is an unobstructed path to that voxel in 3D.

Figures 3, 4 and 5 illustrate the results of our high quality, controlled continuity surface and volume

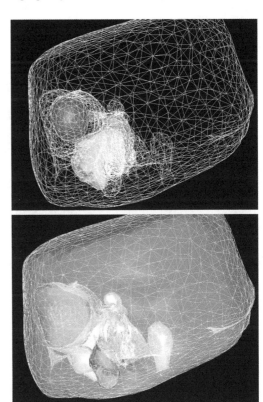

Figure 4. Illustration of surface meshing strategy: wireframe and solid-wireframe rendering of tissues. Top: critical and pathological tissue; bottom: elastic and rigid tissue.

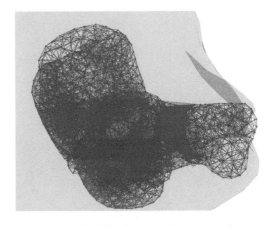

Figure 5. Tetrahedralization example: elastic and tumor tissue.

Figure 3. Illustration of surface meshing strategy: simplex meshing and dual triangulation. From top to bottom: rigid tissue simplex, dual rigid tissue triangulated surface, elastic tissue near pathological and critical tissue or near surgical path, in simplex (row 3) and triangulated (bottom) form.

285

meshing. The mesh size currently is chosen arbitrarily, 8 mm for rigid surface, and 4 mm for other surface and tetrahedral meshing. However, this mesh size will be adapted to simulation requirements, and it is conceivable for example that these tissues will be meshed more sparsely.

5 CONCLUSION

This paper presents a new method for selecting and meshing subvolumes of interest for ENT surgery simulation. It emphasizes the usefulness of surface models in particular, as Fast Marching and Simplex models are applied in tissue subvolume refinement and in meshing respectively. The tissue preprocessing uses the Fast Marching method under various guises to exploit distance to the surgical path and to highly relevant tissues, to exclude from consideration irrelevant soft tissues. The meshing also features Almost-regular Tetrahedralization, and both surface and volume meshing afford explicit control over mesh size.

REFERENCES

Audette, M. A., H. Delingette, A. Fuchs, O. Burgert, and K. Chinzei (January 2007). A topologically faithful, tissue-guided, spatially varying meshing strategy for computing patient-specific head models for endoscopic pituitary surgery simulation. *Journal of Computer Aided Surgery 12*(1), 43–52.

Delingette, H. (1999). General object reconstruction based on simplex meshes. *Internal Journal of Computer Vision 32*(2), 111–146.

Fuchs, A. (2001). Almost regular triangulations of trimmed nurbs-solids. *Engineering with Computers 17*, 55–65.

Lorensen, W. and H. Cline (July 1987). Marching cubes: a high resolution 3d surface construction algorithm. *Computer Graphics 21*(4), 163–170.

MNI (2007). Brain imaging software toolbox. Montreal Neurological Institute, *http://www.bic.mni.mcgill.ca/software/*.

Rouy, E. and A. Tourin (1992). A viscosity solutions approach to shape-from-shading. *SIAM J. Numer. Anal. 29*(3), 867–884.

Sethian, J. (1999). *Level Set Methods and Fast Marching Methods: Evolving Interfaces in Computational Geometry, Fluid Mechanics, Computer Vision, and Materials Science*. Cambridge University Press, 2nd ed.

Computational Vision and Medical Image Processing – João Tavares & Natal Jorge (eds)
© 2008 Taylor & Francis Group, London, ISBN 978-0-415-45777-4

ECG self-diagnosis system at P-R interval

Pedro R. Gomes
Universidade Lusiada, Famalicão, Portugal

Filomena O. Soares & J.H. Correia
Universidade do Minho, Guimarães, Portugal

ABSTRACT: This paper presents a data-acquisition system for monitoring the heart electrical activity in detail with self-diagnosis at P-R interval and application in atrial fibrillation (AF). The hardware acquisition system is based on a custom printed-circuit board with pre-amplifier, filters and interface for short term Ag/AgCl electrodes. The electrodes position follows the vector cardiogram distribution (left and right arms and left and right legs). A MATLAB algorithm was developed for self-diagnosis. A regular signal-acquisition board for personal computer is used as the control of the acquisition. AF is the most common cardiac arrhythmia in clinical practice and has in recent years received considerable research interest since the mechanisms causing its initiation and maintenance are not sufficiently well understood. The main goal of this work is to develop a low-cost electrocardiogram (ECG) system with help-diagnosis functionality.

1 INTRODUCTION

The electrical potentials generated by the heart appear throughout the body and on its surface. The potential differentials are determined by placing electrodes on the surface of the body and measuring the voltage between them.

The ECG records and compares the electrical activity detected in the different electrodes and the result obtained is called "lead". For example, the electrical events detected by the electrodes connected to the left and right arms called "lead I" (Neuman, 1992) (Figure 1).

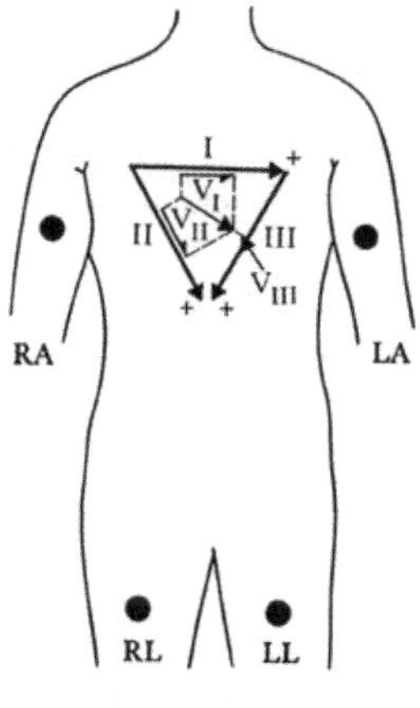

The ECG signal is characterized by six peaks and valleys labeled with different characters: P, Q, R, S, T and U (Stein, 2000) (Figure 2).

The conventional ECG systems record changes in electrical activity by drawing a trace on a moving paper strip (Berbari, 2000). However, these charts must be interpreted by the physicians.

Nowadays, ECG systems for self-diagnosis are very primitive, expensive and not reliable. Therefore, the proposed system is a step further in the ECG analysis, as it allows self-diagnosis in the complete ECG signal and it is a low-cost system.

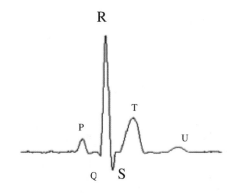

Figure 1.   Vectocardiogram leads.

Figure 2.   Form of ECG signal.

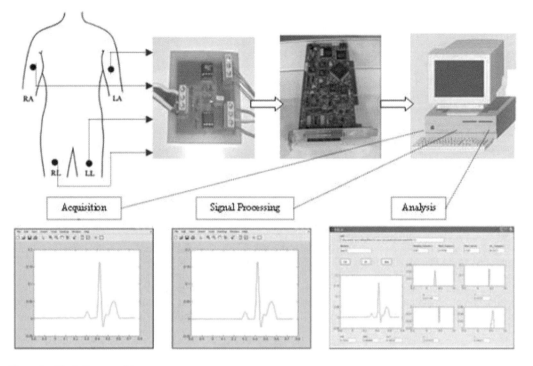

Figure 3.   Block diagram of the system.

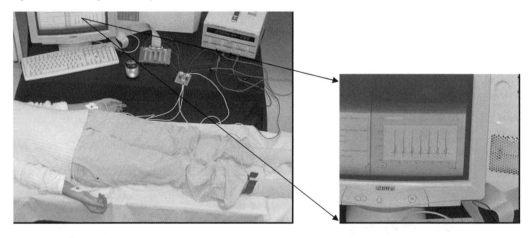

Figure 4.   Experimental set-up.

2   ECG DATA-ACQUISITION SYSTEM

Figure 3 shows a block diagram of the system. The hardware consists of a printed circuit board (PCB), including signal conditioning, filtering and amplification. The electrical activity of the heart is filtered, amplified and converted in a digital signal. A data-acquisition board, NI 6014, set in differential mode is used to control the acquisition hardware and A/D conversion.

The software for acquisition and monitoring was developed in MATLAB. Also, it was developed a database (DB) for storing the signals for future analysis. The signal acquisition can be repeated and recorded with a different identification. For each acquired signal, a new file is created in the DB with all obligatory parameters filled in.

The mandatory parameters are the date of acquisition, the name, age and sex of the patient. The data (voltage magnitude and time) is saved in two arrays.

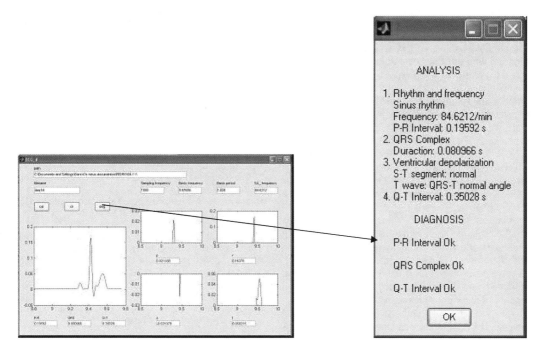

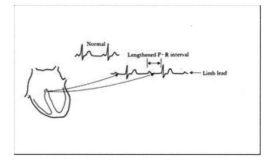

Figure 5. Print-screen of MATLAB software for ECG analysis.

Two Butterworth filters are used, for reducing the 50 Hz noise and for eliminating the DC component of the signal. After, the signal is analyzed in the frequency domain by the Fourier Fast Transformer (FFT). After the selection of the signal it is used a digital filter to achieve the peaks P, Q, R, S, and T without noise and their correspondent position in the array. The basic period is calculated to obtain the length or the array. Each point of the array is got using the basic MATLAB functions to set the positions of the relevant peaks. A "time-window" is setup in order to go through the whole signal acquired and to find possible differences and record them for a later analysis.

Figure 4 shows the experimental setup with the developed ECG data-acquisition PCB system, the patient with the electrodes and the signal displayed in the PC monitor.

3 ANALYSIS AND SELF–DIAGNOSIS

The software developed in MATLAB processes each acquired signal. This processing consists of verifying the period, the sampling frequency and calculating the SA node frequency (sinus rhythm). The software also calculates the intervals P-R, QRS and Q-T, presenting the peaks P, Q, R, S and T.

The user can select an interval, e.g., the P-R, and to obtain information about its contents (Figure 5).

Figure 6. Atrioventricular block.

The first goal of this work was the development of the application of the ECG self-diagnosis system to the P-R interval. During atrial fibrillation (AF), the P-R interval and hence, the ventricular rate, is commonly observed as irregular. However, ventricular activity of this sort is present not only in AF but also in a variety of other arrhythmias, including multifocal atrial tachycardia and atrial flutter with variable atrioventricular (AV) block (Figure 6). Abnormal cardiac rhythms in P-R interval can be detected using these developed algorithms, as the expansion of the atria. Also, AV block can be detected. As an example, in first-degree hearth block, all atrial impulses reach the ventricles, but the P-R interval is abnormally because an increase in transmission time through the affected region (Figure 6) (Petrutiu et al., 2006).

4 CONCLUSIONS

The main goal of this work is to develop an electrocardiogram with self-diagnosis functionality.

An experimental setup was designed. The hardware acquisition system was developed in order to pre-amplify and filter the ECG signal. Also, it includes an interface for the electrodes. A MATLAB-based software was developed for ECG signal acquisition and processing. The algorithms used determine the peaks P, Q, R, S and T and calculate the correspondent intervals P-R, QRS and Q-T. The user can select an interval and obtain information about the contents. The first tests were done for cardiac problems related with the P-R interval.

As future work, the extraction of atrial signal during atrial fibrillation requires nonlinear signal processing since atrial and ventricular activities overlap spectrally and, therefore, cannot be separated by linear filtering. Average beat subtraction is the most widespread, best validated, and clinically applied technique for atrial signal extraction and relies on the fact that AF is uncoupled to the ventricular activity. Principal component analysis (PCA) and independent component analysis (ICA) are other approaches that can be used to separate atrial fibrillation from ventricular activity (Bollman & Lombardi, 2006).

This low-cost ECG system shows good results and can easily be used as rapid and reliable point-of-care diagnosis equipment.

REFERENCES

Berbari, E. J., 2000. Principles of Electrocardiography, The Biomedical Engineering Handbook: Second Edition, Boca Raton: CRC Press LLC.

Bollmann, A. & Lombardi, F., 2006. Electrocardiology of atrial fibrillation, IEEE EMBC Magazine, pp. 15–23, December.

Neuman, M. R., 1992. Medical Instrumentation, Application and design, Webster Editor, Second Edition, USA.

Petrutiu, S., Ng, J., Nijm, G. M., Al-Angari, H., Swiryn, S., & Sahakian, A.V., 2006. Atrial fibrillation and waveform characterization, IEEE EMBC Magazine, pp. 23–24, December.

Stein, E., 2000. Rapid Analysis of Electrocardiograms, Lippincott Williams & Williams Inc., USA.

Computational Vision and Medical Image Processing – João Tavares & Natal Jorge (eds)
© 2008 Taylor & Francis Group, London, ISBN 978-0-415-45777-4

Lung and chest wall structures segmentation in CT images

J.H.S. Felix, P.C. Cortez, M.A. Holanda, V.H.C. Albuquerque, D.F. Colaço & A.R. Alexandria
Federal University of Ceará, Fortaleza, Ceará, Brazil.

ABSTRACT: Computed Tomography (CT) of thorax is nowadays the most accurate image technique for diagnosis of the majority of lung and chest diseases. Despite this fact there are still limitations of CT in diagnosing and specially quantifying lung diseases such as Chronic Obstructive Pulmonary Disease (COPD). The present paper presents a method of automatic classification capable to segment the lungs and the chest wall elements in patients with COPD in supine and prone positions. The technique of binary mathematical morphology was used to segment the lungs and the thoracic cavity using region growing following for negative this image. The lungs, the thoracic cavity and the pulmonary vessels were all successfully segmented with the application of the mathematical morphology and region growing. This method of processing CT images may be a promising tool for qualitative and quantitative studies of chest CT images.

1 INTRODUCTION

Computed Tomography (CT), developed by Hounsfield, allowed him to win the Nobel Prize in 1972. It was initially applied to measure the radiographic density precisely. Nowadays, the major application of pulmonary CT is to diagnose lung and chest diseases (Kak & Slaney 1999, Hounsfield 1973 and Drummond 1998).

Computed Tomography constitutes an excellent tool for quantifying the radiographic density of the lung parenchyma with and without pathologies. Computer pre-processing techniques have being used to turn the human visual analysis of the alveoli more accurate. However, exists a lack of equipments with an appropriated software to analyze the images precisely (Hounsfield 1973 and Drummond 1998).

Computer Vision uses several techniques of Digital Image Processing with the purpose of segmenting, recognizing and identifying details of the region of interest (ROI) of any image. These techniques are used to detect damaged regions of lungs and to measure pulmonary volume, getting a diagnosis more accurate of the case and, consequently, a medical treatment more adjusted (Celli *et al.* 2004).

The main purpose of this paper is to demonstrate a method based on computer programs to segment the aerated lung, the lung tissue and its blood vessels and thoracic cavity both in volunteers with healthy lungs and in patients with Chronic Obstructive Pulmonary Disease (COPD) based on region growing and mathematical morphology techniques.

Developing environment used for such segmentations is the C/C++, in Windows XP platform, regarding to the easiness that this tool offers in the Object-Oriented and Visual Programming.

2 MATERIALS AND METHODS

2.1 Region Growing

Region Growing (RG) is a technique applied in two-dimension digital images. RG carries the grouping of pixels sub-groups or groups in a ROI based on gray scale variation.

Region growing is an applied technique in 2D images to join ROI. Region clustering is determined by the choice of a seed in each analyzed region as, for example, a pixel which has similar characteristics to original image. These characteristics are gotten, normally, through intensity parameters as local average, local variance and others (Jan 2006 and Ballard & Brown 1982).

2.1.1 Region growing algorithm

Initially an Image Region (IR) is determined as the area of interest in which it is necessary to analyze. Then this image can be subdivided in *n* regions IR_1, IR_2,..., IR_n, (Gonzalez & Woods 2002 and Ballard & Brown 1982). The algorithm for region growing is presented briefly below:

$$IR = \bigcup_{i=1}^{n} \tag{1}$$

IR_i is a region connect for $i = 1, 2, ..., n$. $\quad$ (2)

$IR_i \cap IR_j = \varnothing$ for all $i = 1, 2, ...,n$. $\quad$ (3)

$P(IR_i) = True$ for $i = 1, 2, ..., n$. $\quad$ (4)

$P(IR_i \cup IR_j) = False$ for $i \neq j$. $\quad$ (5)

Equation (1) indicates that the segmentation must be complete. Equation (2) points must previously be connected with some definite point. Applying Equation (3) is necessary to split the analyzed region. Region must satisfy the condition in Equation (4) to be segmented and Equation (5) indicates which region IR_i and IR_j are different in sense of predicate P.

In this paper the technique of local growing is used, in which the closest neighbor is grouped in a region formed by pixels with level similar to the one of the seed.

2.2 Mathematical morphology

Researches about mathematical morphology had started in France in the middle of 1964 in the Paris School of Mines, being its precursors Matheron 1975 and Serra 1982. Such technique had the objective to explore the geometric forms in an image.

Basic morphologic operators such as dilation, erosion, anti-dilation and anti-erosion constitute the base of mathematical morphology, therefore, studies of operator decomposition between reticulates are carried (Banon & Barrera 1998).

Morphologic transformation consists in applying a structuring element of varied size on the analyzed image. Such element is a mask that can assume different geometric forms as, for example, circle, square, rectangle, triangle, hexagon, cross point and other polygons (Facon 1996).

2.2.1 Erosion and dilation
Considering sets A and B, being the set A an image in analysis and B the structuring element. Binary erosion for the not empty sets A and B (A⊖B) in Z^2 is defined mathematically as

$$A \ominus B = \{z \mid (B)_z \subseteq A\}, \quad (6)$$

being characterized for the analyzed image reduction (Gonzalez & Woods 2002), and for the binary dilation to the sets A and B

$$A \oplus B = \{z \mid [(\hat{B})_z \cap A] \subseteq A\} \quad (7)$$

where the results are an expansion of the analyzed image.

2.2.2 Opening and closing
Binary opening of a set A by B is the erosion of A by B, followed by dilation of the result by B, being defined as

$$A \circ B = (A \ominus B) \oplus B. \quad (8)$$

This causes the effect of contour smoothing, eliminating small lumps (Gonzalez & Woods 2002).

Inverting the order of the operations which define the opening, in other words, dilation of set A by B, followed by the erosion of the result by B, produces the closing operation, expressed by

$$A \bullet B = (A \oplus B) \ominus B. \quad (9)$$

This results in fulfilling the small holes and/or gaps in the contour of the analyzed image (Gonzalez & Woods 2002).

2.2.3 Mathematical morphology algorithm
Initially the automatic threshold of Fisher is used to binarize the original image of a CT, followed by opening, dilation and erosion operations. Carried through these operations, an intersection is applied between the binarized and eroded image. Finally, a subtraction of the binarized image with the intersection image is done to segment the pulmonary veins.

3 RESULTS

3.1 Region growing results

Region growing is used to segment lungs in supine and prone positions, through the application of the grouping technique with local regions.

Automatic segmentation of lungs easy the pulmonary geometric analysis, lung area and volume, therefore patient with pathology present pulmonary growing. The lung parenchyma segmentation makes the visualization of the lungs easier and may improve the visual detection of parenchyma alterations.

Tests are carried with one seed for each lung, point of origin in analyzed region growing which carries the growing in a band of intensity variation of pixel with 20% tolerance.

In Figure 1a and 1b are shown original image supine end prone positions, respectively. In Figure 1c and 1d are presented the negative images of the pulmonary cavity segmentation in supine and prone position, respectively.

In Figure 1e and 1f, the negative of results of the lungs segmentation are shown (gray region), which is used to calculate area and volume of the lungs. These results can be inferred on some disease that affects the lungs.

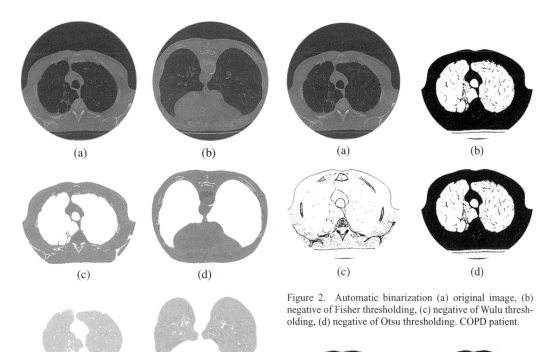

(a)   (b)

(c)   (d)

(e)   (f)

Figure 1. Lung segmentation using region growing (a) original image in supine position, (b) original image in prone position, (c) negative image of the pulmonary cavity in supine position, (d) negative image of the pulmonary cavity in prone position, (e) negative of the segmented image of the aerated lung, (f) negative of the segmented image of the aerated lung in prone position. Patient with COPD.

(a)   (b)

(c)   (d)

Figure 2. Automatic binarization (a) original image, (b) negative of Fisher thresholding, (c) negative of Wulu thresholding, (d) negative of Otsu thresholding. COPD patient.

(a)   (b)

Figure 3. Mathematical morphologic operation (a) negative opening and (b) negative dilation followed by erosion. COPD patient.

3.2 *Mathematical morphology results*

Tests realized in this paper were performed with the binary morphological opening operators with rectangular structuring element with matrix 5×7 and binary dilation and erosion with rectangular structuring element of the 8×10.

It was obtained better results using rectangular structuring element than square and circular ones, because it permits to detect veins structures with more precision than the other elements.

The obtained results with mathematical morphologic application are similar to obtained by human visualization and are efficient in the lungs segmentation and pulmonary veins, separately.

In Figure 2a is shown the original image of lungs in supine position, and Figures 2b, 2c, 2d, 2f are presented the binarized image with automatic threshold of Fisher (Cocquerez & Phillip 1995), WuLu (Wu *et al.* 1998) and Otsu (Otsu 1979), respectively. To apply such threshold is necessary to select the best among

then, and afterwards applied in the binarization of the original image.

Automatic threshold of Fisher and Otsu present better results, when binarizing the original image in supine position than other algorithms, since they preserve significant lung details after binarization.

In Figure 3a is shown the result of the opening application on the image presented on Figure 2b, occuring the elimination of the pulmonary veins which results in fissures assigned by point A. In Figure 3b is presented the result of a dilation application followed by erosion, which eliminate those fissures.

Figure 4a represents the result of intersection between the binarization and erosion of the Figures 2b and 3b, respectively. In resulting image is present the pulmonary veins segmentation, as it is shown in the Figure 4b.

The same automatic binarization algorithms and same morphologic operators applications applied in

293

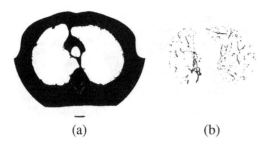

(a)   (b)

Figure 4. Mathematical morphologic segmentation results (a) negative intersection and (b) negative subtraction (segmentation of lung tissue, including blood vessels in COPD patient).

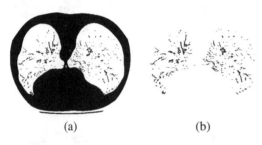

(a)   (b)

Figure 5. Pulmonary veins segmentation (a) negative of Fisher threshold in prone position, (b) negative of pulmonary veins segmented. COPD patient.

supine position were used in pulmonary veins segmentation on prone position.

Figure 5a is shown the original image in prone position and Figure 5b the result of pulmonary veins automatic segmentation in this position.

Segmentation in prone position presents more satisfactory results than the ones obtained in supine position. Thus, segmentation of pulmonary structures and veins present trustworthy results.

4 CONCLUSIONS

The use of region growing and mathematical morphology methods, implemented by on computer programs, makes possible to segment aerated lung, lung tissue and blood vessels, as well as thoracic cavity in patient with COPD.

Region growing reveals to be an efficient tool on pulmonary cavity and lungs segmentation, keeping the details of the pulmonary structures and veins, obtaining accurate results.

Lung tissue including blood vessels segmentation based on binary morphologic operations is an efficient tool, which segment correcting the pulmonary structure and veins in supine position and a satisfactory segmentation in prone position, where occurs small losses in the structures and veins.

ACKNOWLEDGMENTS

The authors would like to thank FUNCAP for their financial support, Department of Teleinformatic Engineering and Laboratory of Teleinformatic (LATIN/LESC) and the Image and Pulmonology Departments of the Walter Cantídio University Hospital.

REFERENCES

Ballard, D.H. & Brown, C.M. 1982. Computer Vision. New Jersey: Pretice-Hall.

Banon, G. J. F. & Barrera, J. (2) 1998. Bases da Morfologia Matemática para a análise de imagens binárias. São José dos Campos: INPE.

Celli, B. & MacNee, W. & committee members. 2004. Standards for the diagnosis and treatment of patients with copd: a summary of the ats/ers position paper. E.R.J. 23: 932–946.

Cocquerez, J.P. & Phillip, S. 1995. Analyse d'images: filtrage et segmentation. Masson. France: 240–245.

Drummond, G.B. 1998. Computed Tomography and pulmonary measurements, British J. of Anaesthesia 80: 665–671.

Facon, J. 1996. Morfologia Matemática: Teorias e Exemplos. Curitiba: Editora Universitria Champagnat da PUC-Paraná.

Gonzalez, R.C. & Woods, R. (2) 2002. Digital Image Processing. EUA: Prentice Hall.

Hounsfield, G.N. 1973. Computerized transverse axial scanning (Tomography): Part 1.description of system. British Journal of Radiology 46: 1016–1022.

Jan, J. 2006. Medical Image Processing, Reconstruction and Restoration: Concepts and Methods, T. F. Group, LLC.

Kak, A.C. & Slaney, M. 1999. Principles of Computerized Tomographic Imaging. IEE Press.

Madani, A., Keyzer, C. & Gevenois, P.A. 2001. Quantitative computed Tomography assessment of lung structure and function in pulmonary emphysema, E. R. J. 18(8): 720–730.

Matheron, G. 1975. Random sets and integral eometry. New York: Wiley.

Otsu, N. 1979. A threshold selection method from gray-level histograms. Systems, Man and Cybernetics 9(1). IEEE Trans: 62–66.

Serra, J. 1982. Image analysis and mathematical morphology. London: Academic Press.

Wu, L., Ma, S. & Lu, H. 1998. An effective entropic thresholding for ultrasonic images. Fourteenth International Conference on Pattern Recognition: 1552–1554.

Computational Vision and Medical Image Processing – João Tavares & Natal Jorge (eds)
© 2008 Taylor & Francis Group, London, ISBN 978-0-415-45777-4

Fuzziness measure approach to automatic histogram threshold

Nuno Vieira Lopes
ESTG-IPL, Leiria, Portugal

Humberto Bustince
UPNa, Pamplona, Spain

Vítor Filipe & Pedro Melo Pinto
CETAV-UTAD, Vila Real, Portugal

ABSTRACT: In this paper, an automatic histogram threshold approach based on a fuzziness measure is presented. This work is an improvement of an existing method. Using fuzzy logic concepts, the problems to find a minima of a criterion function are avoided. Similarity between gray levels is the key to find an optimal threshold. Two initial regions of gray levels, located at the boundaries of the histogram, are defined. Then, using an index of fuzziness, a similarity process is started to find the threshold point. A significant contrast between objects and background is assumed. Previous histogram equalization is used in small contrast images. No pior knowledge of the image is required.

1 INTRODUCTION

Image segmentation represents an important role in computer vision and image processing applications. Segmentation of nontrivial images is one of the most difficult tasks in image processing. Segmentation accuracy determines the eventual success or failure of computerized analysis procedures. Segmentation of an image entails the division or separation of the image into regions of similar attribute. For a monochrome image, the most basic attribute for segmentation is image luminance amplitude (Pratt 2001). This situation can be described as follows:

If I is the set of all pixels and $P()$ is a homogeneity predicate defined on groups of connected pixels, then segmentation is a partition of the set I into connected subsets or image regions $(R_1, R_2, \ldots, R_n)$, such that

$$\bigcup_{i=1}^{n} R_i = I, \text{ where } R_i \cap R_j = \emptyset, \forall i \neq j. \quad (1)$$

The uniformity predicate $P(S_i)$ is true for all regions, and $P(R_i \cup R_j)$ is false, when $i \neq j$ and R_i, R_j are neighbors.

Segmentation based on gray level histogram threshold is a method to divide an image containing two regions of interest: object and background. In fact, applying this threshold value to the whole image, pixels whose gray level is under this value are assign to a region and the rest to the other. Histograms of images with two distinct regions are formed by two peaks separated by a deep valley called bi-modal histograms. In such case, the threshold value must be located on the valley region. When the image histogram doesn't denote a clear separation, several threshold techniques based on minimizing a criterion function might perform poorly. Fuzzy set theory is a powerful tool to deal with human perception and, on the other hand, overcomes the problem of how to find a criterion function minima. Several comparative studies are found in (Sezgin and Sankur 2004; Tizoosh 2005).

In Section 2 a general description of the fuzzy set theory and index of fuzziness measuring is done. The existing method is described in Section 3. The proposed method is presented in Section 4. Limitations and detected problems of the existing method are also discussed. Section 5 shows comparative results to illustrate the effectiveness of the proposed approach and are presented the final conclusions.

2 GENERAL DEFINITIONS

2.1 Basics of fuzzy set theory

Contrarily to crispy sets, fuzzy set theory assigns a membership grade to all elements among the universe of discourse according their potential to fit in some class. The membership grade can be expressed by a mathematical function $\mu_A(x_i)$ that assigns, to each element in the set, a membership grade between 0 and 1. Let X be the universe of discourse and x_i an element of X. A fuzzy set A in X is defined as

$$A = \{(x_i, \mu(x_i))\}, x_i \in X \qquad (2)$$

The S-function is used to modeling the membership grades. This type of function is defined as

$$\mu_{AS}(x) = S(x; a, b, c)$$

$$= \begin{cases} 0, & x \leq a \\ 2\{(x-a)/(c-a)\}^2, & a \leq x \leq b \\ 1 - 2\{(x-c)/(c-a)\}^2, & b \leq x \leq c \\ 1, & x \geq c. \end{cases}$$
$$(3)$$

With parameters a and c the S-function can be controlled. Parameter b is called the crossover point where $\mu_{AS}(b) = 0.5$. The Z-function is also be used with an expression obtained from S-function as follows

$$\mu_{AZ}(x) = Z(x; a, b, c) = 1 - S(x; a, b, c) \qquad (4)$$

2.2 Measures of fuzziness

A reasonable approach to estimate the average ambiguity in fuzzy sets is measuring its fuzziness (Pal and Bezdek 1994). The fuzziness of a crisp set should be zero, as there is no ambiguity about whether an element belongs to the set or not. If $\mu_A(x) = 0.5, \forall x$, the set is maximally ambiguous and its fuzziness should be maximum. Grades of membership near 0 or 1 indicate lower fuzziness, as the ambiguity decreases. Kauffman introduced an index of fuzziness comparing a fuzzy set with its nearest crispy set. A fuzzy set A^* is called the crisp set of A if the following conditions are satisfied

$$\mu_{A^*}(x) = \begin{cases} 0, & \text{if } \mu_A(x) < 0.5 \\ 1, & \text{if } \mu_A(x) \geq 0.5. \end{cases} \qquad (5)$$

This index is calculated by measuring the normalized distance between A and A^* defined as

$$\psi_k(A) = \frac{2}{n^{1/k}} \left[\sum_{i=1}^{n} |\mu_A(x_i) - \mu_{A^*}(x_i)|^k \right]^{1/k} \qquad (6)$$

where n is the number of elements in A, $k \in [1, \infty]$. Depending if $k = 1$ or 2, the index of fuzziness is called linear or quadratic. Such an index reflects the similarity between a set of elements. If a fuzzy set shows low index of fuzziness, the elements have high rate of similarity.

3 EXISTING METHOD

This work is an improvement of an existing method based on a fuzziness measure to find the threshold value in a gray image histogram (Tobias et al. 1885; Tobias and Seara 2002). In order to implement the thresholding algorithm on a basis of the concept of similarity between gray levels, Tobias & Seara made the assumptions that there exists a significant contrast between the objects and background and the gray level is the universe of discourse, a one-dimensional set, denoted by X. The purpose is to split the image histogram into two crisp subsets, object subset O and background subset F, using the measure of fuzziness previously defined. The initial fuzzy subsets object and background, denoted by B and W, are associated with initial histogram intervals located at the beginning and the end regions of the histogram. The gray levels in each of these initial intervals have the intuitive property of belonging with certainty to the final subsets object or background. For dark objects $B \subset O$ and $W \subset F$, for light objects $B \subset F$ and $W \subset O$. These initials fuzzy subsets object and background are modeled by the S and Z membership functions respectively. The parameters of the S and Z functions are variable to adjust its shape as a function of the set of elements.

These subsets are a seed for starting the similarity measure process. A fuzzy region placed between these initial intervals is defined as depicted in Fig. 1. Then, to obtain the segmented version of the gray level image, we have to classify each gray level of the fuzzy region as being object or background. The classification procedure is done by adding to each of the seed subsets a gray level x_i picked from the fuzzy region. Then, by measuring the index of fuzziness of the subsets $B \cup \{x_i\}$

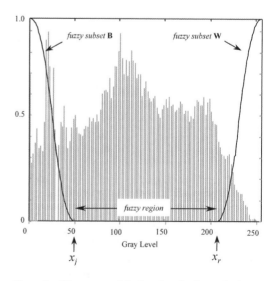

Figure 1. Histogram and the functions for the seed subsets.

and $W \cup \{x_i\}$, the gray level is assigned to the subset with lower index of fuzziness (maximum similarity). Applying this procedure for all gray levels of the fuzzy region, we can classify them into object or background subsets. Since the method is based on measures of index of fuzziness, these measures need to be normalized by first computing the index of fuzziness of the seed subsets and calculating a normalization factor α according to

$$\alpha = \frac{\psi(W)}{\psi(B)} \tag{7}$$

where $\psi(W)$ and $\psi(B)$ are the IF's of the subsets W and B respectively. Fig. 2 illustrate how the normalization works. For dark objects the method can be described as follows:

1. Compute the normalization factor α;
2. For all gray levels x_i in the fuzzy region compute $\psi(B \cup \{x_i\})$ and $\psi(W \cup \{x_i\})$;
3. If $\psi(W \cup \{x_i\})$ is lower than $\alpha.\psi(B \cup \{x_i\})$, then x_i is included in set F, otherwise x_i is included in set O.

4   PROPOSED METHOD

The concept presented above sounds attractive but has some limitations concerning the initialization of the seed subsets. The authors only refer that these subsets should contain enough information about the regions and its boundaries are defined manually. The proposed method aims to overcome some limitations of the existing method. In fact, the initial subsets are defined automatically and they are large enough to accommodate a minimum number of pixels defined at the beginning of the method. This minimum depends on image histogram shape and is a function of the

number of pixels in the gray level intervals $[0, 127]$ and $[128, 255]$ calculated as follows

$$MinPix_{Bseed(Wseed)} = P1. \sum_{i=0(128)}^{127(255)} h(x_i) \tag{8}$$

where $P1 \in [0, 1]$ and $h(x_i)$ denotes the number of occurrences at gray level x_i. After a test period with a set of images, the value $P1 = 0.4$ provided good results for a large number of images. However, in images with low contrast the method performed poorly due to the fact that one of the initial regions represents a low number of pixels. So, previous histogram equalization is done in images with low contrast to provide an image with significant contrast. If the number of pixels belonging in the gray level intervals $[0, 127]$ or $[128, 255]$ is lower than a value P_{MIN} defined by $P_{MIN} = P2.M.N$, where $P2 \in [0, 1]$ and M, N are the dimensions of image, the image histogram is equalized. After some experimental results the value of $P2 = 0.25$ is chosen. Equalization is done using the concept of cumulative distribution function (Gonzalez

Figure 3.   Test images and the corresponding ground-truth images.

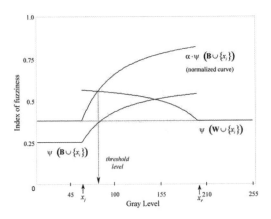

Figure 2.   Normalization step and determination of the threshold value.

297

and Woods 1993). The probability of occurrence of gray level x_i in an image is approximated by

$$p(x_i) = \frac{h(x_i)}{M.N} \qquad (9)$$

For discrete values the cumulative distribution function is given by

$$T(x_i) = \sum_{k=0}^{i} p(x_k) = \sum_{k=0}^{i} \frac{h(x_k)}{M.N} \qquad (10)$$

Thus, a processed image is obtained by mapping each pixel with level x_i in the input image into a corresponding pixel with level $s_i = T(x_i)$ in the output image via Equation 10.

5 EXPERIMENTAL RESULTS AND CONCLUSIONS

To test the performance of the proposed method, images containing small and large objects with clear or fuzzy boundaries were used. A manually generated ground-truth image was defined for each image and used as a gold standard. Original images and their gold standard are illustrated in Fig. 3. Results are compared with two well established methods: the Otsu technique (**OTSU**) and the Fuzzy C-means clustering algorithm (**FCM**). This way, a comparison between fuzzy and non fuzzy threshold algorithms is done and the results of the three techniques are presented in Fig. 4. Performance is defined comparing the gold standard image

Figure 4. Results of three algorithms. For each image, from left to right: Otsu method, Fuzzy C-means algorithm and final improved method.

298

Table 1. Performance of individual methods (%).

Image	OTSU	FCM	IM1	IM2
blocks	94.38	81.18	98.90	99.43
gearwheel	97.99	97.14	95.59	95.59
potatoes	97.57	97.69	97.57	97.57
rice	94.97	85.86	82.06	97.70
shadow	91.03	88.64	95.08	95.08
stones	97.22	96.52	97.81	97.81
zimba	98.05	84.70	96.55	99.11
mouse	49.00	85.87	41.68	57.68
mouse2	73.65	59.09	79.75	79.75
blood	97.49	99.43	85.17	85.17
bird	90.45	78.73	92.14	92.14
moon	26.56	99.97	99.53	91.40
bath	68.46	60.77	84.45	84.45
field	94.89	92.06	99.58	99.58
m	83.69	86.26	88.99	90.89
σ	21.92	13.09	15.26	11.43

with the corresponding image provided by the three different methods. To measure such performance, a parameter η, based on the misclassification error, was used (Tizoosh 2005):

$$\eta_{(\%)} = \frac{|B_O \cap B_T| + |F_O \cap F_T|}{|B_O| + |F_O|} \times 100 \qquad (11)$$

where B_O and F_O denote the background and foreground of the original (ground-truth) image, B_T and F_T denote the background and foreground pixels in the result image, and $|.|$ is the cardinality of the set. This parameter varies from 0% for a totally wrong output image to 100% for a perfectly binary image. The performance measure for every algorithm is listed in Table 1. Average and standard deviation is also presented. The methods indicated by **IM1** and **IM2** represent the improved method without and with histogram equalization. After comparing results, the improved method with histogram equalization provides good results in general, with particular attention in images with imprecise edges.

REFERENCES

Gonzalez, R. C. and R. E. Woods (1993). *Digital Image Processing*. Addison-Wesley Publishing Company.
Pal, N. R. and J. C. Bezdek (1994). Measuring fuzzy uncertainty. In *IEEE Transactions on Fuzzy Systems*, Volume 2.
Pratt, W. K. (2001). *Digital Image Processing* (Third Edition ed.). John Wiley & Sons, Inc.
Sezgin, M. and B. Sankur (2004, January). Survey over image thresholding techniques and quantitative performance evaluation. *Journal of Electronic Imaging 13*(1), 146–165.
Tizoosh, H. R. (2005). Image thresholding using type II fuzzy sets. *Pattern Recognition 38*, 2363–2372.
Tobias, O. J., R. Seara, and F. A. P. Soares (1996). Automatic image segmentation using fuzzy sets. In *Proceedings of the 38th Midwest Symposium on Circuits and Systems*, Volume 2, pp. 921–924.
Tobias, O. J. and R. Seara (2002). Image segmentation by histogram thresholding using fuzzy sets. In *IEEE Transactions on Image Processing*, Volume 11.

Computational Vision and Medical Image Processing – João Tavares & Natal Jorge (eds)
© 2008 Taylor & Francis Group, London, ISBN 978-0-415-45777-4

Measuring critical exponents with a low cost camera in a human standing task

A.S. Fonseca & F. Vistulo de Abreu

Department of Physics, University of Aveiro, Portugal

ABSTRACT: Here we show how a human standing task can be studied with a very simple experimental setup – a conventional laser and a webcam. Subjects are asked to point steadily towards a screen, and movement fluctuations are analyzed. We obtained time series displaying scaling properties on one point distributions, rescaled ranges and power spectra. We analyzed how critical exponents vary when a load is attached to the subject's wrist. Hurst exponents decrease with increasing loads, following what seems to be a subject independent trend.

1 INTRODUCTION

Simple and reliable diagnosis methods have always been important to develop medical science. Finding parameters that reveal a subject's condition can be a matter of experience, requiring the collection of important sets of data. For this reason clinical practice can benefit from reproducible low cost methods that can be easily performed in a clinical context.

Recently it has been proposed that measuring fluctuations in standing postural control can reveal intrinsic features related to a subject neural and muscular condition (Collins 1994, Sabatini 2000). This information can be potentially useful for diagnosis of diseases, like Huntington and Parkinson diseases (Hausdorff 2001, Munoz-Diosdado 2005), the evaluation of a rehabilitation progress or the ageing condition.

Many tasks studied so far have practical disadvantages. Often they may not be easy to implement due to the expensive equipment or because they can be uncomfortable to the patient. For instance, it may require that a subject walks for 1 h (Hausdorff 2001) or balances a stick in a confined space (Cabrera 2002). This may be the reason why some approaches have not been implemented more widely in a clinical context. Consequently, not enough information has been gathered concerning the clinical relevance of these approaches.

Here we developed a study involving low cost instrumentation: it requires only a webcam and a laser pointer. Also, it is non-invasive, requires minimal preparation and lasts only for 2 minutes. Hence, it could be easily implemented in a widespread clinical context.

2 EXPERIMENTAL SETUP

We asked five healthy male subjects with ages between 18 and 30, to hold a laser and point towards a screen, for 2 minutes. Subjects were 30 cm away from the screen. To study the effect of an applied load on muscle control we considered two situations: one, in which subjects pointed freely, and the other, in which subjects held a 1,6 kg weight firmly attached to the wrist. We recorded the image of a red laser pointer on a screen, with a webcam recently introduced in the market that records up to 90 frames per second: PHILIPS SPC 900NC. The webcam was 8 cm away from the screen. A blue screen was used to increase contrast. Experiments were performed in darkness but we verified that results do not depend crucially on this requirement. Images were acquired with a frame rate of 90 frames per second and with a ROI of 120×160 pixels. This corresponded to an area of $\sim 2\,cm^2$ on the screen, and hence typical movements fluctuated within less than $0.1°$.

3 IMAGE PROCESSING

We verified that the laser pointer produced a uniform circular pattern. Consequently, we found appropriate to use a fixed threshold as segmentation technique. Each frame was converted from RGB to Gray Scale as the efficiency of the segmentation improved. A typical result is shown in Figure 1.

4 RESULTS

Typical time series for the laser pointer coordinates are shown in Figure 2. In Figure 3 we also show x-y paths

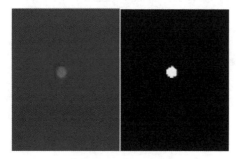

Figure 1. Typical laser pattern recorded and the corresponding segmentation.

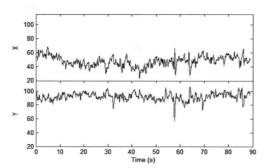

Figure 2. Typical time series for the movement along X and Y, measured in pixels, for a subject without weight.

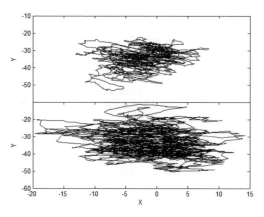

Figure 3. Typical x-y paths during 55s, for experiments with (upper) and without (lower) weight for the same subject.

for the same subject, with and without weight. The range of movements increased in all five subjects when the weight was added. In Figure 4 we show typical time series for the fluctuations in the X and Y coordinates and in Figure 5 we show their histogram. From these two figures it is clear that fluctuations follow a non-gaussian distribution. Instead we have a Lévy-type of

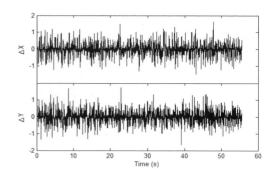

Figure 4. Typical time series for the variation of movement along X (upper) and Y (lower), measured in pixels, for a subject without weight ($\Delta t = 1/90s$).

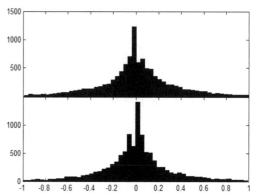

Figure 5. Typical histogram obtained for the variation of movement along X (upper) and Y (lower), for a subject without weight.

distribution which is usually associated in physics to critical phenomena (Sornette 2006).

This type of distribution is characterized by having diverging moments. Quantities like the variance – the second moment of the distribution – are ill-defined and, consequently, in practice, vary considerably from sample to sample. For this reason, other parameters are necessary to characterize reliably a time series. One possibility is the calculation of the critical exponents, which will be described in the next section.

5 DATA ANALYSIS

Critical phenomena are characterized by the existence of long ranged correlations in the system. These can be measured through two-point correlation functions $C(T) = \langle \Delta X(t+T)\Delta X(t) \rangle$, where $\Delta X(t)$ is the fluctuation observed at time t. In the case of long-ranged correlations, the correlations decay slowly, i.e. $C(T) \sim T^{-\beta}$, with $0 < \beta < 1$. In this case the sum of all correlations diverges. This decay can be more easily

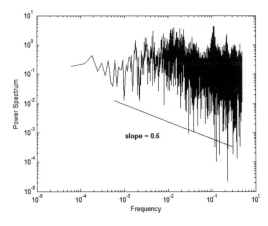

Figure 6. Power Spectrum for data series Pg(1), $S(f) \sim f^{-\alpha}$.

analyzed with the Fourier transform of the correlation function, through the analysis of the power spectrum $S(f)$. This can be straightforwardly shown to decay as $S(f) \sim f^{-\alpha}$, with $\alpha = 1 - \beta$ (Rangarajan 2000, Sornette 2006). The analysis of the power spectrum is nevertheless not a very precise method to extract these exponents. An alternative scheme is to perform a Hurst exponent analysis (Hurst 1965).

The basic idea of the Hurst exponent analysis is to divide the time series of ΔY (or ΔX) in sub-series of length T (Fig. 7a). For each sub-series a rescaled variable is defined, dividing by the standard deviation, S, to obtain samples with unit variance (Fig. 7b). Then, a cumulative series is derived for each sub-series using: $s_i = s_{i-1} + \Delta \tilde{Y}_{i-1}$, $s_0 = 0$, where $\Delta \tilde{Y}$ is the rescaled variable for a sub-series (Fig. 7c). Finally, the mean of the range of cumulative sub-series of length T, R/S_T, is calculated and the procedure is repeated for other window sizes T.

For most stochastic process R/S_T increases with T. If the increments ΔY are derived from Gaussian noise, then $(R/S_T) \sim T^{1/2}$, corresponding to Brownian noise in the $Y(X)$ series. This exponent could also be obtained from a correlated series, if the data had been reshuffled, destroying correlations. For uniform movements (as with constant speed), $(R/S_T) \sim T$. In general, for scale invariant processes we can write:

$$(R/S_T) \sim T^H$$

where H is called the Hurst exponent. Persistent series with long range memory have $1/2 < H < 1$, while long ranged antipersistency leads to $0 < H < 1/2$.

As pointed by Rangarajan (2000), Hurst exponents do not imply the existence of long range memory processes, requiring the confirmation that power spectrum and Hurst exponents are consistent, i.e. that they verify the relation $H = (1 + \alpha)/2$. In anyway, Hurst exponents can always be used to characterize time series.

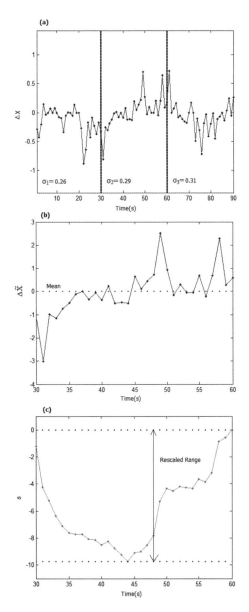

Figure 7. Basic steps in the Hurst exponent analysis. (a) Fluctuations time series, ΔX, are divided in sub-series with window sizes T. (b) For each sub-series the mean and standard deviation is calculated and a normalized sub-series is calculated, with zero mean and unit standard deviation. We show this procedure applied to the window starting at $t_i = 30s$. (c) Then, a cumulative time series s_i is constructed summing fluctuations from $t = t_i$ to $t = t_i + T$. In this example it is clear that while fluctuations are negative, the cumulative time series decreases, and when they become prominently positive, it increases. From s_i the range is calculated. The average over the rescaled ranges for all windows of size T, gives the rescaled range R/S_T. The procedure is repeated for different window sizes T.

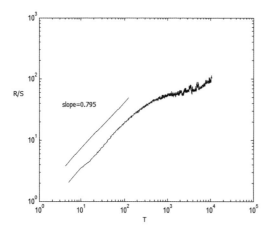

Figure 8. Log-log plot of the rescaled range statistic R/S_T against the window size T, for data series Pg(1) (see Table 1). The markedly linear region shows that the rescaled range scales with T. From the slope we can extract the Hurst exponent, H. The Hurst exponents obtained in the experiments are all above 0.5, suggesting that correlations can be present in the data. Reshuffling the data we obtained exponents close to 0.5 (in this case 0.55), as correlations are destroyed.

Table 1. Hurst Exponents for five subjects: (1) Without weight, (2) With weight.

	ΔX	ΔY
Br(1)	0.846	0.813
Br(2)	0.811	0.701
Pg(1)	0.889	0.839
Pg(2)	0.831	0.748
Fs(1)	0.930	0.844
Fs(2)	0.903	0.847
R(1)	0.952	0.845
R(2)	0.915	0.728
Hg(1)	0.858	0.854
Hg(2)	0.865	0.863

In the present case it was possible to calculate Hurst exponents because R/S_T scaled with T (Fig. 8). In Table 1 we present all Hurst exponents extracted from the data series.

While Hurst exponents seem to vary from subject to subject, two general trends exist. Firstly, fluctuations along X have higher Hurst exponents than along Y and secondly when a load is added to the wrist, Hurst exponents decrease. Hence, muscle effort seems to be clearly related to Hurst exponent variations. We also compared Hurst exponents with power spectrum exponents. Power spectra showed an overall decay with exponents that agree with Hurst exponents, and hence indicate the existence of long ranged correlations in the data. However, we verified that power spectra

possessed some structure, whose significance could also be analysed, but will not be studied here.

6 DISCUSSION

In order to gain a better understanding of these results we formulated a simple one-dimensional model. We assumed that a subject uses a perceptual (cognitive) reference position x_R towards which he attempts to direct the pointer. This reference perceptual position may vary along the time. Within the simplest approximation we considered that it evolved according to a Brownian motion equation:

$$dx_R / dt = \xi_R(t)$$

where $\xi_R(t)$ is Gaussian noise: $\langle \xi_R(t) \rangle = 0$, $\langle \xi_R(t)\xi_R(t') \rangle = \sigma_R^2 \delta(t - t')$. We also considered that the subject brings the pointer position x_P towards x_R, reacting in a non-linear way:

$$dx_P / dt = -\alpha(t)(x_P - x_R)^3 + \xi_m(t)$$

The nonlinearity in this equation is crucial to describe the results, as it leads to the important on-off intermittency described in (Cabrera 2002). This nonlinear response assumes that a subject reacts non-proportionally depending on how far the pointer is from the reference position. The intensity of the reaction depends on $\alpha(t)$, which is related to the muscle and cognitive control in executing the task. $\alpha(t)$ can also fluctuate in time, due to a noisy component, that we modeled with another Gaussian noise component: $\alpha(t) = \alpha_0 + \xi_C(t)$, where $\langle \xi_C(t) \rangle = 0$, $\langle \xi_C(t)\xi_C(t') \rangle = \sigma_C^2 \delta(t - t')$. Fluctuations in $\alpha(t)$ arise from errors following voluntary (controlled) movements. Another source of variability could have an involuntary origin, and is accounted by $\xi_m(t)$ which is also modeled as a zero mean Gaussian noise. In principle, we could expect that exercises requiring higher muscle effort can increase the amplitude of all these sources of variability.

The previous equations can be reduced to a single equation, changing variables according to $X = x_P - x_R$, which leads to:

$$dX / dt = -(\alpha_0 + \xi_C)X^3 + \xi_{add}$$

where $\sigma_{add}^2 = \sigma_m^2 + \sigma_R^2$, where σ_m^2 is the variance associated to $\xi_m(t)$. We should remark that the additive noise can have different physiological origins: involuntary muscle movements and the diffusion of the cognitive reference position.

We simulated this stochastic differential equation, for several parameter ranges. We found regimes in

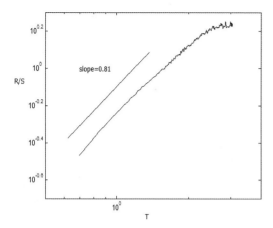

Figure 9. Log-log plot of the rescaled range statistic R/S_T against the window size T, for the theoretical model with $\alpha_0 = 0.1$, $\sigma_R = 0.01$, $\sigma_C = 0.01$.

which histograms of fluctuations of the pointer position become scale invariant as in a Lévy distribution, agreeing with the type of distributions shown previously. Other critical properties, like the existence of Hurst scaling (as shown in Figure 9) and power spectrum decay as a power law, are also exhibited by this model.

An interesting qualitative result is that changing σ_C has no effect on the Hurst exponent. On the contrary, increasing α_0 or σ_R can decrease Hurst exponents significantly, as shown in Figure 10. As a consequence, studying Hurst exponent variations should not provide information concerning how finely a subject controls a movement. However, it should be sensitive to the impact of involuntary movements, as those induced by muscle effort. In fact, this agrees with the empirical results which indicate a decrease of the Hurst exponents when experiments are performed with a load (Table 1). This type of model would also predict that elder people are likely to produce results with smaller Hurst exponents, as observed in (Hausdorff 2001), for a different type of task. This model is nevertheless an over simplification, that is intended to capture how a human subject performing this type of experiment is prone to generate scale invariant data as the one we observed. However, this model does not incorporate long ranged correlations intrinsic to neuromuscular human movements. We remarked that exponents extracted from power spectra and Hurst analysis did not agree with each other. As pointed in Rangarajan (2000) this could be an indication that long ranged correlations are absent in the model derived series. Hence, this simple model is capable of capturing some important task derived features, but lacks some intrinsic subject derived features.

Currently we are working towards improving the model and confronting theory and experiments in a

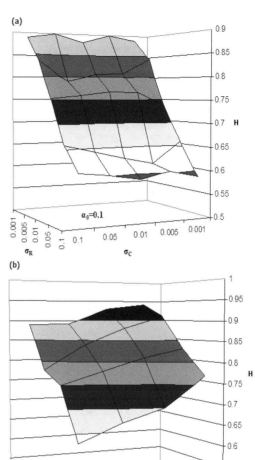

Figure 10. Hurst Exponents obtained from our simplified model in different parameter regions: (a) $\alpha_0 = 0.1$, and σ_R, σ_C are varied; (b) $\sigma_C = 0$, and α_0 and σ_R vary. In the range of parameters described in this figure perfect scale invariant behavior for R/S_T and power spectra was found.

wider range of experimental situations. This model although quite simple, it is also quite rich. It shows that this type of task should be sensitive to perturbations with distinct origins: cognitive and involuntary controlled movements. Hence, suitably designed tasks should be able to access different clinical parameters that may be useful to characterize a subject's condition.

This work points towards the possibility of using low cost imaging techniques to quantify and gain a deeper understanding of human motion control potentially useful in a widespread clinical context.

ACKNOWLEDGMENTS

This work received financial support from FCT, under the grant POCTI/NSE/47379/2002.

REFERENCES

Cabrera, J.L. & Milton, J.G. 2002. On-Off Intermittency in a Human Balacing Task *Physical Review Letters* 89: 158702-6

Chen, Y., Ding, M., Kelso, J.A.S. 2001 Origins of Timing Error in Human Sensorimotor Coordination *Journal of Motor Behavior* 33:3–8.

Collins, J.J. & Deluca, C.J. 1994 Random Walking during Quiet Standing *Physical Review Letters* 73: 764–767.

Rangarajan, G. & Ding, M. 2000 Integrated approach to the assessment of long correlations in time series data *Physical Review E.* 61: 4991–5001.

Hausdorff, J.M. 2001. When human walking becomes random walking fractal analysis and modeling of gait rhythm fluctuations. *Physica A 302* 138:147.

Hurst, H.E. & Simaiki, Y. M. 1965. *Long term Storage: An Experimental Study.* London: Constable.

Munoz-Diosdado, A. 2005. A non linear analysis of human gait time series based on multifractal analysis and cross correlations *Journal of Physics: Conference* Series 23 87–95

Sabatini, A.M. 2000. A Statistical Mechanical Analysis of Postural Sway Using Non-Gaussian FARIMA Stochastic Models. *IEEE Transactions on Biomedical Engineering.* 47:1219–1227.

Sornette, D. 2006. *Critical Phenomena in Natural Sciences: Chaos, Fractals, Self-Organization and Disorder: Concepts and Tools.* Berlin: Springer-Verlag.

Computational Vision and Medical Image Processing – João Tavares & Natal Jorge (eds)
© 2008 Taylor & Francis Group, London, ISBN 978-0-415-45777-4

3D reconstruction of the retinal arterial tree based on fundus images using a self-calibration approach

D. Liu, N.B. Wood & X.Y. Xu
Department of Chemical Engineering, Imperial College London, UK

N. Witt, A.D. Hughes & S.A.McG. Thom
ICCH, Imperial College London, UK

ABSTRACT: Systemic diseases, such as hypertension and diabetes, are associated with changes in the retinal microvasculature. Although a number of studies have been performed on the quantitative assessment of the geometrical patterns of the retinal vasculature, previous work has been confined to 2 dimensional (2D) analyses. In this paper, we present an approach to obtain a 3D reconstruction of the retinal arteries from a pair of 2D retinal images acquired *in vivo*. A simple essential matrix based self-calibration approach was employed for the "fundus camera-eye" system. Vessel segmentation was performed using a semi-automatic approach and correspondence between points from different images was calculated. The results of 3D reconstruction show the centre-line of retinal vessels and their 3D curvature clearly. 3D reconstruction of the retinal vessels is feasible and may be useful in future studies of the retinal vasculature in disease.

1 INTRODUCTION

The retina, which lies at the posterior fundal surface of the eye, has the highest oxygen requirement per unit weight of any tissue in the body (Masters 2004) and this makes it particularly vulnerable to vascular insults impairing oxygen and nutrient supply. Retinal vascular anatomy and network structure are adversely affected by high blood pressure, diabetes mellitus, ageing and atherosclerosis. Diabetic eye disease is one of the commonest causes of blindness in UK. A number of studies have shown that generalized arteriolar narrowing and retinopathy are associated with increased risk of stroke, ischaemic heart disease, heart failure, renal dysfunction and cardiovascular mortality (Wong & McIntosh 2005). Therefore, quantitative assessment of the retinal vascular network is very important.

The geometric patterns of the retinal microvascular network are readily observed *in vivo* using fundal photography (Stanton et al. 1995). However, quantitative analysis of the geometrical patterns requires vessel segmentation and reconstruction. The reconstruction of the retina, especially the area of optic disc has been performed by several researchers (Xu & Chutatape 2006, Kai et al. 2005), but most pathological changes in the microvasculature occur away from this region. D reconstruction of the retinal vascular tree from fundal images is a considerable challenge and only a few such attempts have been made so far

(Martinez-Perez & Espinosa-Romero 2005). When retinal images are obtained with a fundus camera, the intrinsic parameters of the fundus camera-eye system are altered by the relative displacement between camera and the eye of the subject. These changes can be reduced by acquiring retinal images with small displacements of the camera in its original plane which is assumed to be parallel to the surface of the lens. And the intrinsic parameters can be assumed to be fixed. The retinal vessels of interest can be segmented using a semi-automatic approach (Martinez-Perez et al. 2002) and the point correspondence between different images can also be calculated. In order to acquire a metric reconstruction result, self-calibration was performed. For this the pixels of the camera-eye system are assumed to be nearly perfectly rectangular (which means that the aspect ratio is considered to be one and there is no skew) and the principal point of the camera-eye system is assumed to lie at the centre of the final retinal image. With these assumptions the self-calibration approach can be reduced to a simple system and a final metric 3D reconstruction can be recovered.

2 METHODOLOGY

2.1 *Image acquisition*

The retinal images for reconstruction were obtained in a normal subject following mydriasis with tropicamide

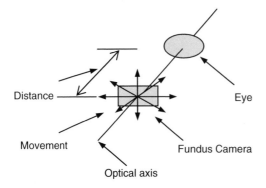

Figure 1. Diagram illustrating the approach to retinal imaging.

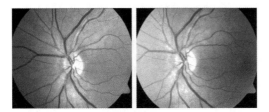

Figure 2. A stereo pair of retinal images for reconstruction.

(1% eye drops) using a commercial retinal fundus camera (Zeiss FF 450plus) with a 30° field of view. Digitized images were captured using a CCD camera and transferred to a PC for analysis. The principle is illustrated in Figure 1 and a stereo (approximate) pair of retinal images is shown in Figure 2. Although the eye and fundus camera are very complex, they are combined and simplified as one single lens in the analysis described here. Because this special system combines the eye and the fundus camera, the displacement between them, such as the change in relative distance and rotation, will alter the intrinsic parameters of the combined eye-camera system. In order to minimize these changes, the distance between the camera and eye was fixed and only a small, near planar displacement of the fundus camera was made when the retinal images were acquired.

Retinal images were acquired with a resolution of 1280 × 103 pixels but were reduced to 499 × 402 pixels prior to analysis in order to reduce the computational time. The pair of retinal images shown in Figure 2 was used for reconstruction.

2.2 Estimation of epipolar geometry

In order to obtain the fundamental matrix F, some corresponding fiducial points (at bifurcations) were obtained first. Matching points were selected by an operator and are marked out in Figure 3. Based on

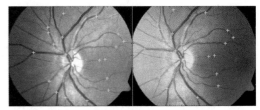

Figure 3. Retinal images with marked corresponding points.

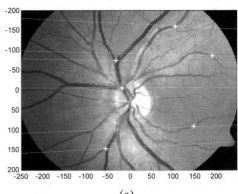

(a)

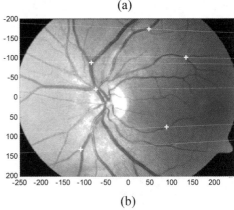

(b)

Figure 4. The epipolar lines of the two retinal images.

these corresponding points, the fundamental matrix F was recovered by applying the gold standard algorithm developed by Hartley & Zisserman (2003), which minimizes the re-projection geometric error:

$$\sum_i d(x_i, \hat{x}_i)^2 + d(x'_i, \hat{x}'_i)^2 \qquad (1)$$

Where $x_i(u_i, v_i) \leftrightarrow x'_i(u'_i, v'_i)$ are the marked correspondence, and $\hat{x}_i \leftrightarrow \hat{x}'_i$ are the estimated correspondence that satisfy $\hat{x}_i F \hat{x}'_i = 0$ exactly for rank-2 estimated fundamental matrix F.

Figure 4 shows several epipolar lines of the two retinal images calculated by the above algorithm.

2.3 Self-calibration

In principle, a projective reconstruction could be obtained without the calibration matrix based on the fundamental matrix, F, but, in practice, due to the ambiguity of projective reconstruction, results were insufficiently accurate. Therefore the 3D reconstruction was made on the basis of a metric projection. It is known that a metric reconstruction of a scene may be computed by using the essential matrix E which could be derived from the calibration matrix K (Hartley & Zisserman 2003):

$$E = K^T F K \qquad (2)$$

where the superscript T represents the transpose of the K. A standard linear camera calibration matrix K has the following entries:

$$K = \begin{bmatrix} f & s & u_0 \\ 0 & \alpha f & v_0 \\ 0 & 0 & 1 \end{bmatrix}$$

where f is the focal length in pixels and α is the aspect ratio. (u_0, v_0) are the coordinates of the principal points, and s is the skew factor which is zero for rectangular pixel.

For the special fundus camera-eye system, a general photogrammetric calibration method, which depends on a calibration object with a known 3-D geometry (Zhang 1999, Xu & Chutatape 2003), is not available. Instead a self-calibration method was employed. Since the image acquisition process was specially designed to minimise alteration of the camera-eye system, we can assume that the intrinsic parameters of camera-eye system were constant. The aspect ratio is considered to be unity, which means that the pixels are nearly perfectly rectangular and there is no skew. The principal point of the camera-eye system is assumed to be at the centre of the final retinal image. Therefore the only unknown parameter of the calibration matrix is the focal length.

It has been recognized that if n_k is the number of intrinsic parameters known in all views and n_f is the unknown but constant intrinsic parameter, the views, m, required for self-calibration will be:

$$mn_k + (m-1)n_f \geq 8 \qquad (3)$$

therefore two views will be sufficient in this case (Mendonca & Cipolla 1999).

A self-calibration method, based on the characteristics of the essential matrix E, was used to recover the unknown focal length: two of the three singular values of E should be identical and the other should be zero (Mendonca & Cipolla 1999). The cost function:

$$C = \omega_{12} \frac{\sigma_1 - \sigma_2}{\sigma_2} \qquad (4)$$

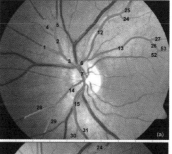

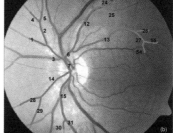

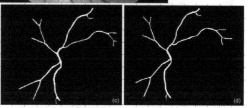

Figure 5. (a), (b) two different retina images with marked vessels; (c), (d) the segmented vessels of (a), (b).

was minimised by a direct search algorithm. $\sigma_1 > \sigma_2$ are the non-zero singular values of $E = K^T F K$, and ω_{12} is the weight factor which equals the number of points used in the computation of the fundamental matrix, F. According to Fusiello this algorithm should have a good global convergence (Fusiello 2001). After obtaining the focal length by this self-calibration method, the essential matrix could be calculated from (2).

2.4 Recovery of the projection matrix

The corresponding points $x_i \leftrightarrow x_i'$ in the 2D images and the unknown 3D points, X_i, on the object have the relationship:

$$x_i = PX_i, \; x_i' = P'X_i \qquad (5)$$

here P and P' are the two projective matrices and are assumed to be $P = K[I \mid 0]$ and $P' = K[R \mid t]$, where R is the rotation matrix and t is the translation vector. If the essential matrix E of the camera-eye system was obtained, the matrices P and P' could be retrieved from E up to scale and a four-fold ambiguity (Hartley & Zisserman 2003). At the same time the extrinsic parameters, the rotation axis l and the angle of rotation α could be obtained as:

$$l = (R_{32} - R_{23}, R_{13} - R_{31}, R_{21} - R_{12})^T \qquad (6)$$

$$\alpha = \arccos\left(\frac{trace(R) - 1}{2}\right) \qquad (7)$$

The projection matrix P and P' are computed according to the essential matrix E obtained above.

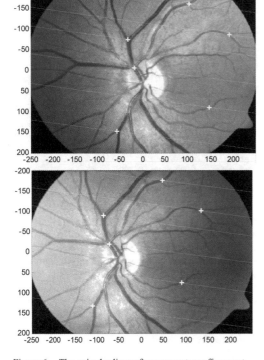

Figure 6. The epipolar lines after parameter refinement.

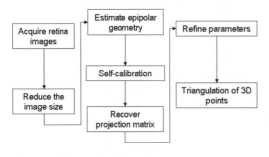

Figure 7. The reconstruction process.

2.5 Vessel segmentation

Because we are concerned with the geometry of the vessels, it is necessary to segment the vessels of interest. Therefore a semi-automatic method (Martinez-Perez et al. 2002) was used to perform vessel segmentation and to calculate the 2D coordinates of the individual pixels corresponding to the centrelines of segmented vessels. If the same vessel were segmented from different images, the correspondence between them could be obtained. The segmented vessels are marked in red in Figures 5a, b, and the segmentation results are shown in Figures 5c, d.

2.6 Parameter refinement

Based on the correspondent points obtained from vessel segmentation, the extrinsic parameters R and t, and the intrinsic parameter f are refined by minimizing the cost function

$$C_r = \sum_i \frac{(x_i'\tilde{F}x_i)^2}{(\tilde{F}x_i)_1^2 + (\tilde{F}x_i)_2^2 + (\tilde{F}^T x_i')_1^2 + (\tilde{F}^T x_i')_2^2} \qquad (8)$$

Here the fundamental matrix $\tilde{F}$ was calculated as

$$\tilde{F} = K^{-T}[t]_\times R'K \qquad (9)$$

where R' is the rotation matrix calculated from Rodrigues' formula (Hartley & Zisserman 2003)

$$R' = \cos\alpha I + \sin\alpha[\frac{l}{\|l\|}]_\times + (1 - \cos\alpha)[\frac{l}{\|l\|}]_\times^2 \qquad (10)$$

After this refinement, the final projection matrix could be obtained from R, t and K.

2.7 Reconstruction of 3D points

Knowing the projection matrices for two images separately, the 3D coordinates of each point, X_i, could be calculated. In order to get a smoother reconstruction, the corresponding points from the segmentation were smoothed and interpolated using a cubic spline.

An iterative linear-eigen (Iterative-Eigen) method (Hartley & Sturm 1997) was used to perform the triangulation of 3D points. Equations in (5) can be written as:

$$\begin{cases} \dfrac{1}{w_{i,j}}(u_i p^{3T} X_{i,j} - p^{1T} X_{i,j}) = 0 \\ \dfrac{1}{w_{i,j}}(v_i p^{3T} X_{i,j} - p^{2T} X_{i,j}) = 0 \\ \dfrac{1}{w_{i,j}'}(u_i' p'^{3T} X_{i,j} - p'^{1T} X_{i,j}) = 0 \\ \dfrac{1}{w_{i,j}'}(v_i' p'^{3T} X_{i,j} - p'^{2T} X_{i,j}) = 0 \end{cases} \qquad (11)$$

where p^{iT} and p'^{iT} are the rows of P and P' respectively. $w_{i,j}$ and $w'_{i,j}$ are the weight factors at the jth step of iteration which have the form:

$$w_{i,j} = p^{3T} X_{i,j-1}, \ w'_{i,j} = p'^{3T} X_{i,j-1} \qquad (12)$$

All calculations were performed using Matlab (The Mathworks) and parameter refinement was based on the codes from the Oxford Brookes toolbox (http://cms.brookes.ac.uk/staff/PhilipTorr/). The reconstruction process is shown in Figure 7.

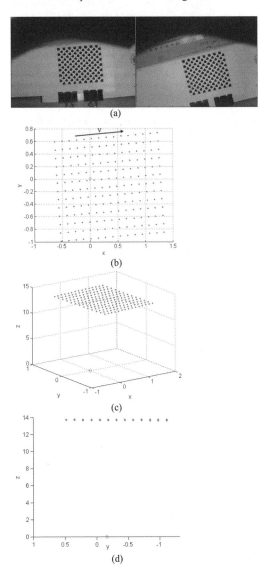

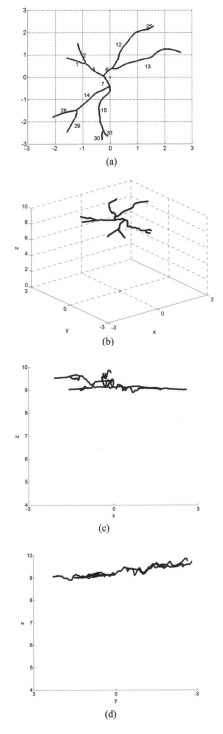

Figure 8. (a) two images of chessboard; (b) top view of the reconstructed corners; (c) the 3D view; (d) side projection along v direction in (b).

Figure 9. 3D reconstruction results of the centre-lines of the marked retinal vessels: (a) top view of the reconstructed vessels; (b) 3D view; (c), (d) side projections.

311

3 EXPERIMENT AND RESULTS

Two images of the same chessboard taken by a normal camera using the method described above were adopted to test the reconstruction procedure. The corners were detected automatically and used for reconstruction. The reconstruction results in Figure 8 show that the procedure proposed here is practicable.

Figure 9 shows the reconstruction results of the segmented vessels in the central region of the image shown in Figure 5. Figure 9a is the 2D view of the recovered centreline points of the marked 3D vessels. The numbers (according to Figure 5a) indicate which segments of the marked vessels are retrieved. Figure 9b is the 3D view of the reconstructed vessel centrelines. Figures 9c, d show the side projections of the centre lines of the reconstructed vessels. These preliminary results demonstrate that 3D reconstruction of the retinal vessels is feasible and may be useful in future studies of the retinal microvasculature in health and disease. Future studies will attempt to validate the 3D reconstruction by using a model eye with known geometry.

REFERENCES

Fusiello, A. 2001. A new autocalibration algorithm: experimental evaluation. *Computer analysis of Images and Patterns 2001, Lecture Notes in Computer Science* 2124: 717–724.

Hartley, R.I. & Sturm P. 1997. Triangulation. *Computer Vision and Image Understanding*. 68(2): 146–157.

Hartley, R. & Zisserman, A. 2003. *Multiple View Geometry in Computer Vision*. Cambridge: Cambridge University Press.

Kai, Z., Xu, X., Zhang, L., et al. 2005. Stereo Matching and 3-D Reconstruction for Optic Disk Images. *CVBIA, LNCS* 3765: 517–525.

Martinez-Perez, M.E., Hughes, A.D., Stanton, A.V., et al. 2002. Retinal vascular tree morphology: A semi-automatic quantification. *IEEE Transactions on Biomedical Engineering* 49(8): 912–917.

Martinez-Perez, M.E. & Espinosa-Romero, A. 2005. Optical 3D reconstruction of retinal blood vessels from a sequence of views. *Proc. of SPIE* 5776: 605–612.

Masters, B. 2004. Fractal analysis of the vascular tree in the human retina. *Ann Rev Biomed Eng* 6: 427–452.

Mendonca, P.R.S. & Cipolla R. 1999. A simple technique for self-calibration. *Proceedings of the IEEE Conference on Computer Vision and Pattern Recognition*: 500–505.

Stanton, A.V., Wasan, B., Cerutti, A., et al. 1995. Vascular network changes in the retina with age and hypertension. *J. Hypertension* 13: 1724–1728.

Wong, T.Y. & McIntosh, R. 2005. Systemic associations of retinal microvascular signs: a review of recent population-based studies. *Ophthalmic and Physiological Optics*: 25–195.

Xu, J. & Chutatape, O. 2003. Comparative study of two calibration methods on fundus camera. *Proceedings of the 25 Annual International Conference of the IEEE EMBS*: 17–21.

Xu, J. & Chutatape, O. 2006. Auto-adjusted 3D optic disk viewing from low-resolution stereo fundus image. *Computers in Biology and Medicine* 36: 921–940.

Zhang, Z. 1996. Determining the epipolar geometry and its uncertainty: a review. *INRIA*: 1–56.

Zhang, Z. 1999. Flexible camera calibration by viewing a plane from unknown orientations. *In Proceedings of the International Conference on Computer Vision*: 666–673.

Computational Vision and Medical Image Processing – João Tavares & Natal Jorge (eds)
© 2008 Taylor & Francis Group, London, ISBN 978-0-415-45777-4

Automatic Couinaud liver segmentation using CT images

D.A.B. Oliveira, R.Q. Feitosa & M.M. Correi

Computing Engineering Department, Rio de Janeiro State University – Rio de Janeiro, RJ, Brazil
Electrical Engineering Department, Catholic University of Rio de Janeiro – Rio de Janeiro, RJ, Brazil
Unigranrio and National Cancer Institute-INCA – Rio de Janeiro, RJ, Brazil

ABSTRACT: This paper presents an algorithm to segment the liver structures on computed tomography (CT) images according to the Couinaud orientation. Our method firstly separates the liver from the rest of the image. Then it segments the hepatic and portal vessel trees inside the liver area using a region growing technique combined with hysteresis thresholding. Finally, the method estimates the planes that best fit each of the three branches of the segmented hepatic veins and the plane that best fits the portal vein. These planes define the subdivision of the liver in the Couinaud segments. An experimental evaluation based on real CT images demonstrated that the outcome of the proposed method is generally consistent with a visual segmentation.

1 INTRODUCTION

Current computed tomography scanners allow for non-invasive high resolution imaging of the human body. A major challenge accompanying this improvement is dealing with the enormous amount of data generated in the form of image sequences.

By and large the CT data analysis is performed visually by a radiologist. This is a time consuming task, whose accuracy depends essentially on the experience of the analyst (Kakinuma et al., 1999; Li et al., 2002). Digital Image Processing techniques can be used to develop methods that automatically perform many of the tasks involved in the CT analysis, improving productivity and the overall accuracy.

The segmentation process is particularly arduous in abdominal CT images because different organs frequently share the same intensity value range and are often near to each other anatomically. Many techniques have been proposed in the literature for the analysis of abdominal CT scans. They can be roughly divided in two main groups: model driven and data driven approaches (Masutani et al, 2005).

Model driven techniques (e.g. Lamecker et al (2004) and Soler et al (2000)) use pre-defined models to segment the desired object from the available images. This kind of technique basically searches the images for instances that fit a given model described in terms of object characteristics such as position, texture and spatial relation to other objects.

Data driven techniques (e.g. Kim et al (2000) and Fujimoto et al (2001)) try to emulate the human capacity to identify objects using some similarity information present on image data, automatically detecting and classifying objects and features in images. Many of them use known techniques such as region growing and thresholding, combined with some heuristic knowledge about the aimed object.

Other important issue consists in segmenting vessels. Inside the liver, the contrast can be insufficient to distinguish minimally the veins from the liver parenchyma. Another problem is that the portal and hepatic veins can be erroneously connected leading to identification as a single object. Kirbas et al (2004) presented a review of vessel extraction, in which many of the available techniques for this purpose can be found.

This paper proposes a data driven algorithm for 3D segmentation of the liver structures based on CT image sequences using the Couinaud orientation (Couinaud, 1957), which is presently used as reference in liver surgical procedures.

The following sections describe our approach in greater detail. Section 2 discusses the details of our segmentation methods, section 3 reports some results, and the main conclusions are presented in section 4.

2 THE 3D SEGMENTATION METHOD

The segmentation method consists of four main steps:

a) segmentation of organs and muscle tissues from the rest of the image based on a dual thresholding.
b) segmentation of the liver using heuristic related to anatomy, such as liver position and relative density of liver's tissue.

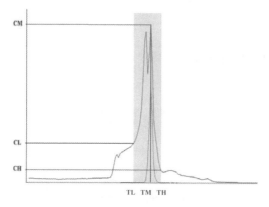

Figure 1.   Histogram analysis for gray level range definition.

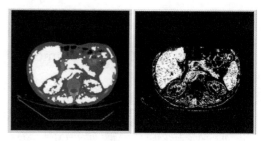

Figure 2.   Resulting images of range selection.

These pixels form connected components that will be used in the later segmentation steps.

c) segmentation of the hepatic and portal vessels in the liver by a region growing technique combined with hysteresis thresholding inspired in Canny's Edge Detector (Canny et al., 1986).

d) Estimation of the planes that define the subdivision of the liver in the eight Couinaud segments.

Each step is described in details in the next subsections.

2.1   Segmentation of organs and muscle tissue

Organs and muscles tissue are the main presence in abdominal images. Typical gray values of these tissues occur around the maximum (CM) of the gray value histogram for the whole CT sequence.

Figure 1 shows the histogram of a sample CT exam, the range of intensities corresponding to organs and muscles and the lower and upper limits TL and TH defining this range.

Let CM be the maximum CT histogram count, TM the corresponding intensity value, and CL and CH the counts corresponding respectively to TL and TH. It has been observed in our experiments that the ratios $RL = CL/CM$ and $RH = CH/CM$ do not significantly change from CT exam to CT exam. In fact these ratios lied around $RL = 0.6$ and $RH = 0.2$ through all our experiments.

This regularity suggests the following procedure to select the lower and higher threshold values:

a) Compute and smooth the histogram of the whole CT exam;

b) Detect the maximum histogram count CM;

c) Multiply CM by the constant factors RL and RH, and obtain the count values CL and CH;

d) Search the smoothed histogram for the intensity values TL and TH closest to TM corresponding to CL and CH, such that $TL < TM$ and $TH > TM$.

Figure 2-a show a CT image in which the pixels with gray levels falling in this range are shown in white.

2.2   Segmentation of the liver

The next step consists in segmenting the liver. Generally the liver appears as homogeneous areas on CT slices, i.e. its intensities are restricted to a narrow gray value interval. This can be observed in Figure 1, where the histogram of the pixels belonging to the liver is drawn in red over the histogram of the whole CT sequence shown in blue.

Our method determines the extreme values of this interval in the following way:

One image of the CT set where the liver is present is selected as the main sample and passed as a parameter to the algorithm. Then, the largest connected component of this slice located on the upper-left side of the image (right side of the human body), is identified and its mean value on the original image is computed.

Using the pixels of organs and muscle tissue previously segmented, a new gray level range is defined following a similar procedure of the subsection 2.1. The histogram count value corresponding to the liver mean value is used as the maximum count value and the range limits are calculated using as limiting ratios the value 0.8 for both cases. The threshold values obtained this way are applied to the regions selected in the previous step. Figure 2-b shows the collection of objects obtained in this fashion in CT slice of figure 1-a. Notice that the kidneys and the muscles were for the most part discarded.

A simple procedure extracts the liver from the remaining objects. Starting on the main sample it is executed on the next adjacent slice upward and downward in the CT image set till all slices have been processed. It consists in three main steps:

a) Select the biggest object in the collection;

b) If its centroid is in the upper left quadrant of the CT image, go to step c, otherwise discard this object from the collection and go back to step a;

c) If the selected object is connected to another object of an adjacent slice previously classified as liver, classify it as liver, otherwise discard the object from the collection and go back to step a;

314

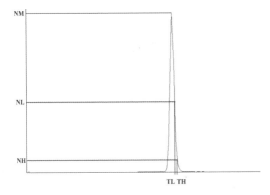

Figure 3.   Liver histogram analysis for vessel segmentation.

Clearly the first iteration does not pass through step c and the object selected in step b is set as liver directly.

2.3   Segmentation of Hepatic and Portal Veins

This section describes our adaptive method inspired on the Canny Edge Detector (Canny et al., 1986) to segment the portal and hepatic vessel trees from the segmented liver.

Initially we select a threshold VH, such that the intensities above it identify unambiguously the vessels. A second threshold VL (VL < VH) is further selected such that intensities below it clearly indicate liver parenchyma. Figure 3 shows these limits and the liver histogram in red.

These two threshold values define three ranges of pixel intensities, namely:

– the strong vessel range, defined by intensities above VH,
– the weak vessel range, comprising intensities between VL and VH, and
– the liver tissue range, covering intensities below VL.

The construction of the vessel tree is performed by a region growing approach consisting of the following basic steps:

a) Build the weak vessel object set defined by the pixels with values above VL.
b) Build the strong vessel object set defined by the pixels with values above VH.
c) Take the strong vessel set computed in the preceding step as the initial vessel tree estimate.
d) Add to the vessel tree estimate all objects of the weak vessel set connected to it.
e) Repeat the previous step using the current vessel tree estimated until it stops growing.

Experiments using the above explained procedure have shown that the selection of the threshold values VL and VH is a key issue for an accurate segmentation.

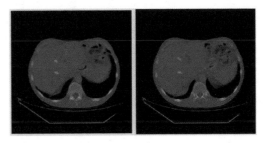

Figure 4.   Vessel tree segmentation.

Appropriate values for these parameters change from image to image depending on the contrast and average brightness level.

We searched appropriate values for VL and VH manually through many experiments using different CT sequences. We observed that the histogram counts for the manually selected values stayed at a roughly constant ratio to the intensity corresponding to the maximum count.

Considering NM the maximum liver histogram count, and NL and NH the counts corresponding respectively to VL and VH , the ratios rl = NL/NM and rh = NH/NM do not significantly change from CT exam to CT exam. These ratios were determined experimentally as rl = 0.5 and rh = 0.2.

This regularity suggests the following procedure to select the lower and higher threshold values:

a) Compute and smooth the histogram of the image region inside the liver;
b) Detect the maximum histogram count NM and the corresponding intensity VM.
c) Multiply NM by the ratios rl and rh, and obtain the count values NL and NH.
d) Search the smoothed histogram for the intensity values VL and VH corresponding to NL and NH, whereby both VL and VH are greater than VM.

Figure 4 shows an example of the results produced by the proposed method. The hepatic vein is shown in red and the portal vein in blue, for two different slices.

2.4   Segmentation of Couinaud regions

The Couinaud paradigm divides the liver into eight independent segments each one having its own vascular inflow, outflow, and biliary drainage. Because of this division into self-contained units, each can be removed without damaging those remaining.

Our method estimates the subdivision of the liver in the eight Couinaud segments, by fitting planes to the portal vein, and to each of the hepatic vein branches. To separate the three main branches of the hepatic vein we apply the k-means algorithm on the 3 dimensional coordinates of the pixels identified in the preceding step as belonging to the hepatic vein. It is assumed that there are three clusters. A restriction for singleton

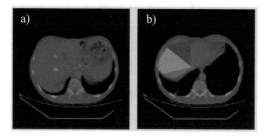

Figure 5. Segmentation by plane fitting – (a) shows the hepatic and portal veins in red and blue respectively; (b) shows the Couinaud segments in different colors.

value is imposed so as to guarantee that no cluster will be empty. This leads to three different objects corresponding to each branch of the hepatic vein.

Then, a least squares based procedure determines the four planes that best fit the points of each branch of the hepatic vein and the portal vein segmented before. These planes divide the liver in the Couinaud segments. Figure 5 shows the Couinaud segments in different colors obtained by this procedure on a sample CT sequence. Figure 5-a shows one CT slice with the hepatic vein in red and the portal vein in blue, and figure 5-b shows another CT slice with the derived segments in different colors.

3   RESULTS

A software prototype implementing the proposed method has been built for validation purpose. Based on the VTK library (Schroeder et al., 1998) the prototype also implements both the surface and volumetric visualization of the structures segmented. It receives as input the segmented structures of each image slice and the thickness of the CT slices available in the DICOM image file header.

Figure 6 shows our segmentation results through a 3D surface model generated by our prototype. In figure 6-a liver, the hepatic vein and the portal vein are shown respectively in gray, blue and red. Figure 6-b shows the Couinaud segments in different colors with the hepatic and portal veins also present. It can be shown that the segments are divided according to the veins orientation, as proposed.

Experiments performed on seven different CT sequences have shown that the results produced by the proposed method are consistent with the visual perception.

4   CONCLUSION

This work proposes an algorithm to segment the liver in computer tomography (CT) images according to the Couinaud classification.

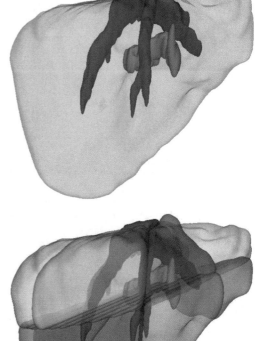

Figure 6.   3D models of segmented structures of the liver.

Experiments conducted on 7 CT series confirm that the proposed method produces a result consistent with the visual analysis. The method has the potential of becoming a useful tool in various applications. It can be used to generate 3D liver representations to aid visual diagnostic and surgery planning. Shape attributes other than volume may also be measured from the 3D model and explored in Computer Aided Diagnostic environments.

The assessment of segmentation accuracy is a major concern for the continuation of this work. This is one of the next steps planned for the continuation of this research.

REFERENCES

American College of Radiology, National Electrical Manufacturers Association, "Digital Imaging and Communications in Medicine (DICOM): Version 3.0", Draft

Standard, ACR-NEMA Committee, Working Group VI, Washington, DC, 1993.

Canny, J., "A Computational Approach to Edge Detection, IEEE Trans. on Pattern Analysis and Machine Intelligence", vol. PAMI-8, no. 6, November 1986, pp. 679–698.

Couinaud, C. "Le Foie: Etudes Anatomiques et Chirurgicales. Masson". Paris. 1957.

Fujimoto H., Gu L., and Kaneko T., "Recognition of abdominal organs using 3D mathematical morphology," Trans. Inst. Electron. Inf. Commun. Eng. D-II, no. 5, pp. 843–850, May 2001.

Kakinuma R., Ohmatsu H., and Kaneko M., "Detection failures in spiral CT screening for lung cancer: Analysis of CT findings." Radiology 212, pg 61–66, 1999.

Kim S.-H., Yoo S.-W., Kim S.-J., Kim J.-C., and Park J.-W., "Segmentation of kidney without using contrast medium on abdominal CT image," in Proc. of ICSP 2000 Fifth Int. Conf. on Signal Processing, Beijing, Aug. 21–25, 2000.

Lamecker H., Lange T., and Seebaß M. "Segmentation of the liver using a 3D statistical shape model". 2004.

Lamecker H., Zachow S., Haberl H., and Stiller M. Medical Applications for Statistical 3D Shape Models. Proc. Computer Aided Surgery Around the Head, volume 17of Fortschritt-Berichte VDI, p. 61, 2005.

Li F., Sone S., Abe H., MacMAhon H., Armato III S., and Doi K., "Lung cancers missed at low-dose helical CT screening in a general population: Comparison of clinical, histopathologic, and imaging findings." Radiology 225, pp 673–683, 2002.

Lorensen W. E., and Cline H. E., "Marching Cubes: A High Resolution 3D Surface Construction Algorithm", Computer Graphics (Proceedings of SIGGRAPH '87), Vol. 21, No. 4, pp. 163–169. 1998.

Masutani Y., Uozumi K., Masaaki Akahane and Kuni Ohtomo, "Liver CT image processing: A short introduction of the technical elements", European Journal of Radiology, Volume 58, Issue 2, Liver Lesions, May 2006, Pages 246–251.

Masutani Y., MacMahon H., and Doi K., "Computerized detection of pulmonary embolism in spiral CT angiography based on volumetric image analysis," IEEE Trans. Med. Imag., vol. 21, no. 12, pp. 517–1523, Dec. 2002.

Oliveira, D.A.B. Mota, G.L.A. Feitosa, R.Q. Nunes, R.A. and "A region growing approach to pulmonary vessel tree segmentation using adaptive threshold". CompIMAGE proceedings, pp. 319–324, 2006, Porto.

Otsu, N., "A Threshold Selection Method from Gray-Level Histograms," IEEE Transactions on Systems, Man, and Cybernetics, Vol. 9, No. 1, 1979, pp. 62–66.

Schroeder W., Martin K., and Lorensen B., "The Visualization Toolkit." Prentice Hall, 2nd edition, 1998.

Soler L. Delingette H. Malandain G. Montagnat J.; Ayache N. Clement J.M. Koehl C. Dourthe O. Mutter D. Marescaux J. "Fully automatic anatomical, pathological, and functional segmentation from CT scans for hepatic surgery," Proc. SPIE Vol. 3979, p. 246–255, Medical Imaging 2000: Image Processing, Kenneth M. Hanson; Ed. 2000.

Computational Vision and Medical Image Processing – João Tavares & Natal Jorge (eds)
© 2008 Taylor & Francis Group, London, ISBN 978-0-415-45777-4

Individualizing biomechanical shoulder models using scaling, morphing and fittings

Germano T. Gomes
Department of Anatomy, Embryology, Histology and Medical Physics, Ghent University, Belgium

Emmanuel A. Audenaert & Lieven De Wilde
Department of Orthopedic Surgery and Traumatology, Ghent University, Belgium

Katharina D'Herde
Department of Anatomy, Embryology, Histology and Medical Physics, Ghent University, Belgium

ABSTRACT: It is increasingly apparent the necessity of subject specific biomechanical models that are aimed at clinical applications, in particular in the design and analysis of orthopedic implants, preoperative planning and interpretation of post operative outcomes. Currently, imaging techniques like MRI and CTs can potentially provide a source for complete individualized models; however, the constraints of clinical reality preclude the creation of full complex models of a region of interest. This gap can be bridged by creating idealized generic musculoskeletal models that can be morphed into patient specific models using limited CT and morphometric data obtained pre-operatively. The transformation from the generic model to a specific one is achieved by a volumetric morphing technique using non uniform scaling, translations, rotations and shear matrices combined with the optimum fitting, in a least squares sense, of spheres, cylinders, ellipsoids, circles, ellipses, planes and general curves, around bony and muscle structures.

1 INTRODUCTION

Painful rotator cuff-insufficient has a tremendous effect on the independence of the aging population. Surgical treatment often includes joint replacement, the functional outcome of which remains unpredictable. The resulting shoulder may have radically modified anatomical and mechanical properties.

Even small anatomic changes have a major impact on the final biomechanical functionality, particularly if portions of the rotator cuff are missing or non-anatomic designs are used. To date optimum prosthetic configuration and technique is not yet defined (Magermans et. al. 2005).

In order to capture anatomical variations medical imagining techniques like MRI and CT scanning can be used but clinical constraints e.g. cost, exposure to unacceptable levels of radiation and available time, prevent, more often then not, the acquisition of information with enough resolution to create detailed individualized biomechanical models capable of revealing the effects these anatomical variations. Even when this information is available, the extraction of important features from the data that can aid the preoperative planning, e.g determining the centre of rotation of a particular joint, cannot be done accurately by simple visual inspection. Therefore a real need exists to use to the fullest the limited data obtained routinely preoperatively and extend its applicability (Kaptein & van der Helm 2004).

To achieve this goal a program as been developed that allows scaling, morph and fitting of geometric shapes to reconstructed 3D bones, using two sets of data: that obtained preoperatively before shoulder surgery and that of a detailed 6 cadaver study.

2 MATERIAL AND METHODS

2.1 Cadaver study

It was hypothised that muscle fibers pahs could be reliable defined by staining them with very thin and therefore flexible metal wires, carefully sutured to them. First a CT analysis of a pork bone-muscle model was performed. Iron wires of thickness 0.9, 1.1 and 1.3 mm; and copper wires of thickness 0.7, 0.9 and 1.1 mm were used. The CT data off this pork bone-muscle model was 3D reconstructed using the software package Mimics (Materialise NV, Heverlee,

Figure 1. Cooper wires sutured to Deltoid muscle.

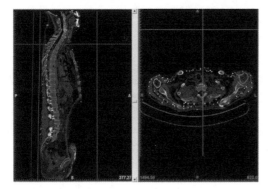

Figure 2. Example of CT slices in sagittal and transverse plain.

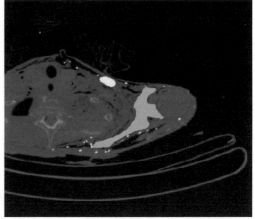

Figure 3. Colour masking of scapula, humerus and clavicle bones.

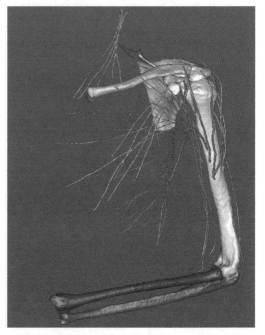

Figure 4. 3D reconstruction of Scapula, clavicle, humerus, radius, ulna and shoulder girdle muscle fiber paths.

Belgium). The proprieties –scattering and visibility versus flexibility- of the different reconstructed metal wires were compared in relation to the pork muscle-bone model.

The 0.7 mm copper wire was selected for application of the method on the human cadaver specimens.

Six cadavers were dissected and in all cadaver specimens the shoulder girdle muscles were sutured with copper wires of 0.7 mm thickness, 18 muscles in total for each side of the body (left and right) with an average of 5 wires per muscle (Figure 1). All dissections and suturing were performed by a professional anatomist and medical students under is supervision.

From the cadaver specimens spiral CTs at 1 mm slice intervals were taken from head to hip level at the University hospital of Gent.

2.2 3D Reconstruction

The Mimics software (Materialise NV, Heverlee, Belgium) was used for the 3D reconstructions from the 2D slices (Figure 2).

The general procedure followed was: importing of the CT images in DICOM format into to Mimics, specification of orientation and spacing of the dataset, thresholding of the bony tissue and cooper wires,

color masking using a mix of semiautomatic (region growth, multislice editing, Boolean operations, hole filling) and manual coloring (Figure 3) and finally creation of 3D models. All bone reconstructions were checked by trained anatomists for possible errors in the segmentation process and corrected when necessary.

The files are exported as 3D meshes in the STL format having in general in excess of 50 000 vertices.

Some post processing operations were performed in order to reduce the file size namely reducing the number of triangles to a more manageable number and smoothing the bone. Post processing was done with the software MAGICS (Materialise NV, Heverlee, Belgium). The STL files are used as an input for the Scaling/Morphing software program.

2.3 Scaling, fitting and morphing

The Matlab programming language was chosen for the creation of the program due to it's wide availability within the academic and industrial communities, excellent prototyping capabilities, availability of a large number of internal toolboxes with many pre made functions and as well the existence of freeware routines written by other researchers.

The concept of the program is that a Graphical user interface (GUI) is the front of the program while Matlab performs the computational intense mathematical calculations on the background.

3 RESULTS

The main features of this program are:

1. Reading of 3D STL files and creation of 3D patchs in the main plot area (Figure 5).
2. Generating Views: top, bottom, right, left, front, back and isometric
3. Display of mesh points and triangles
4. Axes selection and sequence dependence: The axis along which the transformation is applied (X,Y,Z) can be exclusively selected one at a time. The transformations can be applied with or without sequence dependence. That is to say if sequence dependence is on the transformations accumulate on the same bone one after the other and for the different axis selected. With sequence dependence of the bone is always rest to its original state when a different axis or transformation is selected. This is important as in general rotational transformations do not commute.
5. Transformations: these transformation allow to scale and morph the bones along 3 different axis, adding to a total of 12 different possible transformations, when combined in a sequence depend manner, almost any shape of bone even pathological can be obtained. The transformations are: Scaling, Rotations, Translations and Skewing. Values for the transformations can be introduced using a slider or typed numerically.
6. Fitting: the program allows the optimum fitting in a least square sense with statistical analysis of the quality of the fit. Planes, cylinders, spheres, circles, lines, ellipses. These techniques are important in detecting centers of rotation of joints, best axis

Figure 5.   Humerus bone read into GUI.

Figure 6.   Sphere fit to the glenohumeral head to determine the joint centre of rotation.

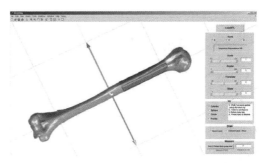

Figure 7.   Cylinder fit to the humeral column to determine best axis of insertion of prosthesis.

for fitting prosthesis and best cutting planes. A minimum of seven points must be picked in the area of interest, the algorithms behind this feature work well up to thousands of picked points, so extremely accurate fittings can be obtained if needed (Figures 6 and 7).

7. Morphing: this encapsulates morphing operations using a variant of the ICP algorithm (Gaojin et al 2006). The basic idea behind this is that having an original ideal bone we are able to morph it automatically into another bone for witch we only have limited CT information. The algorithm rotates,

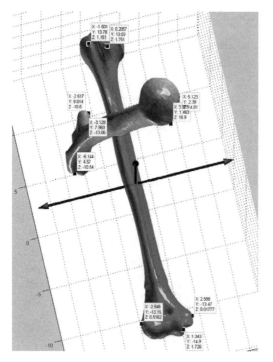

Figure 8.   Selection of 5 points on 2 different bones.

Figure 9.   Morphed bone and original attain reasonable correspondence using only 5 points.

translates and scales the original ideal complete bone in order to do an optimum fitting to the incomplete one. This algorithm works ideally with tens to hundreds of points, to illustrate the robustness of the technique we load two different humerus into the program and we morph one into to the other using only 5 reference points (Figures 8 and 9).

4   DISCUSSION AND CONCLUSIONS

In order to create customized or individualized biomechanical models that aid in preoperative planning of surgery and design of new prosthesis there is a need to take into account anatomical variations. Since at the moment it is realistically impossible to perform both MRI and CT scans of a single patient at a resolution and covering a volume that would allow extracting these individual features, we must find ways to bridge this gap and use morphing of limited available data obtained preoperatively combined with detailed data obtained from cadaver studies.

This study as generated such data including for the first time the muscle fibre paths of all muscles involves in the shoulder girdle. The use of this information with the general transformation techniques presented here allows reconstruction of all the bones involved to fit those of the patient. Furthermore the scaled muscle fibre path curves also allow determining muscle moment arms. Finally the least squares fitting of simple geometric shapes permits us to precisely obtain centers of rotation, axes of bones for prosthesis insertion and best cutting planes.

It remains to be proven quantitatively, on a blind analysis, the precision, accuracy and reliability of this solution. Ongoing work will try giving an answer to these issues.

ACKNOWLEDGMENTS

The authors are grateful to DePuy International (a Johnson and Johnson Company) for the support of Germano T. Gomes throughout this study.

REFERENCES

Kaptein BL & van der Helm FC. 2004. Estimating muscle attachment contours by transforming geometrical bone models. J Biomechanics 37(3):263–73.

Magermans, D.J., Chadwick, E.K.J., Veeger, H.E.J. and. van der Helm, F.C.T. 2005. Requirements for upper extremity motions during activities of daily living. Clinical Biomechanics 20 (6): 591–599.

Gaojin, W., Zhaoqi, W., Shihong, X., Dengming, Z. 2006.Least-squares fitting of multiple M-dimensional point sets. The Visual Computer 22, 387–398.

Computational Vision and Medical Image Processing – João Tavares & Natal Jorge (eds)
© 2008 Taylor & Francis Group, London, ISBN 978-0-415-45777-4

A new semiautomatic method for ICH segmentation and tracking from CT head images

N. Pérez, J.A. Valdés & M.A. Guevara
Center for Advanced Computer Sciences Technologies, Ciego de Ávila University, Ciego de Ávila, Cuba

L.A. Rodríguez
Intensive Care Unit, Morón Hospital, Morón, Ciego de Ávila, Cuba

A.J. Silva
IEETA, Aveiro University, Portugal

ABSTRACT: Recently, computer-aided diagnosis (CAD) has become one of the major research subjects in medical imaging and diagnostic radiology. CAD provide a computer output as a "second opinion" to assist radiologists' image readings. The early evolution and quantitative analysis of the human spontaneous intracerebral brain hemorrhage (ICH) are significant for the treatment and the follow-up of patient's recovery. This paper presents a new method for semiautomatic segmentation and tracking of ICH from computerized tomography (CT) head images, which is based on a suitable combination of digital image processing and pattern recognition techniques. The segmentation and tracking processes involved two main steps: first selection of a query object (ICH) on one study slice (image) and second the automatic object retrieval on the rest of the slices of the patient study. The proposed method had been tested successfully by medical researchers in a representative dataset of 36 ICH CT head images patient studies.

1   INTRODUCTION

Recently, computer-aided diagnosis (CAD) has become one of the major research subjects in medical imaging and diagnostic radiology. The basic concept of CAD is to provide a computer output as a "second opinion" to assist radiologists' image readings (Doi 2005). The early evolution and quantitative analysis of the human spontaneous intracerebral brain hemorrhage (ICH) are significant for the treatment and the follow-up of patient's recovery.

This paper outlines a new method for semiautomatic segmentation and tracking of ICH from computerized tomography (CT) head images. This algorithm is based on a suitable combination of digital image processing and pattern recognition techniques. The segmentation and tracking processes involve two main steps: selection of a query object and object retrieval throughout the selected patient study.

Query object selection (created with minimal user intervention) consists on extracting (in any selected patient study slice) the contour of the recognized ICH region. We used live wire techniques (Liang 2006) to extract the ICH edges, and hereafter based on the selected contour a set of reference points and associated similarity vectors are computed. Reference points and their associated similarity vectors form the query basic elements.

Automatic object retrieval has the aim to track the query object picked over target slices of the patient study. This algorithm attempts to find the points more similar to the current reference points on the slice under processing. The new retrieved points are then used as the reference points to detect the ICH region in the next slice. This process is repeated over all the target slices.

The proposed method had been used by medical researchers to evaluate the behavior and changes in ICH structure (shape, size, etc.) during the disease course and it was validated successfully in a representative dataset of 36 CT head images patient studies with presence of ICH.

The paper is organized as follows: Section 2 describes some previous work related with ICH image analysis. Section 3 outlines the implementation of the new proposed method. In section 4 we describe results obtained with the algorithm application in a representative dataset of patients studies with presence of ICH and in section 5 we present the main conclusions.

2 PREVIOUS WORK

Several works related with segmentation and tracking of ICH from CT head images based on different image analysis techniques have been reported.

Cosic and Loncaric (Cosic 1997; Loncaric 1999) proposed a method based on unsupervised fuzzy clustering and expert system-based region labeling techniques that includes volume measurement of the ICH and the edema regions. Unfortunately the automatic method was not robust to achieve an acceptable routine rate of successful segmentations.

Majcenic and Loncaric (Majcenic 1997) proposed a stochastic method for ICH segmentation based on simulated annealing techniques but the authors recognized the method is computationally complex.

Sucu et al (Sucu 2005) demonstrated the value of XYZ/2 (simple volume estimation method of intracerebral hematoma) to estimate the volume of chronic subdural hematoma, but the applied segmentation and tracking processes were completely manual.

Matesin et al (Matesin 2001) present a method for automatic segmentation and region labeling of CT head images of stroke lesions. This method implement a rule-based expert system focused on the knowledge about symmetry properties of the brain.

Liao et al (Liao 2006) a proposed a model of the deformed midline according to the biomechanical properties of different types of intracranial tissue. Midline shift is an important quantitative feature that clinicians use to evaluate the severity of ICH and other pathologies related with brain compression.

3 PROPOSED METHOD

The new proposed method includes the following steps:

a) Image preprocessing.
b) Query objects selection.
c) Object retrieval.

3.1 Image preprocessing

In order to reduce the computational cost and to improve ICH contrast with respect to background a simple contrast enhancement procedure was introduced in order to emphasize the object's (ICH) edge intensities. We use a linear contrast stretching function to map the grayscale values to new values such that 1% of data is saturated at both intensity extremes.

3.2 Query objects selection

The selection of query object (I_n) is made by the following steps:

a) Extract the object (ICH) contour (CT_n).

b) Extract the reference points (set of points that belong to the query object contour)
c) Create similarity vectors from reference points.

Query object elements are formed by the reference points and their associated similarity vectors. The quantity of reference points and their associated similarity vectors is variable and depend on the specific problem domain knowledge (e.g. for the ICH problem solution we obtained successful results using only 8 reference points).

3.2.1 ICH contour extraction

We used live wire techniques for object contour extraction (segmentation). Live wire was proposed as an interactive boundary tracking technique (Falcão 2000) that share some similarities with snakes and it is generally considered in the literature as a competing snake technique. Like snakes, the idea behind the live wire technique is to allow image segmentation to occur with minimal user interaction, while at the same time allowing the user to exercise control over the segmentation process. This technique shares two essential components: a local cost function that assigns lower cost to image features of interest, such as edges, and an expansion process that forms optimal boundaries for objects of interest based on the cost function and seed points provided interactively by the user (Liang 2006). The obtained curve (object contour) after the live wire application is indeed, a polygonal curve, so we interpolate it with a spline function to obtain a continuous curve, which produce a better object contour approximation. The extracted curve is then used to compute the reference points.

3.2.2 Reference points

Reference points are formed by a set of points that are located close or in the object's contour with direction $i\theta$, where $i : 0..\alpha - 1$, $\alpha = int(360/\theta)$, $0 < \theta \le 180°$ and θ is the displacement angle, taking the object center of mass as the center of the coordinates system. The angle θ will be selected by the user and it will depend on the problem knowledge domain (Fig. 1). Initial reference points are determined by the following algorithm:

Algorithm 1: "Generate Initial Reference Points"
1: Select initial slice $I_{(n)}$
2: Select $\theta, 0 < \theta \le 180°$
3: Select ε: admissible (small) distance between $P_i(x, y)$ and CTn
4: Compute $P_i(X, Y)$: reference points
Coordinates (X,Y) of P_i are obtained by the

$$P_i(X,Y) = C_x + r * \cos(i), C_y + r * \sin(i)$$

expression:
where $r \in Z^+, 1 \le r \le R$ and r increase in 1 on each iteration until $r = R$ and R is the value in which

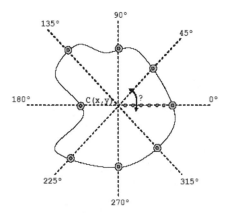

Figure 1. Reference points obtained with $\theta = 45°$, $\varepsilon = 0.5$.

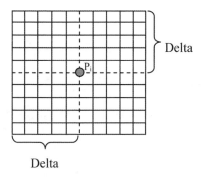

Figure 2. Similarity matrix.

$d_{Pi}(P_i(X, Y), CT_n) \leq \varepsilon$, d is the euclidian distance between $P_i(X, Y)$ and CT_n C_x and C_y are the (X, Y) centroid coordinates

5: *Store new calculated reference points in the* $P_i(X, Y)$ *array*

3.2.3 Similarity vectors

Similarity vectors (S_i) are the basic elements that allow to track the ICH regions on the rest of slices (images) belonging to a selected patient study. The reference points represent the centroid $(C_{x,y})$ of the similarity matrixes (Q_i). Similarity matrixes are MxM windows, where $M = 2*\delta + 1$, δ is a positive integer scalar provided by the user (see Fig. 2). The similarity vectors are obtained from the similarity matrices using the following mathematic formulation:

$$Q_i = \begin{matrix} a_{11} & a_{12} & a_{13} & \cdots & a_{1m} \\ a_{21} & a_{22} & a_{23} & \cdots & a_{2m} \\ a_{31} & a_{32} & a_{33} & \cdots & a_{3m} \\ \vdots & \vdots & \vdots & & \vdots \\ a_{m1} & a_{m2} & a_{m3} & \cdots & a_{mm} \end{matrix}$$

$$S_i = \{a_{1,1} \ldots a_{m,1}, a_{1,2} \ldots a_{m,2}, a_{1,m} \ldots a_{m,m}, a_{1,m+1} \ldots a_{m,m+1}\}$$

where S_i represent the set of similarity vectors corresponding to P_i and Q_i

3.3 Object retrieval

Object retrieval is in general a process devoted to find a query object on a set of images or images databases (Llerena 2004).

In our case object retrieval has the aim to track (extract) the ICH regions on a set of slices selected by the user that belong to the patient under study.

Similarity descriptors are critical for the performance of object retrieval process. Several similarity descriptors were evaluated, but the best results were achieved with the similarity descriptor proposed by Fuertes (Fuertes 1999), which we use to select the new reference points belonging to the slice under analysis:

$$d(P_{i(n)}, P_{i(n+1)}) = \sqrt{(S_{i(n)} - S_{i(n+1)})^t * (S_{i(n)} - S_{i(n+1)})} \quad (1)$$

$S_{i(n)}$ is a similarity vector associated to reference point $P_{i(n)}$ in the I_n image, $S_{i(n+1)}$ is a vector associated to the point $P_{i(n+1)}$ in the I_{n+1} slice, t denote a transpose matrix. The new reference points will be those $P_{i(n+1)}$ where $d(P_{i(n)}, P_{i(n+1)})$ is minimum.

Our algorithm employ the created similarity vectors on query object to find the reference points related to ICH: the points that identify the ICH occurrence on the rest of slices set of patient under study. The object retrieval algorithm developed is outline below:

Algorithm 2: "Retrieval"
1: *Select new slice* $(I_{(n+1)})$
2: *Compute* P_{ik} *points*
    for each i angle
    $P_{ik}(X, Y) = C_{n_x} + k* \cos(i), C_{n_y} + k* \sin(i)$
    where $k : 1..D + \sigma$, σ *is a value selected by the user, D is the euclidean distance between* C_n *and* $P_{i(n)}$
    Build vector S_{ik} *from* P_{ik} *(see 3.2.3)*
    Compute R_{ik}
    $R_{ik} = s(S_{ik}, S_{in})$: s *similarity function*
3: *Select* $m = k$ *where* R_{ik} *is minimum*
4: *Created and store* $P_{i(n+1)}$ *and* $S_{i(n+1)}$
    $P_{i(n+1)} = P_{im}(X, Y), S_{i(n+1)} = S_{im}$
5: *Build the contour* $CT_{(n+1)}$ *based on the* $P_{i(n+1)}$ *array*
6: *Compute the centroid from of* $CT_{(n+1)}$
7: *if (slice set selected is empty) then end else goto 1*

325

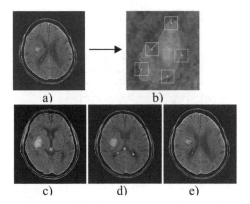

a) b)

c) d) e)

Figure 3. Patient Number 062-06 a) Selected image, b) Query object, c–e) Retrieval images.

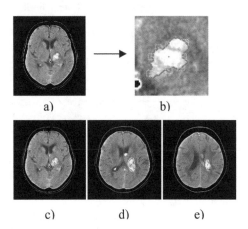

a) b)

c) d) e)

Figure 4. Patient Number 317-06 a) Selected image, b) Query object, c–e) Retrieval images.

4 RESULTS

We have developed a new method for object segmentation and tracking that was applied successfully to segment and track ICH regions from CT head images. Segmentation is performed using live wire techniques while the tracking process is carry out based on a suitable combination of digital image processing and pattern recognition techniques as was described in details on section 3.

In order to test our method we implemented a 1. software prototype in MATLAB, which was validated by medical researchers in a representative dataset of 36 CT head images patient studies. The results obtained indicate that our approach may be 2. particularly accurate and effective for the segmentation and tracking of the ICH regions.

The algorithm was tested on three different set (cases) of parameters settings, such as (θ the

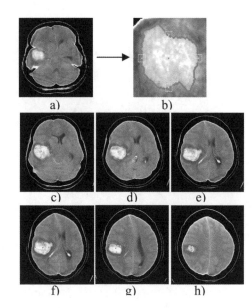

a) b)

c) d) e)

f) g) h)

Figure 5. Patient number 753-06 a) Selected Image, b) Query object, c–h) Retrieval images.

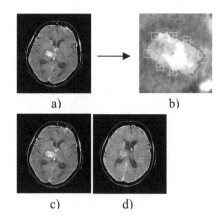

a) b)

c) d)

Figure 6. Patient number 469-06 a) Selected Image, b) Query object, c–d) Retrieval images.

displacement angle, δ the window size for the similarity matrix and ε the admissible euclidean distance from CT_n and selected reference points):

Case 1: $\theta = 45°$, $\delta = 20$ and $\varepsilon = 0.5$
Case 2: $\theta = 72°$, $\delta = 20$ and $\varepsilon = 0.5$
Case 3: $\theta = 90°$, $\delta = 20$ and $\varepsilon = 0.5$

The system answered successfully in 30 cases of 36 patient studies that represent the 83.3% of the total analyzed patient studies. The better segmentation results were obtained with an angle $\theta = 45°$ (case 1).

In Figures 3–8 it is possible to observe successful ICH segmentation and tracking on different patient

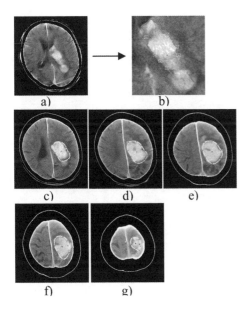

a)   b)

c)   d)   e)

f)   g)

Figure 7. Patient number 138-06 a) Selected Image, b) Query object, c–g) Retrieval images.

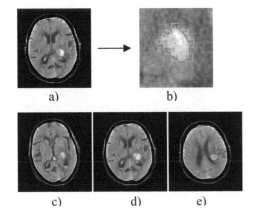

a)   b)

c)   d)   e)

Figure 8. Patient number 674-06 a) Selected Image, b) Query object, c–e) Retrieval images.

studies. The main difficulty of our method resides on sharp deformations between ICH query and the ICH slices shape, which can be observed in figure 9.

5 CONCLUSION

In this paper we proposed a new method for object segmentation and tracking that was applied successfully to segment and track ICH from CT head images. The method efficiency was confirmed on a representative dataset of patient studies. Compared with other

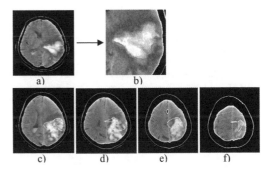

a)   b)

c)   d)   e)   f)

Figure 9. Patient number 566-06 a) Selected Image, b) Query object, c–f) Retrieval images.

approaches devoted to ICH segmentation (Cosic 1997; Majcenic 1997; Loncaric 1999) our approach demonstrated a less computational complexity and higher accuracy due to a tighter user control of whole process.

ACKNOWLEDGMENTS

This work was carried out in the Center for Advanced Computer Sciences Technologies at Ciego de Avila University, Cuba. The authors thanks the Cuban Neurosciences Center for their financial support, as well as the Radiology Department of the Morón Hospital, Cuba for give us a representative set of patient images studies and their participation on the evaluation of the proposed method performance.

REFERENCES

Cosic, D., Loncaric, S. (1997). *Computer System for Quantitative Analysis of ICH from CT head Images*. Proceedings of the 19th Annual International Conference of the IEEE.

Doi, K. (2005). "Current status and future potential of computer-aided diagnosis in medical imaging." *The British Journal of Radiology* **78**(Special Issue): S3–S19.

Falcão, A. X., Udupa, J. K., Miyazawa, F. k. (2000). "An ultra-fast user steered segmentation paradigm: live-wire-on-the-fly." *IEEE Transactions on Medical Imaging* **19**(1): 55–62.

Fuertes, J., M. (1999). Recuperación de Imágenes en bases de datos a partir del color y la forma. *E. T. S. de Ingeniería Informática*. Granada, Universidad de Granada. **PhD**.

Liang, J., McInerney, T., Terzopoulos, D. (2006). "United Snakes." *IEEE Transactions on Medical Image Analysis* **10**: 215–233.

Liao, C.-C., Chiang, I-Jen, Xiao, F., Wong (2006). "TRACING THE DEFORMED MIDLINE ON BRAIN CT." *Biomedical Engineering Applications, Basis & Communications* **18**(6): 305–311.

Loncaric, S., Dhawan, A., P., Kovacevic, D., Cosic, D., Broderick, J., Brott, T. (1999). *Quantitative intracerebral brain hemorrhage analysis*. Proceedings of SPIE Medical Imaging, San Diego, USA.

Llerena, Y. (2004). Técnicas para el almacenamiento y recuperación de información gráfica en bases de datos de imágenes. *CEIS*. La Habana, Centro Universitario José Antonio Echevarria. **Master Sciences**.

Majcenic, Z., Loncaric, S., (1997). *CT Image Labeling Using Simulated Annealing Algorithm*. IX European Signal Processing Conference.

Matesin, M., Loncaric, S., Petravic, D., (2001). *A Rule-based Approach to Stroke Lesion Analysis from CT Brain Images*. Proceedings of 2nd International Symposium on Image and Signal Processing and Analysis, Pula, Croatia.

Sucu, H., K., Gokmen, M., Gelal, F. (2005). "The Value of XYZ/2 Technique Compared With Computer-Assisted Volumetric Analysis to Estimate the Volume of Chronic Subdural Hematoma." *Stroke* **36**: 998–1000.

Computational Vision and Medical Image Processing – João Tavares & Natal Jorge (eds)
© 2008 Taylor & Francis Group, London, ISBN 978-0-415-45777-4

Segmentation of objects in images using physical principles

Patrícia C.T. Gonçalves & João Manuel R.S. Tavares
Mechanical Engineering and Industrial Management Institute – INEGI, Porto, Portugal
Faculty of Engineering of the University of Porto – FEUP, Porto, Portugal

Renato Natal Jorge
Mechanical Engineering Institute, Porto, Portugal
Faculty of Engineering of the University of Porto – FEUP, Porto, Portugal

ABSTRACT: The final goal of this work is to automatically extract the contour of an object represented in an image after manually defining an initial contour for it. This rough initial contour will then evolve until it equals the border of the desired object. The contour is modelled by a physical formulation, using the Finite Element Method, and its evolution to the desired final contour of the object to segment is governed by several forces: internal forces, defined by intrinsic physical characteristics selected for the model; and external forces, defined in function of image features that best represent the desired object. To physically model the considered contour we adopt Sclaroff's isoparametric finite element, and to obtain the evolution of the model towards the object border we use Nastar's methodology that consists in solving the dynamic equilibrium equation.

1 INTRODUCTION

Segmentation, the identification of an object represented in an image, is one of the most common and complex tasks in the domain of Computational Vision. Usually, whenever it is intended to extract higher level information from an image, the used process starts by segmenting it. Thus, image segmentation is one of the areas with more work done in Computational Vision and so it will probably be throughout the times.

There are several methods for segmenting objects represented in images; for example, active contours, level set methods, active shape models and deformable templates. Active contours (also known as snakes), introduced by Kass et al. (1988), use an initial curve that deforms elastically and is immersed in a potential field, function of the image features, that attracts it to the object border. The goal of this method is to minimize the total energy of the curve in order to define the contour of the object to segment.

With level set methods (Sethian 1999), the deformation of the contour is formulated as a propagating wave front that is considered as a specific level set of a function. This function has a term of speed defined by the object in the image that stops the propagation of the wave front as soon as it delimits the object.

Active shape models (Vasconcelos & Tavares 2006) need *a priori* knowledge of the object: the object to segment is sampled by a set of points in each of the images of the training set, as well as by the gray levels around each point. The several sets of points of the training images are aligned to minimize the distance between corresponding points and, analyzing the variation of that distance, we obtain the point distribution model that is used to define an average shape for the object in question and to restrict its movement in the segmentation of that object in new images.

Another method that also uses *a priori* knowledge is that of deformable templates (Carvalho & Tavares 2006). The templates are defined by parameters, which describe the expected geometrical shape of the object to segment, and interact dynamically with the image during the segmentation process. As with snakes, an energy function is defined that attracts the template to the object border, the minimum of this energy corresponding to the best possible segmentation.

The main goal of this work is to extract the contour of an object represented in an image, after the definition of an initial contour for it. This contour, defined roughly by the user of the developed implementation, will evolve throughout an iterative process until it reaches the border of the desired object. For that purpose, it was decided to use deformable models defining their behaviour according to physical principles, as proposed by Nastar (1994). Thus, it is used the dynamic equilibrium equation, also known as equation of motion, applied to an elastic physical model of the contour used in the segmentation process.

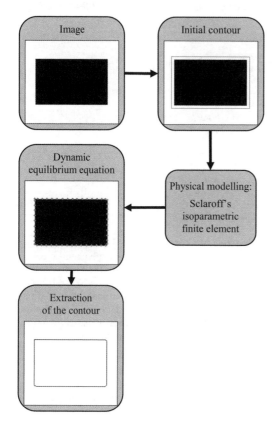

Figure 1. Schema of the methodology used.

The here used methodology is summarily described in Figure 1. The first step consists of drawing a shape on the image with the object that is close to its border, which will be considered as the initial segmentation contour for that object. Next, this contour is modelled according to physical principles using the Finite Element Method, namely Sclaroff's isoparametric finite element (Sclaroff 1995). To move the model to the border of the object to segment, the dynamic equilibrium equation is solved, that describes the equilibrium between the internal and external forces involved. The internal forces are defined by the physical characteristics adopted for the model, determined by the adopted virtual material and the distance between the nodes of the model; and the external forces are defined in terms of the image features that best describe the object to segment.

2 PHYSICAL MODELLING

After defining the initial contour for the object to segment, we model it in physical terms; that is, we assign mass, stiffness and damp to each point of the contour, that is, to each node of the used model.

To model the initial contour and simulate its elastic behaviour, Nastar (1994) used affine interpolation functions together with finite differences. In this work, we employ the Finite Element Method and Gaussian interpolants instead. Namely, we use Sclaroff's isoparametric finite element (Sclaroff 1995) that uses a set of radial basis functions that allows an easy insertion of the data points in the model. Therefore, Gaussian interpolants are used and the nodes of the model do not need to be previously ordered. With this isoparametric finite element, when an object is modelled it is as if each of its feature points are covered by an elastic membrane (Sclaroff & Pentland 1995, Tavares et al. 2000).

For the m points $\mathbf{X}_i(x_i, y_i)$ of the initial contour manually defined by the user, the interpolation matrix $\mathbf{H}$, which relates the distances between the contour points, of Sclaroff's isoparametric finite element is built using, (Sclaroff 1995, Sclaroff & Pentland 1995):

$$g_i(\mathbf{X}) = e^{-\|\mathbf{X}-\mathbf{X}_i\|^2/2\sigma^2}, \tag{1}$$

where σ is the standard deviation, which controls the interaction between the contour points. Then, the interpolation functions, h_i, for the finite element are given by:

$$h_i(\mathbf{X}) = \sum_{k=1}^{m} a_{ik} g_k(\mathbf{X}), \tag{2}$$

where a_{ik} are coefficients that satisfy $h_i = 1$ at node i and $h_i = 0$ at the other m-1 nodes. These interpolation coefficients compose matrix $\mathbf{A}$, and can be determined by inverting matrix $\mathbf{G}$ defined as:

$$\mathbf{G} = \begin{bmatrix} g_1(\mathbf{X}_1) & \cdots & g_1(\mathbf{X}_m) \\ \vdots & \ddots & \vdots \\ g_m(\mathbf{X}_1) & \cdots & g_m(\mathbf{X}_m) \end{bmatrix}. \tag{3}$$

Thus, matrix $\mathbf{H}$ of the interpolants functions will be:

$$\mathbf{H}(\mathbf{X}) = \begin{bmatrix} h_1 & \cdots & h_m & 0 & \cdots & 0 \\ 0 & \cdots & 0 & h_1 & \cdots & h_m \end{bmatrix}. \tag{4}$$

The mass matrix of Sclaroff's isoparametric element, $\mathbf{M}$, is defined as (Sclaroff 1995, Sclaroff & Pentland 1995, Tavares et al. 2000):

$$\mathbf{M} = \begin{bmatrix} \mathbf{M}' & 0 \\ 0 & \mathbf{M}' \end{bmatrix}, \tag{5}$$

where $\mathbf{M}'$ is a sub-matrix $m \times m$ defined as $\mathbf{M}' = \rho\pi\sigma^2 \mathbf{A}^T \mathbf{\Gamma} \mathbf{A} = \rho\pi\sigma^2 \mathbf{G}^{-1} \mathbf{\Gamma} \mathbf{G}^{-1}$, because $\mathbf{A}$ is symmetric, ρ is the mass density of the virtual material adopted for the model, and the elements of matrix $\mathbf{\Gamma}$ are the square roots of the elements of matrix $\mathbf{G}$.

On the other hand, the stiffness matrix, $\mathbf{K}$, is given by:

$$\mathbf{K} = \begin{bmatrix} \mathbf{K}_{11} & \mathbf{K}_{12} \\ \mathbf{K}_{21} & \mathbf{K}_{22} \end{bmatrix}, \qquad (6)$$

where $\mathbf{K}_{ij}$ are symmetric $m \times m$ sub-matrices depending on constants, α, β and λ, that are functions of the virtual material adopted for the contour (Sclaroff 1995, Sclaroff & Pentland 1995, Tavares et al. 2000):

$$\mathbf{K}_{11_{ij}} = \pi\beta \sum_{k,l} a_{ik} a_{jl} \left[\frac{1+\lambda}{2} - \frac{\hat{x}_{kl}^2 + \lambda \hat{y}_{kl}^2}{4\sigma^2} \right] \sqrt{g_{kl}}, \qquad (7)$$

$$\mathbf{K}_{22_{ij}} = \pi\beta \sum_{k,l} a_{ik} a_{jl} \left[\frac{1+\lambda}{2} - \frac{\hat{y}_{kl}^2 + \lambda \hat{x}_{kl}^2}{4\sigma^2} \right] \sqrt{g_{kl}}, \qquad (8)$$

$$\mathbf{K}_{12_{ij}} = \mathbf{K}_{21_{ij}} = -\frac{\pi\beta(\alpha+\lambda)}{4\sigma^2} \sum_{k,l} a_{ik} a_{jl} \hat{x}_{kl} \hat{y}_{kl} \sqrt{g_{kl}}, \qquad (9)$$

where $\hat{x}_{kl} = x_k - x_l$, and $\hat{y}_{kl} = y_k - y_l$.

In this work, we use Rayleigh's damping matrix, $\mathbf{C}$, which is a linear combination of the mass and stiffness matrices with constraints, μ and γ, based upon the chosen critical damping (Bathe 1996, Cook et al. 1989):

$$\mathbf{C} = \mu\mathbf{M} + \gamma\mathbf{K} . \qquad (10)$$

3 DYNAMIC EQUILIBRIUM EQUATION

To estimate the evolution of the physical model built for the contour towards the object edges, and thus to obtain the desired segmentation, we solve the second order ordinary differential equation, commonly known as Lagrange's dynamic equilibrium equation:

$$\mathbf{M}\ddot{\mathbf{U}}^t + \mathbf{C}\dot{\mathbf{U}}^t + \mathbf{K}\mathbf{U}^t = \mathbf{F}^t, \qquad (11)$$

for each time step t, where $\mathbf{U}$, $\dot{\mathbf{U}}$ and $\ddot{\mathbf{U}}$ are, respectively, the displacement, velocity and acceleration vectors, and $\mathbf{F}$ represents the external forces (Gonçalves et al. 2006, Pinho & Tavares 2004). This equation relates the contour physical properties with its displacement, speed and acceleration in order to keep the balance with the external forces applied to the model.

To solve the dynamic equilibrium equation we use Newmark's method (Gonçalves et al. 2006), a direct integration process that has been modified and improved by many researchers.

3.1 External forces

The external forces are usually defined as the gradient of a potential field, $\mathbf{V}$. The external force at point P is given by:

$$\mathbf{F}(P) = -\nabla\mathbf{V}(P). \qquad (12)$$

Nastar & Ayache (Nastar & Ayache 1993, Nastar 1994) define that potential as:

$$\mathbf{V}(P,t) = \frac{1}{2}k\left\|\overrightarrow{P(t)Q}\right\|^2, \qquad (13)$$

where k is a stiffness constant and Q is the point of the image resulting from the use of image filters closest to P, a point of the contour model. To simplify: the external force applied to point P of the contour, in every instant t, is proportional to the displacement of point P:

$$\mathbf{F}(\mathbf{X}_P(t)) = k(\mathbf{X}_Q - \mathbf{X}_P(t)). \qquad (14)$$

In this work, Shen & Castan's edge detection operator (Shen & Castan 1992) is used to build the image from which point Q is obtained.

Because an image with the detected edges can have noisy data that do not concern the object we are trying to segment, an adequate point Q can be difficult to obtain. Thus, determining a line orthogonal to each point of the contour, we determine Q to be the most valuable data point that the orthogonal line intersects.

4 SOME RESULTS

To illustrate the results of the methodology here proposed, to segment an object represented in an image by identifying its contour, consider the images in Figure 2. The first one has the synthetic object to segment in black, and in red the initial contour defined for it. The second image represents the segmentation obtained for a physical model with 37 nodes (points) and made of rubber, and considering $k = 50,000$ N/m. The third image has the segmentation result. In this case, the computational process took 8s to achieve the final result. (In this work, a personal computer was used, with an Intel Pentium D at 3 GHz processor and 2 GB of RAM.)

If the initial contour is modelled with 61 nodes instead of 37, the obtained contour describes better the original object, especially in areas with small curvature, Figure 3, but the computational process takes 22s to finish.

The bigger the constant k, the stiffness constant in the definition of the external force, the faster the segmentation process runs, because bigger external forces

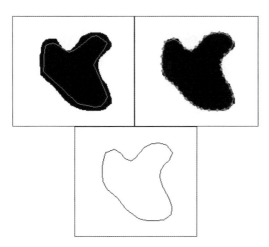

Figure 2.   Initial contour (top left); result of the segmentation with $k = 50,000$ N/m considering a model with 37 nodes and made of rubber (top right); extracted contour (bottom).

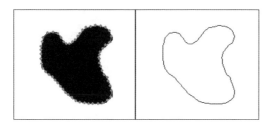

Figure 3.   Result of the segmentation using a model with 61 nodes, made of rubber, and $k = 50,000$ N/m (left); extracted contour (right).

Figure 4.   Initial contour (top left); result of the segmentation with $k = 20,000$ N/m considering a model with 91 nodes and made of rubber (top right); extracted contour (bottom).

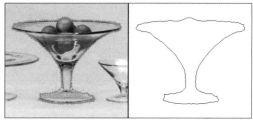

Figure 5.   Result of the segmentation using a model with 181 nodes, made of rubber, and $k = 20,000$ N/m (left); extracted contour (right).

applied to the model mean that it will deform faster. However, the value of k is related with the number of nodes used in the contour model: usually, the bigger the number of nodes, the smaller k has to be so that there are no numerical inconsistencies. For example, for the model in Figure 3, with 61 nodes, k can take values up to 50,000 N/m, but for the model in Figure 2, with only 37 nodes, k can take values up to 70,000 N/m, taking the segmentation process only 6s to end instead of the 8s taken when $k = 50,000$ N/m.

The chosen virtual material for the model built also influences the computational time, because applying the same load to different materials results in different deformations. For example, if the 61 nodes model in Figure 2 were made of aluminium instead of rubber, using $k = 70,000$ N/m would make the computational process take over four hours to end. Because aluminium is more rigid than rubber, it can support a bigger external force, so k can be bigger. Using $k = 1.9 \times 10^7$ N/m in the 61 nodes model made of aluminium, the segmentation is achieved after 327s. Even using the highest k possible for an aluminium model,

the segmentation process takes a lot longer to finish than with a rubber one using the highest possible k for this case, which is consistent with the expected behaviour of real objects, because it is easier and faster to deform rubber than aluminium.

The first image in Figure 4 represents real objects, and, in red, it has the initial contour manually defined. The second image represents the segmentation result using a model with 91 nodes and made of rubber, and $k = 20,000$ N/m. The third image has the contour extracted from the segmentation process. In this case, the computation took 90s to achieve the final result.

Similar to the object in Figures 2 and 3, if the initial contour is modelled using more nodes, the obtained contour describes better the original object, but the computational process takes longer to finish. For example, in Figure 5, the initial contour was modelled with 181 points, and the final result was only obtained after 520s.

In the case of more complex images, such as images with objects overlapped, like the one in Figure 6, the

332

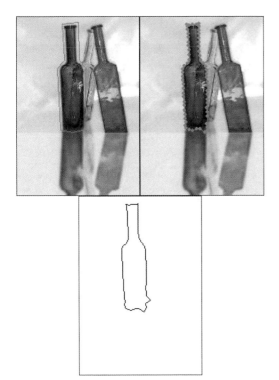

Figure 6. Initial contour (top left); result of the segmentation with $k = 30,000$ N/m considering a model with 63 nodes and made of rubber (top right); extracted contour (bottom).

final result may not be the expected one, because the edges of others objects can be more intense than the edges of the object to segment, consequentially attracting some nodes of the model to the border of the wrong objects.

Another problem arises if some nodes of the initial contour are not close enough to the object border: those nodes are not attracted to the edge of the object to segment and the segmentation result is not the adequate one, as the example presented in Figure 7.

5 CONCLUSIONS AND FUTURE WORK

In this paper, a methodology to segment objects represented in images based on physical principles was presented.

The experimental results obtained using our methodology, some presented in this paper, are quite satisfactory. However, this approach has two major problems:

1) It becomes slower as the number of nodes of the model used for the contour in the segmentation process increases; what can be very inconvenient when a detailed contour extraction is to be accomplished.

Figure 7. Initial contour (top left); result of the segmentation with $k = 50,000$ N/m considering a model with 91 nodes and made of rubber (top right); extracted contour (bottom).

2) The segmentation result can be severely compromised when the image in which the object is represented is very complex, with noisy data or objects overlapped, for instance.

Because of these two major problems, in the near future, some changes to fasten and improve the segmentation process will be introduced, such as trying different approaches for the definition of the external forces, and using other integration methods for the resolution of the dynamic equilibrium equation. The use of adjusted finite elements for large deformations is also a subject to be considered in the following stages of this work, and so is a solution to solve the problem of the nodes not close enough to the contour.

The segmentation of some specific type of images will be also considered. For that, it will become necessary and advantageous the introduction of adequate constraints in the segmentation methodology to improve its robustness.

ACKNOWLEDGMENTS

The presented work was partially done in the scope of the project "Segmentation, Tracking and Motion Analysis of Deformable (2D/3D) Objects Using Physical Principles", with reference POSC/EEA-SRI/55386/2004, financially supported by *FCT – Fundação para a Ciência e a Tecnologia* in Portugal.

REFERENCES

Bathe, K.-J. 1996. *Finite Element Procedures*. New Jersey, USA: Prentice-Hall.

Carvalho, F.J. & Tavares, J.M. 2006. Two Metodologies for Iris Detection and Location in Face Images. *Proc. of the CompIMAGE – Computational Modelling of Objects Represented in Images: Fundamentals, Methods and Applications, Coimbra, Portugal, 20–21 October*.

Cook, R., Malkus, D. & Plesha, M. 1989. *Concepts and Applications of Finite Element Analysis*. New York, USA: John Wiley and Sons.

Gonçalves, P., Pinho, R.R. & Tavares, J.M. 2006. Physical Simulation Using FEM, Modal Analysis and the Dynamic Equilibrium Equation. *Proc. of the CompIMAGE – Computational Modelling of Objects Represented in Images: Fundamentals, Methods and Applications, Coimbra, Portugal, 20–21 October*.

Kass, M., Witkin, A. & Terzopoulos, D. 1988. Snakes: Active Contour Models. *International Journal of Computer Vision* 1(4): 321–331.

Nastar, C. & Ayache, N. 1993. Fast Segmentation, Tracking, and Analysis of Deformable Objects. *Proc. of the Fourth International Conference on Computer Vision, Berlin, Germany, 11–14 May*.

Nastar, C. 1994. *Modèles Physiques Déformables et Modes Vibratoires pour l'Analyse du Mouvement Non-Rigide dans les Images Multidimensionnelles*. Thèse de Doctorat. École Nationale des Ponts et Chaussées. Champs-sur-Marne, France.

Pinho, R.R. & Tavares, J.M. 2004. Dynamic Pedobarography Transitional Objects by Lagrange's Equation with FEM, Modal Matching and Optimization Techniques. *Lecture Notes in Computer Science* 3212: 92–99.

Sclaroff, S. 1995. *Modal Matching: a Method for Describing, Comparing, and Manipulating Digital Signals*. PhD Thesis. Massachusetts Institute of Technology. Cambridge, USA.

Sclaroff, S. & Pentland, A. 1995. Modal Matching for Correspondence and Recognition. *IEEE Transactions on Pattern Analysis and Machine Intelligence* 17(6): 545–561.

Sethian, J.A. 1999. *Level Set Methods and Fast Marching Methods: Evolving Interfaces in Computacional Geometry, Fluid Mechanics, Computer Vision, and Materials Science*. New York, USA: Cambridge University Press.

Shen, J. & Castan, S. 1992. An Optimal Linear Operator for Step Edge Detection. *CVGIP: Graphical Models and Image Processing* 54(2): 112–133.

Tavares, J.M., Barbosa, J. & Padilha, A.J. 2000. Matching Image Objects in Dynamic Pedobarography. *Proc. of the RecPad 2000 – 11th Portuguese Conference on Pattern Recognition, Porto, Portugal, 11–12 May*.

Vasconcelos, M.J. & Tavares, J.M. 2006. Methodologies to Build Automatic Point Distribution Models for Faces Represented in Images. *Proc. of the CompIMAGE – Computational Modelling of Objects Represented in Images: Fundamentals, Methods and Applications, Coimbra, Portugal, 20–21 October*.

Computational Vision and Medical Image Processing – João Tavares & Natal Jorge (eds)
© 2008 Taylor & Francis Group, London, ISBN 978-0-415-45777-4

A didactic application for creating new sights from multiple images using light fields

Brígida Mónica Faria, A. Augusto de Sousa & Luís Paulo Reis
Faculty of Engineering of the University of Porto, s/n Porto, Portugal

ABSTRACT: A Light Field is an imaged based rendering technique based on the eight dimensional Plenoptic Function, simplified to a four dimensional function. It describes the amount of light covering the space, from any point, in any arbitrary direction. It ignores variables like the time and wavelength and assumes that radiance is equal throughout a line in the free space and so it is more efficient than most of the other image based rendering techniques in the literature. Drawbacks appear in the quality of the visualization but may be reduced by the correct use of the technique regarding the desired application. This paper presents a didactic application of this type of imaged based rendering technique that uses multiple sights of a scene through different points of views. The application of light-fields to medical images is thought one of the main objectives of this work and so, a simple application of this image based rendering technique, to this type of images is presented in order to show the usefulness of the approach.

1 INTRODUCTION

In computational terms, a geometrically based scene is a data structure that represents a set of objects. Each object is visible by its boundary or surface that, by its turn, can be curved or, alternatively, a composed by a polygon mesh plane However, the real world generates energy. So, a study about light-fields and its application to biomedical images should start by understanding light behavior, in terms of interaction with objects surfaces, as well as its propagation and its perception by human beings. The creation of images through the evaluation of a model of propagation of light is known by image synthesis where the main objective to be reached is photorealism. One of the related problems consists in simulating illumination. This process should not be slow and the projected 2D images should be realistic. It is intended that the process must be fast and possible to perform in real time. The main objective is to try achieving the visual sensation of the corresponding real scene. Ideally, a complete illumination simulation should be performed so it would make it impossible to distinguish between the real and the virtual scene. Nevertheless, this would result in a too much high computational speed.

Light Field is an image based rendering technique supported by the eight dimensional Plenoptic Function, simplified to a four dimensional function. This function describes the amount of light that covers the space, from any point, in any arbitrary direction; by the other hand it ignores several variables like the time and wavelength and it assumes that radiance is equal throughout a line in the free space. With these simplifications, it is more efficient than most of the other imaged based rendering techniques in the literature. Drawbacks appear in the quality of the visualization but may be reduced by the correct use of the technique regarding the desired application.

The technique, due to its inherent complexity, is not sufficiently known from most computer science researchers and users. This paper intends to present this technique based on a didactic application using medical images as main test cases, in order to show the usefulness of the approach.

The paper is organized as follows. In section 2 it is described the deduction of the radiance equation and the importance of understanding how illumination data is introduced. Section 3 presents the study elaborated by (Adelson 1991) that explains as the world can be analyzed, from camera or points of view, as a set of light rays that form the space. In the following section, considerations are made to the level of definition, representations and construction of light fields. In section 5 a didactic application of light fields and how it can be used in medical visualization of 3D human organs, is presented. Finally, section 6 is dedicated to the work main conclusions, complemented by some perspectives of future work.

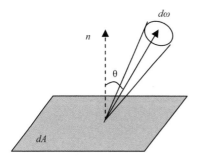

Figure 1. Radiance is the flow for unit of area projected for unit of solid angle.

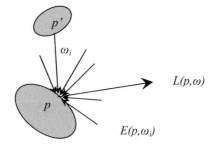

Figure 2. Radiance equation.

2 RADIANCE EQUATION

Radiance (L) is the flow that leaves a surface, for each unit of projected area of the surface, for unit direction of solid angle (Cohen et al. 1993). Being dA the surface area from the energy leaving in direction θ in relation to the normal of dA through the solid angle $d\omega$, the process is exemplified in figure 1:

If the radiance that leaves the area dA is L, the correspondent flow is:

$$d^2\Phi = LdA \cos\theta d\omega$$

To the measure dA and $d\omega$ becomes more worthless $d^2\Phi$ represents the flow throughout a ray in the direction θ. Another way to think radiance is $L(p, \omega)$ that it is the function that is integrated throughout the solid angle and projected area in order to get the radiating power (flow) of the area. The illumination problem would be decided if it was possible to find $\Phi(p, \omega)$ for all the points and directions. Based on study by the scientific community, estimated constraints had been applied in order to simplify the function and applied radiance instead of radiant power, $L(p, \omega)$ instead of $\Phi(p, \omega)$.

The equation deduced and expressed in terms of radiance instead of flow is assigned for Radiance Equation:

$$L(p,\omega) = L_e(p,\omega) + \int_{\Omega} f(p,\omega_i,\omega)L(p,\omega_i)\cos\theta_i d\omega_i$$
$$= L_e(p,\omega) + \int_0^{2\pi}\int_0^{\frac{\pi}{2}} f(p,\omega_i,\omega)L(p,\omega_i)\cos\theta_i \sin\theta_i d\theta d\phi \quad (1)$$

where $L_e(p, \omega)$ is the emitted radiance.

Some solutions have been presented for this equation. An interesting approach is the incorporation of all 2D view points of images that represent the 3D scene. The problem is how to extract the necessary information for the construction of these images. The process of extracting, from the radiance equation, the projected 2D images is called image synthesis or image rendering.

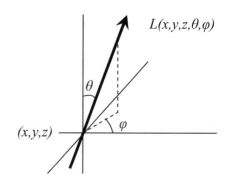

Figure 3. Parameterization of a ray in the space through the position (x, y, z) and direction (θ, φ).

The radiance throughout all the rays in a region of the 3D space, illuminated by a homogeneous set of lights, is called Plenoptic function.

3 PLENOPTIC FUNCTION

World appearance may be analyzed as a dense set of light rays that fill the space and can be observed through the eyes position or chambers in the space. This set of light can be represented through the Plenoptic function, proposed by (Adelson 1991). The rays in space can be parameterized by three coordinated x, y and z and two angle θ and φ; adding the variables time and wavelength, the function becomes with dimension 7.

Let see the example off an image made with a pinhole camera: the light that passes through the camera center of projection correspond is made up of light rays that can also be considered as samples of the Plenoptic function.

The process of Image Based Rendering (IBR) is based on images to produce new images. On the basis of the Plenoptic function, the concept of image based rendering can be defined as a process with two phases: image sampling and image synthesis. Given a continuous function that describes the scene, the image sampling phase consists on the selection of samples extracted through the Plenoptic function, for representation and storage. In the image synthesis phase, the continuous function, Plenoptic function, is reconstructed from the captured samples (Chai et al. 2000).

One of the problems in IBR is the function number of dimensions, seven, that makes the sampling process very difficult. The huge amount of necessary data makes it very difficult to present all the functions in a single representation. Much of the research made in IBR has concentrated in finding out the best estimate to reduce the size of the data samples, keeping a reasonable quality in the phase of rendering.

Many representations of IBR exist in the literature, but usually follow two main strategies to reduce the size of the necessary data. The first consists of constraining the observer's space of visualization; this constraint enables to effectively reduce the dimension of the Plenoptic function, which facilitates the image sampling and synthesis. The second strategy consists of introducing more scene information as it is the case of its geometry; light field techniques do not use this type of solution.

An IBR definitive time representation makes it possible to reduce the data size even more, through the sampling compression. The sampling analysis can supply the information of the minimum number of images/light rays necessary to synthesize the scene with satisfactory quality. On the other hand, the compression can still remove the redundancy inside and between images.

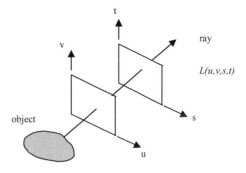

Figure 4. Light Field parameterization.

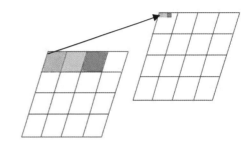

Figure 5. Light Field representation.

that later are calculated by quadrilinear interpolation because there exist light rays in the neighborhood of the image set.

4 LIGHT FIELDS

The most known 4D IBR representations are Light fields (Levoy 1996) and Lumigraph (Gortler 1996). Both ignore the wavelength and time and assume that radiance does not change throughout the line that supports a light ray in a space free of participating media. Although the space parameterization of guided lines is still a problem, the known solutions to overcome this difficulty became being a similar one: one light ray is recorded for its intersections with two parallel planes, respectively indexed with coordinates (u, v) and (s, t): L(u, v, s, t).

In figure 4 an example can be seen, in which the two planes, called camera and focal planes, are parallel to each other.

An example of light ray is also shown in figure 4, indexed as (u_0, v_0, s_0, t_0). The two planes are discretised, so that a finite number of light rays is recorded. An image is obtained if all the discret points, on the focal plane, are connected to the points of camera plane, (2D light rays subset). Therefore, the 4D representation is also a set of 2D images. To create a new sight of the object, the sight is divided in light rays,

4.1 Light Field: definition and representation

A light field is an IBR technique that uses multiple views of a scene corresponding to different view points. The main idea is to capture, in a large array, the light that arrives from the scene. The scene light conditions can be then characterized through the 5 dimensions of the Plenoptic function that describe the flow of light (radiance) in each point of the tridimensional space (corresponding to 3 degrees of freedom) in any direction (more 2 degrees of freedom).

Assuming that the object is perfectly limited and a the environment space is completely transparent (there is no participating media), the radiance doesn't change throughout a light ray. This permits to reduce the problem dimension from 5 to 4 by the use of the two parallel planes. Being so, a light ray is parameterized by the coordinates of its intersection with the two parallel planes, this being the first adopted representation.

By convention, the first plane system of coordinates is (u, v) and the second is (s, t), and a guided line is defined by joining a point in the uv plane to a point in st plane. In practice, the parameters u, v, s and t are normalized to values between 0 and 1, and the resulting points in each plane are constrained to be in a convex quadrilateral. This representation is called light slab. One light slab represents a light beam that enters in a quadrilateral and leaves by another.

This representation allows placing one of the planes in the infinite. This may be convenient, since, in this case, lines can be parameterized only by a point and a direction. In the construction of light fields from orthographic images or images with a fixed field of sight is possible.

A great advantage of this representation is the highest efficiency of geometric calculations involving Linear Algebra. By convention, the camera plane has coordinates (u, v) and the focal plane has coordinates (s, t); the camera is placed in such a way that its optical axis is perpendicular to both planes. The focal plane can be interpreted as the camera plane image, being (s, t) the pixel position. The 4-uple (u, v, s, t) contains the information of the (s, t) pixel color in the sight position (u, v). To get a complete description, that allows an arbitrary position of the vew point, 6 pairs of planes can be used, each one lined up in a side of a cube envolving the scene. Clearly, other positions and directions of camera can be used for the light field acquisition, if the resultant space 4-D is re-sampled in order to coincide with the light field representation.

The above representation is similar to the lumigraph, although this last one has some information of the scene geometry. The light field (in fact a subgroup of this) is then represented by L(u, v, s, t) and discretised by every possible lines between the two planes, defined by rectangular grids imposed in each plane. The plane uv is subdivided into squares, where each vertex forms the center of projection, with a rectangular subset of the st plane, like as a window of the eye plane forms an associated image. It has an image associated with each point of the plane uv, and a radiance value associated to each (u, v, s, t) combination representing lines that intersected the scene convex hull. This may be described as forming one light slab. In order to cover all the possible directions of the rays, six copies – three orthogonal rotations and, in each case, two directions, are necessary.

Once constructed, the light field can be used to synthesize an image through a virtual camera that does not correspond to the set of cameras in the plane uv. A new image can be formed through the sampling of the set of respective lines through the point of view in the required directions (Lin et al. 2000).

The light field approach was developed as a method to create new sights of real scenes images. Assuming that digital photographs under constrained conditions are taken off forming a set of points of view and directions associates with the planes uv this set of images can clearly be used to create a light field. A light field for virtual scenes can be constructed using a different rendering system in order to create images.

4.2 *Representations of Light Fields*

The representation in two parallel planes (2PP) has computational advantages, especially related with the storage, since images are rectangular and with the use of specialized hardware for image synthesis (especially for texture) for the reconstruction.

However, the 2PP representation does not give an uniform distribution of the rays in the 4D space. This means that the quality of the synthesized image is a function of the view point and of the direction of sight: points of view that are close to the center of the plane uv, and with orthogonal directions to this plane will result in better images than others, far from the center or with oblique directions of sight. This is due to the number and distribution of rays that vary for different positions. A deeper analysis of the representation 2PP as a space 4D can be found in (Gu et al. 1997).

Camahort and Fussel (Camahort et al. 1998) presented a study for three alternative representations of light fields: 2PP as above, two spheres (2 SP) and the representation through a direction and point (DPP). The representation 2SP involves to place a sphere around the scene, and to subdivide it according to an uniform grid. Then, links between pairs of vertices form the light field representation. Representation DPP also involves the rank of a sphere around the scene, but with random points over the sphere surface. This defines a vector through the origin until this point. The plane orthogonal to this vector and through the center of the sphere is considered, and limited for the sphere (that is a disk in the origin of the sphere). Choosing a set of points uniformly distributed in the same direction of the original vector, the collection of all these rays will form a set of rays uniformly distributed in the ray space, each ray being represented by its direction and its intersection in the disk through the origin. A different representation using two spheres was introduced in (Ihm Park et al. 1997). In this work, a sphere is placed around a scene and then a set of smaller spheres in the surface of the spheres in the border to represent directions. Also in this work, a method based in wavelets is used for compression of the corresponding data structure.

Camahort and Fussel supplied an in-depth analysis of these different methods (as well as a uniformity quarrel), and shown that the boarding direction and point resulted in little polarization in the image synthesis and so simple corrections need to be applied.

5 DIDACTIC APPLICATION OF LIGHT FIELDS

In order to demonstrate the potentialities of the light field technique, a simple application that allows to explore, in an interactive environment, all the

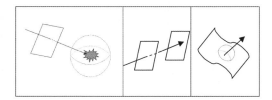

Figure 6. Light Field parameterizations. At left: spherical parameterization, in the middle two planes parameterization; At right: through a point and direction.

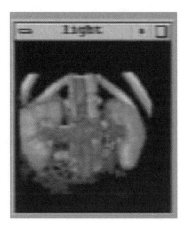

Figure 7. Light Field image with lungs and heart.

potentialities of this technique, was adapted. The initial application was developed in C++ language, in a Linux environment, using as a base, the standard libraries OpenGL and GLUT and initial lightpack 1.0 program. The initial application was developed by (Ohly 1999). In figure 7 the application interface for visualizing medical images, is visible.

Between the parameters that may be configured, are distinguished the parameters that are a disposal made under a fixed measure of struct *LFParam* that are combined with a number of parameters in a struct *LFParams* for extended light field. Its number can be set in compiling time with defined *LF_Maxparms*.

In the implementation, the image synthesis with the library *liblight* brought some advantages to the rendering program of light fields (*lifview*) that, in such a way, shows to the original data, as well as the synthesis of image, of light fields. The reconstruction and the sampling optimization were implemented and studied with the program *panalysis* (parameter analysis) that characterizes the control of the data source for analysis of the parameters effect.

Through the experimentation of different values for diverse parameters, as the options of compression or removal of the planes uv and st, it is possible to an user with basic knowledge of image processing and analysis and computer graphics, to perceive and to use this technique in its multimedia applications.

6 CONCLUSIONS

This article presented the light field's technique for image based rendering. To describe this technique it was firstly essential to introduce the base concepts such as in graphical computing how the environment study is performed. In this way, the equation of the radiance and the Plenoptic Function are fundamental to understand the use of this technique that only demands dimension 4. The representations of light fields were also target of study, as well as the different options to do it.

Since the first publications related to the Light Fieldsubject, the interest for this theory has increased due to the easy way to construct light fields. However, this technique is more appropriate to static scenes than to animate ones and with illumination without high variations. Thus, due to the great number of images necessary to get a good relation in amount/quality, the study on the sampling and compression continues to be an area of great interest and with relative results (Magnor et al. 2000), (Wilburn 2002) and (Chan et al. 2003).

The developed didactic application allows, in an extremely simple way, to a user with basic knowledge of image processing and analysis, as well in computer graphics, to perceive and to use this technique. It also enables its applications to try all the types of parameterizations of light fields in different images and to visualize the results with the aid of diverse indices of performance.

The implemented didactic application needs however to be complemented and made available to the community for a more wider use. In this way, a website with information about light fields and enabling interactive experimentation of the technique is being developed. The availability of this site will allow a more generalized use of the light fields technique in the area of multimedia.

REFERENCES

Adelson, E. & Bergen, J. 1991. The plenoptic function and the elements of early vision, Computational Models of Visual Processing, 3–20.
Camahort, E., Lerios, A. & Fussel, D. 1998. Proceedings of the 9th EUROGRAPHICS workshop Rendering, Uniformly Sampled Light Fields.
Chai, J., Tong, W., Chan, S. & Shum, H. 2000. Plenoptic Sampling, Computer Graphics (SIGGRAPH'00), 307–318.
Chan, S., Chan, K. & Shum, H. 2003. The data compression of Simplified Dynaic Light Fields, ICASSP'03, Hong Kong.

Cohen, M. & Wallace, 1993. Radiosity and Realist Image Sintesis, Academia Press Professional, Harcourt Brace and Company, Pulishers, 13–40.

Gortler, S., Grzeszczuc, R. & Cohen, M. 1996. The Lumigraph Computer Graphics (SIGGRAPH'96), 43–54.

Gu, X., Gortler, S. & Cohen, M. 1997. Proceedings of the 8th EUROGRAPHICS workshop on Rendering Polyedral Geometry and the two Plane Parameterization Render.

Ihm, I., Park, S. & Lee, R. 1997. Proceedings of the 5th Pacific Conference on Computre Graphics and Applications, Rendering of Spherical Light Fields, 59.

Levoy, M. & Hanrahan, P. 1996. Computer Graphics, SIGGRAPH, 96, Light Field Rendering, 31–42.

Lin, Z. & Shum, H. 2000. On the number of Samples Needed in Light Field Rendering with constant deph assumption, Proc. CVPR'00, Hilton Head Island, South Carolina, USA.

Magnor, M. & Girod, B. 2000. Data compression for light field rendering, IEEE Trans. On CSTV, Vol. 10, No. 3.

Ohly, P. 1999. Extended Light Fields and their application to Volume Rendering.

OWilburn, B. & Smulski, M & Lee, H. & Horowitz, M. 2002. The Light Field Video Camera, Proceedings of Media Processors 2002, SPIE Electronic Imaging.

Computational Vision and Medical Image Processing – João Tavares & Natal Jorge (eds)
© 2008 Taylor & Francis Group, London, ISBN 978-0-415-45777-4

Teaching kinematics and dynamics by using virtual reality capabilities

Paulo Flores

Mechanical Engineering Department, University of Minho, Portugal

ABSTRACT: In this work the virtual reality capabilities in teaching kinematics and dynamics is presented and discussed. Over the last few decades several commercially available software tools can be used for enhancing design and analysis tasks related to mechanisms and other dynamics systems. The present work includes development of free-body diagrams and equations of motion of individual components of an elementary slider-crank mechanism driven by a rotational motor. Standard numerical analysis techniques using Excel and virtual prototyping environment provided by Working Model software are used. As expected, both simulation environment yield similar results, however, the visual display of the global motion produced using Working Model provides an excellent correspondence between the abstract mathematics and a realistic animation of the physical reality. In general, students involved with this kind of project teaching have expressed enhancing understating of the subject matter with the integration of the software tools. A second and more complex demonstrative example is used to show the virtual reality capabilities in the kinematics and dynamics of complex mechanical systems, namely, in what concerns to the simulation of a hexapod robotic system.

1 INTRODUCTION

It is well known that over the last few years, a number of commercially available software programs are powerful tools for engineers and designers, being an alternative to the traditional approaches based on analytical analysis (Claro & Flores 2002). Yet, in this paper, the computer capabilities are used in a teaching perspective. In fact, the computer aided analysis software can be used to study kinematic and dynamic characteristics of motion of simple and complex systems. Design software simulation allows testing design performance and simulating the behavior of all components prior to build a physical prototype. This enables the determination of the most critical situation and, consequently, the design of the mechanical components and motion generators needs (Flores 2000, Flores 2005).

The traditional approach of teaching based on *'chalk and talk'* style attempts to transmit knowledge from teachers to a passive recipient. However, during the last years, there is a growing awareness among engineering educators that while this style of instruction is suitable for teaching engineering analysis, it has some limitations when it comes to nurturing creativity, synthesis and engineering design, where different possible solutions have to be tested (Akay 2002, Lima et al. 2007, Teixeira et al. 2007). Thus, it is essential to combine teaching by lectures and active

learning techniques in order to have high motivation and participation of the students and, consequently, to reach better understanding of the issues taught. In addition, the computer packages related to the mechanical systems in general become a powerful alternative to the classic approach, and promote students' creativity. Furthermore, for 2D and 3D complex systems, the traditional mathematical analysis of the equations of motion can be a hard and difficult task, if not impossible for some cases. The modern software simulation programs provide an efficient way to help students in the active learning process (Akay 2003). Moreover, the reform movement in engineering education inspired by Engineering Criteria of ABET (Accreditation Board of engineering Technology) is consistent with this method (Peterson 1997).

In this paper, the virtual reality capabilities are used in teaching kinematics and dynamics. The remainder of the paper is organized as follows. In section two the capabilities of the computational programs are presented. A virtual construction and simulation of a slider-crank mechanism is done in section three. The traditional analysis of the same mechanism is presented in section four. Section five includes the computational modelling and simulation of a more complex system. Finally, in the last section, the main conclusions from this study are drawn and the perspectives for future research are outlined.

2 COMPUTER-AIDED SIMULATION'S CAPABILITIES

In recent years a number of computer-based aids for engineering has emerged, due not only to the development on hardware, but also owing to the improvements on software and mathematical tools. In this context, the computational programs specially dedicated to the kinematic and dynamic analysis of mechanical systems became a powerful alternative to the classic methods. Simulation software allows testing design performance and predicting component's behavior, prior to building a physical prototype (MSC 1999, Flores 2000). These software packages present several advantages, such as (*i*) the possibility of simulating and visualizing the global motion produced by the creation of virtual models, (*ii*) the opportunity of testing, (*iii*) evaluating and correcting different configurations in real conditions, (*iv*) the capability of observing the functionality of the system, (*v*) the flexibility and easiness to process the information and, above all, (*vi*) the economy of time, materials and money spent on its development (Nikravesh 1988, Shabana 1989).

By and large, mechanical simulation requires data inputs on components geometry (which can be created via a CAD system) and mass properties, connections between the elements (i.e., degrees of freedom or restrictions/constraints) and external forces acting on the system (springs, dampers, revolute and linear actuators, etc.). From these inputs, the equations of motion are automatically generated and solved for every component, using numerical methods approach, and displacements, velocities, accelerations and reaction forces are computed (Nikravesh 1988).

In general, the computational programs use numerical methods to allow the solution of the equations of motion of mechanical systems, which are governed, in general, by a set of algebraic differential equations arising from mechanics principles. The main numerical methods used by these computational programs are Newton Raphson's method, Euler's method (the simplest and the fastest method) and Runge-Kutta's method, among others (Burton 1979). The accuracy of results depends on the choice of integration method as well as on the time step selected. The choice of the time step is a critical parameter in fixed numerical integrators, because it affects significantly the speed and accuracy of the result. In general, a small time step produces more accurate simulation results, but requires more computational effort per given time period than a larger integration time step. Additionally, these software packages have automatic collision detection, which is used to simulate body's surface interaction. Results are presented to the designer either in graphic or digital form and the computed data can also be exported to other applications such as

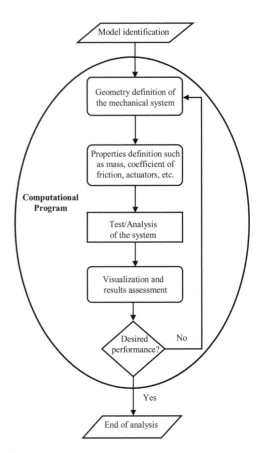

Figure 1. Working steps with computational programs dedicated to perform kinematic and dynamic analysis of mechanical systems {adapted from Flores 2000}.

spreadsheets or Finite Element Analysis (FEA) programs, where additional analysis, evaluation and post processing can be performed.

Computer-Aided programs, usually, offer a complete array of 3D joints and constraints, from motors, actuators, etc. enabling the user to model all types of complex 3D mechanisms. In commercial programs the moving components (part and assemblies) of the mechanical systems are modeled as rigid bodies that are connected to each other in accordance with realistic connections between components with various types of joints (allowing 0–6 degrees of freedom, DOF). The connections provided by these programs are called ideal or perfect connections, that is, the effects of clearance, mass and friction of joints are neglected. Another limitation of the commercial packages is related to the non consideration of the flexibility/deformability of bodies.

Figure 1 depicts the working steps during the analysis of mechanisms when a computational program especially dedicated to perform kinematic and dynamic analysis of mechanical systems is used (Flores 2000). In short, according to the above mentioned issue, computational packages related to the mechanical systems, in general, become a powerful and user-friendly alternative to the classic methods, mainly in complex systems, such as robots, where the traditional approach is very problematic, if not impossible in some cases.

3 TRADITIONAL ANALYSIS OF THE SLIDER CRANK MECHANISM

In this section the analytical analysis, based on the classical approach of the slider-crank mechanism is presented. This methodology is developed based on the free-body diagram of each body that constitutes the mechanism together with the applications of the dynamic equations of motion, also known as the Newton-Euler equations of motion or D'Alembert principle (Flores 2000).

Figure 2 shows the configuration of the mechanism, comprising four rigid bodies that represent the crank, connecting rod, slider and ground, three revolute joints and one ideal translational joint. The inertia properties and length characteristics of each body are given in Table 1.

In the kinematic and dynamic simulation, the crank is the driving element and rotates at a constant angular velocity of 500 rpm clockwise. The initial system configuration corresponds to the top dead point.

Figure 3 shows the free body diagrams of the crank, connecting rod and slider. Thus, applying the dynamic equations of motion to these diagrams yields the following set of equations,

$$F_{34}^x = m_4 a_4^x \tag{1}$$

$$F_{34}^y - W_4 + N_R = m_4 a_4^y \tag{2}$$

$$F_{23}^x - F_{43}^x = m_3 a_3^x \tag{3}$$

$$F_{23}^y - F_{43}^y - W_3 = m_3 a_3^y \tag{4}$$

$$F_{23}^x \frac{l_3}{2} \sin\varphi - F_{23}^y \frac{l_3}{2} \cos\varphi - \\ F_{43}^x \frac{l_3}{2} \sin\varphi - F_{43}^y \frac{l_3}{2} \cos\varphi = I_3 \alpha_3 \tag{5}$$

$$F_{12}^x - F_{32}^x = m_2 a_2^x \tag{6}$$

$$F_{12}^y - F_{32}^y - W_2 = m_2 a_2^y \tag{7}$$

$$M_2 - W_2 \frac{l_2}{2} \cos\theta - F_{23}^y l_2 \cos\theta + F_{23}^x l_2 \sin\theta = 0 \tag{8}$$

By knowing the kinematic data of the slider crank mechanism, equations (1) up to (8) can be solved for the dynamic characteristics of the mechanism, namely for the reaction forces and moments that act on the mechanism. It should be mentioned that for simple systems it is possible to perform this kind of analysis, however, for more complex systems, the formulation

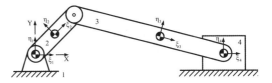

Figure 2. Slider-crank mechanism.

Table 1. Geometric properties of the slider-crank mechanism.

Body Nr.	Description	Length [m]	Mass [kg]	Moment of Inertia [kgm²]
1	Ground	–	–	–
2	Crank	0.050	0.30	1.0×10^{-4}
3	Connecting rod	0.120	0.21	2.5×10^{-4}
4	Slider	–	0.14	–

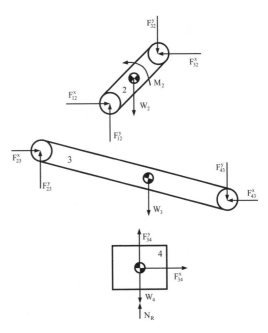

Figure 3. Free-body diagrams of the slider-crank mechanism.

of the equations of motion can be a very hard task since it involves a good deal of mathematical manipulation, if not impossible for some cases, as it will be demonstrated in section 5 of the present paper.

4 VIRTUAL MODELING AND SIMULATION OF THE SLIDER CRANK MECHANISM

Working Model program was used to model and analyze the slider-crank mechanism. Furthermore, a CAD

Figure 4.   Slider-crank mechanism modeled in the Working Model program.

Figure 5.   Mechanical structure of the hexapod robot.

program was used to construct the links of the mechanism. The crank was modeled as a flywheel in order to account for realistic situation. The same considerations were taken into account for the connecting rod and slider. Figure 4 shows the global configuration of the slider-crank mechanism in the Working Model environment and in the case that the system is assembly. This process is quite intuitive for users.

The links that constitute the slider-crank mechanism were, then, constrained by kinematic joints. In the present situation, the mechanism includes three revolute joints and one translational joint. One of the links is considered to be fixed, that is, the ground. An angular motor is attached at the center of mass of the crank in order to actuate the mechanism.

The geometric and inertia properties of the slider-crank mechanism are the same as listed in table 1.

As it is evident from figure 4, it should be highlighted that with this type of commercial programs it is quite useful to teach kinematics and dynamics by using virtual reality capabilities. With this approach it is very easy to teach basic concepts, such as the meaning of degrees of freedom of a system, the physical meaning of the kinematic constraints associated with the different type of joints, just to mention a few of the facilities. Moreover, it is possible to simulate real and different scenarios by including small changes in the mechanism topology, being possible a parametric study, a sensitivity analysis, an optimization process, instead of using the complex and abstract mathematical models based in the traditional approach.

5   APPLICATION OF A HEXAPOD SYSTEM

In this section, a more complex system is presented in order to show the real capabilities of the virtual reality in the analysis of mechanical systems.

Figure 6.   Kinematic configuration of a leg.

Figure 5 shows the mechanical structure of a hexapod robot that consists of one rigid body, load carrying mainframe with six legs, similar and symmetrically distributed. Each leg is composed by four links, interconnected by four revolute joints and attached to the main body by means of a fifth revolute joint. Revolute motors and linear actuators accomplish traction movement and elevation, respectively. The figure 6 illustrates the kinematic configuration of one leg. The foot of each leg is rigidly attached.

Figure 7 shows an animation sequence of the virtual simulation that deals with climbing a standard set of stairs (height $\approx$170 mm, deep $\approx$280 mm). as it is obvious, the study of motion characteristics of this type of systems is quite difficult when the traditional formulations and used, since the system topology is complex and involves contact problems. Conversely, by using the virtual reality capabilities, the analysis of any mechanical system becomes an easy and funny task!

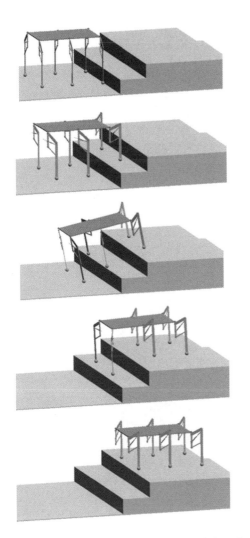

Figure 7. Animation sequence of a virtual simulation of a standard set of stairs climbing.

6 CONLUDING REMARKS

In this paper, an attempt to demonstrate the virtual reality capabilities in the teaching kinematics and dynamics was done. In the process, the traditional approach was presented and used to model a simple mechanical system, namely the slider-crank mechanism. Yet, this methodology is abstract in nature. Thus, the visual display of the global motion produced using Working Model provides an excellent correspondence between the abstract mathematics and a realistic animation of the physical reality. In general, students involved with this kind of project teaching have expressed enhancing understating of the subject matter with the integration of the software tools. Furthermore, a second and more complex demonstrative example was used to show the virtual reality capabilities in the kinematics and dynamics of complex mechanical systems, namely, in what concerns to the simulation of a hexapod robotic system.

REFERENCES

Akay, A., New directions in Mechanical Engineering; Big-Ten-Plus Mechanical Engineering Department Heads; Clearwater Beach, Florida, 2002.

Akay, A., The Renaissance Engineer: Educating Engineers in a Post-9/11 World, *European Journal of Engineering Education*, Vol. 28(2), pp. 145–150. 2003.

Burton, P., *Kinematics and Dynamics of Planar Machinery*, Prentice-Hall, New Jersey, 1979.

Claro, J.C.P. & Flores, P., Influência da Modelização das Juntas na Análise do Desempenho de um Mecanismo, *8as Jornadas Portuguesas de Tribologia*, Universidade de Aveiro, Aveiro, Portugal, (editado por José Grácio, Paulo Davim, Qi Hua Fan and Nasar Ali), Maio, 2002, pp. 215–219.

Flores, P., *Análise Cinemática e Dinâmica de Mecanismos com Recurso a Meios Computacionais*, Trabalho de Síntese integrado nas Provas de Aptidão Pedagógica e Capacidade Científica, Universidade do Minho, Guimarães, Portugal, Abril, 2000.

Flores, P., *Dynamic Analysis of Mechanical Systems with Imperfect Kinematic Joints*, Tese de Doutoramento, Universidade do Minho, Guimarães, Portugal, Fevereiro, 2005.

Lima, R.M., Carvalho, D., Flores, A.M., van Hattum-Janssen, N., A case study on project led education in engineering: students and teachers' perceptions, *European Journal of Engineering Education*, 2007, 32(3), pp. 1–11.

MSC/Working Model *User's Manual*, 1999.

Nikravesh, P. E., *Computer Aided Design of Mechanical Systems*, Prentice Hall, New Jersey, 1988.

Peterson, G.D., Engineering Criteria 2000: a bold new change agent. *ASEE Prism*. September 1997, pp 30–34.

Shabana, A. A., *Dynamics of Multibody Systems*, John Willey & Sons, New York, 1989.

Teixeira, J.C.F., Silva, J.F., Flores, P., Development of Mechanical Engineering Curricula at the University of Minho, *European Journal of Engineering Education*, 2007 (in press).

Computational Vision and Medical Image Processing – João Tavares & Natal Jorge (eds)
© 2008 Taylor & Francis Group, London, ISBN 978-0-415-45777-4

Advanced graphical tools on modelling and control of automation systems

E. Seabra, J. Machado & P. Flores

Mechanical Engineering Department, Enginnering School, University of Minho, Guimarães, Portugal

ABSTRACT: In this paper it is shown, as it is possible, and desirable, the use of a simulation technique in the development and analysis of industrial controllers. Simulation is used in the design process of dynamic systems. The results of simulation are employed for validating a model. The paper concentrates on these aspects and applications of simulation, presents building modules blocks for modelling and simulation and discusses requirements for simulation and validation. In this paper the modelling was performed by using the object-oriented language Modelica and the library for hierarchical state machines StateGraph.

1 INTRODUCTION

In recent years, many software engineering researchers have identified the software process as the key issue to obtain higher quality products, improved productivity, and more controllable projects (Humphrey 1989; Junkermann et al. 1994, Montangero et al. 1994, Lamb et al. 1991).By software process we mean the set of activities, rules, methodologies, tools, and roles that participate in the development of software within a given organization. For this purpose the software engineering community is producing an increasing effort in designing and developing languages and the related support technology to formally describe, assess, and – wherever possible – automate software processes.

Such languages have been designed to allow automatic generation of efficient simulation code from declarative specifications. The Modelica language (Fritzson et al. 1998; Elmqvist et al. 1999; Fritzson et al. 2002) and its associated support technologies have achieved considerable success through the development of specific libraries.

Modelica supports both high level modelling by composition and detailed library component modelling by equations. Models of standard components are typically available in model libraries. Using a graphical model editor, a model can be defined by drawing a composition diagram (also called schematics) by positioning icons that represent the models of the components, drawing connections and giving parameter values in dialogue boxes. Constructs for including graphical annotations in Modelica make icons and composition diagrams portable between different tools. In this paper it is shown, as it is possible,

and desirable, the use of this simulation technique in the analysis of industrial controllers, because the industrial controllers developed will be more robust and much less subject to errors.

To accomplish our goals, in this work, the paper is organized as follows. In Section 1, it is presented the challenge proposed to achieve in this work. Section 2 presents a general presentation of the case study involving a tank filling/emptying system. Further, it is presented the methodology to obtain the controller program deduced from an IEC 60848 SFC specification of the system desired behaviour. Section 3 is exclusively dedicated to the plant modelling, being presented the adopted approach. Section 4 presents and discusses the obtained results on simulation performed with the Modelica Language. Finally, in Section 5, the main conclusions and future work are presented.

2 CASE STUDY

Figure 1 illustrate an example of a tank filling/emptying system, that consists of two tanks, two level analogue sensors (one for each tank) and three onoff valves.

In the normal operation, the system works as follows. Tank 1 is filled by opening valve V1. When the level of the tank 1 becomes high, the valve V1 is closed. After a wait time of ten time units, valve V2 is opened and the fluid flows from tank 1 into tank 2. When tank 1 is empty, valve 2 is closed and after a waiting time of fifteen time units, valve 3 is opened and the fluid flows out of tank 2. Finally when tank 3 is empty, valve V3 is closed.

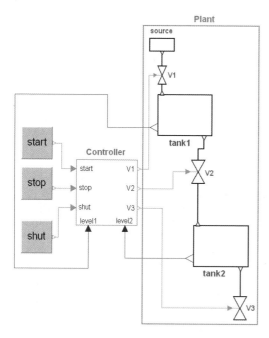

Figure 1. Scheme of the filling/emptying system.

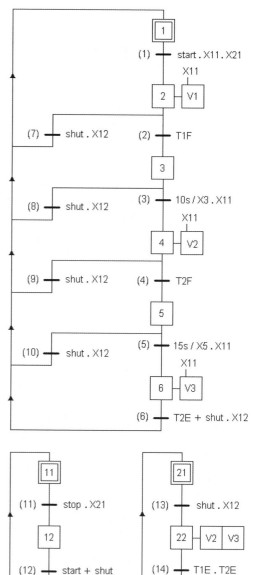

Figure 2. SFC specification of the controller.

The above normal operation can be influenced by three buttons:

- Button "start" starts the above process. When this button is pressed, after a "stop" or "shut" operation, the process operation continues;
- Button "stop" stops the above process by closing all valves. Then, the controller waits for further input (either "start" or "shut" operation);
- Button "shut" is used to shutdown the process, by emptying at once both tanks. When this is achieved, the process goes back to its start configuration. Clicking on the button "start" restarts the process.

2.1 Controller

In order to guarantee the desired behaviour of the system described above, the controller was developed according to IEC 60848 SFC specification, which is presented in Figure 2.

The PLC program which controls the process in closed-loop has input and output variables as described in Table 1.

The tank level is given in % of the fill tank. The Boolean variables T1F (tank1 full) and T2F (tank2 full) were considered true when the level1 and level2 was greater than 0.98, respectively. On the other hand, the Boolean expression T1E (tank1 empty) and T2E (tank2 empty) were assumed true when the level1 and level2 was less than 0.01, respectively.

The controller was modelled using the Dymola program and the object-oriented language Modelica

Table 1. Input/Output variables of the controller.

Input	Output
start – process start	V1 – open valve1
stop – process stop	V2 – open valve2
shut – process shutdown	V3 – open valve3
level1 – % fill tank1	
level2 – % fill tank2	

348

```
model Tank1

    Modelica.Blocks.Interfaces.RealOutput levelSensor;
    Modelica.StateGraph.Examples.Utilities.inflow inflow1;
    Modelica.StateGraph.Examples.Utilities.outflow outflow1;
    Real level "Tank level in % of max height";
    parameter Real A=1 "ground area of tank in m²";
    parameter Real a=0.2 "area of drain hole in m²";
    parameter Real hmax=1 "max height of tank in m";
    constant Real g=Modelica.Constants.g_n;
equation
    der(level) = (inflow1.Fi - outflow1.Fo)/(hmax*A);
    if outflow1.open then
        outflow1.Fo = sqrt(2*g*hmax*level)*a;
    else
        outflow1.Fo = 0;
    end if;
    levelSensor = level;

end Tank;

connector Modelica.Blocks.Interfaces.RealOutput =
                output RealSignal "'output Real' as connector";

connector Modelica.Blocks.Interfaces.RealSignal
    "Real port (both input/output possible)"
    replaceable type SignalType = Real;

    extends SignalType;

end RealSignal;

connector Modelica.StateGraph.Examples.Utilities.inflow

    import Units = Modelica.SIunits;

    Units.VolumeFlowRate Fi "inflow";
end inflow;

connector Modelica.StateGraph.Examples.Utilities.outflow

    import Units = Modelica.SIunits;

    Units.VolumeFlowRate Fo "outflow";
    Boolean open "valve open";
end outflow;
```

Figure 3. Modelica program code for the model of tank1.

(Elmqvist and Mattson, 1997) with the library for hierarchical state machines StateGraph (Otter, 2005).

3 PLANT MODELLING

The plant was modelled as the controller using the Dymola program and the object-oriented language Modelica.

Figure 3 presents the program code for the tank1 model. The Modelica code for modelling the tank2, is similar to the code obtained for the tank1 model because is equal the numbers of fill sources. Additionally the Figures 4 and 5 present the program code used for modelling, respectively, the valves and the source.

4 SIMULATION RESULTS

In order to perform the simulation, it is necessary to define the parameters, start and stop time of the simulation, the interval output length or number of output intervals and the integration algorithm. In the present work, in all simulations performed, the Dass algorithm (Basu, 2006) with 500 output intervals was used.

```
model valve
    Modelica.Blocks.Interfaces.BooleanInput valveControl;
    inflow inflow1;
    outflow outflow1;
equation
    outflow1.Fo = inflow1.Fi;
    outflow1.open = valveControl;
end valve;

connector Modelica.Blocks.Interfaces.BooleanInput =
                input BooleanSignal "'input Boolean' as connector";

connector Modelica.Blocks.Interfaces.BooleanSignal =
                Boolean "Boolean port (both input/output possible)";

connector Modelica.StateGraph.Examples.Utilities.inflow

    import Units = Modelica.SIunits;

    Units.VolumeFlowRate Fi "inflow";
end inflow;

connector Modelica.StateGraph.Examples.Utilities.outflow

    import Units = Modelica.SIunits;

    Units.VolumeFlowRate Fo "outflow";
    Boolean open "valve open";
end outflow;
```

Figure 4. Modelica program code for the model of the valves.

```
model Source

    outflow outflow1;
    parameter Real maxflow=1 "maximal flow out of source";

equation
    if outflow1.open then
        outflow1.Fo = maxflow;
    else
        outflow1.Fo = 0;
    end if;
end Source;

connector Modelica.StateGraph.Examples.Utilities.outflow

    import Units = Modelica.SIunits;

    Units.VolumeFlowRate Fo "outflow";

    Boolean open "valve open";
end outflow;
```

Figure 5. Modelica program code for the model of the source.

In order to study the system behaviour different values for physical variables of the plant were used. Table 2 shows the variables considered in the simulation of the plant model.

It was performed three simulations with the purpose of verifying if the SFC of the controller system (Fig.2) modelled with Modelica language with the library for hierarchical state machines StateGraph simulated correctly the system, respectively, in their start, stop and shutdown process. The values for the plant variables considered in these simulations were $Q = 1$, $G1 = G2 = 1$, $Ht1 = Ht2 = 1$, $A1 = 0.2$ and $A2 = 0.05$.

Table 2. Variables of the plant.

Plant	Variable
source	Q – flow rate [m^3/s]
tank1, tank2	$G1, G2$ – ground area [m^2]
	$Ht1, Ht2$ – height [m]
	$A1, A2$ – drain hole area [m^2]

Table 3. Time actuation of the buttons.

	Period time actuation [s]		
Simulation	Start	Stop	Shut
Start process	[0–1]		
Stop process	[0–1, 24,25]	[11–12]	
Shut process	[0–1]	[11–12]	[15–16]

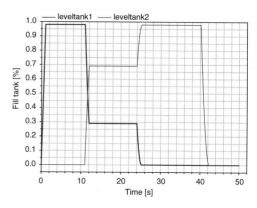

Figure 7. Level tanks in function of time in stop operation process of the system.

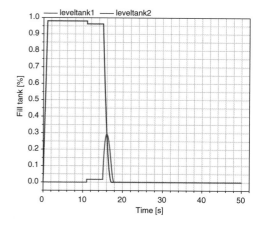

Figure 8. Level tanks in function of time in shutdown operation process of the system.

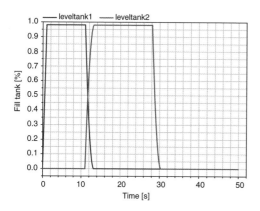

Figure 6. Level tanks in function of time in normal operation process of the system.

The Table 3 depicts the buttons actuation period times used in the three simulations performed.

Figure 6 shows the level tanks results of the first simulation that they were not pressed the buttons "stop" and "shut" during the production cycle, which corresponds to the normal system operation.

Observing Figure 6 it can be concluded that the system is properly simulated by the developed Modelica program code, since during the time specified by the SFC specification the tanks remain filled and empty.

On the other hand, Figure 7 shows results of the second simulation performed with the actuation of the "stop" button during the production cycle.

Analyzing Figure 7 it can be also concluded that the stop operation (closing all valves) is properly simulated by the proposed program. Because it can be verified, that after the actuation of the "stop" button (time 11s) the tanks levels stay stationary until being pressed the "start" button (time 24s). After that time, by opening the appropriate valves, the filling/emptying process continues.

Figure 8 shows results of the third simulation performed with the actuation of the "shut" button after the actuation of "stop" button during the production cycle.

Observing Figure 8 it can be also concluded that the shutdown operation is properly simulated by the proposed program. Because it can be verified, that after the actuation of the "stop" button (time 11s) the tanks levels stay stationary until being pressed the "shut" button (time 15s). After that time, the solution present in the tank1 is immediately drained for the tank2 and later emptied by the opening of valves 2 and 3.

5 CONCLUSIONS

The simulation used to evaluate the controller and plant behaviour has been developed and proposed in this paper.

The present research proved to be successful using the Modelica programming Language to obtain plant models. Also it will be proved that is adequate the use of the Dymola program with the library for hierarchical state machines StateGraph for the analysis of industrial controllers, because it is possible to eliminate a set of program errors of some, possible system behaviours in reduced intervals of time. This approach implies that the industrial controllers developed will be more robust and much less subject to errors.

Moreover, the simulation techniques allow us to test different delays of the plant functioning and to see if a property, for different considered delays, is still true or if different delays imply that a property is true and after is false.

In conclusion, using this approach a manufacturer of industrial automated systems does not need the physical part of the machine for later perform tests and simulation of the system controller. In consequence allows to reduce the times of production of the automated systems.

ACKNOWLEDGEMENTS

This research project is carried out in the context of the SCAPS Project supported by FCT, the Portuguese Foundation for Science and Technology, and FEDER, the European regional development fund, under contract POCI/EME/61425/2004 that deals with safety control of automated production systems.

REFERENCES

Basu S., Pollack R., Roy M. 2006. Algorithms in Real Algebraic Geometry, In Springer (Eds), *Algorithms and Computation in Mathematics.*, vol. 10, 2ª edition.

Elmqvist E., Mattson S. 1997. *An Introduction to the Physical Modelling Language Modelica*. Proceedings of the 9th European Simulation Symposium, ESS'97. Passau, Germany.

Elmqvist, Hilding, Mattsson S., Otter M. 1999. *Modelica – a language for physical system modeling*, visualization and interaction. Proceedings of the IEEE Symposium on Computer-Aided Control System Design. August, Hawaii.

Fritzson, Peter, Vadim E. 1998. *Modelica, a general object-oriented language for continuous and discrete event system modeling and simulation*, 12th European Conference on Object-Oriented Programming (ECOOP'98). Brussels, Belgium.

Fritzson, Peter, Bunus P. 2002. *Modelica, a general object-oriented language for continuous and discrete event system modelling and simulation*. Proceedings of the 35th Annual Simulation Symposium. April, San Diego, CA.

Otter M., Årzén K., Dressler I. 2005. *StateGraph – A Modelica Library for Hierarchical State Machines*. Modelica 2005 Proceedings.

Humphrey W.S. 1989. Managing the Software Process. In Addison-Wesley (Eds), *SEI Series in Software Engineering*.

Junkermann G., B. Peuschel, W. Schäfer, and S. Wolf. 1994. In Research Studies Press Ltd, Merlin: Supporting Cooperation in SoftwareDevelopment Through a Knowledge-Based Environment. In A. Finkelstein, J. Kramer, and B. Nuseibeh, editors, *Software Process Modelling and Technology*, 103–130.

Lamb C.W., G. Landis, J.A. Orestein, and D.L. Weinreb. 1991. *The Object Store Database System*. Communications of the ACM, 34(10), October.

Montangero C. and V. Ambriola. 1994. OIKOS: Constructing Process-Centered SDEs. In A. Finkelstein, J. Kramer, and B. Nuseibeh, editors, Research Studies Press Lda, *Software Process Modelling and Technology*, 187–222.

Computational Vision and Medical Image Processing – João Tavares & Natal Jorge (eds)
© 2008 Taylor & Francis Group, London, ISBN 978-0-415-45777-4

Medical and biological imaging with optical coherence tomography

Carla Carmelo Rosa
Universidade do Porto, Porto, Portugal
INESC-Porto, Instituto Nacional de Sistemas e de Computadores, Porto, Portugal

Michael Leitner
Universidade do Porto, Porto, Portugal

Adrian Gh. Podoleanu
University of Kent, Canterbury, United Kingdom

ABSTRACT: A versatile optical imaging system is presented that provides imaging resolutions down to the micrometer range. The system is built for time domain optical coherence tomography, with versatility in the scanning regime to be employed when scanning samples in the transverse and depth directions, thus generating cross-section images (B-scans) by using either transverse priority or depth priority. The system is targeted for eye fundus imaging but is easily adapted for the imaging of other biological samples, in vivo, by using its non-invasive property.

1 INTRODUCTION

1.1 Optical Coherence Tomography

Optical coherence tomography (OCT) is an imaging technique based on some optical interferometry configurations illuminated by a light source of low temporal coherence, such as large spectrum superluminescent diodes (SLD) (Huang, 1991; Tearney 1996; Fercher, 2003). The interferometer's scanning depth resolution, typically 10–20 micrometers, is enhanced by using larger bandwidth sources. To produce an image equivalent to an in depth section of the sample, the beam has to be laterally scanned at its surface while the interferometer's reference path is varied, producing coherence gating along the sample depth direction. Imaging speeds thus depend on the type of scanning and its speed, pretended image size, total depth, and finally signal acquisition.

OCT imaging has been most often used in ophthalmology during the last decade (Moreira, 2003), and in other medical specialties, such as dermathology (Weizel, 2001), dentistry (Colston, 1998; Feldchtein, 1998; Baumgartner, 2000), endoscopy (Tadrous, 2000).

The recent developments of OCT, such as Fourier domain (Leitgeb, 2003) and spectral OCT (Wojtkowski, 2003), have been targeted to the improvement of *in vivo* high resolution imaging of

biological samples towards video rate time scales. This is of foremost importance since movement artefacts are one of the causes for image degradation by blurring.

At the same time, ultrahigh resolution OCT (Fujimoto, 2003) allows resolution of the order of the micrometer by using ultra wide spectrum sources, and enhancing transverse resolution by using adaptive optics techniques (Drexler, 2001; Roorda, 2002). These are more difficult to conciliate with spectral domain techniques, and time domain configurations taking advantage on the development of fast scanning delay lines still need to be used.

In this paper a versatile time-domain OCT system with medical and biological imaging capabilities is presented. It generates cross-sectional images, known as B-scans, by using either transverse priority or depth priority scanning regimes. In time-domain OCT the detected interferometric signal amplitude is related to the reflectivity of the sample in the object path, at a certain sample depth, thus each *tomogram* is obtained without any digital processing. The images may be obtained in adjacent lines of space, allowing post-processing of the acquired cross-sectional images for the reconstruction of the whole 3D volume being imaged and other post-processing tasks, such as features extraction and automatic detection of forms.

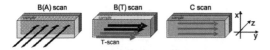

Figure 1. Different scanning regimes. Sample depth and incident light beam is represented along the z coordinate. The arrows indicate the priority scan direction, for successive line scans.

2 EXPERIMENTAL CONFIGURATION

2.1 Scanning regimes

Most of the commercial devices available today generate longitudinal images, or B-scans, from a set of fast depth scans of the object under test, referred to as A-scans. Adjacent depth profiles are combined as horizontal or vertical B(A)-scans, accordingly to the movement of an XY scanner pair, see Figure 1. A different scanning scheme is possible, where the B-scan image is generated by grouping transversal scans (T-scans) acquired at different successive depths, by slowly varying the optical path difference in the OCT interferometer. In this way a B(T)-scan image is generated, where the fastest movement is along the direction transversal to the incident beam axis. A C-scan (or en-face) has the familiar orientation of microscopy images, but it should be noted that, although the axial scanning is at rest, such C-scans do not necessarily represent constant depth sections; the depth where-from C-scans are acquired depends on the localization of the coherence surface within the object, hence it depends on the object shape, its optical properties, and the transverse scanning modality.

The core of the system relies on an original transmissive fast scanning delay line (Rosa, 2005), allowing identical scanning speeds for both the transverse (B(T) and C) and the longitudinal (B(A)) scanning regimes.

2.2 Imaging Transverse Resolution

The transversal imaging resolution in OCT is defined by the optics interfacing with the sample, or object, under investigation. It is ultimately limited by diffraction properties of the optics used on the focusing and collection of light from the sample being imaged.

The transverse resolution is given by the diffraction limit of the interfacing optics, by:

$$\Delta x = 0.50 \frac{\lambda}{NA} \tag{1}$$

and corresponds to the diameter of the Airy disk, at wavelength λ and with optics of NA numerical aperture. This value also applies to the case of conventional microscopy, when each point of the object is imaged as a diffraction limited spot.

An alternative method for higher resolution imaging is confocal microscopy, where spatial resolution enhancement is achieved by using both diffraction limited illumination of the sample for high transverse resolution, and very small aperture detection to achieve depth selectivity, e.g by using a pin-hole. Both illumination and detection need to be scanned along the sample. The pinhole at the detection will slightly improve the transverse resolution as compared to traditional microscopy, to:

$$\Delta x_{confocal} = 0.72 \, \Delta x = 0.36288 \frac{\lambda}{NA} \tag{2}$$

On the case of retina imaging, the transverse resolution is ideally of the order of $10 \, \mu m$, but in real life this value can increase due to aberrations introduced by the shape of the eye due and its morphology.

2.3 Axial Resolution

The axial (or depth) resolution is defined by the optical properties of the source. The temporal coherence of the source can be expressed in terms of the optical path delay between the reference and object paths in the interferometer as:

$$\Delta z = \frac{2\ln 2}{\pi} \frac{\lambda_0^2}{\Delta\lambda} \tag{3}$$

where λ_0 is the central wavelength of the optical source, and $\Delta\lambda$ its full width at half maximum, and Δz the sample depth. The wider the source, the better the axial resolution. In this way, coherence defines a time (depth scan) gate providing the high spatial resolution of the OCT technique.

In confocal imaging with ideal pinholes, the confocal gate, or depth along which the beam irradiance does not vary more than 50%, is given by:

$$\Delta z_{confocal} = 0.89 \frac{\lambda n}{(NA)^2} \tag{4}$$

where n is the refractive index of the imaging medium. The depth resolution is achieved either by increasing the numerical aperture of the imaging system, or by increasing the refractive index of the medium between the optics and the sample. This technique is used for cell imaging, where resolutions below $1 \, \mu m$ are needed, and using high n immersion oils. For skin imaging, depth resolutions from 10 to $1000 \, \mu m$ are needed, and these can be achieved by traditional high NA objectives. For the particular case of the eye, there is no way to control both the refraction index, typically $n \sim 1.36$, and the aperture of the imaging lens, the eye lens. Using typical values of the eye, the confocal depth resolution in no better than $200 \, \mu m$, and this value exceeds largely the typical dimensions of small structures within the retina.

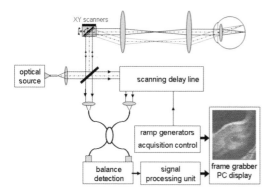

Figure 2. Multiscan OCT system configuration.

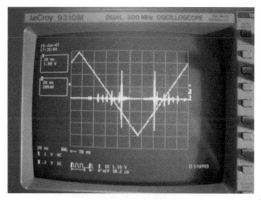

Figure 3. Test sample for OCT system calibration.

Optical coherence tomography should be considered as a complimentary method for *non-invasive* imaging of micro-scale structures within living tissues and biological samples. The depth resolution can be tuned by using suited optical sources of broad light spectrum, and it does not require any contact with the sample.

2.4 Setup

The configuration of the OCT system used in this study is shown in Figure 2. The optical source is either an SLD (Superlum, Russia) emitting light with spectrum centred at 810 nm and a FWHM of 19 nm, or the light from a femtosecond laser, built in house (CLOQ/Femtolab). The incoming light is split between reference and object arms of a Michelson interferometer. A pair of XY galvanometric scanners supported by some interfacing optics direct the beam into the sample under study. The light back-reflected from the sample is launched into a broadband single mode directional coupler where it combines with the light exiting a transmissive fast scanning spectral delay line. With interference occurring at the coupler, a balanced receiver is used to bring detection's noise closer to shot-noise limited regime. The output of the signal-processing unit is passed to a frame grabber controlled by triggers driven by the ramp generators driving the transverse scanners and the scanning delay line.

Safety issues associated with the use of these systems have been recently studied (Rosa, 2007).

2.5 OCT Signal

The photocurrent at each photodetector is derived considering a Michelson interferometer configuration with a sample mirror, resulting in:

$$i_D(z) = i_0 \left(k_{obj}^2 + k_{ref}^2\right)\left(1 + \frac{2k_{obj}k_{ref}}{k_{obj}^2 + k_{ref}^2} \text{Re}\left\{\gamma\left(\frac{2z}{c}\right)\right\}\right) \quad (5)$$

where k_{obj} and k_{ref} are the total reflectance in the object and reference arms, respectively, z is the imbalance between reference and object paths, c is the speed o light, i_0 the amplitude of the photocurrent, and $\gamma()$ is the temporal coherence function describing the optical source.

Transparent tissue or tissue composed of weakly reflecting layers, flat over the cross section of the probing beam, may be approached by a one-dimensional model where response is given solely in terms of the z depth position local reflectivity $R(z)$. With this assumption, the photocurrent may be rewritten as:

$$i_D(z) = i_0 k_{obj}k_{ref} \sqrt{R(z)} * \text{Re}\{\gamma(z)\} \quad (6)$$

with $*$ denoting the convolution of the reflectivity profile with the coherence function, and suggesting an alternative representation in the spectral domain:

$$i_D(z) \propto F^{-1}\{R(k)H(k)\} \quad (7)$$

Function $R(k)$ is the Fourier transform of the square of the reflectivity $R(z)$, and $H(k)$, obtained from the coherence function, may be thought as the point spread function of the OCT system.

3 RESULTS

3.1 Imaging Test Samples

In Figure 3 a depth scan along a sandwich of five microscope cover glass slides leaning against a mirror may be seen, showing the typical peaked structure of the raw OCT signal. The ramp signal on the scope represents the driving signal to the fast scanning delay line in the reference arm. The signal to noise ratio was 80 dB, and the total scan depth of 3.3 mm. The highest peak corresponds to the reflection from the ending mirror at the object, and the small peaks to the reflection in each cover glass slide. There are two additional peaks

355

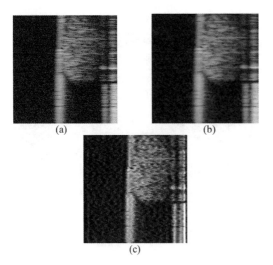

(a)   (b)

(c)

Figure 4. (a) original B(T) image: depth and transverse; (b) wiener filter to reduce noise; (c) deconv estimating PSF from reflection at the first glass slide interface reflection.

of lower amplitude, one next to the mirror peak, and the other ending the signal train, produced by reflections at the cell tape holding the slides sandwich together and close to the mirror.

Measuring the distance between secondary maxima, corresponding to the reflection at each cover slide boundary, a value of $160 \mu m$ is obtained, which is in good agreement with typical values.

3.2 *Image enhancement*

The OCT images can be post-processed to enhance the representation of the real sample. Equation 7 suggests a deconvolution operation in order to retrieve the reflectivity profile. This procedure is illustrated in the images of Figure 4. These resulted from probing a biological sample placed between microscope cover glass slides. Probe/axial direction is the horizontal direction. Fig. 4(a) shows the raw image. Fig. 4(b) resulted from low-pass filtering the raw image with an adaptive Wiener Matlab filter. Since the glass covers in the sample introduce step transitions in the refraction index of the medium, the PSF of the optical system can be estimated by extracting a line of pixels centred at the air-glass interface, as shown in Figure 5.

The deconvoluted image, in Fig. 4(c), shows enhanced boundaries and a better definition of the specimen between the microscope glass slides.

The intensity value for each pixel is scaled with the logarithm of the envelope of the photocurrent signal.

4 CONCLUSIONS

We present imaging capabilities of a time domain optical coherence tomography system implemented

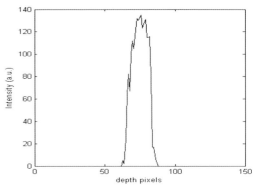

Figure 5. Estimated point spread function used in the deconvolution on Fig. 4.

for broadband operation. A comparison with confocal and traditional microscopy is made showing that OCT presents advantages for imaging micrometric structures within samples, namely the non-invasive approach and the capability to perform *in vivo* imaging, *in real time*. With a simple approach, we demonstrate that further enhancement can be achieved by processing the raw images, where knowledge on the optical performance of the system is introduced.

Here, we applied a simple deconvolution technique, using a point spread function obtained from experimental measurements of the impulse response of the system along the depth directions, resulting on a final reflectivity map image with clearer boundaries definition. This procedure can also be applied simultaneously with a PSF including both the depth and the transverse response.

The system is currently being fully characterized and will evolve towards ultra-high resolution imaging by using the ultra wide spectrum of a femtosecond laser. More results will be presented at the conference site.

ACKNOWLEDGMENTS

Michael Leitner acknowledges the support of the Marie Curie Early Stage research Training network "High resolution optical measurement and imaging", sponsored by the European Commission MEST-CT-2005-020353, in the Faculdade de Ciencias da Universidade do Porto, Porto, Portugal.

REFERENCES

Baumgartner A., Dichtl S., Hitzenberger C.K., Sattmann H., Robl B., Moritz A., Fercher A.F., Sperr W., "Polarization-Sensitive Optical Coherence Tomography of Dental Structures", *Caries Research*; 34:59–69 (2000)

Colston B., Sathyam U., DaSilva L., Everett M., Stroeve P., and Otis L., "Dental OCT," Opt. Express 3, 230–238 (1998)

Drexler W., Morgner U., Ghanta R. K., Kartner F. X., Schuman J. S. and Fujimoto J. G., "Ultrahigh-resolution ophthalmic optical coherence tomography", *Nature Medicine*, 7, 502–507 (2001).

Feldchtein F., Gelikonov V., XIksanov R., Gelikonov G., Kuranov R., Sergeev A., Gladkova N., Ourutina M., Reitze D., and Warren J., "In vivo OCT imaging of hard and soft tissue of the oral cavity," Opt. Express 3, 239–250 (1998)

Fercher, A.F., Drexler, W., Hitzenberger, C.K. and Lasser, T. Optical coherence tomography – principles and applications Rep. Prog. Phys., 2003 , 66 , 239–303

Fujimoto J. G., "Optical coherence tomography for ultrahigh resolution *in vivo* imaging", *Nature Biotechnology* 21, 1361–1367 (2003)

Huang, D., Swanson, E. , Lin, C. , Schuman, J., Stinson, W. , Chang, W. , Hee, M. , Flotte, T. , Gregory, K. , Puliafito, C. and Fujimoto, J. G. 1991. Optical Coherence Tomography, *Science* 254: 1178–1181

Leitgeb R., Hitzenberger C. and Fercher A., "Performance of fourier domain vs. time domain optical coherence tomography," Opt. Express 11, 889–894 (2003)

Moreira R. O., Trujillo F. R., Meirelles R. M. R., Ellinger V. C. M. and Zagury L., "Use of Optical Coherence Tomography (OCT) and Indirect Ophthalmoscopy in the Diagnosis of Macular Edema in Diabetic Patients", International Ophthalmology, 24(6) (2001)

Tearney, G. J. and Bouma, B. E. and Boppart, S. A. and Golubovic, B. and Swanson, E. A. and Fujimoto, J. G. 1996. Rapid acquisition of in vivo biological images by use of optical coherence tomography, *Opt. Lett.* 21: 1408–1410

Rosa, C. C. , Rogers, J. and Podoleanu, A.Gh. 2005. Fast Scanning Transmissive Delay Line for Optical Coherence Tomography, *Optics Letters* 30:3263–3418

Rosa C. C., Rogers J., Pedro J., Rosen,R. & Podoleanu A. 2007. Multiscan time-domain optical coherence tomography for retina imaging, *Applied Optics* 46(10):1795–1808

Roorda A., Romero-Borja F., Donnelly W. J., Queener H., Hebert T. J. & Campbell M. C. W. 2002. Adaptive optics scanning laser ophthalmoscopy, *Opt. Express* 10:405–412

Tadrous P. J. " Methods for imaging the structure and function of living tissues and cells: 1. Optical coherence tomography", The Journal of Pathology, pp 115–119, 2000

Welzel J., "Optical coherence tomography in dermatology: a review", Skin Research and Technology, Volume 7, Number 1, February 2001, pp. 1–9(9)

Wojtkowski M., Bajraszewski T., Targowski P. and Kowalczyk A., "Real-time in vivo imaging by high-speed spectral optical coherence tomography," Opt. Lett. 28 1745–1747 (2003)

Computational Vision and Medical Image Processing – João Tavares & Natal Jorge (eds)
© 2008 Taylor & Francis Group, London, ISBN 978-0-415-45777-4

3D reconstruction of pelvic floor for numerical simulation purpose

Fátima Alexandre
IDMEC – Polo FEUP, Faculty of Engineering of University of Porto, Portugal

Rania F. El Sayed
Radiodiagnosis Departement – Faculty of Medicine of University of Cairo, Egypt

Teresa Mascarenhas
Faculty of Medicine of University of Porto, Portugal

R.M. Natal Jorge, M.P. Parente & A.A. Fernandes
IDMEC – Polo FEUP, Faculty of Engineering of University of Porto, Portugal

João Manuel R.S. Tavares
Faculty of Engineering of University of Porto, Portugal

ABSTRACT: Female pelvic floor disorders (stress urinary incontinence, fecal incontinence, pelvic organ prolapse) affect approximately 60% of woman over 60 years old [1]. The real geometry and architecture of female pelvic floor and connective tissues are complex and difficult to visualize from two-dimensional (2D) images. To facilitate the understanding of pelvic floor geometry, in this work 3D models were building. A 3D helpful model of pelvic floor could aid the understanding of the anatomy and physiology of this complex part of the female body. These models were constructed from 2D medical images, obtained by magnetic resonance imaging (MRI). The purpose of this study was to help the identification of pelvic disorders by using MRI scans and reconstructed 3D models. Three women were studied. Pelvic organs and their structures were manually segmented, namely: bladder, urethra, vagina, levator ani (puborectalis, iliococcygeus and coccygeus), obturator internus and pubic bone. 3D models of female pelvic floor were created from a combination between the individual organs, muscles and bones. Three 3D model created were compared and differences were noted between the pelvic floor 3D model of a woman with disorders and the 3D models without disorders.

Keywords: Women Pelvic Cavity, Numerical Simulation, Image Segmentation, 3D Reconstruction

1 INTRODUCTION

The pelvis, pelvic floor musculature and the associated structures comprise one of the most complex regions of the human anatomy.

Recent advances in magnetic resonance imaging (MRI) have allowed us to study the appearance of these regions in normal women and in the women with pelvic floor dysfunction (Fig 1).

The main objective of this work was to use a combination of axial MRI and three dimensional (3D) models to describe these differences.

Three different women were studied, two Egyptians and one Caucasian woman.

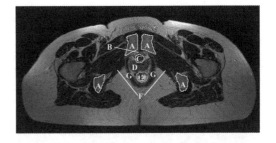

Figure 1. Axial MR Image of the pelvic cavity.
A – bones, B – bladder, C – urethra, D – vagina, E – rectrum, F – obturator internus, G – levator ani

There were three steps to do this study. The purpose of the first step was to obtain familiarity with the anatomy of the levator ani subdivisions and determine characteristic features for each subdivision in each scan plan (Fig. 1). In a second step differences between the 3D models of the levator ani in obtained from the MRI scans of three women were identified. The construction of 3D model female pelvis constituted the third step.

2 MATERIALS AND METHODS

The pelvic floor 3D model construction was based on the three pairs of levator ani: the pubococcygeus and the puborectalis and the iliococcygeus muscles, which described in the existing literature [4–6].

The levator ani muscles and muscular structures of the urethra comprise a system. These muscles are recruites during a cough to help prevent urine loss during stress. The coordinated action of these elements depends upon the central nervous system [2].

The pubococcygeus and the puborectalis form a U shape (Fig. 2) as they originate from the inner surface of the public bone on either side of the middle and insert into the vagina. The iliococcygeus muscle forms a horizontal sheet that spans the opening in the posterior region of the pelvis between the pubococcygeus muscle, the posterior ilia and sacrum.

This information was used to construct five individual model muscle bans.

Two dimensional (2D) MRI has been used to assess the anatomy of the female pelvic floor.

The MRI data were manually segmented where relevant outlines of levator ani muscles were digitized from consecutive stretch (used the CAD Sofware – Autodesk Inventor 11).

The imported profiles were connected to do the render (lofted). The manual segmentation time was, approximately, 2 hours for each subject [3].

Three-dimensional surface models were generated (Fig 3) and the computer time for 3D model generation was approximately one hour per subject [3].

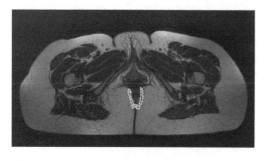

Figure 2.   Pelvic floor muscles and structures.

Three models (3D) of pelvic floor muscles was built (Fig. 3, 4, 5).

In the model of the Fig. 4 and 6 we can see an asymmetry in the levator ani.

Slice-by-slice was segment manually from each subject structures (Fig. 6 – a, b, c, d, e, f).

The MRI data of the Caucasian woman (slices with thickness of 5 mm) was chooses to construct 3D surface models with levator ani muscles, obturator

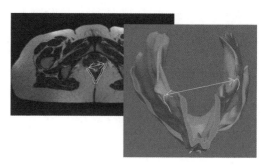

Figure 3.   2D MRI and 3D Model of the levator ani (slices with thickness of 2 mm), asymmetry in the levator ani (Egiptian woman) – white arrow.

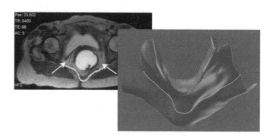

Figure 4.   2D MRI and 3D Model of the levator ani (slices with thickness of 5 mm), asymmetry in the levator ani (Caucasian woman) – white arrows.

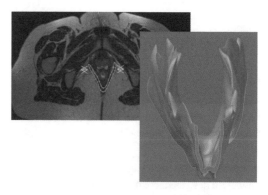

Figure 5.   2D MRI and 3D Model of the levator ani (slices with thickness of 5 mm), symmetry in the levator ani (control Egiptian woman) – white arrows.

internus muscles, urethra, vagina, rectrum, blader and bones (Fig. 7).

3 RESULTS AND DISCUSSION

The 3D anatomy and geometry of healthy female pelvic floor derived from MRI images shows consistent signal intensity.

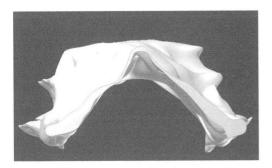

Figure 6a. Bones.

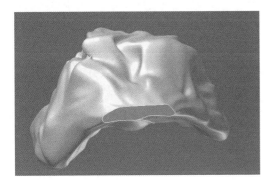

Figure 6b. Bladder.

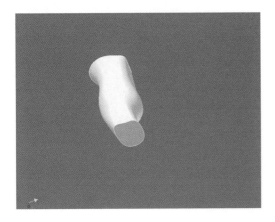

Figure 6c. Urethra.

3D model reconstruction is feasible and can be used in a research setting to evaluate complex anatomy relationships which may help to identify pelvic floor dysfunction.

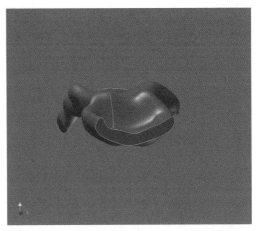

Figure 6d. Vagina.

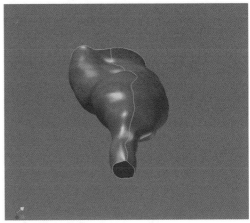

Figure 6e. Rectrum.

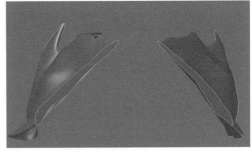

Figure 6f. Obturator internus muscles.

361

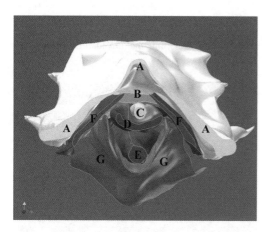

Figure 7. Female pelvic floor 3D model. Color scheme: white = bones (A), yello (flat) = bladder (B), yello = urethra (C), purple = vagina (D), gray(dark) = rectrum (E), brown = obturator internus muscles (F), red = levator ani muscles (G)

This methodology can also be used to evaluate levator muscle morphology and volume in a patient.

A larger study can be conducting to validate this technique and to better understanding the relationship between pelvic floor geometry and pelvic floor dysfunction.

4 FUTURE WORK

Neurophysiology of the pelvic floor is not completely understood yet. The importance of its symmetry and asymmetry of innervation has been pointed out. These facts have clinical relevance in case of pelvic floor trauma or incontinence surgery. Better reconstruction techniques of 3D model and 3D finite element models are necessary to help the specialists to confirm correlations between symptoms development and asymmetry of sphincter innervation.

It is proposed also as future case study, using the 3D reconstruction, the racial differences in pelvic morphology and geometry.

ACKNOWLEDGMENTS

The presented work was partially done in the scope of the project "Segmentation, Tracking and Motion Analysis of Deformable (2D/3D) Objects Using Physical Principles", with reference POSC/EEA-SRI/55386/2004, and for the project with reference PTDC/SAU-BEB/71459/2006, both financially supported by FCT – Fundação para a Ciência e a Tecnologia in Portugal.

REFERENCES

[1] Parikh, M., Rasmussen, M., Brubaker, L., Salomon C., Sakamoto, K., Evenhouse, R., Ai, Z., Damaser, M. S., "Three Dimensional Virtual Reality Model of the Normal Female Pelvic Floor", *Annals of Biomechanics Engineering, Vol. 32 , No. 2 (2004) pp. 292–296*

[2] James A. Ashton-Miller, Denise Howard, John O. L. DeLancey "The Funcional Anatomy of the Female Pelvice Floor and Stress Continence Control System" *Scand J. Urol Nephrol Supp. 2001: (207): 1–125*

[3] Lenox Hoyte, Julia R. Fielding, Eboo Versi, Charles Mamisch, Carl Kolvenbach, Ron Kikins "Variations in Levator Ani volume and geometry in normal living subjects, versus prolapse, and genuine stress incontinence: The application of MR based 3D reconstruction in evaluating pelvic floor dysfunction".

[4] Kuo-Cheng Lien, Brian Mooney, John O. L. DeLancey, James A. Ashton-Miller "The levator ani muscle and nerve supply to the puborectalis muscle" *Obstet Gynecol. 2004 (January); 103(1): 31–40*

[5] Roberts WH, Harrison CW, Mitchell DA, Fischer HF "The levator ani muscle and nerve supply to the puborectalis muscle" *Clin Anat. 1988; 1: 256–88.*

[6] Lawson J.O. "Pelvic Anatomi. I. Pelvic floor muscles" *Ann R. Coll Surg Engl. 1974; 54: 244–52*

Computational Vision and Medical Image Processing – João Tavares & Natal Jorge (eds)
© 2008 Taylor & Francis Group, London, ISBN 978-0-415-45777-4

An Application of Hough Transform to identify breast cancer in images

E.M.P. deF. Chagas & D. L. Rodrigues

Programa de Pós-Graduação em Engenharia Mecânica / MEA – Pontifícia Universidade
Católica – PUC MINAS, Av. Dom José Gaspar, Coração Eucarístico – Belo Horizonte, Brasil

J.M.R.S. Tavares

Instituto de Engenharia Mecânica e Gestão Industrial, Laboratório de Óptica e Mecânica Experimental,
Faculdade de Engenharia da Universidade do Porto, Departamento de Engenharia Mecânica e Gestão
Industrial Rua Dr. Roberto Frias, Portugal

A.M. Reis & D. Miranda

Unidade Local de Saúde Matosinhos – Hospital Pedro Hispano, Departamento, Imagiologia, Portugal

R. Duarte

Centro Hospitalar V. N. Gaia / Espinho, Departamento Imagiologia, Portugal

ABSTRACT: The present work focus on the identification of malignant neoplasias in breast by using the Hough Transform in digital images. A detector of circumference was implemented based on the Hough Transform once the pattern to identify is circular. The detector presented here was developed using MatLab, and was applied to detect circumferences in tomography images.

1 INTRODUCTION

World-wide, the breast cancer is the more prevalent malignant tumor among women. It corresponds to 23% of all the malign tumors, with 1.15 million new cases in 2002 (Parkin et al 2005). More than half of the cases occurred in the industrialized countries and the estimates for the United States of America in 2007 are of 178.480 new cases (26%) and 40,910 deaths, whilst in Europe, in 2006, had 429.900 (28,9%) of new cases resulting in 132,000 deaths (Jemal et al 2007; Ferlay 2006).

The mortality rate of breast cancer is the second highest that affects women, only beaten by lung cancer (Landis et al 1998). Currently, an efficient method for prevention the breast cancer does not exist. Thus, the best form to reduce mortality passes by early detention, followed of adequate treatment (Chan et al 1999).

2 MAMMOGRAPHY

The survival from breast cancer depends, basically, on the size of the tumor and the presence or absence of metastasis in lymphatic nodes. Small tumors detected through the mammogram and without involvement of lymphatic nodes have survival rates of 90% (Brant &

Helms 2006). The goal is to carry through a precocious diagnosis, contributing directly for a better prognosis, allowing a less radical surgery and providing better indices of free intervals of disease. Currently, the mammogram is the most efficient method in the detection of clinically occult illness, being the only image-based method recommended for breast cancer screening. The technological improvements in the mammogram field allow to precocious lesions detection. The introduction of organized screening programs of mammography contributed significantly to the reduction of the mortality rate of breast cancer, namely in industrialized countries (Tyczynski et al 2004; Levi et al 2005; Paci et al 2002). Esteems are that mortality in Europe associated with this type of cancer had an annual reduction of about 2% between 1995 and 2000 (Levi et al 2005).

The breast cancer can be presented under several patterns, whose characteristics influence in a general way the suspicion of malignancy or benignity of one determined lesion. However, the definitive diagnosis of breast cancer is only confirmed after the cytological or histological analysis.

Some types of breast cancer can present in the mammogram as a mass that causes parenchymal distortion, classically with indistinct or ill-defined borders. However, this is the presentation form of only 20% of non-palpable cancers (Lewin Jr., D'Orsi & Hendrick

2004). The tumor type that most frequently manifests this way is the invasive ductal carcinoma, followed by the tubular carcinoma and lobular. Masses with well-defined margins generally represent benign lesions, despite about 5% of these can be malignant (Marsteller & Paredes 1989).

The microcalcifications (calcifications <0.5 mm), associated or not to a mass, represent the first mammography malignancy signal. These suggestions of malignancy generally present a great variability in its form (pleomorphics) and size. Normally, they are elongated, numerous and have trend to form groups. The ductal carcinoma is frequently detected by mammography under the form of branching microcalcifications. The morphology of the microcalcifications cannot, however, be used to predict the type of carcinoma in cause. The microcalcifications are visualized in about half of the breast cancers cases and in about one third they represent the only mammographic finding (Sickles 1986). However, about 25 to 35% of the biopsied microcalcifications represent malign lesions (Brant & Helms, 2006).

An architectural parenchymal distortion or density asymmetries are indirect signs of malignancy that many times leads to a misinterpretation of the mammogram. In the architectural parenchymal distortion, mammary focal area appears distorted because the infiltration of the tumor leads to an interruption and spacing of the parenchyma before the formation of a visible mass. The infiltrating lobular carcinoma often manifests such this, representing about 22% of the false negatives cases (Majid et al. 2003). Separately, these subtle indirect signs have a low positive predictive value of malignancy, but this is increased when associated to masses or microcalcifications.

3 COMPUTER AIDED DIAGNOSTICS

However mammogram is not a perfect screening tool because its high sensibility is opposed by a low specificity (Morton et al. 2006).

Several studies have shown that interpretation errors are frequent in breast cancer screening (Burhenne et al. 2000; Vitak, 1998).

Based on these studies, about 20 to 30% of cancers could have been detected on earlier screening mammograms (Karssemeijer et al. 2003).

Studies of missed cancers that were later detected on screening have demonstrated that most (57%–75%) of missed lesions have some finding visible on mammograms in retrospect (Burhenne et al. 2000; Birdwell, Bandodkar & Ikeda 2005; Ikeda et al. 2004).

In this direction, the computer-aided detection systems (CAD) had been conceived to try to reduce the number of false negatives of breast cancer screening. These systems intend to assist the radiologists indicating suspicious areas in the mammogram. CAD systems are supported in the theory of when an area is pointed by the system, the radiologist will analyse with extreme care, minimizing the risk of a lesion area passing unobserved (Karssemeijer et al. 2003).

Depending on various studies enclosing this substance, the sensitivity of CAD systems varies of 65% to 86% for masses and 78.4% to 100% for microcalcifications (Morton et al. 2006; Birdwell, Bandodkar & Ikeda 2005; Baker, Delong and Floyd 2004).

Some studies had already been considered to determine the impact of CAD systems in breast cancer screening. In 2001 through the analysis of 12.860 mammograms, Freer et al related in its study that the use of CAD systems contributed for an increase the cancer detection rate by 19.5% in early stages (Free & Ulissey 2001). In the same direction, Birdwell et al, through the study of 8.682 women, reported an increase breast cancer detection rate of 7.4% (Birdwell, Bandodkar & Ikeda 2005). These data still had been corroborated in 2006 by Morton et al, in a study involving 18,096 women, describing an increase of the detention rate of 7.62% (Morton et al. 2006). However, in the last prospective study at the time of preparing this article that enclosed 222,135 women, Fenton et al defended that the use of CAD systems is associated with a reduction of the acuity in the detention of breast cancer, opposing the above-mentioned studies (Fenton et al. 2007).

The relatively lack of consensus to this subject is the starting point for the accomplishment of this work, intending to contribute to the research of new techniques applied for CAD systems.

4 HOUGH TRANSFORM

Frequently, the Hough Transform is used to identify parametric patterns, as lines, circles and ellipses, presented in digital images. The main idea is to map one specific space composed by image's pixels into the respective space of parameters, which is organized in form of an m-dimensional accumulator (where m corresponds to the number of parameters considered). By using this transformation, the issue is to focus in the parameters' space, where relations that connected points of the space of data, that is the image, in order that the search of one analytical form for the pattern to be detected or to be reconstructed is simplified. Its basic principle consists in getting points from an image through the utilization of gradient and threshold transforms (Grimson & Huttenlocher 1990; Duda 1972).

In this direction, a transformation is applied in the digital image in order that all points that belong to the same curve are mapped in one, and only one, point of the new parameters' space defined for the curve being detected. Thus, in this work, each edge of the image is transformed, by the mapping process, to detect cancer cells in the parameters' space, which are indicated in this space through the respective accumulator. These accumulators are incremented and in the end the ones with higher values will indicate the localizations of the desired shape in the input image (Pistori, Pistori & Costa, 2006; Pereira 1995).

5 METHODOLOGY

During this work, some experimental tests were done using mamographic images with presence of tumor selected from the current literature. First of all, to remove some noise that the input mamographic images have, they need to be filtered; with this, we make the signal/noise relationship better (Bauab 2005; Caldas et al. 2005). With the purpose to improve the distinction between the background of the images and the desired objects in the same, the micro calcifications, a threshold method was used (Castleman 1998; Jain 1989; Gonzalez & Wintz 1987) to convert them to two states. This method consists in the choice of the threshold value T using the histogram of gray levels. After the binarization process, it is necessary the segmentation of the image, that is the identification of the objects to be used in the subsequent processing steps. The segmentation is carried through using the Transformed of the Negative. Usually, segmentation algorithms are based on one or two basic properties of the gray levels: similarity and discontinuity. In this work we choose to use the first option that is based in thresholding and region growing.

The main concern in this work was not to test the use of the Hough Transform to detect circles in a large number of images, but to analyze in terms of which conditions these circles are correctly detected.

6 IMAGE SEGMENTATION

Image segmentation is necessary to classify an image, by dividing it in its constituent parts. The main idea of this operation is to characterize interest regions in order to get information on edges as well as of interior points of them. The data considered can be represented by borders or by complete regions. The representation by borders is adjusted when the main interest concentrates in characteristics of external shapes, like rounded, oval, tubular, and others. The representation by regions make it possible to concentrates in internal properties also, such as texture, bright, color, contrast, and others (Castleman 1998; Gonzalez & Wintz 1987).

In mammographic images, the representation by borders has as objective the identification of injuries through its format, making possible, at a first moment,

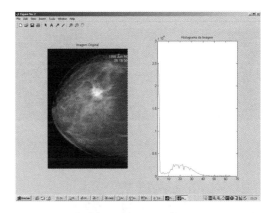

Figure 1. An original image and its histogram.

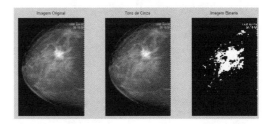

Figure 2. Original image (left), filtered image (center) and binary image (right).

the characterization of a malignant or benign tumor, simply analyzing its structural form (Pereira et al. 2006; Le Gal, Chavanne & Pellier 1984). In this work, the segmentation done is based on the application of five algorithms implemented: thresholding, contrast, negative, region growing and edges detection.

The well known technique thresholding is here used to identify modal peaks in the data histogram, in order to transform the original image into a binary image. This capacity is important for the automatic election of the threshold values in situations in which the characteristics of the input images can vary along a wide band of intensity distribution. Usually, the threshold process does not present satisfactory results in images with illumination not uniform or low contrast, (Gonzalez & Wintz 1987). Figure 1, presents an example of an original mammographic image and its histogram. Figure 2, shows the original image, the image converted into gray levels and the threshold image.

Frequently, the contrast technique is used to increase the dynamic scale of the gray levels of an input image. Figure 3, shows the use of this transformation.

Negative transformation just inverts the gray level of the original image, (Gonzalez & Wintz 1987). Figure 4, illustrates the use of this transformation.

The technique of region growing is characterized by grouping pixels or sub-regions in regions of bigger

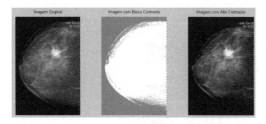

Figure 3. Original image (left), Low contrast image (centre) and high contrast image (right).

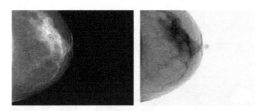

Figure 4. Original image (left), inverted image (right).

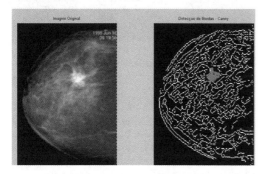

Figure 5. Original image (left), and edges detected by using the Canny's operator (right).

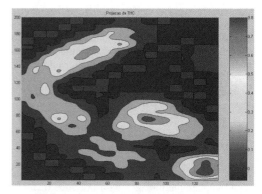

Figure 6. Projection of the Hough Transform for circles (Radius = 10 and Resolution = 8).

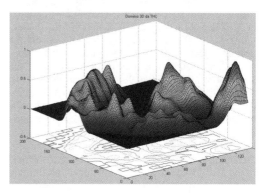

Figure 7. 3D domain of the Hough Transform for circles (Radius = 10 and Resolution = 8).

dimension. Here, this technique is used to make possible the detention of injuries in the mammograms. In accordance with (Castleman, 1998; Gonzalez & Wintz 1987), this technique initiates with the definition of points known as "seeds" and to grow from them the regions that present similar properties.

Finally, the Canny's operator was used to detect the edges of intensity in the images (Vale & Poz, 2006). Figure 5, illustrates this usage.

7 RESULTS

In order to validate the efficiency of developed algorithms, they had been applied in images, first simulated and later in real mammography's images. The application of the algorithms in simulated images is very important to evaluate the system in all its complexity, since the real images do not have so diversified and controllable space. Based on this and considering elements with characteristics previously known, the system is ready to a larger range validation.

The tests done along this work had been carried through on mammograms available in current literature. In total, they had been applied in 6 mammograms. After the acquisition of the mammograms, was carried through the image pre-processing by opening the contrast, in order to correct the low deriving contrast of the acquisition process. Later, the Canny's edge detector was applied in each image (Vale & Poz, 2006). Finally, the Hough transform was applied to obtain the recognition of shapes presented in the original images. The implemented Hough Transformed to detect circles in images needs the following parameters: radius of the circles to be detected in the input image and resolution of the matrix defined for votes accumulation. Figures 6, 7 and 8, shown examples of the projection of the Hough Transform, 3D domain of the Hough Transform, and the circles detected with radius = 10 and resolution = 8 in an input image, respectively.

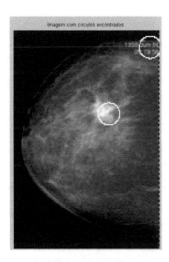

Figure 8.   Original image with the circles founded indicated (Radios = 10 and Resolution = 8).

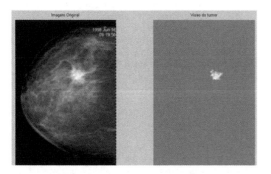

Figure 9.   Original image (left) and tumor enhanced visualization obtained using image processing techniques (right).

8   DISCUSSION

The used thresholding technique was based on local statisticians of the gray levels values, having as final objective the threshold of one determined region present in the input image. For each region of a mammogram it was calculated the value of "threshold", varying from 255 (maximum value) to 0 (zero, minimum value), allowing the specialist to extract the anatomical structure from the input images. Analyzing the results obtained with this technique, we can observe its higher sensibility in the detention of small injuries present in the input images.

The use of the negative transform allowed improving the visual quality of the original images. With this, important attributes of shape, contour and texture had been visually identified.

9   CONCLUSION AND FUTURE WORKS

The algorithms considered here had been implemented to work as support tools to assist medical doctors in the detection of mammary anomalies, malignant or benign, and promoting the accurate diagnostic from images. Despite the techniques proposed present good results, we intend to improve the thresholding technique, in order to automatically adjust the ideal threshold value.

Even though the results obtaining have presented satisfactory quality, it is clear that for the success of a computational system detect and track breast cancer, a more detailed analysis is demanded, including a large number of test images and the utilization of higher quality images.

In future, we intend:

- Improve the algorithms based on Hough transform that had allowed the detection of circles in images;
- Do experimental tests with a large number of representative images of breast cancer;
- The validation of the results obtained by medical specialists;
- Consideration of artificial intelligence techniques in the recognition of patterns in mammography images that had been associated to breast tumors;
- Implement a CAD system that makes possible the precocious diagnosis of the breast cancer from images.

REFERENCES

Alecrin, I.N.; Taniguchi, C.K.; Calvoso JR., R. et al. 2001. Calcificações mamárias: quando biopsiar?. *Rev. Assoc. Med. Bras.*, ene./mar.47(1):10–11. ISSN 0104–4230.

Baker, J.; Lo, J.; Delong, D. & Floyd C. 2004. Computer-aided Detection in Screening Mammography: Variability in Cues. *Radiology* 233:411–417.

Bauab, S.P. 2005. Mamografia digital: um caminho sem volta. *Radiol Bras*, May/June 38(3):iii–iv. ISSN 0100–3984.

Birdwell, P.; Bandodkar, R. & Ikeda, D. 2005. Computer-aided Detection with Screening Mammography in a University Hospital Setting. *Radiology* 236:451–457.

Brant, W. & Helms, C. 2006. *Fundamentals of Diagnostic Radiology*, 3rd edition. Lippincott Williams & Wilkins.

Burhenne, L.J.; Wood, S.A.; D'Orsi, C.J., et al. 2000. Potential contribution of computer-aided detection to the sensitivity of screening mammography. *Radiology* 215:554–562.

Caldas, F.A.A.; Isa, H.L.V.R.; Trippia, A.C. et al. 2005. Quality control and artifacts in mammography. *Radiol Bras*, July/Aug. 38(4):295–300. ISSN 0100–3984.

Castleman, K.R. 1998. *Digital Image Processing*. New Jersey: Prentice-Hall, Englewood Cliffs.

Chan, H.P. et al. 1999. Improvement of radiologists Characterizaton of mammographic masses by using computer-aided diagnosis: an ROC study. *Radiology* 212:817–827.

Duda, R. & Hart, P. 1972. Use of the Hough transformation to detect lines and curves in picture. *Comm. of ACM* 1:11–15.

Fenton et al. 2007. Influence of computer-aided detection on performance of screening mammography. *New England Journal of Medicine* 356:1399–1409.

Ferlay, J.; Autier, P.; Boniol, M.; Heanue, M.; Colombet, M. & Boyle, P. 2007. Estimates of the cancer incidence and mortality in Europe in 2006. *Annals of Oncology* 18: 581–592.

Freer, T.W.; Ulissey, M.J. 2001. Screening mammography with computer-aided detection: prospective study of 12,860 patients in a community breast center. *Radiology* 220:781–786.

Gonzalez, R.C. & Paul W. 1987. *Digital Image Processing*. Addison-Wesley Publishing Company, EUA.

Grimson, W.E.L. & Huttenlocher, D.P. 1990. On the sensitivity of the hough transform for object recognition. *IEEE Transactions on Pattern Analysis and Machine Intelligence, PAMI* 12(3):255–274.

Huynh, P.T.; Jarolimek, A.M. & Daye, S. 1998. The False-negative Mammogram. *RadioGraphics* 18:1 13–l 154.

Ikeda, D.; Birdwel, R.; O'Shaughnessy, K.; Sickles, E. & Brenner, R. 2004. Computer-aided Detection Output on 172 Subtle Findings on Normal Mammograms Previously Obtained in Women with Breast Cancer Detected at Follow-Up Screening Mammography. *Radiology* 230:811–819.

Inca. 2006. *Câncer de mama*. Available at: http://www.inca. gov. br/conteudo_view.asp?id=336. Accessed in: 10 dez. 2006.

Jain, A.K. 1989. *Fundamentals of Digital Image Processing*. EUA: Prentice-Hall., 1989.

Jemal, A.; Siegel, R.; Ward, E.; Murray, T.; Jiaquan, Xu & Thun, M. 2007. 2007 Cancer Statistics, 2007. *CA Cancer J Clin* 57:43–66.

Karssemeijer, N.; Otten, J.; Verbeek, A.; Groenewoud, J.; Koning, H.; Hendriks, J. & Holland, R. 2003. Computer-aided Detection versus Independent Double Reading of Masses on Mammograms. *Radiology* 227: 192–200.

Landis, S.H.; Murray, T.; Bolden, S. & Wingo, P.A. 1998. Cancer statistics 1998. *CA Cancer J Clin* 48:6–29.

Le, G.M; Chavanne, G. & Pellier, D. 1984. *Valeur diagnostique dês microcalcifications groupées découvertes par mammographies*. Masson , Paris.

Leitch, A.M.; Dodd, G.D.; Costanza, M. et al. 1997. American Cancer Society guidelines for the early detection of breast cancer: update 1997. *CA Cancer J Clin* 47: 150–153.

Levi, F.; Bosetti, C.; Lucchini F. et al. 2005. Monitoring the decrease in breast cancer mortality in Europe. *Eur J Cancer Prev* 14 (6): 497–502.

Lewin, J.R.; D'Orsi, C.J. & Hendrick, R.E. 2004. Digital mammography. *Radiol Clin North Am* 42: 871–884.

Majid, S.; Paredes, E.; Doherty, R.; Sharma, R. & Salvador, X. 2003. Missed Breast Carcinoma: Pitfalls and Pearls. *RadioGraphics* 23:881–895.

Marsteller, L. & Paredes, E. 1989. Well defined masses in the breast. *Radiographics* 9: 13–37.

Mclaughlin, R.A. 1998. Randomized hough transform: improved ellipse detection with comparison. *Pattern Recognition Letters* 19(2):299–305.

Morton, J.M.; Whaley, D.; Brandt, K. & Amrani, K. 2006. Screening Mammograms: interpretation with Computer-aided Detection-Prospective Evaluation. *Radiology* 239(2): 375–383.

Paci, E; Duffy, S.W.; Giorgi, D., et al. 2002. Quantification of the effect of mammographic screening on fatal breast cancers: the Florence Programme: 1990–1996. *Br J Cancer* 87:65–69.

Parkin, D.; Bray, F.; Ferlay, J. & Pisani, P. 2005. 2002 Global Cancer Statistics. *CA Cancer J Clin 2005* 55:74–108.

Pereira, A.S. 1995. Processamento de imagens médicas utilizando a Transformada de Hough. *PhD Thesis*. Institut of Fisic of the São Carlos, 432p., may.

Pereira, A.S.; Marranghello, N.; Yokota, C.S. & Nikuma, K.H. 2006. *Detecção de microcalcificações de bordas lisas e agrupamentos em formação, para auxílio ao diagnóstico médico de câncer de mama*. Available at: http://www.sbis.org.br/cbis9/arquivos/291.doc. Accessed in: 09 dez. 2006.

Pistori, H.; Pistori, J. & Costa, E.R.H. 2006. *Hough-Circles: um módulo de detecção de circunferências para o ImageJ*. Available at: http://rsb.info.nih.gov/ij/plugins. Accessed in: 26 nov. 2006.

Sickles E.A. 1986. Mammographic features of 300 consecutive nonpalpable breast cancers. *AJR Am J Roentegenol* 146: 661–663.

Tyczynski, J.E.; Plesko, I.; Aareleid, T. et al. 2004. Breast cancer mortality patterns and time trends in 10 new EU member states: mortality declining in young women, but still increasing in the elderly. *Int J Cancer* 112 (6): 1056–1064.

Vale, G.M. & Poz, A..P.D. 2006. *Processo de detecção de bordas de canny*. Available at: http://calvados.c3sl.ufpr.br/ojs2/index.php/bcg/article/viewFile/1421/1175. Accessed in: 12 out.

Vitak B. 1998. Invasive interval cancers in the Ostergotland mammographic screening programme: radiological analysis. *Eur Radiol* 8:639–646.

Computational Vision and Medical Image Processing – João Tavares & Natal Jorge (eds)
© 2008 Taylor & Francis Group, London, ISBN 978-0-415-45777-4

A simple mathematical model for the woman's breast geometry and its use in the breast characterisation

Paulo G. Macedo, Elisabete A. Mota & Vera M. Sá

Dept. of Applied Mathematics – Faculdade de Ciências da Universidade do Porto, Portugal

ABSTRACT: In this work, a simplified two dimensional model for the woman's breast geometry is proposed. Using a variational method, the breast of the young woman is modeled as an elastic bag with a given volume and outer area filled with a homogeneous incompressible fluid subjected to gravity as the only external force acting on it.

A characterisation of the woman's breast profile in the erect position is also proposed. This is achieved, in a similar way to what is done in Astrophysics, by using a two-dimensional diagram which we have called the Breast Diagram. Using simple results of variational calculus in two dimensional plane geometry we establish allowed and forbidden regions for the existence of breast profiles on this diagram. A quantitative parameter for measuring the breast fall tendency is proposed. A method for studying the evolution of the woman's breast with age using this diagram is also proposed (A more detailed version of the contents of this presentation can be found in Macedo et al. (submitted for publ.)).

1 INTRODUCTION

The use of plastic surgery to change or enhance the woman's breast is a common medical practice today. However, there is only an empirical knowledge of the correlation between the size of the implant and the end result in the shape of the implanted woman's breast.

The woman's breast profile is such that, once its Perimeter P and its Area S are given, the shape of the profile curve $y = y(x)$ is uniquely determined and can be obtained using variational calculus [Macedo et al. (2007)]. As we shall see further on, since gravity has a given value in the Earth's surface, this means that each breast profile is uniquely characterised by specifying the corresponding values of these two parameters P and S (see Fig. 1).

The purpose of this work will be to develop a simplified mathematical model for the breast geometry. Due to the complexity of a three-dimensional model, we shall use a bi-dimensional approach which corresponds to the breast profile geometry analysis. Inspired by the Hertzsprung-Russell diagram [see Jaschek (1990) & Kaler (1989)] used in Astrophysics, we shall also propose a new two dimensional diagram which we shall call the Breast Diagram (or S-P diagram) as a quantitative way of classifying the woman's breast profile. As a consequence, we shall introduce a new parameter to characterise the tendency for the breast to fall with age, which we shall call the "Breast Fall Parameter", δ.

We believe this may, in the future, be a first step to help plastic surgeons to better calculate the implant size to be used in the breast enhancement surgery, knowing what type of breast shape is sought as the end result.

For a breast with a given skin area, the bigger its volume, the rounder its shape. The breast diagram may help in determining the volume of implants to be used in breast enhancement surgery in order to obtain a desired final breast shape.

2 MATHEMATICAL FORMULATION OF THE PROBLEM

2.1 Simplifying assumptions about the breast profile geometry

To simplify the problem, we shall neglect the shape of the breast's nipple. We shall also assume that the upper and lower intersection points of the breast profile with the bone and muscles structures are separated by a distance of x_1. Since all we are interested in, is the shape of the breast, we shall use this distance as our unit length. In this way, all breast profiles will be normalized to this unitary length. I.e. $x_1 = 1$.

On the other hand, although being often unrealistic, we shall assume that the line which unites these two referred points is vertical, like pictured in Figure 1.

Furthermore, to simplify the problem, we shall also assume that the breast profiles to which this model

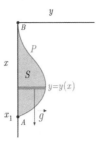

Figure 1.   Breast profile curve represented by $y = y(x)$.

applies are such that the profile curve can be described by a function $y = y(x)$. This geometrically corresponds to the fact that no other point of the profile is lower than the point of inferior intersection $x = x_1$.

2.2   The breast profile geometry as a result of a variational principle

Most Nature's physical systems are such that is possible to describe their configuration or temporal behaviour using variational principles. This means that they have a configuration or temporal evolution that minimises some functional of a function of the variables that describe them [Clegg (1968)]. This is well known namely in Mechanics where the movement of a mechanical system can be studied by defining a function called the Lagrangian of the system which is a function of the generalised co-ordinates that characterize the system. The evolution of the system is such that minimises a functional of the Lagrangian which is called the "Action" for the system. Using variational calculus methods, one can therefore obtain a description of the way in which these co-ordinates vary with time (that is, a description of the system movement).

Something similar also happens in field theory were the field equations can be obtained in a similar way, by a variational method, from extremal conditions on a functional [Landau & Lifshitz (1959)].

In a similar way, we can assume that the breast geometry, can be obtained by a variational principle. We shall assume that, at a given time in the woman's life, the geometry of the breast can be modeled as that of a bag filled with water. It was shown [see Macedo et al. (2007)] that, under the two following assumptions:

1) For a given moment in the woman's life the volume and surface skin area of the breast have a given fixed value,
and that:
2) The only external force acting on the breast is gravity,

out of the many shapes the breast profile could take, nature chooses the one which minimises its total potential energy. This energy being the sum of the elastic potential energy of the tissues involved plus the gravitational potential energy. Since we have assumed the breast skin tissue to have a given fixed area this means that the elastic potential energy of the skin tissues does not vary and therefore the only varying potential energy is the gravitational one. This is a result from our above mentioned assumption that, for the purpose of this model, the breast skin would behave as being inelastic.

On the other hand, we shall assume that the breast skin tissues span an area which, at a given moment in the woman's life, is fixed. This in turn means that the breast profile has a given fixed perimeter. We shall also assume that the breast volume, at a given moment in the woman's life, is fixed. This in turn means that the breast profile has a given fixed area. This, in turn, means that, at a given time, each breast profile is characterized by fixed values for its perimeter P and for its area S. This implies that the breast profile geometry will be such that minimizes the gravitational potential energy of the tissues which constitute the breast.

To simplify matters we shall also assume that the tissues in the breast interior are homogeneous (both in their density ρ and in their mechanical parameters).

Looking at figure 1, we can see that, in the presence of the gravitational field $\vec{g}$ that produces a gravity acceleration g, the gravitational potential energy of the tissues which constitute the breast will be given by a functional U of the breast shape function $y(x)$:

$$U = \rho g \int_0^{x_1} yx\,dx \tag{1}$$

The above mentioned assumption, (that the geometry of breast shape is such that it minimizes its potential energy), means that the variation of the functional U of $y(x)$ is null for the required function $y = y(x)$ which corresponds to the actual breast profile:

$$\delta U = 0 \tag{2}$$

On the other hand, the conditions that, for a specific moment in the life of a woman, the area S and the perimeter of her breast profile P are fixed, constrains the function $y(x)$ with the two following conditions:

$$S = \int_0^{x_1} y\,dx = const. = c_1 \tag{3}$$

and

$$P = \int_0^{x_1} \sqrt{1 + \left(\frac{dy}{dx}\right)^2}\,dx = const. = c_2 \tag{4}$$

The problem of finding the function $y = y(x)$ that satisfies the condition (2), under the constraint conditions

(3) and (4), is equivalent to the problem of finding an extremal function $y(x)$ for the functional I given by [see Clegg (1968)]:

$$I = \int_0^{x_1} \left(\rho g y x + \lambda y + \mu \sqrt{1 + \left(\frac{dy}{dx}\right)^2} \right) dx \qquad (5)$$

were λ and μ are Lagrange multipliers such that:

$$\delta I = 0. \qquad (6)$$

If we call $y\prime$ the derivative $\frac{dy}{dx}$ and L the Lagrange function for this system, given by:

$$L = \rho g x y + \lambda y + \mu \sqrt{1 + y'^2} \qquad (7)$$

then we can write the functional I given by equation (5) in the following way:

$$I = \int_0^{x_1} L \, dx \qquad (8)$$

In this case, the condition (6) will be equivalent to the condition given by the following Euler-Lagrange's equation:

$$\frac{d}{dx}\left(\frac{\partial L}{\partial y'}\right) = \frac{\partial L}{\partial y}. \qquad (9)$$

Substituting on this equation the Lagrangian function L given by (7), and choosing a units such that the length x_1 has unitary value ($x_1 = 1$) and the gravity field such that $g = 1$, one obtains the following differential equation:

$$\frac{d}{dx}\left(\mu \frac{y'}{\sqrt{1+y'^2}}\right) = x + \lambda. \qquad (10)$$

Integrating this equation one obtains:

$$\mu \frac{y'}{\sqrt{1+y'^2}} = \frac{x^2}{2} + \lambda x + c, \qquad (11)$$

where c is an integration constant. Solving this first order differential equation, one obtains the following solutions:

$$y(x) = \pm \int \frac{\frac{x^2}{2} + \lambda x + c}{\sqrt{\mu^2 - \left(\frac{x^2}{2} + \lambda x + c\right)^2}} dx + k, \qquad (12)$$

were k is another integration constant.

The parameters λ, μ, c and k can be obtained by solving the system of the four equations constituted by the boundary conditions:

$$y(0) = 0 \qquad (13)$$

$$y(1) = 0, \qquad (14)$$

along with the restrictions obtained by fixing the breast area (eq.3) and the perimeter (eq.4).

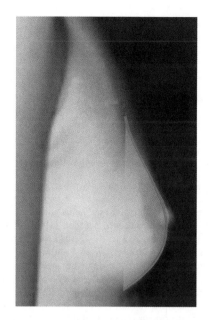

Figure 2. Overlay of the breast profile calculated in Maple for $S = 0.15$ and $P = 1.16$ with the photograph from which the values were taken.

3 A NUMERICAL SOLUTION AS A VALIDATION TEST

Using the integration series method (grade five) we solved numerically the above mentioned system of equations.

We started by taking a photograph of a woman's breast and using the software Geometry Sketchpad we drew a set of points along its profile, generating then a polygon (represented in 2 by the shaded area) then, using that same software, we calculated its area S and perimeter P.

In this case, the breast profile area was $S = 0.15$ and its perimeter was $P = 1.16$. The numerical solution we obtained is the one corresponding to grey line in Figure 2. When compared to the breast profile determined from a photo we consider to have obtained a good match, as can be seen in Figure 2.

4 THE BREAST DIAGRAM

4.1 The breast diagram and its allowed and forbidden zones

The Breast skin surface Area and the Breast Volume translate in the two dimensional sagital plane projection into Breast profile perimeter P and the Breast profile area S. These two parameters determine uniquely the geometrical shape of the breast profile.

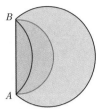

Figure 3. Three examples from the family of curves which maximise the subtended area.

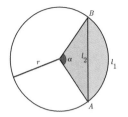

Figure 4. The Fence problem.

Therefore, one can use these two parameters to represent the shape of the breast in a two dimensional diagram (this was inspired by the use of the H-R diagram in Astrophysics [see Jaschek (1990) & Kaler (1989)]). Each breast profile corresponds to a point in this diagram with coordinates S and P.

Since in this work we shall be interested in the shape of the breast and not on its size, we normalise all breast sizes by taking the length of the line segment $\overline{AB}$ (which unites the upper and lower insertion points of the breast profile) as our unit length.

We shall call Breast Exclusion Curve (or BEC) the curve C in the breast diagram which corresponds to the locus of the family of arc circles containing the two fixed points A and B (as represented in Figure 3).

Since the circle is the 2D geometric curve which, for a given enclosed area, minimises its perimeter [see Clegg (1968)], no other curve can enclose a bigger area with the same perimeter. Therefore, any breast profile with a given area would change its shape if its perimeter is shrinked and when reaching the curve C it would burst.

In the next section, we shall therefore determine the equation for the curve in the S-P diagram corresponding to that locus.

4.2 The breast exclusion curve equation

In what follows, we shall determine the equation for the curve C using a parametric representation of the form $S = S(\alpha)$ and $P = P(\alpha)$ where the parameter α is to the angle corresponding to the arc circle considered above.

We shall start by considering the perimeter P and the area S subtended by an arc circle which is the solution of the classical and well known isoperimetrical problem in variational calculus called the "fence problem" (see [2]). A line segment $\overline{AB}$ of length l_2 is considered uniting two fixed points A and B and we shall then consider the family of all the arcs of circumference, of varying radii r, with varying perimeter l_1 which contain these two points.

The perimeter of the arc circle which we are interested in is the length l_1 represented in Figure 4.

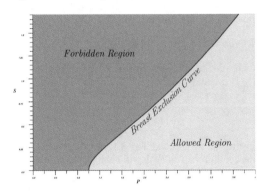

Figure 5. Representation of the Breast Exclusion Curve in the S-P diagram. The curve separates two regions in the diagram. One corresponding to an allowed region and one which is a forbidden region for the existence of breast profiles in them.

We shall call α the angle AOB measured in radians. The length l_1 is given by $l_1 = \alpha r$. The length l_2 of the segment $\overline{AB}$ is given by $2r \sin\left(\frac{\alpha}{2}\right)$.

In order to calculate the value of these two parameters, P and S, we shall consider two cases:

In both cases, the curve C can be given in a parametric representation by the following equations:

$$S = \frac{\alpha - sin(\alpha)}{8\,sin^2(\alpha/2)} \tag{15}$$

and

$$P = \frac{\alpha}{2\,sin(\alpha/2)} \tag{16}$$

where α is the parameter.

Using the Maple software, we ploted this curve C given in the parametric form by the two above equations in the breast diagram. The result is the curve represented in Figure 5.

4.3 The breast fall tendency

It is well known that women's breasts tend to become flaccid and fall with age. The question is how can this fall tendency of the breast be quantitatively measured?

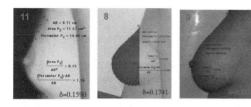

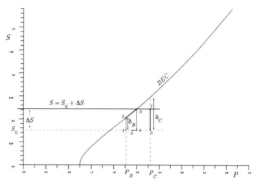

Figure 6. Data for three women's breasts used in this study.

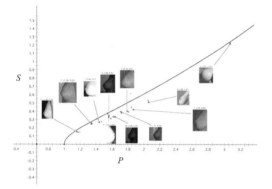

Figure 7. S-P Breast Diagram.

Figure 8. Expected breast track in the S-P diagram during the pregnancy and breast feeding period.

An answer to this question can be obtained using the above mentioned breast diagram.

For a given breast perimeter P we shall define the "Breast Fall Parameter" as the parameter δ given by the square root of the difference ΔS between the value of $S(P)$ for the BEC and the value of the breast profile area S for that breast. I.e. $\delta = \sqrt{\Delta S}$. This is shown in Figure 8. The square root is used in order for the "Breast Fall" to have the dimension of a length.

Looking at Figures 6, 7, one can see that the parameter δ is a good candidate to quantitatively describe the fall (or lack of roundness) of a given breast.

5 SAMPLE AND DATA ANALYSIS

In Figure 6 we present the measured values for the area S, perimeter P and the parameter δ, from some women in our sample.

Taking our measured sample of women's breasts, the points with coordinates (P, S) along as the corresponding breast photos were plotted in the breast diagram. The result is the following Figure 7.

As one would expect, looking at Figure 7, one can see that when represented in the breast diagram, all the points corresponding to the breasts in our sample lie below the Breast Exclusion Curve (BEC). The distance between each point and the BEC is a measure of the lack of roundness, or the fall tendency, for the corresponding breast.

6 THE BREAST DYNAMICS USING THE S-P DIAGRAM

Women's breasts change their volume and skin area along the women's life. This translates in the fact that each breast profile, when represented in the breast diagram, follows a path along the woman's life which is determined by the presence of gravity, on the one hand, and on the other hand by the stretching of the skin tissues.

One of most evident changes occurs during the pregnancy and breast feeding period [see Cox et al. (1999)].

In Figure 8 we suggest a theoretically expected track path in the S-P diagram of two women's breast with different breast fall parameters undergoing the process of pregnancy and subsequent breast feeding. This is in no way an experimental result. Instead it is the expected result of this model, which we would like to propose for experimental test by medical groups. Taking into account the changes in volume and skin area, the expected breast track in the S-P diagram during the pregnancy and breast feeding period could be represented as shown in the above Figure 8.

We shall suppose that a given woman's breast has area S_0 before she becomes pregnant. During pregnancy, in the first stage, the breast volume tends to grow due to the growth of the mammary glands in order to be able to produce a milk volume, ΔV_L, which is the amount the child needs to drink on each breast feed. This volume growth translates, in the 2-dimensional S-P diagram, into an increase of the breast profile area given by, ΔS_L. Depending on the value of the breast fall parameter, δ, for the specific breast when compared with $\sqrt{\Delta S_L}$, two situations can arise. These are represented in the S-P diagram of Figure 8 by the cases of two different breasts namely B and C with the same starting profile area S_0 but different perimeters P_B and

P_C before the corresponding pregnancies represented in the diagram by the starting points 1 and 6:

1. When $\delta > \sqrt{\Delta S_L}$ (case of breast C). – In this case, the breast volume can grow without the need to enlarge the skin area. The volume growth therefore stops before the breast point in the S-P diagram reaches the BEC (represented by the track from 6 to 7).

2. When $\delta < \sqrt{\Delta S_L}$ (case of breast B). – In this case, the breast volume growth can not be complete without the enlargement of the breast skin area. The volume growth reaches the BEC (represented by the track from 1 to 2), and therefore can not carry on along a vertical line in the S-P diagram. In order to reach the total volume growth ΔV_L, the breast must follow the BEC until the intersection point of the BEC with the horizontal line $S = S_0 + \Delta S_L$. This is represented in Figure 8, by the track 2 to 3.

Once the breast feeding process is finished, the breast no longer produces milk and therefore starts to reduce its volume. On the other hand the breast skin can also reduce its area (depending on whether or not its skin has stretched and on the skin elasticity). The end result shall be that, after a while, the point representing the breast B in the S-P diagram will move to point 5 (as represented in Figure 8). On the other hand, we expect that since breast C has not undergone any skin stretching can simply return to its starting point 6 (or a nearby point).

7 DISCUSSION AND CONCLUSIONS

The S-P breast diagram seems to us a particularly useful instrument to study, in a quantitative way, both the geometry and the dynamics of the breast evolution with age. It can also be useful in its study for specific time intervals like the pregnancy and lactation as well as the menstrual cycle.

Looking at Figure 7 one can notice examples of the good correlation between the intuitive perception of the breast fall tendency and its mathematical description using the parameter δ. For example, one can see that a breast with clear fall like the one numbered as 5 has a much bigger δ than the one numbered as 11. An extreme example as the one numbered as 4 although its large volume shows "great roundness" and this is in accordance with the fact that δ is minimal (its corresponding point in the diagram is almost touching the BEC). Even between two breasts apparently similar, like the ones numbered as 8 and 9, one can see that the parameter δ distinguishes the two by showing that number 9 has a bigger fall than number 8. This is in accordance with the clearly bigger visual roundness of number 8 when compared with number 9.

REFERENCES

[1] Macedo, G. Paulo; Mota, Elisabete & Sá, Vera. A simple Mathematical Model for the Woman's Breast Profile Geometry and its Characterisation. Submitted for publication in the *International Journal for Computation Vision and Biomechanics*.

[2] J. C. Clegg 1968. Calculus of Variations. Edinburgh: Oliver & Boyd.

[3] L. Landau & E. Lifshitz 1959. The Classical Theory of Fields. London: Pergamon press.

[4] Jaschek, Carlos & Jaschek, Mercedes 1990. The Classification of Stars, 2nd edition. New York, NY: Cambridge University Press. ISBN 0-521-26773-0.

[5] Kaler, James B. 1989. Stars and Their Spectra: An Introduction to the Spectral Sequence. New York, NY: Cambridge University Press. ISBN 0-521-30494-6.

[6] Cox, D.B.; Owens, R.A.; Kent, J.C.; Mitoulas, L. & Hartmann, P.E. 1999. Lactation in women breast volume and milk production during extended. *Exp Physiol.* 84:435–447.

Computational Vision and Medical Image Processing – João Tavares & Natal Jorge (eds)
© 2008 Taylor & Francis Group, London, ISBN 978-0-415-45777-4

Matching contours in images using curvature information

Francisco M. Oliveira
Faculdade de Engenharia da Universidade do Porto, Porto, Portugal

João Manuel R.S. Tavares
Instituto de Engenharia Mecânica e Gestão Industrial,
Faculdade de Engenharia da Universidade do Porto, Porto, Portugal

ABSTRACT: The work here described consists in searching for the optimum global matching between contours of two objects represented in images, which are sampled by equal or different number of points. Thus, to determine the optimum global matching between the points of two contours, it is used curvature information, that is totally invariant to rigid transformations. For the case of contours sampled by different numbers of points, two approaches are proposed to exclude, from the matching process, the additional points. In the last section of this paper, a method for the determination of the rigid transformation associated to two contours that is based in the same solution considered in our matching process is also described.

1   INTRODUCTION

The determination of correspondence between the data of two objects represented in images is a topic of raised importance and hard research in Computational Vision (Maciel 2002, Tavares 2001). In image processing and analysis, the applications that need the determination of correspondence between objects are numerous. Some examples that can be referred are: tracking and movement analysis (Bastos et al. 2006, Pinho et al. 2005, 2007, Tavares et al. 2000, Tavares 2001), 3D reconstruction (Azevedo et al. 2006, Maciel 2002), recognition (Tavares et al. 2000, Vasconcelos et al., 2006), registration (Vasconcelos et al., 2006), etc.

In this paper, a methodology to determine the matches between points of contours represented in images is presented. The referred methodology uses curvature information and optimization of the global matching cost.

2   MATCHING OF CONTOURS

2.1   *Equal number of points*

As it is well known, rigid transformations or, using the mathematical definition, transformations of similarity, can produce changes in an object, as in its position, scale or orientation, but they do not imply changes in its form. An invariant feature that persists between two objects, when one is obtained from the other by a rigid transformation, is the curvature information.

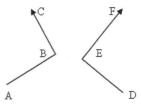

Figure 1.   In our matching methodology the direction in which the contours are being analysed is considered and, therefore, considering the parts of two contour represented, we have $A\hat{B}C = 270°$ and $D\hat{E}F = 90°$; instead, if the corrected direction was not considered, we have $A\hat{B}C = 90° = D\hat{E}F$.

Being two objects to be matched represented by a set of points that define the polygon associated with the contour of each one, the curvature information along the same can be estimated using the angle defined for each collection of three consecutive points. To calculate the angle associated to each point, it is also important to consider the direction along the contour in which the curvature is being analyzed, otherwise, strange situations can happen, as exemplified in figure 1. Thus, the assumption that the points to be matched are correctly ordered is important, see figure 2.

Let us considerer that to point i of a *contour 1* has associated the amplitude of angle α_i, and to point j of a *contour 2* correspond the amplitude of angle θ_i. Consequently, along these two contours to be matched,

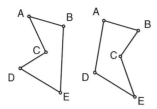

Figure 2. Two different contours configurations that are defined by the same set of points.

it can be considered the association between each point and the angle of its curvature. It should be noted that the sequences of angles obtained for each contour depends only on its shape, and not on its scale or position in the original image.

The following step of our matching methodology consists in searching for the correspondences between angles of the two contours with the minor global matching cost associated. That is, the goal of this step is to minimize the sum of absolute values of the angular differences between matched points.

Thus, we have an assignment problem, but with a fundamental constraint that prevent the application of the traditional assignment algorithms: the order of the points that defines the two contours should be maintained. Therefore, the methodology proposed in (Maciel 2002), for instance, can not be used. For that reason, it is necessary to develop a new matching algorithm that we explain next.

Let us consider that the two contours to be matched are sampled by n points. Using our approach, we obtain two sequences of angular amplitudes associated with the curvature along each contour to be matched:

$$\alpha_1, \alpha_2, ..., \alpha_n \text{ and } \theta_1, \theta_2, ..., \theta_n.$$

The next step is to test the n matching hypotheses successively, figure 3, saving the associated matching values in the following matrix of costs:

$$\begin{bmatrix} |\alpha_1 - \theta_1| & |\alpha_1 - \theta_2| & ... & |\alpha_1 - \theta_n| \\ |\alpha_2 - \theta_2| & |\alpha_2 - \theta_3| & ... & |\alpha_2 - \theta_1| \\ ... & ... & ... & ... \\ |\alpha_n - \theta_n| & |\alpha_n - \theta_1| & ... & |\alpha_n - \theta_{n-1}| \end{bmatrix}.$$

Therefore, the sum of the elements of the first column of the matrix of costs represents the angular cost of the first global matching; the sum of the elements of the second column represents the angular cost of the second global matching, and so forth. With this approach, the best matching will be given by:

$$\min\left\{\sum_{i=1}^n |\theta_i - \alpha_i|; \sum_{i=1}^n |\theta_{i+1} - \alpha_i|; ...; \sum_{i=1}^n |\theta_{i+n-1} - \alpha_i|\right\},$$

with $\theta_{i+n} = \theta_i$.

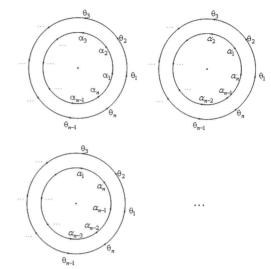

Figure 3. Example of searching for the best global matching between two contours sampled by equal number of points.

Thus, in order to use our method, the points of each contour to be matched should be previously ordered, and the two contours should be analyzed in a common direction. Thus, reflections between contours to be matched are not possible.

Notice that the point i of the *contour 1* does not have to correspond to the point i of the *contour 2*. The only condition that has to be respected is that, for example, if point i of *contour 1* corresponds to point j of *contour 2*, then its subsequent neighbour in *contour 1* (point $i+1$) has to correspond to subsequent neighbour of point j in *contour 2* (point $j+1$). Thus, this restriction of order reduces the hypothesis of our matching problem from $n!$ to n.

In figure 3, a graphical representation of our matching searching methodology is made. The associated idea is the following: the two contours are "overlapped" and one is "rotated" successively in order that the best fit between the curvature angles is found.

In figure 4, we can see an adequate matching found between two contours using our methodology. In this example, one of the contours is rotated 175° relatively to the other one.

2.2 Different number of points

If a contour has more points than the other one, for example, *contour 1* has only n points and *contour 2* has m points, with $n < m$, then an approach to reject the $m - n$ extra elements from the angles' sequence of *contour 2*, $(\theta_i)_{i=1,2,...,m}$ is necessary.

In this work, we developed two approaches to discard the angles associated with the exceeded points.

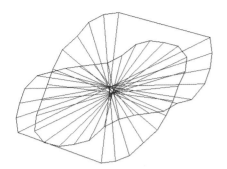

Figure 4. Matching obtained between two contours that have a rotation of 175° between them (lines between contours represent the matches found).

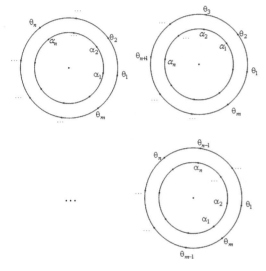

Figure 5. Example of searching for the best global matching between two contours sampled by different number of points.

In both, we use the same methodology described previously to get the angular differences between the possible correspondences.

Thus, first, the matrix of angular costs for the m possible global matches is obtained:

$$
\begin{bmatrix}
|\alpha_1 - \theta_1| & |\alpha_1 - \theta_2| & \cdots & |\alpha_1 - \theta_{m-n}| & \cdots & |\alpha_1 - \theta_m| \\
|\alpha_2 - \theta_2| & |\alpha_2 - \theta_3| & \cdots & |\alpha_2 - \theta_{m-n+1}| & \cdots & |\alpha_2 - \theta_1| \\
\cdots & \cdots & \cdots & \cdots & \cdots & \cdots \\
|\alpha_n - \theta_n| & |\alpha_n - \theta_{n+1}| & \cdots & |\alpha_n - \theta_m| & \cdots & |\alpha_n - \theta_{n-1}|
\end{bmatrix}
$$

Once again, the main idea is to imagine that the sequences of angles are "overlapped" and one of them is "rotating", finding, this way, the best global correspondence between them, figure 5.

If necessary, in the matching procedure adopted, points can be eliminated from the contour that presents more data points. This elimination, and consequent angles recalculation, might be necessary, because, the angles' amplitudes can be rather different between the two contours, and therefore, the probability of bad matches can be increased considerably. Thus, the geometry of the contour that presents more points will be simplified until a good global matching can be found. Two approaches to perform this elimination are described in the next subsections.

In both approaches, in case of a good matching to be verified before eliminating all exceeded points, the process will be interrupted.

2.2.1 Approach I

Let us consider as the best possible global matching between the angles' amplitude of the *contour 1* with the ones of *contour 2*, the case for which is minimum the global angular cost for the n angles matched in the m possible hypotheses; that is:

$$
\min\left\{\sum_{i=1}^{n}|\theta_i - \alpha_i|; \sum_{i=1}^{n}|\theta_{i+1} - \alpha_i|; \ldots; \sum_{i=1}^{n}|\theta_{i+m-1} - \alpha_i|\right\},
$$

where $\theta_{i+m} = \theta_i$.

In the next step, from the sequence of angles $(\theta_i)_{i=1,2,\ldots,m}$ is considered the angle that provokes major angular difference and has at the same time an amplitude of around 180°. This last condition guarantees that the curvature of the contour in that point is irrelevant; that is, that point is in an almost straight part of the contour and, because of that, it can be considered insignificant for the description of the associated object's shape. Thus, the point of *contour 2* that corresponds to that angle is discarded from the same one.

Until the second angles' sequence has length equal to n, the search for the optimum matching is repeated, discarding each data point in the conditions previously enunciated. In this searching process, for the best matching possible the one that has the minor total angular cost is selected again, figure 3.

2.2.2 Approach II

In this approach, it is considered for the best matching case the one that has more pairs of angles correctly matched. Being considered as correctly matched, if the angular difference between the angles involved is less than a predefined threshold value. Next, from the result of this matching process, we find the set of angles wrongly matched. From this set, the angle that is closest to 180° is considered, which corresponds to a point of the contour wrongly matched and has curvature that can be considered as negligible. Then, this point is discarded.

Until the second angles' sequence has a length equal to n, the process used to search for the best global matching is repeated and each point found in the conditions previously enunciated is discarded. Finally, the

377

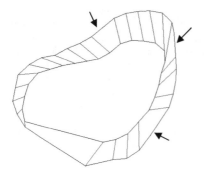

Figure 6. Result of the proposed matching approach with two contours sampled by a different number of points.

best matching is selected by the minor global angular cost.

In figure 6, an example can be visualized in which the exceeded points, that were not matched, are precisely the points whose curvature is small. In this example, one contour is sampled by 25 points and the other by 28. In the same figure, the arrows drawn indicate the 3 data points that were excluded from the matching process.

After this matching process, the points in excess can be matched easier using a second matching criterion.

2.3 Observations

In the case of matching two contours sampled by equal number of points, the developed methodology always makes the one-to-one correspondence that minimizes the global angular cost and generally obtains good results.

If the contours to be matched are sampled by different numbers of points, both approaches considered for the exclusion of the exceeded points, had shown adequate efficient. In the several experimental tests done, the second approach proposed has demonstrated to be slightly more efficient.

3 ESTIMATION OF THE RIGID TRANSFORMATION

3.1 Translation, scale and rotation

For the rejection of the angles and associated points wrongly matched, a new process that is based on the rigid transformation involved between the contours is being developed. In this way, it will also be possible to consider as a matching criteria the distance between each pair of candidate points.

Essentially, the method in development is based on the following steps. The centers of the two contours to be matched are determined using a procedure in which the weight associated to each point depends on the incidence of contour's data in its neighborhood. Thus, for instance, in a region with high concentration of points, the weight associated to each one of those points is reduced.

Next, to determine the scale involved between the two contours, the respective perimeters are calculated, being the quotient the desired value. After that, both contours are centered in the origin and the scale determined is applied on the second contour. This way, two contours centered at the origin and with similar dimensions are obtained.

The problem consists now in finding the optimum estimator for the involved rotation angle. When the contours are sampled using a significant number of points, two sets of samples with equal dimension, one for each contour, are consider. With this procedure, the fact that local contour's details can have exaggerated influence in the global geometry is prevented, as it is only considered the polygon base of each contour. Later, using a process equivalent to the one described previously, the matching with minor global angular cost is obtained. From this matching, only the pairs of angles with angular difference smaller than a predefined threshold value are considered. Then, using each selected pair of points matched, the rotation angle that minimizes the distance between them is obtained; that is, considering the point (x_1, y_1) of *contour 1* and its matched point (x_2, y_2) of *contour 2*, we intended to determine β, such that:

$$f(\beta) = (x_1 - x_\beta)^2 + (y_1 - y_\beta)^2,$$

be minimum, with:

$$\begin{bmatrix} x_\beta \\ y_\beta \end{bmatrix} = \begin{bmatrix} \cos\beta & -\sin\beta \\ \sin\beta & \cos\beta \end{bmatrix} \begin{bmatrix} x_2 \\ y_2 \end{bmatrix}.$$

Later, the weighed average of the obtained rotation angles is considered as the rotation angle involved between the two contours; that is, if the angular difference between two angles considered matched is large, its weight in the final contribution for the global rotation angle is small. Finally, the rotation estimated is applied on *contour 2*.

In figure 7, some examples of experimental results obtained using the presented methodology for the estimation of the rigid transformation involved between two contours are presented.

3.2 Observations

The process proposed in this paper for estimating the existing rigid transformation between two contours represented in images, has shown to be very efficient when the differences between them are essentially of rigid type.

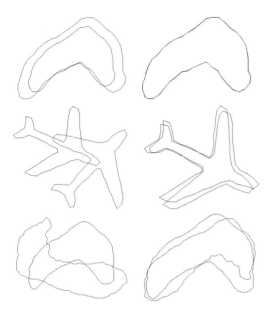

Figure 7. Some examples of results obtained in the determination of the rigid transformation involved between two contours (In each case, in the left are the original contours and in the right are the contours after applying an one of them the rigid transformation estimated.).

The process proposed also shows to be very promising even when is accentuated the non-rigid component involved. In these cases, the process proposed allows the adequate estimation of the rigid part of the global transformation in cause, as it can also be verified from the examples presented in figure 7.

4 CONCLUSIONS

In this paper, a methodology that allows the adequate matching of contour that are represented in images and sampled by points was presented. The referred methodology is based on the curvature information along each contour and in the optimization of the global matching costs.

With the presented methodology, the matching of contours sampled by equal or different number of points is possible.

The experimental results obtained, some presented in this paper, validate the adequateness of the methodology proposed.

In this paper, a solution for the estimation of the rigid transformation involved between two contours, based also on the searching methodology for the minor global angular cost was also presented. The experimental results obtained with this solution are very promising, even in cases that the transformation involved has a significant non-rigid component.

As future works, beyond other possible tasks, we expect to apply the methodology proposed in contours represented along image sequences, to allow the matching of type "one with several" and "several with one" between points, and to quantify contours' similarity.

ACKNOWLEDGEMENTS

This work was partially done in the scope of project "Segmentation, Tracking and Motion Analysis of Deformable (2D/3D) Objects using Physical Principles", with reference POSC/EEA-SRI/55386/2004, financially supported by *FCT – Fundação para a Ciência e a Tecnologia* from Portugal.

REFERENCES

Azevedo, T., Tavares, J., Vaz, M. (2006). Development of a Computer Platform for Object 3D Reconstruction using Active Vision Techniques, *VISAPP 2006 – First International Conference on Computer Vision Theory and Applications*, Setúbal, Portugal

Bastos, L., Tavares, J. (2006). Matching of Objects Nodal Points Improvement using Optimization, *Inverse Problems in Science and Engineering*, Vol. 14, No. 5, 529–541

Maciel, J. (2002). PhD Thesis: Global Matching: Optimal solution to correspondence problems, Instituto Superior Técnico, Portugal

Pinho, R., Tavares, J., Correia, M. (2005). Human Movement Tracking and Analysis with Kalman Filtering and Global Optimization Techniques, *II International Conference On Computational Bioengineering*, Lisbon, Portugal

Pinho, R., Tavares, J., Correia, M. (2007). An Improved Management Model for Tracking Missing Features in Computer Vision Long Image Sequences, *WSEAS Transactions on Information Science and Applications*, Vol. 4, No. 1, 196–203

Tavares, J., Barbosa, J., Padilha, A. (2000). Matching Image Objects in Dynamic Pedobarography, *RecPad 2000 – 11th Portuguese Conference on Pattern Recognition*, Porto, Portugal

Tavares, J. (2001). PhD Thesis: Análise de Movimento de Corpos Deformáveis usando Visão Computacional, Faculdade de Engenharia da Universidade do Porto, Portugal

Vasconcelos, M. Tavares, J. (2006). Methodologies to Build Automatic Point Distribution Models for Faces Represented in Images, *CompIMAGE – Computational Modelling of Objects Represented in Images: Fundamentals, Methods and Applications*, Coimbra, Portugal

Kinematic analysis of human locomotion based on experimental data

Fernando Meireles
MSc Student in Biomedical Engineering, University of Minho, Portugal

Miguel Silva
Mechanical Engineering Department, IDMEC-IST, Technical University of Lisbon, Portugal

Paulo Flores
Mechanical Engineering Department, University of Minho, Portugal

ABSTRACT: A general methodology for kinematic analysis of human locomotion based on experimental data is presented and discussed in this work. The kinematic equations of motion and the human biomodel description are developed by using Cartesian coordinates. The trajectories of the bodies that constitute the biomodel are obtained from experimental data acquisition, in which relevant anatomical points of reference are followed, as they represent the human natural gait motion. These points are typically used to represent the natural joints. After obtaining the data relative to the human gait motion, the points are interpolated in order to define the necessary analytical expressions that represent the guiding constraint equations. These constraints guide all the degrees-of-freedom of the biomodel. In order to describe the constraints in closed-form expressions, cubic interpolation spline functions are used, which consist of polynomial pieces on subintervals, joined together according to certain smoothness conditions. For this purpose, the degree selected for the polynomial functions is 3, being the resulting splines named cubic splines. The reason for that is due to the fact that the cubic polynomial functions are joined together in such a way that they have continuous first and second order derivatives. Since the constraints are defined, they are implemented in a computational code devoted to the kinematic analysis of multibody systems.

1   INTRODUCTION

The study of human body motion as a multibody system is a challenging research field that has undergone enormous developments over the last years (Silva & Ambrósio, 2002). Computer simulation of several human capabilities has shown to be useful in many research and development activities, such as: analysis of top athletic actions, to improve different sporting performances (Raasch *et al.*, 1997); optimization of the design of sportive equipment (Zappa *et al.*, 1995); ergonomic studies, to assess operating conditions for comfort and efficiency in different aspects of human body interactions with the environment (Rasmussen *et al.*, 2002); orthopedics, to improve the design and analysis of prosthesis (Andriacchi & Hurwitz, 1997); occupant dynamic analysis for crashworthiness and vehicle safety related research and design (Ambrósio & Dias, 2007); and gait analysis, for generation of normal gait patterns and consequent diagnosis of pathologies and disabilities (Winter, 1991).

It is known that computational simulation of the human motion requires the implementation of mathematical models that correctly describe the behavior of the human body and its interaction with the surrounding environment. In a broad sense, there are essentially two major classes of biomechanical models: (*i*) the ones described using finite elements methods; (*ii*) and the ones described using multibody formulations. Finite element models are applied in cases where localized structural deformations or soft tissues need to be described and analyzed in detail (Bandak *et al.*, 1996), while multibody models are usually applied in cases where gross-motions are involved and when complex interactions with the surrounding environment are expected (Wismans, 1996). Gait as a gross-motion simulation is usually described using multibody formulations.

The main purpose of this work is to develop a general methodology to perform a biomechanical analysis of the kinematics of human gait, using a multibody systems formulation. The proposed model is a two dimensional model. Being a sagittal-plane model, this model cannot be used to study movements that occur in frontal and transverse planes. Although, the six major determinants of gait, i.e., hip, knee and ankle

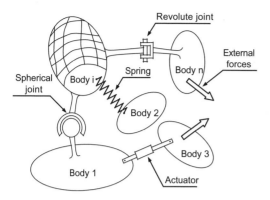

Figure 1. Representation of a general multibody system.

flexion-extension, pelvic rotation, pelvic list, and lateral pelvic displacement, (Norkin, 2001), can only be study in three dimensional models (Pandy, 2001), the development of two dimensional models is far more simple, and can be interesting concerning joint flexion-extension movements and joint moments (Alkjaer et al., 2001).

2 MULTIBODY SYSTEMS FORMULATION

2.1 Multibody concept

A typical multibody system (MBS) is defined as a collection of rigid or flexible bodies that have their relative motion constrained by kinematic joints and that are acted upon forces. The forces applied over the system components may be the result of springs, dampers, actuators or external forces. A generic description of a multibody model is represented in Figure 1. In the formulation presented in this work only rigid bodies are considered. In the two-dimensional space, when the bodies are considered to be rigid, they present three degrees-of-freedom (DOF). There are different coordinates and formalisms that lead to suitable descriptions of multibody systems, each of them presenting relative advantages and drawbacks. The application of Cartesian coordinates has the advantage that the formulation of the equations of motion is straightforward (Nikravesh, 1988).

2.2 Cartesian coordinates

The multibody system geometric configuration is characterized by a set of variables which entirely define a body location an orientation. In the two-dimensional space, the configuration of the rigid body can be defined through three independent coordinates: two coordinates describing the body local system axis localization and one coordinate describing the body orientation in relation to global system axis. The global system is fixed as it can be seen in Figure 2. The

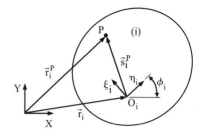

Figure 2. Definition of the Cartesian coordinates.

position of the body with respect to global coordinate system XY is defined by the coordinate vector $\mathbf{r}_i = [x\ y]_i^T$ that represents the location of the local reference frame $(\xi\eta)_i$. The orientation of the local frame with respect to the global system axis is represented by ϕ_i. Therefore, the vector of coordinates that describes completely the rigid body i in the two-dimensional space is,

$$\mathbf{q}_i = \{x\ y\ \phi\}_i^T \tag{1}$$

Thus, once defined these three coordinates, the global position of any point located in the body can be expressed as a function of these three coordinates. A point P on body i can be defined by the position vector $\mathbf{s}_i^P$, which represents the location of point P with respect to the body-fixed reference frame $(\xi\eta)_i$, and by the global position vector $\mathbf{r}_i$, that is,

$$\mathbf{r}_i^P = \mathbf{r}_i + \mathbf{s}_i^P = \mathbf{r}_i + \mathbf{A}_i \mathbf{s}_i'^P \tag{2}$$

where $\mathbf{A}_i$ is the transformation matrix given by,

$$\mathbf{A}_i = \begin{bmatrix} \cos\phi_i & -sen\phi_i \\ sen\phi_i & \cos\phi_i \end{bmatrix} \tag{3}$$

Notice that the vector $\mathbf{s}_i^P$ is expressed in global coordinates whereas the vector $\mathbf{s}_i'^P$ is defined in the body i fixed coordinate system.

2.3 Constraint equations

A kinematic joint imposes certain conditions on the relative motion between the adjacent bodies that it comprises. When these conditions are expressed in analytical form, they are called constraint equations. In a simple way, a constraint is any condition that reduces the number of degrees of freedom in a system.

In this work planar revolute joints are used and a driving constraint equation that guides the body's center-of-mass (CM) along the time is developed.

The revolute joint is a pin and bush type of joint that constrains the relative translation between the two bodies i and j, allowing only the relative rotations, as it is

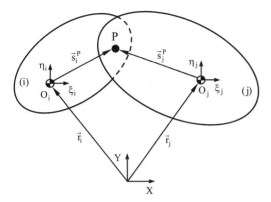

Figure 3.   Planar revolute joint connecting bodies i and j.

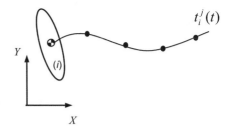

Figure 4.   Body trajectory.

illustrated in Figure 3. The kinematic conditions for the revolute joint require that two distinguish points, each one belonging to a different body, share the same position in space all the time. This means that the global position of a point P in body i is coincident with the global position of a point P in body j. Such condition is expressed by two algebraic equations that can be obtained from the following vector loop equation,

$$\mathbf{r}_i + \mathbf{s}_i^P - \mathbf{r}_j - \mathbf{s}_j^P = \mathbf{0} \tag{4}$$

which is re-written as,

$$\mathbf{\Phi}^{(r,2)} \equiv \mathbf{r}_i + \mathbf{A}_i \mathbf{s}_i'^P - \mathbf{r}_j - \mathbf{A}_j \mathbf{s}_j'^P = \mathbf{0} \tag{5}$$

Thus, there is only one relative DOF between two bodies that are connected by a planar revolute joint.

In general, the motion of one or more bodies in a MBS is specified, e.i, the system is guided typically by rotational or translational actuators. Other type of guiding constraints are those associated with the known trajectories of the bodies CM, which can be obtained by experimental data acquisition, as it is illustrated in Figure 4. The mathematical equation that represents the guiding constraint can be written as,

$$\mathbf{\Phi}^{(g,3)} \equiv \begin{bmatrix} x_i - t_i^x(t) \\ y_i - t_i^y(t) \\ \phi_i - t_i^\phi(t) \end{bmatrix} = \mathbf{0} \tag{6}$$

where, $t_i^j(t)$ ($j = x, y, \phi$) represents the trajectory described by the CM of body i.

2.4   Kinematic analysis and computational strategy

The kinematic analysis is the study of the motion of a system, independently of the forces that produce it. It is performed by solving the set of equations that result from the kinematic and driver constraint setup.

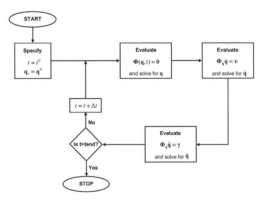

Figure 5.   Computational procedure for kinematic analysis.

When the configuration of a MBS is described by n Cartesian coordinates, then a set of m algebraic kinematic independent holonomic constraints $\mathbf{\Phi}$ can be written in a compact form as (Nikravesh, 1988),

$$\mathbf{\Phi}(\mathbf{q}, t) = \mathbf{0} \tag{7}$$

where $\mathbf{q}$ is the vector of generalized coordinates and t is the time variable, in general associated with driving elements.

The velocities and accelerations of the system elements are evaluated using the velocity (8) and acceleration (9) constraint equations,

$$\mathbf{\Phi}_{\mathbf{q}} \dot{\mathbf{q}} = \upsilon \tag{8}$$

$$\mathbf{\Phi}_{\mathbf{q}} \ddot{\mathbf{q}} = \gamma \tag{9}$$

where $\mathbf{\Phi}_{\mathbf{q}} = \partial \mathbf{\Phi}/\partial \mathbf{q}$ is the Jacobian matrix of the constraint equations, $\dot{\mathbf{q}}$ is the vector of generalized velocities, $\upsilon = \partial \mathbf{\Phi}/\partial t$ is the right hand side of velocity equations, $\ddot{\mathbf{q}}$ is the acceleration vector and γ is the right hand side of acceleration equations, i.e., the vector of quadratic velocity terms, which contains the terms that are exclusively function of velocity, position and time.

The kinematic analysis of a multibody system can be carried out by solving the set of equations (7), (8) and (9). The necessary steps to perform this analysis are summarized as sketched in Figure 5. For details in this formulation, the interested reader is referred to the work of Nikravesh (1988).

3 BIOMODEL OF HUMAN LOCOMOTION

3.1 *Model description*

In this section a biomechanical model of the human body, used in kinematic analysis of the stride period, is presented. This model is developed within a general purpose multibody code that uses the methodologies described in the previous section.

The model is defined using 7 bodies, corresponding to 7 body segments. A description of the 7 anatomical segments and their corresponding bodies is presented in Table 1 (Winter, 2005).

The bodies are connected by 6 revolute joints, as it is illustrated in Figure 6. The model has 9 DOF that correspond to 6 rotations about revolute joints, plus 3 DOF that are associated with the free body translations and rotation of the base body (½ HAT, acronym for head-arms-trunk).

3.2 *Used Methodology*

With the purpose to develop a general approach to perform kinematic analysis of the human locomotion it is necessary to know the bodies' trajectories. For this,

the Biomechanical laboratory of Technical University of Lisbon was used to capture human gait. The motion data of the system consists of the trajectory of a set of anatomical points located at the joints and extremities of the subject under analysis, as depicted in Figure 7. Furthermore, a pressure platform was used to obtain the ground reaction forces.

In the present work, the bodies' trajectories are obtained through a digitization process in which the images collected by two video cameras (frontal and sagittal planes) are used to reconstruct the two-dimensional coordinates of the anatomical points. From these anatomical points' coordinates, the $CM(x, y)$ position are calculated, using anthropometric relations between the segment length and the CM proximal location (Winter, 2005), as well as the body orientation, ϕ. Table 2 shows typical results for the Cartesian coordinates variation along time based on this approach.

The next step consists on the interpolation of the Cartesian coordinates' variation along time in order to achieve bodies trajectory curves $(t_i^j(t))$. Based

Table 1. Geometric properties of the human model.

Rigid body Nr.	Description	Length [m]
1	½ Right HAT	0.250
2	Right upper leg	0.314
3	Right lower leg	0.425
4	Right foot	0.122
5	Left upper leg	0.314
6	Left lower leg	0.425
7	Left foot	0.122

Table 2. Cartesian coordinates variation along time for ½ Hat.

Time	CMx [m]	CMy [m]	CMφ [rad]
0.0000	0.4718	1.0812	1.4866
0.0145	0.4929	1.0804	1.4964
0.0290	0.5131	1.0818	1.5035
...	...	...	...

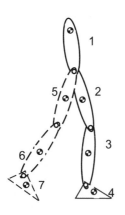

Figure 6. Two-dimensional biomechanical model.

Figure 7. Sagittal data acquisition showing the set of anatomical points' location.

on these curves, the analytical constraint equations, given by expression (6), can easily be obtained. This interpolation is performed employing cubic splines, since higher-order polynomials tend to swing through wild oscillations in the vicinity of an abrupt change, whereas cubic spline provides much more smooth transitions (Chapra, 1989). Further, the use of cubic splines is useful to guarantee the continuity of the first and the second derivatives (velocity and acceleration respectively), property that is extremely important in the kinematic analysis.

4 RESULTS AND DISCUSSION

The methodology proposed in this work to model the guiding constraints associated with bodies' trajectories was implemented in KAP FORTRAN code (KAP, acronym for *Kinematic Analysis Program*), which is a computational program developed by Nikravesh (1988) with the purpose to perform kinematic analysis of multibody systems. Thus, a new subroutine, named *'GUID.for'* was created to evaluate constraint equation violations, the right sides of the velocity and accelerations equations, as well as the entries in the Jacobian matrix for the system's guiding constraints. This subroutine reads the input values of the Cartesian coordinates and performs the necessary calculations with the aid of specific routines for *cubic spline interpolation* and *cubic spline derivative evaluation* available in the *ISML Fortran Numerical Libraries*.

With the intent of validating the implemented methodology in the KAP program, a known input data was prescribed. Then the corresponding output data was evaluated mainly to verify the kinematic consistency of the cubic splines first and second derivative calculation. For this purpose, the function *sin(t)* was chosen, since it is quite easy to calculate its first and second derivatives. The domain selected varies from 0 to π radians.

In the first numerical simulations a total of 219 data points were used to obtain the cubic splines interpolation and the corresponding derivatives. The output values present a good behavior concerning the cubic spline interpolation as well as in evaluation of the first derivative. However, there are some numerical difficulties in the calculation of the second derivative, as it is illustrated in Figure 8, where some oscillation around the values of the second derivative is observed (*219-Input data points*). In order to overcome this problem, two different approaches were performed. The first one consisted of testing several subroutines available in *ISML Libraries*. The second one, based on cubic spline theories, the number of input data points were reduced, from 219 to 42, yet that output frequency is exactly the same, which depends on the user specification. This second approach represents different frame rates of signal acquisition of anatomical

points, that is, different sampling frequencies. When the input data number was reduced, the results obtained are smoother, as it is shown in Figure 9 (*42-Input data points*).

Figure 10 evidences the main difference of the results obtained and illustrated in figures 8(c) and 9(c) when the X and Y-axis are re-scaled. Again, when the

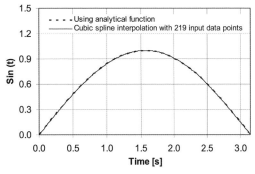

(a) Evaluation of function *sin(t)*.

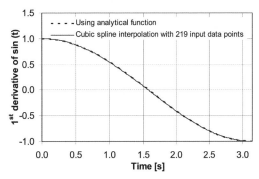

(b) Evaluation of the first derivative of function *sin(t)*.

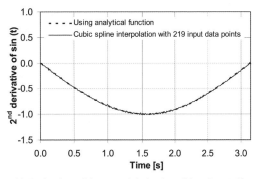

(c) Evaluation of the second derivative of function *sin(t)*.

Figure 8. Evaluation of the function sin(*t*), first and second derivative by using analytical function and cubic splines interpolation with 219 input data points.

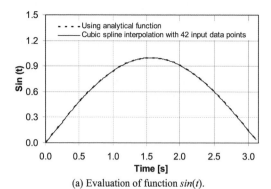

(a) Evaluation of function $sin(t)$.

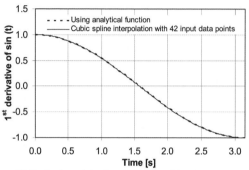

(b) Evaluation of the first derivative of function $sin(t)$.

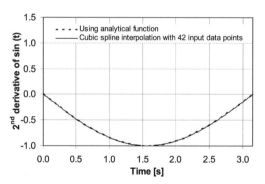

(c) Evaluation of the second derivative of function $sin(t)$.

Figure 9. Evaluation of the function $sin(t)$, first and second derivative by using analytical function and cubic splines interpolation with 42 input data points.

input data number is reduced, the results obtained are smoother, as it is shown in Figure 10b.

5 CONCLUDING REMARKS

In this work, a general approach for kinematic analysis of human locomotion based on experimental data has been presented. The proposed model is a two dimensional model. In the process, the main aspects related

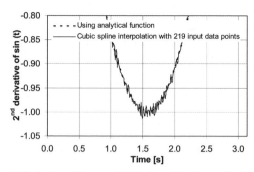

(a) Evaluation of the second derivative of function $sin(t)$ with 219 input data points.

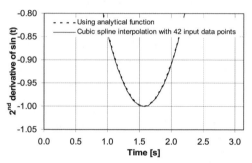

(b) Evaluation of the second derivative of function $sin(t)$ with 42 input data points.

Figure 10. Results obtained for the second derivative of function $sin(t)$ by using cubic splines interpolation with 219 and 42 input data points.

to the formulation of the kinematic analysis of multibody systems were revised. The kinematic equations of motion and the human biomodel description were developed employing Cartesian coordinates.

The data points that constitute the bodies' trajectories were interpolated using cubic splines and with these the guiding constraint equations that guide the bodies during the kinematic simulation were defined. The reason for using such approach is due to the fact that the cubic polynomial functions are joined together in such a way that they have continuous first and second derivatives in the analysis domain. This methodology was validated through numerical simulations performed with different interpolation splines and different number of input data points. It was observed that when the input data points are reduced, the quality of the results obtained is better.

At future work, a natural follow-up is to test the methodology proposed with a general gait motion, as well as to include it in a computational code for dynamic analysis of human locomotion. For this to be achieved additional experimental work is required.

REFERENCES

Alkjaer, T., Simonsen, E.B., Dyhre-Poulsen, P. 2001. Comparison of inverse dynamics calculated by two- and three-dimensional models during walking, *Gait and Posture* 13: 73–77.

Ambrósio, J. & Dias, J. 2007. A Road Vehicle Multibody Model for Crash Simulation based on the Plastic Hinges Aprproach to Structural Deformations, *IJCrash,* 12(1): 77–92.

Andriacchi, I. & Hurwitz D. 1997. Gait Biomechanics and the Evolution of Total Joint Replacement, *Gait and Posture,* 5: 256–264.

Bandak, F., Eppinger, R. & Onimaya, A. 1996. Traumatic Brain Injury, Bioscience and Mechanics, *Mary Ann Liebert. Inc.* New York, NY.

Chapra S.C & Canale, R.P. 1989. *Numerical Methods for Engineers,* 2nd Ed, McGraw-Hill.

Nikravesh, P.E. 1988. *Computer-Aided Analysis of Mechanical Systems,* Prentice Hall, New Jersey.

Norkin, C.C. & Levangie, P. K. 2001. *Articulações – estrutura e função: uma abordagem prática e abrangente,* 2nd Ed., Rio de Janeiro.

Pandy, MG. 2001. Computer Modeling and Simulation of Human Movement, *Annu. Rev. Biomed. Eng.* 3:245–73.

Raasch, C., Zajac, F., Ma, B. & Levine, S. 1997. Muscle Coordination of Maximum-Speed Pedaling. *Journal of Biomechanics,* 30(6) 595–602.

Rasmussen, J., Damsgaard, M., Christensen, S. & Surma, E. 2002. Design Optimization with Respect to Ergonomic Properties, *Structural and Multidisciplinan Optimization,* 24(2):89–97.

Silva, M.P.T., Ambrósio, J.A.C. 2002. Kinematic Data Consistency in the Inverse Dynamic Analysis of Biomechanical Systems. *Multibody System Dynamics,* 8(2): 219–239.

Winter, D. 1991. *The Biomechanics and Motor Control of Human Gait: Normal, Eldery and Pathological.* University of Waterloo Press, Waterloo, Canada., 1991.

Winter, D. 2005. *Biomechanics and Motor Control of Human Movement,* 3rd Ed., John Wiley & Sons, Inc.

Wismans, J. 1996. Models in Injury Biomechanics for Improved Passive Vehicle Safety. *Crashworthiness of Transportation Systems: Structural Impact and Occupant Protection.* J. Ambrósio, M. Pereira and F. Silva. (eds), *Kiuwer Academic Publishers.* Dordrecht.

Zappa B., Casolo, F. & Legnani, G. 1995. Analysis and Synthesis of 3D Motion for Multi-Body Systems with Regards to Sport Performances. *Ninth World Congress on the Theory of Machines and Mechanisms,* Milano, Italy.

Image processing on the Poisson ratio calculation of soft tissues

P. Martins, R. Natal Jorge & A. Fernandes
IDMEC-Polo FEUP, Faculdade de engenharia da Universidade do Porto, Porto, Portugal

A. Ferreira
DEMEGI, Faculdade de engenharia da Universidade do Porto, Porto, Portugal

T. Mascarenhas
Faculdade de Medicina da Universidade do Porto/Hospital de S. João, Portugal

ABSTRACT: An image processing based method of calculating Poisson's ratio is proposed. This approach combines pure geometrical analysis with an optimization based tracking technique. The images were obtained from the video recordings of uniaxial tension tests. The tested materials were prolapsed human vaginal tissues, which are soft, highly hydrated tissues.

Due to the lack of distinctive characteristics on the sample's face (acting as tracking marks), an algorithm that extracts information from the RGB codes of a given region (mask) and tracks it's subsequent evolution was implemented. There were no additional markings (purposely made) on the samples. The image processing software solution that implements the algorithm, was developed using MATLAB® and some functions from the image processing toolbox (*IPT*) were used.

1 INTRODUCTION

The acquisition of data from biological soft tissue experiments, poses important technical problems that act as conditioning factors for the acquisition of significant and meaningful scientific knowledge.

The present work has the objective of constructing an algorithm to perform the Poisson's ratio (v) calculation using low resolution video. The videos are the recordings of the uniaxial tension tests of human prolapsed vaginal tissues. The problem of genital prolapse has a special incidence on post menopausal women (Tarnay and Dorr 2003), therefore peroperative tissue used to perform the mechanical tests reported in the current work, were excised exclusively from post menopausal women, during prolapse correcting surgeries. Previous works concerning the mechanical properties of prolapsed vaginal tissues (Goh 2002; Cosson et al. 2004), provided an important basis for the current work.

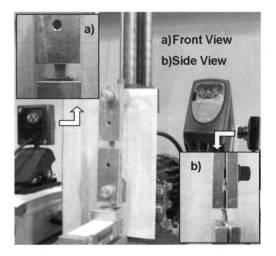

Figure 1. Uniaxial tension test.

1.1 *Uniaxial tension test*

The experimental protocol followed, has been adapted for the study of biological highly hydrated soft tissues. The uniaxial tension tests were performed using the experimental apparatus shown in figure 1. The testing solution used was developed by the authors purposely for the study of hyperelastic materials Martins et al. 2006). The main components of the system are the testing machine, the automatic control module, the sensor array and the acquisition system. A video record of

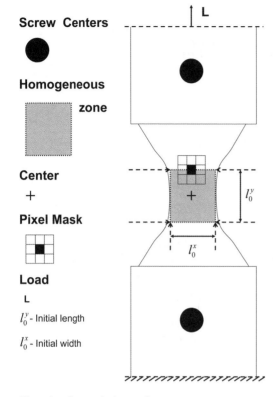

Screw Centers

Homogeneous

zone

Center

+

Pixel Mask

Load

L

l_0^y - Initial length

l_0^x - Initial width

Figure 2.   Geometric Approach.

every test is maintained, mainly for control purposes. The chosen view is identified in figure 1 by the label **a)**.

1.2   *Poisson's ratio*

In a uniaxial tension test the change in length, can be related with the change in width (case of a square section bar) or the change in diameter (case of a circular section bar). As the tensional load (**L**, figure 2) increases on the bar, the initial values for length (l_y^1) increase. The width (or diameter) initial value (l_x^1) decrease. To evaluate the changes in length and width, two strain quantities ε_y and ε_x can be defined according to equations (1) and (2). The ratio between ε_x and ε_y is Poisson's ratio, ν (3). The ratio ν is constant if the strains are small (Lai et al. 1996). A typical value of ν for steel is 0.3. As for a (quasi)incompressible material, like rubber, ν value is close to 0.5. On the case of highly hydrated soft tissues, due to the high water content of the material ($>70\%$), the expected value for ν is also close to 0.5. Poisson's ratio may depend on the orientation of the specimen in which case the material is called anisotropic. Anisotropy is related to some degree of organization of the material structure. For an isotropic material, Poisson's

ratio is constant independently of sample's cutting direction.

$$\varepsilon_y \;=\; \frac{l_y - l_y^1}{l_y^1} \tag{1}$$

$$\varepsilon_x \;=\; \frac{l_x - l_x^1}{l_x^1} \tag{2}$$

$$\nu \;=\; -\frac{\varepsilon_x}{\varepsilon_y} \tag{3}$$

2   MATERIALS AND METHODS

Using the experimental apparatus presented in figure 1, a video of the view labeled as **a)** was taken during the uniaxial tension test. A freeware video recording application, Virtual VCR 2.6.9 was used with the following settings:

- Frame rate: 30 fps
- Resolution: 320×240 (CMOS sensor native resolution)
- Video Compression Codec: DivX 5.0.5

The acquired videos were processed using *Virtual-Dub 1.6.9*. First, a crop operation selected only the relevant image areas, corresponding to the scheme in image 2. The video frames were extracted into separate image files. The frames comprising the beginning of (visible) tissue matrix damage to experiment terminus were discarded.

The second part of the process uses a MATLAB® script implementing the algorithmic scheme proposed.

2.1   *Algorithm for poisson's ratio (ν) determination*

In order to calculate ν, the image files were processed with a MATLAB® script. The first image (first frame to consider) is analyzed to acquire the initial geometry. The structures shown in figure 2 are identified sequentially from the original images (figure 3 **a**). On the first stage of the algorithm, the images are processed using functions from Matlab's image processing toolbox (*IPT*), according to the techniques described in (Gonzalez et al. 2004). A sequence of segmentation (figure 3 **b**) and morphological operations (figure 3 **c-d**) enable the calculation of the screw center coordinates. Using this two points, the other geometric elements seen in figure 2 can be obtained by simple geometric calculations.

Having the initial length (l_y^1) and width (l_x^1) values, the subsequent images (frames) will be processed using the optimization scheme. The objective of this procedure is to follow the evolution of the

a) Original Picture

b) edge detection via 'canny' filter

Morphologic operations

c) Image close

d) Image subtract

Figure 3. Image Processing.

pixel mask (defined in figure 2) through subsequent frames, calculating the new values for length (l_y^k) and width (l_x^k).

2.2 Tracking algorithm

The tracking algorithm does not require any mark on the sample surface. The mask can be of an arbitrary $m \times n$ size, allowing (in principle) the application of the algorithm to images of different resolutions. In any frame k, the *RGB* color indexes of the mask pixels, were used to build the matrix $\mathbf{M}_{ij}^k$ (4):

$$\mathbf{M}_{ij}^k = R_{ij} + G_{ij} + B_{ij} \tag{4}$$

$$i = 1,\ldots,m$$

$$j = 1,\ldots,n$$

The algorithm does the minimization of a cost function O defined by the difference between $\mathbf{M}_{ij}^{k-1}$ and $^c\mathbf{M}_{ij}^k$. The matrix $\mathbf{M}_{ij}^{k-1}$ provides the information acquired on the previous frame and $^c\mathbf{M}_{ij}^k$ contains the information of all the candidate masks of the present frame. The set of possible candidates $\{c_1,\ldots,c_l\}$ is defined through geometric considerations.

Finally the choice of the new mask for the kth frame is made using (5)

$$O(c) = \sum_{i=1}^{m} \sum_{j=1}^{n} \left(\mathbf{M}_{ij}^{k-1} - {}^c\mathbf{M}_{ij}^k \right)^2$$

$$c \in \{c_1,\ldots,c_l\} \quad : \quad min\{O(c)\} \tag{5}$$

The following scheme summarizes the complete procedure used to calculate Poisson's ratio (ν) from the experiment video:

1. Use *VirtualDub* to:
   a) Crop video
   b) Split video frames
   c) Remove frames showing material damage

2. Geometric analysis of the 1st frame (figure 2):
   a) Detect screw centers
   b) Identify the homogeneous zone
   c) Define the 1st pixel mask: $\mathbf{M}_{ij}^1$
   d) Calculate initial length (l_y^1) and with (l_x^1)

3. Use optimization algorithm

```
1:  for k ← 2, step, n do
2:      define mask candidates {c₁,...,cₗ}
3:      for all c ∈ {c₁,...,cₗ} do
4:          ᶜMᵏᵢⱼ
5:          O(c) = Σᵢ₌₁ᵐ Σⱼ₌₁ⁿ (Mᵏ⁻¹ᵢⱼ − ᶜMᵏᵢⱼ)²
6:      end for
7:      c : min{O(c)}
8:      Mᵏᵢⱼ ← ᶜMᵏᵢⱼ
9:      Compute (lᵏₓ, lᵏᵧ)
10:     Compute νᵏ
11: end for
```

3 RESULTS

To test the scheme proposed, five uniaxial tension tests were considered. The frames extracted from the video files, were processed according to the procedures presented in the previous section. The Matlab® script that implements the algorithm is able to accommodate a *step* parameter which indicates the frame processing frequency. For $step = 1$ every frame is processed, for $\{step = a : a \in \mathbb{N}\}$ one out of every a frames is processed.

The other parameters fed into the script define the mask size ($m \times n$), which can be of arbitrary size, limited only by the images resolution.

3.1 Step choice

The authors found that using $step = 1$ or $step = 5$, does not affect the overall behavior of the algorithm. Of course the higher step as a diminishing effect in computational time (a smaller number of cycles are performed). This fact means that there is no practical advantage in using the more computationally expensive $step = 1$ instead of $step = 5$.

The tests performed with $step = 10$, provided the best results. A stabilizing effect was noticed by the authors in all experiment recordings analyzed. For

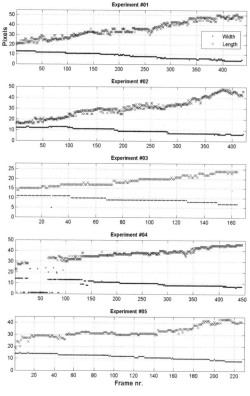

Figure 4. Width and length measurements using the algorithm.

step ≥ 15 an important degradation of the results was noticed, therefore the usage of step sizes in this range is not advisable.

According to the tests performed, the *step* choice for all the results presented was *step* = 10.

3.2 *Mask choice*

The algorithm can accommodate arbitrary mask sizes. The first mask tested had 5 × 3 pixel size, but the lack of homogeneity of the results obtained, lead to the search of another size suitable for a quality increase on the results. Since the CMOS sensor used for video acquisition has low definition (320 × 240 pixels), only a decrease in mask size made sense, therefore, a 3 × 3 pixel mask was used with a considerable increase either in performance and result stability.

Figure 4 shows the algorithm results for the computations of length (l_y^k) and width (l_x^k). The evolution of l_y^k (augmenting) and l_x^k (diminishing) with increasing values of k, follows the expected pattern. Some fluctuations of length and width can be noticed on every graph in figure 4. This effect is particularly evident at the beginning of experiment #4.

Table 1. Minimum and maximum values of length (l_y^k) and width (l_x^k) in pixels.

	l_x^k		l_y^k	
Exp. #	Min.	Max.	Min.	Max.
1	4	13	20	51
2	6	13	16	48
3	5	11	14	24
4	1	13	23	46
5	8	15	18	43

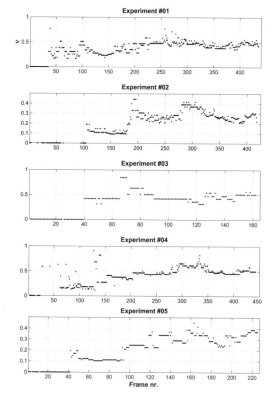

Figure 5. Poisson ratio for each frame.

Table 1 presents the maximum and minimum values for l_x^k and l_y^k in pixels. The small values obtained, are due to the law resolution of the CMOS sensor. This fact is specially relevant for the width (l_x^k) measurement because with such small values ($1 \leq l_x^k \leq 8$) the measurement errors can be significant. The Poisson ratio was computed using (3). Figure 5 shows graphically the v^k results obtained for each analyzed experiment. There is a strong fluctuation of the results (mainly) at the beginning of the experiments.

392

Table 2. Mean value ($\overline{\nu^k}$) and standard deviation (*StD*) for Poisson's Ratio.

Exp. #	$\overline{\nu^k}$	*StD*
1	0.4550	0.0725
2	0.2884	0.0553
3	0.4737	0.1065
4	0.5000	0.0727
5	0.3016	0.0564

A possible estimation of Poisson's Ratio is the average $\overline{\nu^k}$, calculated taking into consideration the individual values of ν^k obtained from frames belonging to the stability interval. For example, considering experiment #1, the frame interval (*k* values) taken into account was, $160 \leq k \leq 435$. The ν values for all the considered experiments are summarized in table 2. On the case of experiments #{1, 3, 4}, the ν estimations are close to 0.5, which is the expected value for incompressible materials (Lai et al. 1996). Since the vaginal tissues have high water content they are assumed to be incompressible. The results for experiments #{2, 5} deviated from the expected range of values. In fact, they are closer to the known values of ν for steel (0.3).

4 CONCLUSIONS

The authors developed a computational scheme, for the calculation of Poisson's coefficient (ν) of biological soft tissue. The computational scheme includes an image processing phase, followed by a tracking algorithm based on the optimal value of a cost function. The cost function is built upon image color information (RGB indexes).

The method was adapted successfully to uniaxial tension tests (simple tension).

It was possible to measure (estimate) the length (l_y^k) and with (l_x^k) during the mechanical test (increasing k; excluding damage).

The estimated values of l_y^k and l_x^k shown in figure 4 suffered some fluctuations, specially at the beginning of the experiment. This is a conditioning factor to the method's efficiency. There may exist several factors contributing to this situation, however, the low resolution of the original video (320×240 pixels) is most certainly responsible for this fluctuations. The

small dimensions (in pixel) of l_y^k and l_x^k (table 1) are a direct consequence of the low resolution and may be an important source for measurement errors.

The Poisson ratio was computed taking an average of the individual values obtained from a set of frames (*k* values) in a range where ν^k results were stable (figure 5). Three (#{1, 3, 4}) of the five considered experiments enabled ν estimations close to the predicted value for incompressible materials, 0.5. The results for the remaining experiments (#{2, 5}), are close to 0.3, the reference value for steel. One possible explanation might be difference in water content of different soft tissue samples. However, the strong fluctuations observed on the graphics 2 and 5 in figure 5 incline the authors to exclude that possibility.

ACKNOWLEDGEMENTS

The support of Ministério da Ciência, Tecnologia e Ensino superior (FCT and FSE, Portugal) and the funding by FEDER under grant PTDC/SAU-BEB/71459/2006.

REFERENCES

Cosson, M., E. Lambaudie, M. Boukerrou, P. Lobry, G. Crépin, and A. Ego (2004). A biomechanical study of the strength of vaginal tissues: Results on 16 post-menopausal patients presenting with genital prolapse. *European Journal of Obstetrics Gynecology and Reproductive Biology* 112(2), 201–205,

Goh, J. (2002). Biomechanical properties of prolapsed vaginal tissue in pre- and postmenopausal women. *International Urogynecology Journal and Pelvic Floor Dysfunction* 13(2), 76–79,

Gonzalez, R. C., R. E. Woods, and S. L. Eddins (2004). *Digital Image processing using MatLAB*. Pearson Prentice Hall.

LAi, W. M., D. Rubin, and E. Krempl (1996). *Introduction to Continuum Mechanics* (Third ed.). Butterworth Heinemann.

Martins, P. A. L. S., R. M. N. Jorge, and A. J. M. Ferreira (2006). A comparative study of several material models for prediction of hyperelastic properties: Application to silicone-rubber and soft tissues. *Strain* 42(3), 135–147.

Tarnay, C. and C. Dorr (2003). *Relaxation of pelvic supports*, Chapter Current Obstetric and Gynecologic Diagnosis and Treatment, 9th ed., pp. 776–797. Lange Medical Books. New York: McGraw-Hill.

Computational Vision and Medical Image Processing – João Tavares & Natal Jorge (eds)
© 2008 Taylor & Francis Group, London, ISBN 978-0-415-45777-4

Author index